T0296116

LONDON MATHEMATICAL SOCIETY LECTURE NOTE SERIES

Managing Editor: Professor Endre Süli, Mathematical Institute, University of Oxford, Woodstock Road, Oxford OX2 6GG, United Kingdom

The titles below are available from booksellers, or from Cambridge University Press at www.cambridge.org/mathematics

"It is exciting to read this important book. It is a marvelous panorama of the life, work, and inspiration of Robert Langlands, as he discovered and developed his grand ideas, and as he guided his students. These ideas now largely shape the broad architecture of representation theory and automorphic forms, creating a grand bridge between analysis and arithmetic – with connections to mathematical physics. The contributors to this volume offer us – and will surely offer future historians of our mathematical age – a splendid introduction to, and overview of, the early days of Langlands' program."

– Barry Mazur, Harvard University

Robert Langlands was born in 1936, in New Westminster, British Columbia. He received his bachelor's and master's degrees in mathematics at the University of British Columbia in 1958, and his Ph.D. from Yale University in 1960. He not only passed his oral examination but also submitted his Ph.D. thesis at the end of his first year at Yale.

The inception of the Langlands Program can be dated to 1967. In 1972, he was appointed professor at the Institute for Advanced Study in Princeton, New Jersey, USA, where he has been a Professor Emeritus.

Professor Langlands has received numerous prestigious awards, such as the 1996 Wolf Prize (with Andrew Wiles) and the 2007 Shaw Prize (with Richard Taylor). He is also the recipient of the 2018 Abel Prize, which has often been regarded as the equivalent of the "Nobel Prize" in mathematics.

Robert Langlands and his wife, Charlotte, have been living in Canada since March 2020.

The Genesis of the Langlands Program

Edited by

JULIA MUELLER
Fordham University, New York

FREYDOON SHAHIDI
Purdue University, Indiana

CAMBRIDGE
UNIVERSITY PRESS

University Printing House, Cambridge CB2 8BS, United Kingdom

One Liberty Plaza, 20th Floor, New York, NY 10006, USA

477 Williamstown Road, Port Melbourne, VIC 3207, Australia

314-321, 3rd Floor, Plot 3, Splendor Forum, Jasola District Centre, New Delhi - 110025, India

103 Penang Road, #05-06/07, Visioncrest Commercial, Singapore 238467

Cambridge University Press is part of the University of Cambridge.

It furthers the University's mission by disseminating knowledge in the pursuit of
education, learning and research at the highest international levels of excellence.

www.cambridge.org
Information on this title: www.cambridge.org/9781108710947
DOI: 10.1017/9781108591218

© Cambridge University Press 2021

First published 2021

A catalogue record for this publication is available from the British Library

ISBN 978-1-108-71094-7 Paperback

Contents

Contributors

S. Ali Altuğ
Boston University

James Arthur
University of Toronto

Matthew Emerton
University of Chicago

Steve Gelbart
Weizmann Institute of Science

Dorian Goldfeld
Columbia University

Alan Goodacre
Dominion Observatory, Ottawa (retired)

Thomas Hales
University of Pittsburgh

Hervé Jacquet
Columbia University

Jean-Pierre Labesse
Marseille Institute of Mathematics

Claude Levesque
Laval University

Julia Mueller
Fordham University

Claude Pichet
University of Quebec at Montreal

Dinakar Ramakrishnan
California Institute of Technology

Derek W. Robinson
Australian National University

Cihan Saçlioğlu
Sabanci University

Yvan Saint-Aubin
University of Montreal

Freydoon Shahidi
Purdue University

Diana Shelstad
Rutgers University

Thomas Spencer
Institute for Advanced Study

Preface

Robert Langlands' contributions to modern number theory, which is generally referred to as the "Langlands Program," has been visionary and ground breaking. The present volume is intended as homage to his early work in which the seeds of the program were planted. This volume covers the period from 1960 to 1967, during which Langlands defined many of the concepts that are basic to the program; at this time, he introduced some of his revolutionary ideas including the functoriality principle and his foundational work on Eisenstein series and Artin's L-functions. The period concludes with Langlands' famous letter to André Weil.

The book is divided into five parts. Parts I, II, and III are a gathering of personal reflections written by a number of his friends and former students; Julia Mueller's *Bulletin of the American Mathematical Society* interview with Langlands is also included. Part IV is a collection of surveys on different aspects of Langlands' work that took place over the period, including different concepts. Part V consists of two articles on Langlands' work in physics that took place later in his career.

The scope and depth of the Langlands Program will keep mathematicians busy for many exciting years, and we hope that this volume will be a useful introduction to it.

James Arthur
Steve Gelbart
Freydoon Shahidi

Introduction

This volume of contributed chapters is assembled as a tribute to Robert Langlands' prodigious accomplishments over the years from 1960 to 1967, from the age of 24 to 31. Surveys of his work during that period are included in Part I and Part IV of this volume. The two nontechnical chapters in Part I are contributed by Julia Mueller and Steve Gelbart. A few reflections from some of his students and friends are contained in Part II and Part III. The chapters in Part IV are survey articles contributed mostly by experts of the Langlands program, and Part V contains two survey articles of Langlands' work in physics.

The Langlands program was launched soon after Langlands' discovery of his automorphic L-functions. This important discovery together with his principle of functoriality can be regarded as pillars of the Langlands program. This far-reaching principle was first proposed in his manuscript "Problems in the theory of automorphic forms." It is a well-formulated conjecture and can be explained using the basic materials of the program. We have an excellent survey article by James Arthur in Part IV, where his idea is to motivate functoriality the way Langlands originally presented it, with Artin L-functions and principal (Godement–Jacquet) L-functions as the background.

The story of Langlands' discovery of his automorphic L-functions started with his foundational work on the most general Eisenstein series (1962–1964). He had to overcome many obstacles faced by others in order to develop the theory in its most general form over number fields. Roughly speaking, his new ideas in this piece of work were rooted in the works of Selberg, Harish-Chandra, and Gelfand, and most importantly, in harmonic analysis and representation theory.

It was pointed out by Langlands on several occasions (see Sections 14.1 and 14.6 of Shahidi's chapter in Part IV, and Mueller's chapter in Part I) that it was his calculations of the constant terms of those most general Eisenstein series that led him to the discovery of a number of central objects in the Langlands program such as the L-groups and his automorphic

L-functions. Without those objects, the formulation of many of his conjectures would not have been possible. The calculations appeared as a monograph "Euler Products," which he presented in his James Whitmore lectures at Yale. Moreover, his calculations, which were also the centerpiece of his famous letter to André Weil in 1967, explained the duality he discovered and many concepts that were involved with the duality. In short, it is his calculations that allowed him to have a formulation of his functoriality principle. His paper "Problems in the theory of automorphic forms," which appeared later, extended his ideas from the Euler products manuscript to the general setting.

It is well known that L-groups played a central rule in Langlands' discovery of his automorphic L-functions. One of their features is in connecting the arithmetic and analytic properties of Langlands' L-functions via Langlands' isomorphism. Since L-groups have no direct connection with the "Satake parameter," it is clear that "Langlands' isomorphism" is a significant and conceptually challenging extension of the "Satake parameter." More discussions on this topic can be found in the chapters of Mueller and Shahidi in this volume. Shahidi's treatment of this topic is both axiomatical and through examples, while Mueller's presentation is less technical. It should be noted here as well that Mueller's chapter in Part I originally appeared in the *Bulletin of the AMS* (Vol. 55, No. 4, October 2018, pages 493–528).

The theory of Eisenstein series is briefly, but masterfully, surveyed by Jean-Pierre Labesse. His chapter in Part IV is an important contribution to this book. In it, Labesse discusses the spectral decomposition for which Eisenstein series are central in accounting for both the continuous and the residual part of the spectrum.

It was Langlands' attempts to formulate a nonabelian reciprocity law that was among the most appealing yet challenging parts of the program to number theorists. They were formulated as conjectures, together with the principle of functoriality, again in his manuscript "Problems in the theory of automorphic forms." These arithmetic aspects are surveyed in this volume by Matthew Emerton's contribution in Part IV. His chapter touches upon Artin's L-functions, on which Langlands had produced a long manuscript with more or less complete local proofs of the existence of root numbers for these L-functions. A complete proof of the existence was later given by Pierre Deligne using local–global methods.

The most basic of all the automorphic L-functions are those studied, in particular, by a method generalizing Tate's thesis. They are usually addressed as "Principal L-functions" for the general linear group $GL(n)$. This basic but important case of automorphic L-functions is surveyed in a chapter in Part IV, contributed jointly by Dorian Goldfeld and Hervé Jacquet.

Langlands has acknowledged in several writings and interviews the influence that Harish-Chandra has had on his work. In particular, Harish-Chandra's study of orbital integrals for real reductive groups played an important role in the harmonic analysis that entered Langlands functoriality. This has influenced Diana Shelstad's early career, and her chapter in Part IV contains her presentation on the early history of functoriality.

Langlands' Ph.D. thesis at Yale, which has only been minimally published, is surveyed by Derek Robinson, with a delightful introduction, in his chapter in Part IV. As he explains, Langlands' thesis included a number of brilliant ideas addressing and providing full proof for density of analytic vectors and the Garding space for general continuous representations of any Lie group. Let us mention that Langlands' Ph.D. thesis is prior to his work on the "Langlands Program." Here are a few words from Langlands himself:

> I will tell you a story about my thesis. I wrote it on my own in the summer between my first and second years at Yale. Charlotte typed it and I submitted, but it was independently written and no-one understood it. The stair in the mathematics building was circular and that allowed me to overhear a conversation between Felix Browder and Kakutani. Kakutani wanted to decline it because no-one understood it. Browder, who was more familiar with me than Kakutani, wanted to accept it. He fortunately prevailed. I had, about the same time, a conversation with Ed Nelson, as I was taking a tourist visit to the Institute with friends from Yale. This conversation led, with no application whatsoever, to a position as instructor at Princeton. The thesis was never published but sometime later it was incorporated (and redone) in a book by Robinson. So, for me, there is considerable history in Robinson's article. It belongs to my first years and there is no connection to Hecke and automorphic forms. That began at Princeton, but my two years at Yale had made an analyst of me. The thesis is by the way impossible for me to read now.
>
> *Robert Langlands, 2020*

Some of Langlands' work in physics is presented in the two survey articles, with an excellent overview, in Part V by Thomas Spencer and Yvan Saint-Aubin.

Finally, we wish to express our sincere gratitude to all the contributors whose participation has given life to this book.

Julia Mueller
Freydoon Shahidi

PART I

Remembrance of Things Past

1

A Glimpse at the Genesis of the Langlands Program

Julia Mueller

1.1 Introduction

This chapter[1] serves as an introduction to the volume as a whole, and it is aimed at a general mathematical audience. Our presentation of the early life and work of Robert Langlands, creator and founder of the Langlands program, is narrated, to a large extent, by Langlands himself. We focus on two of Langlands' major discoveries: automorphic L-functions and the functoriality conjecture. Langlands' desire to communicate his excitement about his newly discovered objects resulted in his famous letter to André Weil in January 1967, and the Langlands program was launched soon afterwards.

Section 1.2 of this chapter focuses on Langlands' early years, from 1936 to 1960, and the material is taken from an interview given by Langlands to a student, Farzin Barekat, at the University of British Columbia (UBC), Langlands' alma mater, in the early 2000s. A copy of this interview is available on Langlands' website at http://publications.ias.edu/rpl/.

Section 1.3 is an overview of Langlands' early work and professional life from 1960 to 1967. It contains a brief descriptive account of the essential events that led him to his discoveries of automorphic L-functions and the functoriality conjecture. Our source material in this part is taken largely from our correspondence and interviews with Langlands himself over the past few years. Those correspondence can also be found on Langlands' website.

1.1.1 A Summary of Langlands' Early Life (1936–1960)

Let us list a few important events in Langlands' early life which, in our view, shaped his professional life later on.

---

[1] An earlier version of this material appeared in the *Bulletin of the AMS* (Vol. 55, No. 4, October 2018), with a slight modification of the introduction here. The idea of writing that article originated from the preparation of this book.

(1) Toward the end of his twelfth grade (see Section 1.2.4), an English teacher, Crawford Vogler, took up an hour of class time to explain to him, in the presence of all the other students, that it would be a betrayal of God-given talents for him not to attend university. Langlands had no intention of attending college because none of his classmates did. At the time, very, very few students, at the best one or two, did so in any year. School meant little to him. However, Langlands was flattered by his teacher's comments and his ambition was aroused.

(2) Taking the aptitude tests at UBC in the fall of 1953 (see Section 1.2.5) was a decisive event in his early life. From the results of his tests, the university counselor suggested at first that he might want to become an accountant, but Langlands rejected that idea right away. The counselor then suggested mathematics or physics, but cautioned him that this would require a master's degree or even a PhD. Langlands did not know what a PhD was, but he decided on the spot that he would become a mathematician or a physicist.

(3) Another important event was the discovery of his natural talent for languages (see Section 1.2.6). Sometime during the year of 1954, Langlands spoke to Professor S. Jennings about his intention to choose honors in mathematics. Professor Jennings declared that to be a mathematician, one had to learn French, German, and Russian. Taking Professor Jennings' advice, Langlands became fluent not only in French, German, and Russian, but also in several other languages, including Italian and Turkish.

(4) Let us take a brief look at Langlands' performance as a student at UBC. What stood out was his fierce independence in conducting his research. However, the result was not completely successful. For example, the topic of his master's thesis was chosen by himself, but unfortunately, the theorem he was trying to prove was false (see Section 1.2.7). There was some question whether the thesis could be accepted, but the committee decided to let him go on to Yale.

(5) Langlands' unique and profound talent in mathematics was first successfully revealed during his first year as a graduate student at Yale University (see Section 1.2.8). He not only completed all the required courses and examinations for the degree, but also submitted his PhD thesis at the end of his first year. Consequently, he was completely free to pursue his own research during his second year as a graduate student at Yale.

While all things seemed to point toward an illustrious career in mathematics, it is important to note that Langlands, as a young man, harbored ambitions of a much more general nature. He did not envision a career as a mathematician. Here he is, in his own words:

> I began my mathematical, or better my university, studies more than sixty years ago just before my seventeenth birthday, after a childhood and adolescence that were in no sense a preparation for an academic career, or a career in any sense. Various

influences came together that suggested that I attend university not so much to prepare myself for a profession but to become what one might think of as a savant. A major influence was perhaps Ernest Trattner's *The Story of the World's Great Thinkers'*, a book that was very popular in the late thirties, so that there are still many copies available on the used-book market at very modest prices. What I envisioned, after enrolling in the University of British Columbia in 1953, was myself, in a jacket with leather elbow-patches and perhaps a pipe in my mouth, gazing over the lawns and the trees while reflecting on a still undetermined, but certainly abstruse, topic. That topic became mathematics, not out of a strong preference for that subject, but because it pretty much required no preparation, only native ability.

1.1.2 A Summary of Langlands' Early Works (1960–1966)

During the first six years in the 1960s, Langlands' research output was prodigious. The important ideas in the discoveries of his automorphic *L*-functions and the functoriality conjecture were rooted mainly in the works of Selberg, Harish-Chandra, and Gelfand; as well as the formulation of adèlic structure of groups by Godement, Tamagawa, and Satake (see Section 1.3.12). Of course, Artin's reciprocity law (see Section 1.3.11.5) also played an important role. Let us highlight some of his important contributions.

(1) Langlands' spectral theory (see Section 1.3.8) was certainly one of his first major contributions; particularly, the most general Eisenstein series he constructed in this piece of work which played an important role in the discovery of his automorphic *L*-functions later on; for example, it became the archimedean component π_∞ of the "automorphic representation π." Let us note that his success in generalizing Selberg's result on spectral theory (see Section 1.3.7) from $SL_2(R)$ to arbitrary reductive groups was very much helped by his knowledge in harmonic analysis and representation theory that he acquired while working on his PhD thesis.

(2) Langlands' calculation of the constant term of his generalized adèlic Eisenstein series in the fall of 1966 (see Section 1.3.14) was a decisive factor in his discovery of automorphic *L*-functions. It not only led him to create the *L-group*, but also to introduce the *Langlands isomorphism*. Please note that *Langlands isomorphism extends* the Satake isomorphism and connects the Hecke algebra with the representation ring of the *L*-group while the Satake isomorphism only relates the Hecke algebra with a certain polynomial ring (see Sections 1.3.15.2 and 1.3.15.3). In Langlands' own words: "The expression 'Satake parameter' is completely inappropriate and could very well be replaced by 'Frobenius–Hecke parameter'."

(3) An Artin *L*-function is defined as an Euler product which does not have analytic properties, but when the number field extension F over $\mathbf{Q}$ is abelian, Artin's reciprocity law allows his *L*-function to acquire "nice" analytic

properties from Hecke's degree-1 L-series (see Sections 1.3.11.3 to 1.3.11.5). However, in a nonabelian number field extension, Artin was unable to provide such analytic properties to his L-functions.

Langlands automorphic L-functions $L(s,\pi,\rho)$, as Langlands discovered, depend on both the *automorphic representation* π and the *finite-dimensional representation* ρ, where π is an isomorphism from a "generalization" of Hecke's degree-2 L-series to the L-group, and ρ is linked, via the L-group, with the Frobenius classes of Artin's L-functions. Hence, Langlands L-functions simultaneously generalize Hecke's L-functions as well as those of Artin.

(4) We learn (see Section 1.3.16) that Langlands was contemplating the problem of establishing "nice" analytic properties for those automorphic L-functions when he created the functoriality conjecture. Apparently, one of the major moments in his mathematical career was the first hint of the genesis of functoriality. That happened during the Christmas vacation of 1966, while looking through the leaded windows of his office distractedly, he was suddenly struck by the epiphany: "The Artin reciprocity law has to be replaced by what I now refer to as functoriality." (See Section 1.3.16.2.)

There are 17 subsections in Section 1.3. A large part of the last section, Section 1.3.17, is a presentation of an interesting and illuminating example of Langlands, which gives a rough sketch on how the functoriality conjecture was discovered.

We hope that by prefacing the story of the evolution of Langlands' mathematical landscape with a summary of crucial life events, we provide the reader with greater insight into both the man and his work.

1.2 Langlands' Early Years (1936–1960)

My time as a student at UBC (University of British Columbia) was the major transitional phase of my life, and it is not possible to appreciate its significance for me without understanding what I brought to it.

1.2.1 Preschool Years

I was born in October 1936, in New Westminster, British Columbia, Canada. My first years were not, however, spent there. They were spent on the coast in a hamlet south of Powell River and slightly north of Lang Bay. It was a very small group of houses, to the best of my recollection, three summer cottages. My father had found work at the nearby Stillwater dam.

One of the other two cottages, perhaps better described as cabins, was occupied by an elderly woman and a child, who I recall was her granddaughter. That they had a goat has also fixed itself in my mind. I also remember a field surrounding their house, but it could have been small. The other house and its inhabitants I do not recall. It may have been occupied only in the summer. Like ours, it was bounded on one side by the beach and the Straits of Georgia, and on the other by a forest or swamp. The only vegetation that fixed itself in my mind was skunk-cabbage.

The five or six years there must have been broken occasionally, by trips to New Westminster to my parents' families. But I have no recollection of such excursions, simply of my mother, my father, a younger sister, and towards the end a second baby sister.

1.2.2 Families in New Westminster

When time came for me to attend school, and perhaps for other reasons, we returned to New Westminster, then a small, in my memory, delightful city with chestnut-lined streets, the trees planted in what was called a boulevard, between the sidewalk and the curb.

New Westminster, during the war years was agreeable for a child. There were very few automobiles. The available area for me was from the north at 8th Ave. to the south at Columbia St., and from Queen's Park in the east to 12th St. in the west. So I had about one-half a square mile in which to freely roam, although I was not very nomadic. I had grandparents and many cousins close to 12th St., or on the other side of 12th, but it never occurred to me to visit them on my own.

Although we were only a short time in New Westminster, it was the period when I made the acquaintance of my many aunts and uncles. My mother's family large, my father's smaller but substantial, so that all in all there were thirteen aunts and uncles together with their husbands or wives and children.

In England my grandparents may have been Methodists but in Canada they were members of the United Church. My father's parents were not so much stern as pious. In my father there were only un-reflected remnants of the piety. I have seen a photo of my father and two sisters, all infants because he was a twin, with their mother before a tent in the forest, probably on Vancouver Island, but he himself never once mentioned those early years. As a child I had found my father's parents and his sisters and brother, in comparison with my mother's family, a little colorless and distant.

We were used to parents, especially fathers, with little schooling. My own father had eight years. His arithmetic was excellent but his reading was on the whole confined to the sport pages. My mother's family was, at least for a child, much warmer. Her parents, had drifted west from Halifax, trying their hand at land-grant farming in Saskatoon before reaching New Westminster.

My grandmother was already a widow when I was born and even before my grandfather's death had to assume a good deal of responsibility in difficult financial circumstances for her ten surviving children. She had many grandchildren, a few older than me, several of about my age, and a good number younger. Most of the latter I never met. My grandmother had affection and energy enough for all of us, although she cannot have been well. Her heart failed after a bus-ride to White Rock and a walk up a steep, dusty hill on a very warm day as she was coming to visit mother when I was ten.

Since I left British Columbia I have seen my aunts, uncles, and cousins on both sides only infrequently, but it has always been a pleasure. I still correspond occasionally with my mother's only surviving sister.

I was enrolled by my mother, a Catholic, in the parochial school, St. Ann's Academy. It was a school, taught partially by nuns, especially in the first two or three years. They were young, pretty I suppose, and encouraging, so that I enjoyed these first years, taking three years in two, or four in three, but then I was moved for the later years of the elementary school to the companion academy, St. Peter's I believe, which was less friendly, with a morose beadle, prompt to resort to his baton, and I grew restive. By the time I reached 6th grade, just before my 10th birthday, we had moved again, to White Rock, where I learned less, indeed pretty much nothing at all, but which also gave me a great deal.

1.2.3 White Rock – the Town of my Youth

I have not visited White Rock for a very long time, several decades. I would suppose that the town of my youth was quite different than the town of today. The logs washed up along the shore were there and for me, for all of us, part of the natural environment, to be enjoyed like the sea itself.

My first two or three years in the town had been carefree. I was a little older than in New Westminster, a little more independent than I had been there. The sea and the shore were immediately accessible; the pier was, at least after I learned to swim and dive, an attraction. I found my way there when free of responsibility.

On Sunday, the main street, which ran along the shore, was devoted entirely to cruising. In my early adolescence, I admired these people, their clothes, above all their freedom, their independence, but I was too young to imitate them. By the time I might have been in a position to do that, I had other goals. Each of us is a child only once, so that it is difficult to distinguish what is particular from the atmosphere in which we grew up from what was particular in us.

My parents did succeed for some years in establishing and maintaining a business, millwork and builders supplies, so that we were better off than many. After the age of twelve or thirteen I worked afternoons, weekends and summers there, and continued working there in the summer even after I began university. It was typical then, and may still be now, for students at UBC to work in the summer to earn enough for room, board and tuition during the university's relatively short fall and winter terms.

There were other sources of income, a newspaper route for the Vancouver's *The Province* to which I was faithful for a year before I tired of the inevitable loss of freedom. For six days of every week, one and a half hours of the afternoon had to be given over to collecting and delivering the newspaper. I also remember changing the marquee at the movie theater, next to the dance hall, but visited only by local residents, so that the featured movie ran just for two days, changing three times each week. This means three visits a week at 10:30 in the evening to the theater in exchange for free access. At first I was delighted with that, but as I was then old enough that the obligatory visits interfered with my social life, I soon abandoned the marquee even though the theatre was only a few hundred yards from our home and business.

School, now a public school, was mixed. I enjoyed that. Except that it was a place frequented by girls and my friends, meant little to me. For a short period I was large for my age, I was also younger than almost all my classmates and maladroit. I had little success with the limited athletic possibilities.

I was probably the despair of the teachers, who, perhaps from the results of IQ tests, were aware that I had considerable untapped academic potential, from which I refused to profit.

1.2.4 Grade 12 – a Very Special and Important Year

In my last year, in grade 12, we had an excellent teacher, Crawford Vogler, with a newly designed textbook, and a newly designed course on English literature. He is one of the people to whom I owe most and for a very specific reason. Toward the end of the year, he took up an hour of class time to explain to me, in the presence of all the other students, that it would be a betrayal of God-given talents for me not to attend university.

I had had no intention of doing that. None of my classmates did: at the time very, very few students, at the best one or two, did so in any year. I was flattered by his comments, my ambition was aroused, and I decided then and there to write the entrance examinations. I worked hard and was successful, even winning a small fellowship from the University.

There was another factor in my changing stance. I have acknowledged elsewhere that I was tainted even at a fairly early age by ambition, but could never satisfy it.

I also was not incapable of intellectual or moral passion. I had decided at the age of seven or eight, not long after beginning to attend the parochial school, that I would become a priest, building myself an altar with improvised paraphernalia in my bedroom. It probably delighted my mother to see this early sign of a vocation, but it soon passed.

In the course itself, Vogler singled me out as one of the students who could usefully present a report on a novel and assigned me Meredith's novel: *The ordeal of Richard Feverel*. He had overestimated me. I read the novel but had no idea what might be said about it, or perhaps I hesitated to express my feelings. It is, after all, a novel of young love.

Unfortunately, by the time I undertook to thank him personally, it was too late. Although still alive after a successful teaching career in the lower mainland, he was, as I learned from his son, no longer in any condition to appreciate expressions of gratitude.

By the time I read Meredith, I had met the girl, Charlotte, whom I was to marry and to whom, although she is no longer a girl, I am still married many years later.

In retrospect, I am astounded that by good luck, certainly not with any foresight, I found in a little town at an early age and with no guidance, someone who could give me so much, in so many ways, in so many different circumstances, and for so long, without sacrificing her independence or totally neglecting her talents.

Her father, having lost his mother as an infant, grew up in a Gaelic-speaking foster family on Prince Edward Island, but went off to the logging camps at the very early age of twelve. When I met him he had spent the best part of his life in Ontario and the Lower Mainland, some of it during the Depression, unemployed and drifting.

My wife's father had only one year of schooling. It was at the age of 30, during the Depression, that he learned how to read, and in principle to write, at the classes that the parties on the left organized for the unemployed. He also acquired a small library, but he could not really read with ease. By his books, however, in particular one with biographies of those savants who were heroes to the socialists, Marx, Freud, Hutton, Darwin and several others, I was inspired with the ambition to be a savant. It is odd, but this ambition has never faded. It is the ambition with which I arrived at the University.

1.2.5 The Aptitude Tests at UBC (1953) – a Decisive Event

At the University, I took, as was common at the time and as was perfectly appropriate for anyone with my lack of academic experience, aptitude tests.

The results were predictable. In those domains, mathematics and physics, where, at least in the context of such tests, only native talent matters I did extremely well. In the others I also did well, but not so well. So the university counselor, whom one was encouraged to consult after the tests, suggested at first that I might want to become an accountant, even a chartered accountant. This lacked all glamour. So he then suggested mathematics or physics, cautioning me that this would require a master's or even a PhD. The latter meant nothing to me, but I did not acknowledge this. I decided on the spot that I would become a mathematician or physicist. The PhD whatever it was, could take care of itself.

As soon as I returned to White Rock from the University, I looked up my future father-in-law, found him in bed, and learned from him what a PhD was. Having set out to become a mathematician and a savant, I was as systematic as possible, even buying a copy of Euclid in the Everyman's edition to remedy, as I thought, my neglect of elementary geometry.

1.2.6 A Talent for Languages Revealed via Mathematics

At some point in the first or second year, because of my intention to choose honors in mathematics, I spoke with Professor S. Jennings. For me, he became later a somewhat comical figure, dapper but a little soiled, who reminded me of the Penguin in Batman comics, although I cannot say for certain that he always carried an umbrella with a crook-handle.

He gave me a piece of advice for which I am grateful to this day. He declared that to be a mathematician one had to learn French, German and Russian. The romantic desire to penetrate the present and the past of the enchanting, mysterious or seductive tones of a tongue not sung at my cradle, was planted in me as a student. This desire came to me simultaneously with mathematics.

Through the advice of S. Jennings, thus through mathematics, and through opportunities given me as a mathematician, I was eventually led to the very words and the very sounds not only of Gauss, Galois or Hilbert but of Thomas Mann,

Proust, Pasternak, or even Giuseppe Tomasi di Lampedusa and Almet Hamidi Tanpinar, not to speak of Robert de Roquebrune or Michel Tremblay, or Mommsen and Michelet. I am afraid such possibilities are no longer available. I myself have not yet reached the classical languages or any truly exotic language, except in an inadequate way. I like to think there is time left to remedy this.

Although it may appear that I quickly abandoned the desire to become a physicist or even to acquire some understanding of physics, the desire persisted and I took a large number of courses that I enjoyed, both as an undergraduate and during my year as a graduate student at UBC.

My experience suggests nevertheless that it is easier to learn mathematics on one's own than physics. It also suggests that my natural aptitude for mathematics was greater than my natural aptitude for physics. I fear, however, that I was more attracted to the mathematical explanations than to the physical phenomena themselves.

It was not possible at the time I was an undergraduate to take books from the library at UBC – there was only the main library – or to visit the stacks. It was nevertheless my first encounter with a library after that on Carnarvon Street in New Westminster and I profited from it as much as whatever knowledge I had would allow. It was a pleasure to handle the books.

1.2.7 An Incomplete Master's Thesis

During my year as a candidate for a master's degree at UBC I had probably tried to read both Weyl's book on algebraic number theory and Lefschetz's *Introduction to Topology*. The first is, of course, a difficult book and God only knows what I understood, but I did come away with a clear notion that the law of quadratic reciprocity, which never appealed to me as an undergraduate, was in the context of the theory of cyclotomic equations genuinely a thing of beauty. Weyl opened my eyes. Lefschetz's book would have been my introduction to topology. I never did take to topology. Whether Lefschetz is to blame or my intellectual limitations I cannot say.

My master's thesis, which was influenced by an undergraduate seminar at UBC, was not a successful undertaking because I discovered just as I was submitting it that I was trying to prove a false theorem. I could not recover much useful from what I had written. There was, I believe, some question about whether it could be accepted. My guess is that the committee was generous, gave me credit for independence and enterprise, and let me go on to Yale and the next step, for which I am still grateful.

1.2.8 Yale University – an Excellent Choice

I was eager to finish the Master's degree as soon as possible and to continue with what seemed to me genuine graduate work. I applied to three institutions: Harvard, Wisconsin and Yale. I was accepted by all three.

Wisconsin was without aid and I would have had to teach. I had discovered in my year as a candidate for a master's degree at UBC that teaching interfered with learning mathematics. So I did not hesitate to decline Wisconsin.

Yale offered a fellowship that would, with almost no help from my family, support both me and my wife, who would not be allowed to work in the USA. Besides I had some familiarity with the mathematics of the faculty at Yale, above all of Hille and Dunford, thus functional analysis, but informed by classical analysis. I was accepted at Harvard but with no support. So the choice was evident.

In retrospect, it was extremely fortunate for me that Harvard did not offer a fellowship. I would have gone there and missed in one way and another a great deal. At Harvard, I would have had to deal with fields that were both popular and extremely difficult and with fellow students who were already initiated into them. That would have taken an incalculable toll.

1.2.9 A Stressful Oral Examination

At Yale I was on my own and allowed to follow my own inclinations. I finished at Yale in two years, indeed the thesis was written in one, so that I had a great deal of time after it was finished in which to think about various problems and to learn various techniques.

I did not prepare for the oral examinations to take place at the end of the first year. In some sense I took them cold and it began badly because I could not prove the simplest things about Noetherian rings. I had obviously read Northcott's book on ideal theory too quickly and too superficially.

I had certainly read with some care the first edition of Zygmund's book on Fourier series available at the time in a Dover edition. I had also read in Burnside's book on finite groups, with dreams of solving the famous conjecture that all simple groups were of even order. Fortunately for me, when Shizuo Kakutani, one of the examiners, discovered that I knew something about Fourier series he began to question me closely about interpolation theorems, which take up a certain amount of space in Zygmund's book and are probably still popular among Fourier analysts, but otherwise little known. Having recently spent considerable time reading the book, and with considerable pleasure, I could respond quickly and correctly to his examination. Thanks to this – so far as I know – I was saved. I am particularly grateful to him for discovering that, even if I did not know what I should have known, I did know something.

Formally, the director of my thesis was Cassius Ionescu Tulcea. The first half of my thesis was a solution of an open problem in the somewhat obscure domain of Lie semi-groups and their representations. It became my first paper, appearing in the Canadian Journal of Mathematics. The second half of my thesis, never properly published, but available, because Derek Robinson incorporated it into his book *Elliptic Operators and Lie Groups*. It had in another way an important effect, because it drew me to the attention of Edward Nelson, then an assistant professor at Princeton, and on his recommendation alone, with no application, with no information whatsoever about me, the department appointed me as an instructor.

1.2.10 A Fateful Future

What I really hoped to do when I completed my PhD was to stay at Yale. I had fallen in love with the atmosphere there: I had a freedom to study and think that

I had never had elsewhere. Several of the faculty encouraged me to stay, but my appointment was blocked, probably by Kakutani, who had, for reasons I never completely understood, taken a dislike to me. His disfavor was a great favor.

I accepted the offer from Princeton, where I had the great fortune to meet Salomon Bochner, whose encouragement had decisive, concrete consequences. I am not sure that Bochner ever understood how much he had done for me. I was a timid young man and he was a genuinely timid old man, so that there were some feelings that were never expressed.

Robert P. Langlands is currently Professor Emeritus at the Institute for Advanced Study in Princeton, New Jersey, where he has been a professor since 1972.

1.3 Overview of Langlands' Early Work (1960–1967)

I was tainted even at a fairly early age by ambition, but could never satisfy it. I also was not incapable of intellectual or moral passion.

1.3.1 The Letter to André Weil (January 1967)

When asked where those ideas in his conjectures came from Langlands often answered "Those ideas were in the air, and I was in the right place and at the right time."

In the early 1960s, Langlands was a junior faculty member at Princeton University. Emil Artin's influence was certainly in the air even though he had left Princeton University in 1958. Artin was noticeably the most respected algebraic number theorist of his time and an authority on abelian class field theory. Artin and his school had seriously investigated the possibility of a nonabelian class field theory but was not able to create one.

John Tate, one of Artin's eminent students showed (in his famous Princeton university thesis (1950)), how to use an adèlic treatment to reformulate and reprove Hecke's complicated results about his L-functions.

One of the "desirable" and "fashionable" research topics at that time, especially in Princeton, was to generalize Tate's thesis from $GL(1)$ to $GL(n)$. The leaders of that group were Godement and Tamagawa. Even though they both were striving for the same goal, they worked totally independently of each other.

The leading mathematician of the other "desirable" topic at that time was Atle Selberg, who was a professor at the Institute for Advanced Study in Princeton, New Jersey. From the 1950s to the 1960s, Selberg's work focused on investigating Eisenstein series for the group $SL_2(R)$. What he was looking for, around 1960, was a generalization of $SL_2(R)$ to other groups.

Let me just remind you that there was a sequence of developments in the 1930s, 1940s and 1950s that would inspire a lot of the later work : Hecke's theory; Siegel's many papers on reduction theory and related matters; Maass's extension of Hecke's ideas to non-holomorphic forms on GL(2) and his introduction of the associated spectral theory of which the Eisenstein series form a part; Selberg's solution of a problem raised by Maass but not solved by him, the construction of the continuous part of the spectrum with the help of the Eisenstein series, and his introduction of the trace formula.

I learned something about this from Selberg's paper, not only for GL(2) but also for groups of higher dimension, but I no longer know exactly what. I had also, for other reasons, read some papers on domains of holomorphy for functions of several variables. Putting this together, I had proved some theorems about the analytic continuation of Eisenstein series in several variables. I would guess that Selberg had also proved them, but one never knew with Selberg.

Many mathematicians from both groups tried their hands at one or the other of these topics obtaining partial results. But it was Langlands who, in 1966 at the age of 30, amalgamated and vastly generalized the essential ideas from his contemporaries as well as the recent past:

* the ideas of Selberg, Harish-Chandra, and Gelfand which are rooted in Eisenstein series, harmonic analysis, and representation theory of certain classes of noncompact groups;
* the usage of adèlic structure on groups championed by Godement as well as Tamagawa and Satake; and
* Artin's legacy on class field theory and his quest for a nonabelian class field theory.

Among what Langlands had created was a series of interlocking conjectures which later became the foundation of the Langlands program. Those conjectures seek to connect deep arithmetic questions with the highly structured theory of infinite-dimensional representations of Lie groups. The latter is part of harmonic analysis. Those visionary conjectures have exposed, quite unexpectedly, the deeply entwined nature of several seemingly unrelated branches of mathematics. In 1967, when those conjectures were first introduced to the public, they encountered some resistance. However, during the past half century, mathematicians around the world have been inspired to ask questions and solve problems, and their solutions have altered the landscapes of (a) number theory (b) arithmetic geometry, as well as (c) representation theory. As an example, let us mention that according to a theorem of Kenneth Ribet, the celebrated Fermat's Last Theorem, an unsolved problem over 300 years, follows from a conjecture of Shimura–Tanijama–Weil (proved by Andrew Wiles and Richard Taylor), which could be viewed as a special case of the general Langlands' conjectures. In addition, the Langlands–Tunnell theorem, proving another case of the Langlands' conjectures, was an important step in the Wiles–Taylor proof.

These conjectures had a profound effect, especially on number theorists, on how they thought about their subject. The conjectures suggested a way to obtain arithmetic information from the rich and rigid structure of algebraic groups and their infinite-dimensional representations which were mostly unfamiliar to number theorists at that time.

In January 1967, Langlands formulated a series of conjectures in his famous letter to André Weil who was, at that time, a world-famous mathematician and a professor at the Institute for Advanced Study in Princeton, New Jersey.

We find, in that famous letter, in a hastily written dense format, an impressive collection of novel and ingenious ideas which in 1967 were far ahead of their time.

(1) A general and comprehensive notion of automorphic L-functions attached to automorphic representations (plus some additional representation theoretic datum) simultaneously generalizing Hecke's L-functions as well as those of Artin. Such a generalization had been out of reach since Hecke's work in 1936.

The notion of L-group[2] (and the dual group), which underpins both the automorphic representations and their L-functions, was a deciding factor in Langlands' discovery of his automorphic L-functions.

(2) The principle of functoriality and nonabelian class field theory, which explains, albeit conjecturally, the connections between automorphic representations on groups whose L-groups are related via an appropriate notion of morphism. Moreover, a special case of the functoriality conjecture generalizes Artin's abelian reciprocity law, and asserts, conjecturally, Artin's reciprocity conjecture for nonabelian field extensions which may be viewed as a general formulation of nonabelian class field theory.

Topic (2) is linked to topic (1) and they were discovered almost simultaneously. Both topics will be revisited in later sections.

The automorphic L-function, $L(s, \pi, \rho)$, as Langlands discovered, depended not only on the automorphic representation π but also on an auxiliary finite-dimensional representation ρ of a canonically associated dual group or L-group.

Langlands originally named his L-function the Artin–Hecke series because it significantly generalized both of these L-functions into a single definition, via the representations ρ and π. This is quite remarkable in the sense that Langlands' L-function is defined as an Euler product akin to Artin's L-function and hence it is arithmetic in nature; but it also possesses analytic characteristics, which are more aligned with Hecke's L-function. This unique feature of combining algebra and analysis plays a central role in Langlands' L-function.

---

[2] The name L-group was suggested by Hervé Jacquet.

From the fall of 1960 till the spring of 1967, Langlands, a promising young mathematician, worked and lived in Princeton, except for the academic year 1964–1965, which he spent in Berkeley, California.

For the rest of Section 1.3, we have Langlands' own recollections, from that period, on some aspects of his professional life.

1.3.2 Salomon Bochner – an Exceptional Mentor (1960–1967)

In the first few months, I was, as a very junior mathematician, invited to speak in the analysis seminar conducted by Robert Gunning. The talk was about a theorem that I had proved as a graduate student while studying Selberg's paper. I believe that Salomon Bochner was very favorably impressed by my independence.

I had worked alone as a graduate student but, more importantly, on at least two different subjects. He encouraged me in a number of ways. He not only urged me to read Hecke's work on the Dedekind zeta-function and related Euler products, but also had me moved one step, and later more steps, up the academic ladder.

Above all, he suggested two or three years later that I offer a course on class field theory. I was scared stiff. Bochner's encouragement and suggestions played a decisive role in my first years at Princeton.

We have learned from Langlands himself that Bochner had been an encouraging and caring mentor. The other mathematician who had an important influence on Langlands' work, at that time, was Atle Selberg. As we know, Langlands had started learning about Selberg's work as a graduate student at Yale University, mostly by reading Selberg's papers. As for their personal contact in Princeton, as far as we know, Selberg was reserved and aloof.

1.3.3 A Memorable Meeting with Atle Selberg (1961)

Selberg, I am sure, had invited me to his office at the suggestion of Salomon Bochner. I was able to follow Selberg's oral presentation of the proof of the analytic continuation of the Eisenstein series for discrete subgroups of SL (2, R) with quotient of finite volume, in my first, and surprising as this may be, my last mathematical conversation with him, since our offices were, many years later, essentially side-by-side for a good long time.

The ideas involved are just those of spectra theory for second-order self-adjoint equations on a $1/2$-line, but I had never really seen these before and certainly not in the hands of a master.

It was a great pleasure to speak, for the first time in my life, with a strong mathematician about serious mathematical matters, or rather to listen to him, and I was tremendously impressed. It was a defining experience. I went away with a reprint of his paper and began to study it carefully, especially the trace formula.

What Selberg was lecturing on at that meeting was the Selberg spectral theory on $SL_2(R)$. This topic is the subject of Section 1.3.7, and Langlands' generalization of Selberg's work is the subject of Section 1.3.8. In Sections 1.3.4 to 1.3.6, we list the topics that are relevant to the works of both Selberg and Langlands

1.3.4 Eisenstein Series

An Eisenstein series of weight $2k$ on a lattice Λ in C is defined by

$$E_{2k} = \sum \lambda^{-2k}, \ \lambda \in \Lambda, \ \lambda \neq 0, \ k \text{ a positive integer.}$$

It is invariant under integral linear change of lattice basis and homogeneous with respect to complex rescaling of the lattice. It is also holomorphic i.e. it satisfies the Cauchy–Riemann equation.

The *real-analytic* Eisenstein series, which was introduced by Maass, is an Eisenstein series without the constraint of being holomorphic; instead, it is an eigenfunction of a non-Euclidean second-order Laplace operator. It played a central role in Selberg's spectral theory on $SL_2(R)$, and later was vastly generalized by Langlands in his spectral theory.

By their definition, Eisenstein series converge absolutely only in some right half-plane and are not even square-integrable. In applications it is therefore necessary to extend their convergence to the entire complex plane by meromorphic continuation, and this was an important and challenging undertaking for both Selberg and Langlands in their investigations on spectral theory.

1.3.5 Automorphic Forms

Since Maass introduced nonholomorphic modular forms of one variable, there was the need to deal with this together with the holomorphic theory.

Gelfand (together with Fomin) translated the notion of modular form to a class of vectors in certain Hilbert space. In this new setting, the theory of holomorphic modular forms is on the same footing with the nonholomorphic modular forms discovered by Maass; that is, the only difference is their Laplace eigenvalues. Also, the moderate growth condition translates into being square integrable. The result is that a modular (cusp) form becomes identified with an element of an infinite-dimensional Hilbert space.

The resulting notion became known as an automorphic form. In this sense, Gelfand started the spectral theory of modular forms.

The modern analytic theory of automorphic forms has its origins largely in the work of Hecke and Siegel: Hecke with his L-functions and Siegel who worked with a large number of groups – general linear groups, orthogonal groups, and symplectic groups – and, in general, brought the nineteenth century into the twentieth.

This – according to my understanding and reading, but this was never systematic – laid the foundation for the modern theory. That one decisive step was the extension of the reduction theory to general reductive groups by Borel and Harish-Chandra [in Arithmetic subgroups of algebraic groups *Ann. Math.* 1962].

The general L^2-theory, thus the theory for general reductive groups, especially the notion of cusp form, seems to have had its origins in two papers, one by Godement [in *Sem. H. Cartan*, 1957/58, Ex-8, pp. 8–10] and the other by Harish-Chandra [Automorphic forms on a semisimple Lie group, 1959].

1.3.6 The Rise of Automorphic Representation

From the 1940s to the 1950s there was a flurry of development of modular forms in two separate directions. On the one hand, Maass and Selberg established the theory of real-analytic modular forms. On the other hand, Siegel, with his reduction theory of many variables, had successfully generalized the theory of modular forms to several complex variables. One of the sticky points was that Siegel modular forms were necessarily analytic functions. This meant that transferring their definitions over to groups was not straightforward, since there may not even be a complex analytic structure. Those two points of view were amalgamated with the work of Harish-Chandra, whose use of Harmonic Analysis allowed him the flexibility to work in both higher dimensions like Siegel, and also with relaxed analytic conditions like Selberg. These simultaneous relaxations, together with the contributions of Gelfand, gave rise to the theory of automorphic representation.

1.3.7 Selberg's Spectral Theory on $SL_2(R)$ (1950s–1960s)

The Hecke theory and thus the possibility of generalizing it was in the air at the time, in part because the possibility of Hasse–Weil L-functions was in the air. In part because Hecke's work was becoming more familiar to mathematicians. This may have been encouraged by the ideas of Maass, taken up by Selberg. Hecke's L-functions and the converse theorems worked for imaginary quadratic extensions.

It was Maass, not Selberg, who created a theory that functioned for real quadratic extensions. Maass was however a weaker analyst than Selberg, and neither he nor his students were able to establish a theory for the continuous spectrum in the general case (of discrete subgroups of SL(2)). Selberg undertook this, but as far as I know with knowledge of Maass's papers the continuous spectrum is given by series that for subgroups of SL(2, Z) are Eisenstein series in the classical sense. For other discrete subgroups of SL(2, R) Maass, I believe, was not able to establish the necessary analytic continuation. That was done by Selberg, who perhaps did not give enough credit to Maass. Roelcke, a student of Maass, also examined the pertinent spectral theory. By the way Maass was influenced by

Hecke, indeed a student of Hecke. For Hecke the L-functions came first. This was probably the case for Maass as well.

We learned from Langlands that the central analytic problem of establishing the spectral theory, continuous and discrete, was broached first by Maass and through his influence but with stronger results by Selberg, who was looking for a spectral decomposition of the space $L^2(\Gamma \backslash G)$, where Γ is a congruence subgroup of G, $G = \mathrm{SL}_2(\mathbb{R})$, and $L^2(\Gamma \backslash G)$ is an infinite-dimensional Hilbert space of square-integrable automorphic forms.

It was Selberg who first appreciated that the classification problem of automorphic forms was essentially an eigenfunction problem for a certain operator. In the case of real-analytic Eisenstein series, this operator is a non-Euclidean second-order Laplace operator which gave both discrete and continuous spectra. Selberg knew that the continuous spectrum can be described completely by the real-analytic Eisenstein series, which is an eignfunction of the Laplacian operator. In general, the discrete spectrum is a mystery. Selberg's way of treating the discrete spectrum was to introduce the Selberg trace formula.

In summary, Selberg succeeded in finding a complete spectral basis for the non-Euclidean Laplace operator on $\mathrm{SL}_2(\mathbb{R})$, and hence paved a way to construct every possible square-integrable automorphic form on $\mathrm{SL}_2(\mathbb{R})$.

In those same two years, I had begun to study the Hecke theory. Gunning with whom I often spoke had written very convenient and accessible notes on it. Also sometime in those first months or years in Princeton, I acquired and began to read the various Paris seminars of Cartan, Godement and others inspired by the work of Siegel, Hecke, Selberg and the work of the pioneers of representation theory Gelfand and coauthors, Bargmann and Harish-Chandra.

I also began to think about the general theory of Eisenstein series taking advantage of the many results on theta series of various kinds in the papers of Siegel. I would occasionally mention these to Selberg, but he always replied that it was a general theorem that was needed. When the general theorem was offered, he said nothing, but the general theorem came later.

It is, by the way, curious that Selberg himself never could deal with the general theory. He himself was never clear about what he knew and did not know.

My guess is that he never understood the notion of a cusp form or the mutual relations between parabolic subgroups. This meant that he could not separate the new problems posed in several variables from those already settled as a consequence of the theory for one variable.

Selberg was extremely interested in the generalization of his result on $\mathrm{SL}_2(\mathbb{R})$ to higher-rank groups; however, he was prevented by many complicated obstructions. He is reputed to have said: "Eisenstein series are Dirichlet series in r complex variables. It seems very difficult to attack the problem of continuation of these."

1.3.8 Langlands' Spectral Theory (1962–1964)

It was Langlands who undertook the project to resolve the problem of the "general theorem" that Selberg had in mind. It was the higher-rank problem for arbitrary reductive groups. Langlands not only gave a complete description of the spectral decomposition of $L^2(\Gamma\backslash G)$ but also succeeded in identifying the spectral basis of $L^2(\Gamma\backslash G)$ for arbitrary reductive groups G, where Γ is a discrete subgroup of G such that $\Gamma\backslash G$ is of finite volume.

This project required a drastic change in language: from a complex analytic framework to a Lie group-theoretic setting. The necessary tool to deal with the latter was mainly provided by Harish-Chandra's work. At the same time, Langlands had to lay the foundation of the theory of general Eisenstein series for general groups.

Eisentein series themselves have a certain domain of convergence, and unfortunately the spectrum resides outside of this range of convergence. Thus, constructing a spectral basis is tantamount to meromorphically continuing Eisenstein series.

Langlands initially encountered the problem of meromorphic continuation of real-analytic Eisenstein series on $SL_2(R)$ as a graduate student at Yale from reading Selberg's paper on the topic. He had some initial success in extending Selberg's ideas to several variables. In doing so, it was clear to him that the problem of several variables was more complex than previously expected.

The theory of Eisenstein series on higher-rank groups turned out to be significantly more involved than on $SL_2(R)$. Langlands' PhD thesis (see Section 1.2.9) had prepared him to handle the challenging situation here. Langlands succeeded in decomposing $L^2(\Gamma\backslash G)$ into both the continuous spectrum and the discrete spectrum. Attempts to parameterize the continuous spectrum by Eisenstein series presented its own problems because the continuous spectrum broke up into a descending tree, where each level represents a factor of the previous one, reducing the problem to lower rank, making them amenable to induction arguments. Roughly speaking, Langlands' proof was an elaborate induction on the rank of the group. However, its significant layer of combinatorial complexity was one of the complications facing Langlands. Moreover, some extra attention is needed when some Eisenstein series actually arise as the "residues" of some other Eisenstein series on larger groups. This phenomenon had not been recognized until then.

> I read Gelfand's address to the ICM in Stockholm, finally understood correctly the notion of a cusp form in general. Since, as I observed, I had some passive experience with the spectral theory of self-adjoint operators and with holomorphic forms of several variables, several months with my nose to the grindstone and a refusal to be discouraged by temporary setbacks – for the proof presented a good number of unexpected obstacles – gave me in the spring of 1964 a complete proof of analytic continuation.

I was exhausted and, moreover, quite dissatisfied with the account of the proof but with no energy and no desire to revise the exposition. If Harish-Chandra had not taken time from his own researches to work through and present at least a part of my paper – that pertaining to Eisenstein series associated to cusp forms – no one may have taken me seriously. To Bochner and Harish-Chandra I owe an enormous amount.

So what did I offer or have available that was not available to Selberg? Of course, one thing I had was an understanding of the style of Harish-Chandra and the context in which he worked. Another, was a clear notion of cusp-form.

At this point Langlands had worked solely with the archimedean place. However, this would be the archimedean component π_∞ of what later become the "automorphic representation π." The nonarchimedean parts of an automorphic representation were also being investigated, around the same time, mainly by Tamagawa and Satake (see Section 1.3.12).

Langlands' spectral theory was undoubtedly his first major achievement.

1.3.9 UC Berkeley – a Disappointing Year (1964–1965)

In the fall of 1964, I went to Berkeley with my wife and three children, the last child being born just after our return to Princeton, an event which allows me to fix the dates. I was unable to initiate any new project. I had hope, having established the analytic continuation of the Eisenstein series, to turn immediately to the trace formula, but it was too daunting.

Although, as I appreciate in retrospect, it was not an entirely unsuccessful year. There are, I now understand, several useful things left over from the year: the proof of the Weil Conjecture on Tamagawa numbers for Chevalley groups and implicitly for quasi-split groups, although the latter had to await the thesis of K.-F. Lai a formula for the inner product of truncated Eisenstein series; and a conjecture inspired by calculations of P. Griffiths and proved by W. Schmid on the cohomological realization of the discrete series representations constructed by Harish-Chandra not long before. I ran a seminar together with P. Griffiths on abelian varieties, but in the end he did much more with the material than I did. Nevertheless I did not attach much importance to them and was discouraged. I did not have the feeling that things were working out.

1.3.10 The Boulder Conference in Boulder, Colorado (Summer 1965)

In the summer of 1965, Langlands attended the Boulder conference in Boulder, Colorado, which was organized by Borel and Mostow. His reflections suggested that it was an important event to him.

I came of age at a time when Hecke, Siegel, Maass and Selberg had revived a theory that goes back to Eisenstein, Dirichlet, and many others and which would now be referred to as the theory of automorphic forms. At the same time Artin and Chevalley had not been able to create a non-abelian class field theory and Artin had even written that he had come to believe that there was no such theory.

At the Princeton University Bicentennial Conference on the Problems of mathematics held in 1956, on p. 5 of a pamphlet published on that occasion, Brauer reported his proof of Artin's conjecture about induced characters. Brauer's proof asserts that characters known to be rational combinations of certain special characters are in fact integral rational combinations. Brauer's result represents a decisive step in the generalization of class-field theory to the non-abelian case, which is commonly regarded as one of the most difficult and important problems in modern algebra.

Artin stated

My own belief is that we know it already, though no-one will believe me – that whatever can be said about non-abelian class field theory follows from what we know now, since it depends on the behavior of the broad field over the intermediate fields – and there are sufficiently many abelian cases. The critical thing is learning how to pass from a prime in an intermediate field to a prime in the large field. Our difficulty is not in the proofs, but in learning what to prove.

We have Langlands' comment on Artin's quotation:

It is evident that Artin's notion of a non-abelian theory is not ours. His notion is closely tied to contributions of the initial creators of the abelian theory; ours lies more in Artin's own contributions.

The Boulder conference had inspired Langlands to investigate the following two problems.

I participated in the Boulder conference and learned, somewhat belatedly, to think in terms of reductive algebraic groups. After having been introduced, in one way or another, in the early 1960s to the papers of Siegel, Selberg, Hecke, and Harish-Chandra and to class-field theory, I had the background to reflect on the problems (1) to create a general theory of L-functions, thus for all possible groups G appearing in the theory of automorphic forms, or better, automorphic representations in general, but along the line of the Hecke L-functions; and (2) to find a non-abelian class field theory, even though my understanding of the abelian theory was weak.

As I recall, in the early sixties, a number of mathematicians had created a structure theory for groups over p-adic fields. These were described in the very successful Boulder conference. Some of the representation theory for real groups, created by Harish-Chandra, by the Russian school and, to a lesser extent, by others, had been extended to groups over p-adic fields, in particular the theory of spherical functions to which among others, Satake had contributed.

The first problem was popular at the time and many unsuitable approaches were discovered. The solution of the second problem was the unfulfilled goal of Emil Artin.

Section 1.3.11 on L-functions is a supplement to the above mentioned problems (1) and (2)

1.3.11 L-functions – Euler, Riemann, Dirichlet, Frobenius, Artin, Hecke, and Langlands

We have mentioned in Section 1.3.1 that the distinct feature of Langlands' automorphic L-functions is their dual nature – both arithmetic and analytic. As we'll see they are vast generalizations of classical L-functions attached to the names listed above, from Euler to Artin and Hecke.

1.3.11.1 The Euler–Riemann ζ-Function

The Euler–Riemann ζ-function was first studied by Euler for real values of s only. He showed, using only the fundamental theorem of arithmetic, that the series had an Euler product expression over all prime numbers p that is

$$\zeta(s) = \sum_n n^{-1} = \prod_p (1 - p^{-s})^{-1} R(s) > 1 \quad n = 1, \ldots, \infty,$$

An immediate result is that the number of primes p is infinite.

Later, Riemann, with the theory of complex analysis, was able to extend the domain of s from real values to complex values. He also showed that $\zeta(s)$ has "nice" analytic properties:

1. analytic continuation to a meromorphic function of s in the entire complex plane, with a simple pole at $s = 1$; and
2. functional equation relating values at s and $1-s$.

Those two analytic properties together with the Euler product will be the *characteristic (or nice) feature* of any future L-functions (or L-series).

1.3.11.2 Dirichlet L-Series

Sometime later, around 1840, Dirichlet introduced complex valued functions $\chi(n), n$ is a positive integer, such that $\chi(nm) = \chi(n)\chi(m), \chi(n+N) = \chi(n)$, and $\chi(n) = 0$ if gcd $(n, N) > 1$ Such a χ is called a *Dirichlet character* with modulus **N**.

By generalizing Riemann's construction, Dirichlet introduced his L-series and showed that it has an Euler product expansion over all primes p

$$L(s, \chi) = \sum_n \chi(n)n^{-s} = \prod_p (1 - \chi_p p^{-s})^{-1} n = 1, \ldots, \infty. \quad (1.3.11.1)$$

Dirichlet showed that $L(s, \chi)$ has analytic continuation to a meromorphic function on the complex plane and a functional equation. In fact, the Dirichlet

L-series behaves very much like the Riemann ζ function, and Dirichlet showed that the number of primes in an arithmetic progression is infinite.

1.3.11.3 Artin L-Functions – Galois Representations and Frobenius Class

Let E be a Galois extension of $\mathbf{Q}$ such that E is the splitting field of a monic, irreducible, integral polynomial $f(x)$ of degree n. The Galois group $G = \mathrm{Gal}(E/\mathbf{Q})$ of E (or f) is the group of automorphisms of E leaving $\mathbf{Q}$ fixed. Since the elements of G permute the roots of f, we have an injective homomorphism of G into the symmetric group S_n on n elements. We say $\mathbf{E}$ (or $\mathbf{f}$) is *abelian* if $\mathbf{G}$ is an *abelian group*.

A *representation* of a group G is simply a homomorphism σ of G into the group of invertible linear transformations of a vector space $\mathrm{GL}(V)$ σ: $G \to \mathrm{GL}\,(V)$.

A *subrepresentation* is one that stabilizes a subspace of V. A representation is *irreducible* if it cannot be written as a direct sum of proper subrepresentations.

Representations distinguish abelian groups from nonabelian groups, because of the following:

Schurs' Lemma Every irreducible representation of a group G is one-dimensional if and only if G is abelian.

It is clear that abelian groups can be described by characters (one-dimensional representations) alone while representations of nonabelian groups are necessarily higher dimensional.

Artin was interested in relating prime numbers p to elements of the Galois group of an irreducible polynomial f of degree n by considering the way in which f factors modulo p. That is

$$f(\mathrm{x}) \equiv f_1(\mathrm{x}) \cdots f_r(\mathrm{x})(\mathrm{mod}\,p), \tag{1.3.11.2}$$

where each f_i is an irreducible polynomial of degree n_i, and $n = n_1 + n_2 + \cdots + n_r$.

An *n-dimensional Galois representation* is a homomorphism r of the Galois group $\mathrm{Gal}(E/\mathbf{Q})$ into $\mathrm{GL}_n(C)$, i.e. $r : \mathrm{Gal}(E/\mathbf{Q}) \to \mathrm{GL}_n(C)$, where E is a finite Galois extension of $\mathbf{Q}$. If E is an abelian extension over $\mathbf{Q}$, and r is irreducible, then r is a character $\sigma : \mathrm{Gal}(E/\mathbf{Q}) \to C^\times$.

It was Frobenius who introduced a canonical way of turning primes into conjugacy classes in Galois groups over $\mathbf{Q}$. He had shown that there is an unique element g of the Galois group $\mathrm{Gal}(E/\mathbf{Q})$, defined by $g(\mathrm{x}) = \mathrm{x}^p$ (mod P), for all integers x of E, where p is an unramified prime and P is a prime ideal lying over p in E, which when considered as an element of S_n is defined by the partition

$$\Pi p = \{n_1, \ldots, n_r), \text{ where } n_1 + \cdots + n_r = n$$

and it is *well-defined up to conjugation.*

We denote the *conjugacy class* of g associated to p by Frob_p and it is called the *Frobenius class*. Any element of the conjugacy class is called a *Frobenius element of p*.

Remarks
(1) $\mathrm{Frob}_p = 1$ if and only if $f(x)$ splits into linear factors modulo p in (1.3.11.2)
(2) Suppose E is the Nth cyclotomic field, then its Galois group over $\mathbf{Q}$ is abelian and isomorphic to $(Z/NZ)^x$, and conjugacy classes in the Galois group are just elements. Now suppose p does not divide N. Then the Frobenius class is $p \bmod N$. This example suggests not only that the distribution of Frobenius conjugacy classes in Galois groups over $\mathbf{Q}$ generalizes Dirichlet's result about primes in arithmetic progressions, but also that the Frobenius class is linked with the Dirichlet character

Artin was inspired by the two notions mentioned above: the first is the notion of a *group representation* that is distinct from a character and that could distinguish between abelian and nonabelian extensions The second is the Frobenius class which is well defined in nonabelian extensions. Historically, Artin's L-functions were developed by Artin in part as an attempt to understand nonabelian class field theory.

A central ingredient in Artin's construction of his local L-function was an n-dimensional Galois representation $r \colon \mathrm{Gal}(E/\mathbf{Q}) \to \mathrm{GL}_n(C)$, where the Frobenius conjugacy class Frob_p will be mapped to a semisimple conjugacy class $r(\mathrm{Frob}_p)$ inside $\mathrm{GL}_n(C)$. Semisimple conjugacy classes in $\mathrm{GL}_n(C)$ have a particularly simple representation: a diagonal matrix with the roots of the characteristic polynomial of the conjugacy class along the diagonal.

Artin defined its local L-function in $\mathbf{Q}$ as

$$L_p(s,r) = [\det(1 - r(\mathrm{Frob}_p)p^{-s}]^{-1} \quad p \text{ unramified,} \qquad (1.3.11.3)$$

in term of the associated characteristic polynomial, where $r(\mathrm{Frob}_p)$ is a semisimple conjugacy class in $\mathrm{GL}_n(C)$.

The global Artin L-function, defined for unramified primes in $\mathbf{Q}$ only, is the Euler product

$$L(s,r) = \Pi_p L_p(s,r) = \Pi_p [\det(1 - r(\mathrm{Frob}_p)p^{-s}]^{-1}. \qquad (1.3.11.4)$$

Artin worked from the beginning with finite Galois extensions E over an arbitrary number field F, a finite field extension of $\mathbf{Q}$, rather than $\mathbf{Q}$ itself. Algebraic number theory assures us that the definitions we have so far can be extended from $\mathbf{Q}$ to F

Let S be the set of prime ideals for F that ramify in E, then for any prime ideal $P \notin S$, there is a canonical conjugacy class Frob_P in $\mathrm{Gal}\,(E/F)$.

From the n-dimensional representation $r \colon \mathrm{Gal}\,(E/F) \to \mathrm{GL}_n(C)$, Artin defined his local and global L-functions in F as follows:

$$L_F(s,r) = \prod_P L_{F,P}(s,r) = \prod_P [\det(1 - r(\text{Frob}_P)(NP)^{-s}]^{-1} \quad P \notin S$$

(1.3.11.5)

where NP is the number of elements of $\mathcal{O}/P$, and $\mathcal{O}$ is the ring of algebraic integers of F.

Artin conjectured that $L_F(s,r)$ has nice analytic properties. Since it is defined as an Euler product, it is fundamentally an arithmetic object and the kind of analysis Dirichlet applied to his L-function was not suitable for $L_F(s,r)$. However, when E/F is an abelian extension, Artin was able to link his L-function with Hecke's degree-1 L-series which has nice analytic properties. This was achieved via the Artin reciprocity law. Let us introduce Hecke's L-series before stating the Artin reciprocity law.

1.3.11.4 Hecke L-Series

1.3.11.4.1 Hecke's Degree-1 L-Series A significant generalization of the works of Riemann and Dirichlet was taken up by Hecke who generalized the Dirichlet L-series to number fields. A Hecke L-series for an abelian number field extension F over $\mathbf{Q}$ takes the form

$$L(s,\chi) = \sum_U \chi(U) N(U)^{-s} = \prod_P L(1 - \chi(P)N(P)^{-1})^{-1}, \quad (1.3.11.6)$$

where the Hecke character χ is a character on the idèle class group, U is an (ordinary) ideal and P is an unramified prime ideal of the ring of algebraic integers of F. Hecke showed that $L(s,\chi)$ has a degree-1 Euler product and it possesses desirable analytic properties such as analytic continuation and a functional equation.

We remark that both Dirichlet's L-series and Dedekind's ζ-function are special cases of Hecke's degree-1 L-series.

1. Taking the number field F to be $\mathbf{Q}$ and the character χ to be nontrivial, then $L(s,\chi)$ specializes to the Dirichlet L-series.
2. Taking the character χ to be the trivial character, then one recovers the Dedekind ζ-function.

1.3.11.4.2 Heck's Degree-2 L-Series – Ramanujan, Mordell, and Hecke Operators Ramanujan, in 1916, was the first to observe that modular forms could have L-series attached to them. He actually conjectured that the classical Delta-function has a degree-2 L-series with Euler product expansion. A year later, Mordell proved this by introducing a series of operators known as Hecke operators (an early version). Hecke operators as we now know them are a vast generalization of Mordell's version by Hecke in 1936.

Relying on the Hecke operators, Hecke showed that given a modular cusp form for $SL_2(Z)$, $f(z) = \sum a_n e^{2\pi i n z}$, $n = 1, \ldots, \infty$, of weight k and level N, its associated *Dirichlet L-series* has a *degree-2 Euler product* expansion if f is an eigenfunction for all the Hecke operators $\mathbf{T}_p$ with $\mathbf{T}_p \mathbf{f} = \mathbf{a}_p \mathbf{f}$ for all primes p, that is

$$L(s, f) = \sum_n a_n n^{-s} = \prod_p (1 - a_p p^{-s} + p^{k-1-2s})^{-1}. \qquad (1.3.11.7)$$

Hecke was also able to show that $L(s, f)$ could be meromorphically continued to the entire complex plane and possessed a functional equation. Hecke's degree-2 L-series shed a new light on the theory of automorphic forms and their L-functions. Interestingly, neither Hecke nor Artin was aware of the other's work on L-functions even though they were working not far from each other. Their L-functions were finally linked by Langlands as a special case of his functoriality conjecture.

1.3.11.5 The Artin Reciprocity Law and Class Field Theory

I will tell you a story about the Reciprocity Law. After my thesis, I had the idea to define L-series for nonabelian extensions. But for them to agree with the L-series for abelian extensions, a certain isomorphism had to be true. I could show it implied all the standard reciprocity laws. So I called it the General Reciprocity Law and tried to prove it but couldn't, even after many tries. Then I showed it to the other number theorists, but they all laughed at it, and I remember Hasse in particular telling me it couldn't possibly be true. Still, I kept at it, but nothing I tried worked. Not a week went by – for three years! – that I did not try to prove the Reciprocity Law. It was discouraging, and meanwhile I turned to other things. Then one afternoon I had nothing special to do, so I said, "Well, I try to prove the Reciprocity Law again." So I went out and sat down in the garden. You see, from the very beginning I had the idea to use the cyclotomic fields, but they never worked, and now I suddenly saw that all this time I had been using them in the wrong way – and in half an hour I had it.

(– Emil Artin, as recalled by Mattuck (in Recountings: Conversations with MIT Mathematicians 2009).)

The Artin Reciprocity Law Suppose E is an abelian extension over F and let S be the set of prime ideals for F that ramify in E. Then for any Galois representation r over F, $r: \text{Gal}(E/F) \to C^x$, there is a Hecke character χ over F such that

$$r(\text{Frob}_P) = \chi(P), \quad P \notin S, \qquad (1.3.11.8)$$

and this lead to

$$L_F(s, r) = L(s, \chi). \qquad (1.3.11.9)$$

Both (1.3.11.8) and (1.3.11.9) are called the *reciprocity law.*

Artin's reciprocity law identifies Artin's abelian L-function $L_F(s,r)$ in (1.3.11.5) with Hecke's abelian L-series $L(s, \chi)$ in (1.3.11.6). This identity enables $L_F(s,r)$ to acquire nice analytic properties such as analytic continuation and a functional equation from $L(s, \chi)$.

Let us list a few consequences of Artin's reciprocity law.

(a) The reciprocity law (1.3.11.8) informs us that abelian Galois extensions E over F are limited to those attached to Hecke characters χ. From now on let E be an *abelian* extension over **Q**, then (1.3.11.8) simplifies to

$$r(\text{Frob}p) = \chi(p), \qquad\qquad (1.3.11.10)$$

where p is a prime in **Q** and χ is a Dirichlet character for **Q** with modulus N.

Let us also recall that $\text{Frob}_p = 1$ if and only if $f(x)$ splits into linear factors in (1.3.11.2).

(b) Let $S(E/\boldsymbol{Q})$ be the set of unramified primes p such that $\text{Frob}_p = 1$. Then $S(E/\boldsymbol{Q})$ is also the set of primes that splits completely in E. From Tate's work (T, p. 165) we know that the map $E \to S(E/\boldsymbol{Q})$ is injective. So we have a criterion which limits the number of abelian Galois extensions E of **Q**.

(c) Suppose $f(x)$ in (1.3.11.2) is a quadratic polynomial Then (1.3.11.10) tells us that *a prime is a sum of two squares if and only if it is congruent to 1 modulo 4.*

(d) Artin's reciprocity law implies the *Kronecker–Weber theorem: Any finite abelian extension of* **Q** *is contained in the cyclotomic extension of Nth roots of 1, for some N.*

The Artin reciprocity law is the essence of *class field theory*. How should we describe class field theory? Let us take a look at what Langlands has to say:

> Class field theory, which establishes a decisive connection between abelian extensions of number fields on one hand and their ideal class groups on the other hand, classifies and constructs all abelian extensions of number fields. So problem arose after the completion of the theory to classify and construct (in some sense) all finite extensions. In 1956 at the bicentennial conference of Princeton University, Artin suggested that perhaps all we could know and all we needed to know in general was implicit in our knowledge of abelian extensions, so that there was in fact little left to do, although it was not clear what it might be.

1.3.11.6 Emil Artin and the Pursuit of Nonabelian Class Field Theory

In the second of his *Two Lectures on Number Theory, Past and Present*, André Weil summarized the state of affairs up to the mid 1930s, after the Abelian Reciprocity Law had been proven, by saying:

> Artin's reciprocity law, which in a sense contains all previously known laws of reciprocity as special cases, deals with a strictly commutative problem. It establishes a relation between the most general extension of a number field with a commutative Galois group on the one hand, and on the other hand the

multiplicative group over that field. Where do we go from there? Well – of course we take up the noncommutative case.

The Artin school had hoped in vain that their theory of simple algebras, and the resulting nonabelian cohomological theory would lead to a nonabelian class field theory, but this never materialized. A generalization of Artin's reciprocity law to nonabelian extensions of number fields is the Langlands' functoriality conjecture, which, in a special case, is Artin's reciprocity conjecture, and identifies, conjecturally, Artin's nonabelian L-functions with Langlands' automorphic L-functions which are analytic in nature. In this respect, Langlands' functoriality conjecture represents a general formulation of nonabelian class field theory – a theory that had eluded Artin's grasp more than half a century ago.

1.3.11.7 Langlands' Automorphic L-Functions

We have seen that Hecke's degree-1 L-series was the analytic counterpart of Artin's abelian L-functions. The analytic counterpart of Artin's nonabelian L-functions is Langlands' automorphic L-functions $L(s, \pi, \rho)$. This identity enables Artin's nonabelian L-functions to have analytic continuation and functional equation, albeit conjecturally.

A novel feature of Langlands L-function $L(s, \pi, \rho)$ is the presence of its dual representations: the automorphic representation π and the finite-dimensional representation ρ. *Automorphic representations* are the nonabelian generalizations, largely due to the introduction of adèlic theory, of *Hecke characters*. The finite-dimensional representation ρ is, roughly speaking, an extension of the Galois representation associated with the Artin L-function, and the Frobenius classes play an important role. In this sense, *Langlands L-functions* represent an unprecedented generalization and fusion of ideas of, among others, Dirichlet, Frobenius, Hecke, *and* Artin. This topic will be revisited in Section 1.3.15.

Why should L-functions deserve so much attention? Well, besides their applications to number theoretical problems, L-functions are particularly relevant to the principle of functoriality. For example, take two L-functions, which are concrete objects to study; their equality is equivalent, via the functoriality conjecture, to the identity of their attached objects. Those objects could appear quite different and hence much harder to handle.

1.3.12 The Adèlic Theory

The adèlic theory played an important role in Langlands' discovery of his L-functions. Chevalley had introduced the adèles into number theory with the intent to streamline notions in class field theory. An important application of the adèles came in Tate's Princeton university thesis (1950). Tate recast Hecke's work about his L-functions using the theory of adèles.

In 1963, Tamagawa, and independently Godement, extended Tate's work on Hecke's L-functions from GL(1) to an inner form of GL(n). In particular, he showed that there is a Hecke algebra for which one can produce (in a similar vein to Hecke) a degree n, pth Hecke polynomial. Further, Tamagawa was systematically introducing adèlic methods and the theory of spherical functions into the theory of automorphic forms. In particular, Satake (1964) translated Hecke theory into the language of spherical functions (the local analogue of Hecke operators), and he also localized Tamagawa's arguments with spherical functions and generalized them from GL(n) to general p-adic groups.

> Spherical functions were in the air and I used that. These, I suppose, relied on a more or less obvious transfer of the theory of spherical functions from the reals to the p-adics that is implicit in the theory of Eisenstein series. It is difficult to transfer oneself back to that time, but this transition was made more or less unconsciously.

1.3.13 A Promising Career in Jeopardy (Fall 1965–Summer 1966)

Langlands started to think seriously about the two problems mentioned in Section 1.3.10 soon after the Boulder conference, but he was not successful. He grew more and more discouraged to the point that he was seriously considering abandoning mathematical research. Consequently, some months later, he decided to spend a year in Ankara Turkey.

> There were in the 1960s still residues of romantic notions of British imperialism, embodied in figures like Gertrude Bell or T. E. Lawrence. So one could dream, even with a wife and four small children, of escaping into the life and language of some exotic land and beginning anew. I did; my wife, more generous than wise, did not discourage me and we made plans to spend a year, at first several years, but the department in Ankara only agreed to a provisional 1-term appointment. The specific choice of Turkey was the result of accidental factors.

The decision had been taken by the summer of 1966. Langlands was, no doubt, attempting to liberate himself. The decision itself freed him and all ambitious projects were dropped. He took up again the study of Russian, which he had abandoned for many years.

1.3.14 Constant Term Calculations – an Unexpected Path to Success (Fall 1966)

In the fall of 1966, Langlands did not spend much time with mathematics. His interest was in the studies of Russian and Turkish.

> I took a beginner's course in Turkish and a course in Russian with which I already had a little experience. The teacher was Valentine Tschebotaref-Bill. I liked her as

a teacher and I think she was fond of me. However, I still had a little time for mathematics but no serious goals. As far as I can recall, I began idly, simply to fill the time, to calculate the constant term of the Eisenstein series associated to maximal parabolics of split groups. I had no goal in mind, just nothing better to do. Suddenly, without any effort on my part, the results suggested the form to be taken by the Euler products for which I had been looking in vain.

The constant terms were quotients of Euler products and these Euler products could be continued to the entire plane as meromorphic functions, already a convincing beginning. The ideas must have matured over the course of the summer and fall of 1966, my thoughts quickening towards the year's end. The strange way in which mathematical insights arise.

Langlands realized that the constant term he was calculating possessed features that could be related to the automorphic L-function he was searching for. This revelation both inspired and energized him, and in the next few months he worked feverishly to reach his goal.

I abandoned the Turkish course and the Russian course from one day to the next, although I was already committed to Turkey. This angered Ms. Tschebotaref-Bill and with good reason. She had been generous to me. I was unable to explain the situation to her satisfactorily.

The unexpected success from his constant term calculations was a total surprise, perhaps not only to Langlands himself but also to everyone who was trying to find an acceptable definition for an automorphic L-function.

I had not recognized in 1966, when I discovered after many months of unsuccessful search for a promising definition of automorphic L-function, what a fortunate, although, and this needs to be stressed, unforeseen by me, or for that matter anyone else, blessing it was that it lay in the theory of Eisenstein series. What is to be very much stressed is that there is no intrinsic reason that there should or must be a connection between the local factors of the Euler product and the representation theory of reductive groups. There is, so far as I know, absolutely no reason that this is so. There is no comparable phenomenon for the group GL (1) alone. The phenomenon cannot be discovered before one has passed – in the spirit of Siegel and his nineteenth-century precursors – to automorphic forms on reductive groups. It is a phenomenon for which neither the Bourbaki nor the Artin–Chevalley school were prepared, and which they still cannot readily accept. These schools were enormously successful and their members are still very influential.

1.3.15 A Successful Search for Automorphic L-Functions (Winter 1966),the Satake Isomorphism and the Langlands Isomorphism

The Eisenstein series that Langlands was working with was the most general Eisenstein series which was generalized by him in his construction of

Langlands' Spectral Theory (Section 1.3.8). The new feature of this Eisenstein series, in 1966, was its adèlic format. Langlands was calculating the p-factors of the Euler products in the constant term which arose in the functional equation of his adèlic Eisenstein series. They appeared in the form

$$\prod \xi_i(\alpha_i s)/(\xi_i(\alpha_i s + 1) \quad i = 1, \ldots, r, \tag{1.3.15.1}$$

where $\xi_i(\alpha_i s)$ is a polynomial. They were defined on $F/\mathbf{Q}$, for reductive groups, where F is a finite field extension over the rational numbers $\mathbf{Q}$

> If I had not searched assiduously for a general form of the theorems of Hecke and of the founders of class field theory, or had not been familiar with various principles of non-abelian harmonic analysis as it had been developed by Harish-Chandra, in particular with the theory of spherical functions, I might have failed to recognize the importance or value [of (1.3.15.1)]. It is the relation expressed [by (1.3.15.1)] that suggests and allows the passage from the theory of Eisenstein series to a general notion of automorphic L-function that can accommodate not only a non-abelian generalization of class-field theory but also, as it turned out, both functoriality and reciprocity. It was the key to the suggestions in the Weil letter.

Langlands was intrigued by two unexpected features in these polynomials: (a) the degree of the polynomial in the numerator of (1.3.15.1) was not the dimension of the representation of the group G as he had expected, instead, it was the dimension of the representation of the dual of that group; and (b) these polynomials in (1.3.15.1), as characteristic polynomials of semisimple conjugacy classes on a dual group, shared some of the same features as the characteristic polynomials in Artin's construction of his L-function via Frobenius conjugacy classes.

Langlands' observation (a) led him to focus not only on the dual group G but also on the link between the local factors of the Euler product and the representation theory of reductive Lie groups. Such a link was established via the creation of the L-group $^L G$. A conjugacy class $c(\pi_p)$ in the L-group is called the "Langlands class."

1.3.15.1 The L-Group

The L-group $^L G$ is defined usually over field extensions K/F as the semidirect product of the dual group $\hat{G}$ with the Galois group $\mathrm{Gal}(K/F)$ (or the Weil group $W(K/F)$),

$$^L G = \hat{G} \rtimes \mathrm{Gal}\,(K/F), \tag{1.3.15.2}$$

where K/F is a Galois extension and $F/\mathbf{Q}$ is a finite field extension with $\mathbf{Q}$ the field of rational numbers.

We note that when $K = F$, then we have $^L G = \hat{G}$. Since the dual group of a reductive group was not established at that time, Langlands' creation of both the dual group and the L-group were two of his major achievements.

I then realized not only that these functions were quotients of Euler products, but also that the numerator, whose form was similar to that of the denominator, could be described in terms of representation theory. The relevant representations were algebraic representations of a complex algebraic group – the L-group. I not only understood the form of these functions, I was also able to prove – thanks to the general theory of Eisenstein series – that they admit meromorphic continuations.

It was Langlands' creation of the L-group together with his observation (b) about the similarity between his characteristic polynomials and those of Artin's, that eventually led him to the definition of the automorphic L-function $L(s, \pi, \rho)$.

1.3.15.2 The Satake Isomorphism

Aside from the L-group, the other essential ingredient in the construction of Langlands' local L-functions is the Satake isomorphism.

> When and how I recognized the role of what is now called the L-group I do not know. The structure of the algebra of spherical functions on a general p-adic group had been established by Satake, as an extension of the known structure theorem of Harish-Chandra for K-biinvariant differential operators.
>
> [The Satake isomorphism] is suggested immediately by the analogous lemma for spherical functions over a real field. The step from it to what I have called the Frobenius–Hecke conjugacy class (i.e. the Langlands class) rather than the Satake parameter – a term often used by others – is technically minute, but entails a fundamental conceptual change. Without any sign of the L-group and without the desire to find an adequate notion of automorphic L-function, there was no need for it.

Let us take a closer look at Langlands' words in the above quote, especially about "The step [that] entails a fundamental conceptual change."

(a) Satake was motivated to construct the Satake isomorphism from his interest in the Euler factors appearing in Tamagawa's L-function associated to automorphic representation on an inner form of GL(n).

(b) Satake describes the contribution of his paper Theory of Spherical Functions on reductive algebraic groups over p-fields (*Publ. Math. IHES*, 18 (1963)), in the following words: "Then our main theorem asserts that $\mathcal{L}(G, U)$ is isomorphic to the algebra of all w-invariant polynomials functions on Hom $(M, C^*) \cong C^a, \ldots$, thus $\mathcal{L}(G, U)$ is an affine algebra of (algebraic) dimension a over C."

The algebra $\mathcal{L}$ (G, U) is the algebra of spherical functions. That is, however, suggested immediately by the analogous lemma for spherical functions over the real field.

(c) Satake proved the nonarchimedean analogue of the Harish-Chandra isomorphism, relating the Hecke algebra with a certain polynomial ring. The similarity of Satake's target ring to that of a general representation ring was striking. However, Satake had no motivation to suspect that it was a representation ring because "without the desire to find an adequate notion of automorphic L-function, there was no need for it."

(d) The conjugacy class $c(\pi_p)$ (i.e. the Langlands class), for unramified p, was an object in the L-group. It was Langlands who established a bijection between $c(\pi_p)$ and the unramified local automorphic representation π_p which is a character on the Hecke algebra H_G. This bijection was *the step* taken by Langlands which *entails a fundamental conceptual change*.

By what we have just seen, the following questions come up naturally.

* How can the Satake isomorphism, "without any sign of the L-group", link the Hecke algebra with a conjugacy class in the L-group $^L G$?
* Since a conjugacy class in the L-group has no direct connection with the Satake isomorphism, what to the "Satake parameter" referring to?

> The expression 'Satake parameter' is completely inappropriate and could very well be replaced by 'Frobenius–Hecke' parameter. It is an extension of the Frobenius parameter adapted to the Hecke theory both in the original and extended forms. This parameter as an element of the definition of automorphic L-functions was discovered in the context that is the continuation of contributions of Frobenius and Hecke.

1.3.15.3 The Langlands Isomorphism

Let us give a very brief account on Langlands' construction of the bijection in (d).

* The Satake isomorphism, as we have seen, only relates the Hecke algebra H_G with a certain polynomial ring.
* Langlands realized, from his constant term calculation and observation 1.3.15 (a), that the Satake target ring was in fact the representation ring of a certain complex algebraic group.
* In searching for that "certain" complex algebraic group, Langlands created the L-group.
* With the L-group in place, a bijection was established between the Satake's target ring and the representation ring of the L-group, denoted by $R\,(^L G)$.
* Linking this bijection with the Satake isomorphism, one gets an isomorphism relating the Hecke algebra H_G with the representation ring $R\,(^L G)$:

$$H_G \approx R(^L G), \qquad\qquad (1.3.15.3)$$

We call the isomorphism in (1.3.15.3) the *Langlands isomorphism*.
* The local automorphic representation π_p is a character on the Hecke algebra H_G. However, via the Langlands isomorphism (1.3.15.3), one can view π_p as a character on $R\,(^L G)$.

* Essentially by definition, characters of a representation ring correspond bijectively to conjugacy classes of the underlying group; therefore, π_p corresponds bijectively to the conjugacy class $c(\pi_p)$ inside the L-group, and this establishes the desired bijection in 1.3.15.2(d). We refer to $c(\pi_p)$ as the conjugacy class associated with π_p for unramified p.

1.3.15.4 Definitions of Langlands' L-Functions

Langlands' definition of L-functions actually presupposed local factors of ramified primes as well as the archimedean prime π_∞. However, the unramified L-functions $L_p(s,\pi_p,\rho_p)$ remain the most important component. In our version of Langlands' definitions of $L_p(s,\pi_p,\rho_p)$ and $L(s,\pi,\rho)$ we restrict ourselves to dealing with unramified primes p in $\mathbf{Q}$ only.

Since it is desirable to have the conjugacy classes $c(\pi_p)$ in a general linear group $G_n(C)$, rather than the complex (disconnected) group $^L G$, where $^L G = \tilde{G} \rtimes \text{Gal}(F/\mathbf{Q})$ and $F/\mathbf{Q}$ is a finite Galois extension, we let ρ_p be a finite-dimensional representation $\rho_p: {}^L G \to \text{GL}_n(C)$.

This step is in analogy with Artin's construction of his L-functions which had been attached to the finite dimensional representations of a finite Galois group. The image $\rho_p(c(\pi_p))$ is a semisimple conjugacy class inside $\text{GL}_n(C)$: a diagonal matrix with the roots of the characteristic polynomial of the Langlands class $c(\pi_p)$ along the diagonal.

We express $\pi = \otimes_p \pi_p$ as a (restricted) tensor product where p is an unramified prime. Langlands defined his L-functions as follows.

Definition (a) $L_p(s,\pi_p,\rho_p) = [\det(1 - \rho_p(c(\pi_p))p^{-s}]^{-1}$ p unramified,
(b) $L_p(s,\pi,\rho) = \prod_p L_p(s,\pi_p,\rho_p) = \prod_p [\det(1 - \rho_p(c(\pi_p))p^{-s}]^{-1}$,
which converges in some right half-plane.

We remark that p^{-s} will be replaced by $N(p)^{-s}$ in the above definitions when $\mathbf{Q}$ is replaced by a finite extension of $\mathbf{Q}$.

One of the reasons that we are engaged with automorphic representations is because the family $c(\pi)$ attached to π is a good object to study: It is believed that $c(\pi)$ is fundamentally connected to the arithmetic world, and that it carries concrete and analytic data. For example, any family $\{c(\pi_p) : p \text{ unramified}\}$ attached to $\{\pi_p : p \text{ unramified}\}$ determines a homomorphism from a Hecke algebra H_G into C. Elements in H_G are called Hecke operators.

The conjugacy classes $\{c(\pi_p) : p \text{ unramified}\}$ thus provide eigenvalues of Hecke operators.

Langlands' success in attaching his L-function to an automorphic representation was an assertion that automorphic representations rather than automorphic forms were desirable objects to study.

Special cases of $L_p(s,\pi,\rho)$
(a) Let $G = \{1\}$ be the trivial group. Then $\pi_p = 1$ and $\rho_p = 1$, and
$$L_p(s,1,1) = (1 - p^{-s})^{-1}$$

is the p-Euler factor of the Riemann ζ-function.

(b) Let G be trivial and let $^L G$ be the Galois group over some unramified extension.

Then

$$L_p(s, 1, \rho) = [\det(1 - \rho(\text{Frob}_p)p^{-s})]^{-1}$$

is the local Artin L-function over $\mathbf{Q}$, where Frob$_p$ is the Frobenius class.

(c) Let $G = \text{GL}_2$, then π_p is just a Hecke character and the local Langlands' L-function over $\mathbf{Q}$ is the local Hecke L-function over $\mathbf{Q}$ (see equation (1.3.11.7)).

1.3.16 The Genesis of the Functoriality Conjecture (Winter Break 1966)

Almost immediately after the definition of automorphic L-function was found, Langlands turned his attention to the problem of establishing his L-functions' analytic properties such as meromorphic continuation to the entire complex plane and the functional equation relating the values at s and $(1 - s)$.

We learned from Langlands (see Section 1.3.14) that the constant terms he was calculating were quotients of Euler products and these Euler products could be continued to the entire plane as meromorphic functions. To him, that was a convincing beginning.

> What I held in my hands was a satisfactory definition of the function
> $L(s, \pi, \rho) = \ldots$ However, the function L_v were still unknown at a finite number
> of places.[3] At this point I had no idea how to define them. The element $\gamma(\pi_v)$
> belongs to a conjugacy class in a certain complex Lie group that is nowadays
> denoted by $^L G$. Even though I had proved the existence of a meromorphic
> continuation for a significant number of these functions, I had no idea how to prove
> the existence of such a meromorphic continuation in general, or whether their
> analytic continuation could be proved at all.

1.3.16.1 The Tamagawa Lecture

This is what I thought about standing by the window. Suddenly – everything was immediately present in my mind, at least according to my own recollection: *Tamagawa has on some occasion that could not have been too long before December 1966, but I am not sure, delivered a lecture in the auditorium of the old Fine Hall in which he discussed the function $L(s, \pi, \rho)$ where G is an inner form of GL(n) and ρ is the defining representation of $^L G = GL(n, C)$, and had treated the problem of analytic continuation. I saw no reason that his proof shouldn't also be valid for GL(n), as indeed it does, as later shown by Godement - Jacquet [2].*

---

[3] Ramified primes and the archimedean prime.

1.3.16.2 Langlands' Epiphany and his Letter to André Weil

This insight quickly led Langlands to the next step, which was to replace Artin's reciprocity law with functoriality. We have the following vivid recollection of Langlands during those last few days at the end of 1966.

> The mathematics department at Princeton University was housed in the old Fine Hall, now Jones Hall, at the time a lovely building. As you enter it, on the right was a small office. It was a beautiful office. On the right was a seminar room, with a large table, chairs, and a large blackboard. The windows in both rooms were leaded, in a medieval style, with views of the presidential gardens. Since I lived on Bank Street, a few steps from the corner of Nassau street and University Place, and there were four children at home, I often worked there evenings and on holidays.

The next paragraph deserves our attention.

> In particular, working there during the Christmas vacation of 1966, looking out the window, it occurred to me that what was needed was a generalization of Artin's original idea that his abelian L-functions were equal to the L-functions defined by ideal class characters. I realized that the conjecture I was in the course of formulating implied the Artin conjecture. Thus, what was needed was that the Artin's reciprocity law has to be replaced by what I now refer to as functoriality. It was certainly one of the major moments in my mathematical career.

While looking out the window, the epiphany of Langlands at that moment was what we might call a turning point in the history of mathematics. It was the moment that the genesis of Langlands most celebrated creation – functoriality – was first revealed. Finally, Langlands' one last decisive insight was his reflection on the Tamagawa's lecture.

> The last, culminating insight came on reflecting how the analytic continuation might be proved in general, using the analytic continuation for $G = \mathrm{GL}(n)$ and the defining representation of $^{L}G = \mathrm{GL}(n, C)$ as the standard L-functions to which all others are to be compared, just as one uses the L-functions of Dirichlet type together with the Artin reciprocity law to establish the continuation of the Artin L-functions. The Artin reciprocity law has to be replaced by what I now refer to as functoriality.
>
>   I suppose I was convinced immediately that I had found what I had been searching for, but I do not remember being especially eager to communicate this to anybody. Who was there? Weil, although he might seem in retrospect to be a natural possibility, turned up by accident.
>
>   At all events, on January 6, 1967, we found ourselves pretty much alone and together in a corridor of the Institute for Defense Analysis, having both arrived too early for a lecture of Chern. Not knowing quite how to begin a conversation, I began to describe my reflections of the preceding weeks. He suggested that I send him a letter in which I described my thoughts. Ordinarily the letter never arrives. Mine did, but with Harish-Chandra, who was then a colleague of Weil and to

whom I was closer, as an intermediary. Harish-Chandra perceived its import, but
Weil, so far as I have since understood, did not.

Whether Weil took the letter seriously or not, the Langlands program was
launched soon afterwards.

At the inception of the Langlands program, with the introduction of
Langlands' newly discovered *automorphic L-functions* attached to general
reductive groups, one of the aims of the program was to search for a list of
properties these *L*-functions were expected to satisfy. This list constituted
a bulk of the original Langlands conjectures, particularly the *functoriality
conjecture* (Section 1.3.17), which can be regarded as a foundation of the
Langlands program.

It is not surprising that the nonabelian theory was not developed until the
late 1960s. The theory of infinite-dimensional representations of reductive
groups was first developed actively in the 1950s, and the *p*-adic case was not
studied intensively until the 1960s; both those theories provided the necessary
tools for Langlands' formulation of functoriality conjecture in January 1967.

1.3.17 The Functoriality Conjecture – from Hecke to Langlands

Let us see what James Arthur [1] has to say about functoriality.

> The principle of functoriality was introduced by Langlands as a series of
> conjectures in his original article [3]. Despite the fact it is now almost fifty years
> old, and that it has been the topic of various expository articles, functoriality is still
> not widely known among mathematicians.
>
> The principle of functoriality can be regarded as an identity between
> automorphic *L*-functions for two groups. Let G and G' be connected quasisplit
> groups over $\mathbf{Q}$, and let φ be an *L*-homomorphism from $^L G'$ to $^L G$, which is
> compatible with the maps $\rho: {}^L G'$ to $\mathrm{Gal}(K/F)$ and $\rho': {}^L G$ to $\mathrm{Gal}(K/F)$, then
> functoriality asserts that for every automorphic representation π' of G', there is an
> automorphic representation π of G such that $c(\pi_p) = \varphi(c(\pi'_p))$.

1.3.17.1 From Hecke's *L*-Series to Langlands' Functoriality

Artin's reciprocity law (1.3.11.9) identifies Artin's abelian *L*-function with
Hecke's degree-1 *L*-series, and Langlands' epiphany was to replace Artin's
reciprocity law with functoriality. In this section, we present an example of
Langlands from [4] which shows, if only as a rough sketch, how he arrived at
functoriality – a sublime mathematical insight – from injecting his novel ideas,
step by step, into a degree-2 Hecke *L*-series (see Section 1.3.11.4).

If I replace s by $s + k - 1$ in the Hecke form

$$\prod_p [(1 - \alpha_p p^{-s})(1 - \beta_p p^{-s})]^{-1}, \alpha_p \beta_p = p^{k-1},$$

then

$$\alpha_p \to \alpha'_p = \alpha_p / p^{k-1}, \beta_p \to \beta'_p = \beta_p / p^{k-1}, \alpha'_p \beta'_p = 1,$$

and the functional equation is between s and $1 - s$.

Let γ_p be the diagonal matrix with α'_p and β'_p in the diagonal which may be treated as a conjugacy class in GL(2, C) and the Hecke form may be written as

$$\prod_p [\det(1 - \gamma_p p^{-s})]^{-1}$$

More generally, although Hecke did not do so, we may replace γ_p by $\rho(\gamma_p)$, where ρ is any finite-dimensional representation of GL(2, C):

$$L(s, \pi, \rho) = \prod_p [\det(1 - \rho(\gamma_p) p^{-s})]^{-1}. \qquad (1.3.17.1)$$

Here $\gamma_p = \gamma_p(f) = \gamma_p(\pi)$, where the automorphic form f, an eigenform of the Hecke operators, is replaced in the notation by the representation π it determines.

For any reductive group G/F and any automorphic representation of $G(A_F)$, the theory of spherical functions allows us to define $\gamma_p(\prod) = \gamma(\prod_p)$ as a conjugacy class in a finite-dimensional complex group, the L-group, for almost all p. The calculations on Eisenstein series I described give Euler products that can all be expressed as the right side of (1.3.17.1), i.e. $\prod_p [\det(1 - \rho(\gamma_p) p^{-s})]^{-1}$. They suggest that this function can be analytically continued and has a functional equation of the usual kind. This question posed, a way to answer it suggests itself.

At this point the Tamagawa lecture (see Section 1.3.16.1) was a decisive insight that led Langlands to connect his Hecke series with his ideas in the next step.

Then in analogy with the Artin reciprocity law, all we would need to do to show the analytic continuation of $L(s, \pi, \rho)$ is to establish the existence of an automorphic representation of GL(n), $n = \dim \rho$, such that

$$\left\{ \rho \left(\gamma_p \left(\prod \right) \right) \right\} = \left\{ \gamma_p \left(\prod \right) \right\}$$

for almost all p.

It is a small step – at least conceptually – from this possibility to the possibility of functoriality in general.

Taking the above ideas one step further, Langlands in [5] succeeded in formulating his definition of functoriality.

I might equally well think that given two reductive groups H and G over a number field F, and given a map $\varphi : {}^L H$ to ${}^L G$ compatible with the maps $\rho : {}^L H$ to Gal(K/F) and $\rho' : {}^L G$ to Gal(K/F), and an automorphic representation $\pi_{H,v}$, then there also exists an automorphic representation $\pi_{G,v}$ such that we have

$$\varphi(\Upsilon(\pi_{H,v})) = \Upsilon(\pi_{G,v}) \qquad (1.3.17.2)$$

for almost every place v.

Let us point out that equation (1.3.17.2) is the *essence of functoriality*. To state functoriality in terms of L-functions, we take the following steps.

(1) $\Upsilon(\pi_{H,v})$ and $\Upsilon(\pi_{G,v})$ are conjugacy classes in ${}^LG_H$ and ${}^LG_G$ attached to $\pi_{H,v}$ and $\pi_{G,v}$ respectively. The above quotation of Langlands informs us that we are not only given the condition $\rho'(\varphi(\Upsilon(\pi_{H,v}))) = \rho(\Upsilon(\pi_{H,v}))$, but also the assumed existence of $\pi_{G,v}$ together with condition (1.3.17.2). An immediate consequence is the following:

$$\rho'(\Upsilon(\pi_{G,v})) = \rho(\Upsilon(\pi_{H,v})). \tag{1.3.17.3}$$

(2) The L-function given by (1.3.17.1) was Hecke's degree-2 L-series reformulated by Langlands. In the context of reductive groups H and G, the local L-function in (1.3.17.1) takes the form

$$L_p(s,\pi_{H,p},\rho_p) = [\det(1 - \rho_p(\Upsilon(\pi_{H,v})p^{-s}))]^{-1}$$

and

$$L_p(s,\pi_{G,p},\rho'_p) = [\det(1 - \rho'_p(\Upsilon(\pi_{G,v})p^{-s}))]^{-1}$$

Langlands' Functoriality Conjecture (Local and Global)

(a) Let $\pi_{G,p} = \pi_{G,v}$ and $\pi_{H,p} = \pi_{H,v}$. Suppose we are given φ,ρ,ρ' and $\pi_{H,p}$ as before, and assume there exists an automorphic representation $\pi_{G,p}$ with condition (1.3.17.2) for each unramified p. Then from (1.3.17.3) we have

$$L_p(s,\pi_{G,p},\rho'_p) = L_p(s,\pi_{H,p},\rho_p)$$

(b) Let $L(s,\pi_G,\rho') = \prod_p L_p(s,\pi_{G,p},\rho'_p)$, and $L(s,\pi_H,\rho) = \prod_p L_p(s,\pi_{H,p},\rho_p)$, then under the same assumptions as in (a), we have

$$L(s,\pi_G,\rho') = L(s,\pi_H,\rho). \tag{1.3.17.4}$$

Both (1.3.17.2) and (1.3.17.4) are known as Langlands' functoriality conjecture.

1.3.17.2 Some Consequences of Functoriality

We conclude by mentioning two immediate consequences of functoriality.

(1) A result of Godement and Jacquet [2] states that the L-functions of $GL(n)$ have nice analytic properties provided the irreducible unitary representations of $GL(n)$ are automorphic.

(2) Using (1) together with the expression in (1.3.17.4) imply that functoriality reduces the study of L-functions for arbitrary reductive groups G to the known results of $GL(n)$.

(3) Artin's reciprocity conjecture is a special case of the functoriality conjecture.

Suppose $H = \{1\}$ (the trivial group) and $G = GL(n)$. Then $\varphi : {}^L H \to {}^L G$ reduces to the map $\rho : \text{Gal}(K/F) \to GL(n)$.

Since π_H is trivial and $\rho'\colon \mathrm{GL}(n) \to \mathrm{GL}_n(C)$ is the "standard" representation, it follows from functoriality and condition (1.3.17.4) that there exists an automorphic cuspidal representation $\pi(\rho)$ such that

$$L(s, \pi(\rho)) = L(s, \rho). \qquad (1.3.17.5)$$

Expression (1.3.17.5) is an assertion of Artin's reciprocity conjecture.

In summary, we repeat that one of the original aims of *abelian class field theory* was to establish nice analytic properties for Artin's abelian L-functions.

A satisfactory solution was provided by Artin's reciprocity law, which identifies Artin's arithmetic L-functions with Hecke's analytic L-functions.

For nonabelian field extensions, Artin's reciprocity conjecture identifies Artin's L-functions, with Langlands' automorphic L-functions, which are analytic in nature. This identity enables Artin's L-functions to have analytic continuation and functional equation, albeit conjecturally. In this respect, Langlands functoriality conjecture, which has Artin's reciprocity conjecture as a special case, represents a general formulation of nonabelian class field theory.

Acknowledgements

This article could not have been written without the cooperation of Professor Robert Langlands. I am deeply grateful for his generosity in sharing his recollections of his early life with me. I would also like to express my sincere appreciation for the help and encouragement that I have received, during the writing of this article, from James Arthur, Herve Jacquet, Barry Mazur, Cris Poor, and in particular the referee, whose detailed and thoughtful comments are extremely helpful. Finally, I wish to thank Dr. Michael Volpato, who was my collaborator at a preliminary stage of the book project *The Genesis of the Langlands Program*, for his contributions.

Figure 1.1 Robert P. Langlands. Courtesy of the Simons Foundation.

42 Julia Mueller

Figure 1.2 Robert P. Langlands in his office at the Institute for Advanced Study, Princeton, New Jersey, USA. It was the office of Albert Einstein from 1947 to 1955. Photographer: Anthony V. Pulid.

References

[1] J. Arthur, *Functoriality and the trace formula* available at www.math.toronto.edu/arthur/
[2] R. Godement and H. Jacquet, Zeta function of simple algebras, *Lecture Notes in Mathmatics*, Vol. 260, Springer-Verlag, Berlin-New York, 1972. MR0342495
[3] R. P. Langlands, Problems in the theory of automorphic forms *Lectures in Modern Analysis and Application*, III, Springer, Berlin, 1970, pp. 18–61. *Lecture Notes in Math.*, Vol. 170. MR0302614
[4] R. Langlands, *The genesis and gestation of functoriality*. TIFR, Mumbai, Feb. 2005 available at www.math.tifr.res.in/sites/default/files/maths/TheGenesis,pdf
[5] R. Langlands *Funktorialitat in der Theorie der atomotphen Formen: Ihre Entdeckung und ihre Ziele*, p.16 (translated by Dr. Hans-Joachim Hein), available at http://publications.ias.edu/rpl/
[6] J. T. Tate, *Global class field theory*, Algebra Number Theory (Proceedings of the Instructional Conference, Brighton, 1965), Thompson, Washington, DC., 1967, pp. 163–203. MR0220697

2

The Early Langlands Program – Personal Reflections

Steve Gelbart

This chapter will not be a survey of the early Langlands Program. Instead, it will be an account of what the early Langlands Program – and Langlands himself – has meant to me.

I thank Julia Mueller for the idea of writing this chapter, and Moshe Baruch and Freydoon Shahidi for several helpful suggestions.

2.1 The Years 1967 to 1972

Langlands was a faculty member at Princeton University from 1960 to 1967, went to Turkey for a year and returned to the USA in 1968 as a Professor at Yale. In the fall of 1967, I was a beginning graduate student at Princeton; I had finished my undergraduate degree at Cornell, and my fields of interest were functional and harmonic analysis. After my first year in graduate school, I began to write my thesis under Eli Stein. I knew almost nothing about automorphic functions. Nevertheless, it was my Ph.D. with Stein that (eventually) brought me into the automorphic realm of Langlands.

Stein, in his paper of the 1967 *Annals of Mathematics*, found the missing irreducible unitary representations of the complex group $GL(n, C)$. My thesis was to develop *partly* Stein's work for the real case. The slightly modified Fourier operator F* on matrix space $M(n, R)$ once again led to a central operator, which on the representation side was a multiplication by a function m that I had to compute. The connection between this problem and the much larger problem concerning L-functions will soon be explained.

When the problem with the function m was solved in 1969, Stein showed me a letter he received from Roger Godement just after Stein had published his 1967 paper. The letter proposed a plan for p-adic groups and representations that would also do what some of Stein's paper had done for $GL(n, C)$. This was a first turning point in my work. Until 1969 I would scan *Mathematical*

Reviews by looking at the relevant sections on Fourier Analysis; after 1969, my attention was focused more on the *p*-adic numbers and algebraic number theory.

A second turning point also came in 1969. This was the publication of the English translation of *Group Representations and Automorphic Forms* by Gelfand, Graev, and Piatetski-Shapiro (originally published in 1966, in Russian). Together with Stein, I read the Introduction of that book in the old Fine Hall Common Room of Princeton. These three authors thanked Godement for his help. They also mentioned Langlands and his new theory of Eisenstein series. Although I had been in graduate school for two years, and had known about Langlands having been on the faculty at Princeton, this was really the first time I was hearing about him and his work from other mathematicians. I was impressed that Stein took him very seriously. I was also impressed that Joe Shalika – an Assistant Professor in the Department – had already had an intense interest in Langlands' work for several years. Shalika helped me with the *p*-adic part of my Ph.D. thesis, which was finished in spring of 1970.

Much of the towering work of Langlands was done from 1965 to 1970. In early 1967, he had written his now-famous letter to André Weil sketching Langlands' monumental program. I knew nothing of this before 1969 and nothing had really been published until 1970. In 1970 Langlands produced "Problems in the theory of automorphic forms" (in *Lecture in Modern Analysis and Applications, III*, Springer Lecture Notes in Mathematics, Vol. 170) and – together with Hervé Jacquet – a 548-page book *Automorphic Forms on GL(2)* (Springer Lecture Notes in Mathematics, Vol. 114). The first work dealt with wide-ranging questions for a reductive group G; the second with what these conjectures implied for $G = GL(2)$. In Princeton, Harish-Chandra thought highly of Langlands' work but still only a few lectures about it had been given. One exception was Shalika, who learned of it from Jacquet (who was a member of the Institute from 1967 to 1969), with the help of Bill Casselman. Another was Godement, who visited the Institute from September 1969 to April 1970. In the last three months of his stay, Godement gave a series of lectures on Jacquet–Langlands' Notes (which were soon to be published informally) that I faithfully attended. But – as a student who never took even one formal course in number theory – I did not get too much from them until a little later.

In early spring of 1970, I was not really aware of the tenuous connection between my thesis and a small part of Jacquet–Langlands' work. But by the time my thesis was published in 1971, I was. This is described below

Why was *Automorphic Forms of GL(2)* so important? First of all, it was almost the first time that properties of such forms were treated wholly in terms of their place in representation theory and over local and global fields. This point of view was encouraged by Godement, the advisor of Jacquet; Godement was the sole mathematician thanked by the authors in their Introduction to the book.

To Langlands, there was also a second reason (as well as many more). Langlands felt that the Main Theorem of Section 16, relating automorphic forms on a quaternion algebra to automorphic forms on $GL(2)$ (first partially proved by Shimizu), should not be proved by L-functions nor by theta-functions but rather by "a beautiful illustration of the power and ultimate simplicity of the Selberg trace formula" (from the Introduction to their book). Going from automorphic forms on a quaternion algebra to automorphic forms on $GL(2)$ is a simple example of Langlands' "functoriality principle." This states that a general correspondence should map automorphic forms on a group G to automorphic forms on G' – or the corresponding L-functions on G to L-functions on G'. Ever since the late 1960s, Langlands has stuck to this belief: the "functoriality principle" should be proved in general by the *trace formula*. As we know now, this belief is becoming more and more confirmed

Now I return to Shalika. Motivated by him in 1969, I studied the adèles by focusing on the $GL(1)$ analysis that Tate worked on in his thesis. I remember also following the treatment of this in the book by Gelfand, Graev, and Piatetski-Shapiro, which I thought was beautiful. Shalika taught me to consider the local and global theories as equally important: in this case, the global theory handed us an abstract functional equation, whereas the local theory led to knowledge of the concrete Dirichlet L-functions. Shalika also talked about his own work where he used a "Weil representation" to construct certain irreducible representations of $SL(2)$, as did Jacquet and Langlands.

In the summer of 1970, my wife Mary and I traveled to Paris to see relatives and friends, and to Nice for the 1970 International Congress of Mathematicians. Stein gave an hour-long Plenary address at the Congress, and I was still very interested in his various results. Langlands gave a non-Plenary lecture, but for some reason, I did not go – I don't know why! Some months earlier, in the spring of 1970, Stein suggested thinking about making an explicit theory of special functions for the homogeneous space $SO(n)/SO(n-m)$, generalizing the classical theory of spherical harmonics. This work had an immediate application to the representation theory of certain noncompact groups. Indeed, the so-called "Weil representation" of the real symplectic group $Sp(m, R)$ in $L^2(M_{2m,m})$ is known to be highly reducible, and the theory of harmonics makes it possible to decompose this representation into holomorphic discrete series representations. This is the same type of "Weil representation" as Jacquet–Langlands used for $GL(2)$. Later, the subject of $SO(n)/SO(n-m)$ shows up as a small part of Langlands' discussion related to Roger Howe's earlier theory of *dual reductive pairs*; see Langlands' letter to Howe of 1975, which appeared in *"Functoriality"*, *The Work of Robert Langlands*.

In 1970–1971 I stayed on at Princeton as an Instructor of Mathematics. When some of Stein's suggested problems on $SO(n)/SO(n-m)$ were solved, I began to think more about Langlands' ideas. In particular, I learned a lot

from Shalika about the Jacquet–Langlands' zeta-integrals in their book and I
could now explain the connection between my thesis and a tiny portion of their
work. I saw that a small part of their local (real) representation theory was
more or less equivalent to a large part of my thesis for $GL(2, R)$. For the sake
of definiteness, take $G = GL(2, R)$ and suppose π is an irreducible unitary
representation in the discrete series of G. For nice f in matrix space $M(2,2)$,
the function $f(x)|\det(x)|^s$ belongs to $L^2(G)$ whenever $\mathrm{Re}(s) > 1$, and the
zeta-integral

$$\zeta(f, \pi, s) = \int f(x)\pi(x)|\det(x)|^s \, dx$$

defines a holomorphic function of s in this range. Using the argument of my
1975 book *Automorphic Forms and Adèle Groups* (chapter 6, Further Notes),
one indeed shows that the analytic continuation and functional equations for
$\zeta(f, \pi, s)$ follow from the computation of m of the modified Fourier operator
F*. This is the slight connection between my thesis and the automorphic realm
of Jacquet–Langlands alluded to earlier.

Finally, as an Instructor, I was lucky enough to enjoy the friendship of a
fresh new Ph.D. of Langlands, Jim Arthur; he helped make my first year fun.
I also got to know Serge Lang, who cotaught the Honors Couse in Calculus
with me; two years later he and Jim went to Yale where Langlands had been,
and where I made many stops in the years to come.

In August 1971, I returned to Cornell as an Assistant Professor. Since
I was finishing up my papers on $SO(n)/SO(n - m)$ and the symplectic group
I became free to turn my attention to the book of Jacquet and Langlands. In the
fall I taught a course in group representation theory, pretty much like the course
Stein taught me in my second graduate year; in the spring, I gave a seminar
where I highlighted some of theorems of Jacquet–Langlands. At the end
I explained how one could use a "Weil representation" to construct a continu-
ous series representation of $SL(2, R)$ that occurs discretely in $L^2(\Gamma \backslash SL(2, R))$
for a particular congruence subgroup Γ of $SL(2, Z)$. I also decided it was time
to write Langlands himself (with whom I had not yet exchanged letters nor
seen). This letter was to introduce myself, and to ask for his "Notes on Artin
L-functions." He wrote back that these Notes were still in preparation, but if
I had some questions he would be glad to try and answer them. At Cornell,
I wrote some Notes about my lectures (called "Notes on automorphic forms
and representations of adèle groups"), and in May I sent them out to several
people, including Langlands. One sign of my being a beginner in this field was
that Maass' important paper (*Math. Ann.* 121, 1949) was not mentioned in the
bibliography of my Notes; Langlands kindly pointed this out to me in a second
letter. In August of 1972, at a Summer Research Institute of the AMS, I spoke
on this continuous representation of $SL(2, R)$ again, at a seminar organized
by Godement. This time, Maass' discovery of real analytic cusp forms was

mentioned; I also cited Shalika (and Tanaka) for generalizing Maass' work by using a "Weil representation" similar to that of Jacquet–Langlands.

2.2 The Year 1972 to 1973

This is the year I went to the Institute for Advanced Study in Princeton and met Langlands. The year could not have been more exciting – some high points were the following.

(1) I was actually supposed to go to the Institute in 1971 and to begin my Assistant Professorship at Cornell the following year. But it turned out that Cornell wanted me to come in 1971, and to put off my year at Princeton until 1972. I was a little disappointed at this, having to move up to Ithaca and then back again in a year. But when I heard that Langlands was to accept an offer of permanent Professor at the Institute in 1972, my feelings changed completely. I think that's what made me work harder than ever and write him a first letter from Cornell. In the fall of 1972, after arriving at the Institute, I met Langlands for the first time, and we slowly became friends.

(2) Before the winter Langlands and I had arranged a weekly handball game in Princeton University's Dillon Gymnasium that continued into the spring. It was about a 20-minute walk to the gym, and we used this time for talking about Labesse–Langlands, L-indistinguishable representations of $SL(2)$, Jacquet's Part II of *Automorphic Forms on GL(2)*, how exactly one computes the multiplicity of the reducible principal series for $SL(2, R)$ in $L^2(\Gamma \backslash SL(2, R))$ where $\Gamma \backslash SL(2)$ is compact, and so on. In the gym, we stuck to handball. In 2017 I asked him who usually won at handball; he didn't know, and I don't remember either!

Figure 2.1 Langlands in front of Dillon Gym (1973). Photo courtesy of Steve Gelbart.

(3) In the spring of 1973, Langlands asked me to give a series of lectures at the Institute on Eisenstein series, the continuous spectrum for $GL(2)$, and the trace formula. Most of this material is to be found in Jacquet–Langlands. I gave complete proofs for $GL(2)$, by following Jim Arthur's as yet unpublished manuscript of the Selberg trace formula for real rank-one groups.

(4) At about the same time, Langlands showed me the pages he had written in his referee's report for a now well-known paper of Shimura in the 1973 *Annals* dealing with modular forms of half-integral weight. At the end of his paper, Shimura had suggested that someone should rewrite his paper in terms of representations; Langlands' report went further, strongly suggesting a generalization for forms of half-integral weight to be representations of the adèlic metaplectic group, and eventually, like Shimura, suggesting some kind of Jacquet–Langlands theory for the two-fold cover of $GL(2)$. I was as impressed by Langlands' projected metaplectic generalization as I was by Shimura's result. It was then that I started thinking about metaplectic groups. During the next 40 years, many very good mathematicians working on automorphic forms made the knowledge of metaplectic groups grow immeasurably (even though they needed new definitions of Langlands' concepts – for example, metaplectic groups are *covering* groups, not linear groups, and the concept of an L-group is just recently being understood for such groups). But in 1973, they were still in their infancy; until then one had mostly the Eisenstein series and theta-functions of Siegel, Weil, and Kubota, the identity between them, and Shimura.

(5) In June 1973 I left the Institute having made – or strengthened – many good friendships. For example, with the metaplectic group coming from all directions, Roger Howe had sketched his new (G, G') oscillator theory to me, the beginnings of his theory of dual reductive pairs. Without a doubt, though, it was Langlands who made this year really special.

2.3 The Years 1973 to 1983

When I got back to Cornell in 1973, Stein suggested that I expand my Cornell "Notes on automorphic forms" for publication in the *Annals of Mathematics Studies* and add to that what I had lectured on at the Institute in the spring. I finished the writing by December 1973, and in the Preface to the book I thanked Langlands for his help and his inspiration.

For a distinguished great mathematician, Langlands had an unusual style of openness with younger people, even Ph.D. students. When he spoke or wrote to them about new problems, he did so with the expectation that younger newcomers could help to solve them. There was no competitiveness here, or feeling that it was a waste of time. I think that people like Eli Stein and Peter Sarnak also had this same way with young mathematicians; that's why they too, for example, have a whole school of young followers. When Langlands had a Ph.D. student who went straight to the top – like Jim Arthur – he felt genuinely pleased; more generally, he would take time to talk to young visitors at the Institute and encourage them. This kind of interaction is greatly appreciated and, when combined with his much larger Langlands Program, makes for a formidable force.

In July of 1975, Borel gave a Bourbaki Seminar talk entitled "Formes automorphes et series de Dirichlet d'apres R. P. Langlands." In it he gave the first attempt at describing the Langlands philosophy. It goes a long way towards making this program doable by proceeding by approximation. It also fixes some terminology; the complex group associated by Langlands to each connected reductive group G is – by a suggestion of Jacquet – denoted by ^{L}G, and called the L-group of G.

Aside 1 Some people have asked when (and where) the terminology "Langlands Program" came from. Here is a guess. In the third sentence of his Bourbaki talk, Borel writes (in French) that "Langlands' conjectures and questions define rather a vast *program* called the Langlands philosophy and this *program* ties together classical and recent results in a very dramatic fashion." After 1975, the term "Langlands Program" started appearing slowly in print, but I don't know yet which author did it *exactly* first.

The metaplectic group I was working with teamed up with the Langlands Program in the following way. In 1975, Shimura proved that integration of a certain real analytic Eisenstein series $E(z,s)$ of half-integral weight multiplied against a theta-series times a nice form $f(z)$ of integral weight equals the "symmetric square" (third-degree) L-function $L(s, f, \text{Sym}(2))$ attached to f. Keeping in mind the Langlands' "functoriality principle," this "had" to be the L-function of a form F on $GL(3)$.

At the same time, Jacquet had been finishing his work with Piatetski-Shapiro and Shalika on $GL(3)$ cusp forms, and had corresponded with me about his own Eisenstein series on the metaplectic group. At the very end of 1975, Jacquet and I suggested that we pool our efforts to prove $L(s, \pi, \text{Sym}(2)) = L(s, \Pi)$ with Π an automorphic cuspidal representation of $GL(3)$), i.e. there is a "lifting from $GL(2)$ to $GL(3)$."

In December of 1975, in a letter to Serre, Langlands sketched his own *reciprocity conjecture* – and hence Artin's conjecture – for a wide class of two-dimensional Galois representations. The modular ingredients included: (i) the theory of "base change for $GL(2)$"; (ii) the theory of lifting from $GL(2)$ to $GL(3)$; and (iii) the "converse theorem for $GL(3)$" and L-functions on $GL(3) \times GL(3)$ due to Jacquet, Piatetski-Shapiro, and Shalika. Ingredient (ii) was left undone at that time. In early February 1976, I wrote to Langlands that Jacquet and I had been working together to show (ii), and by late spring it was proved.

That summer, Mary and I again traveled to Europe to see family and friends. The highpoint professionally was the International Conference on Modular Forms, which was held in Bonn. I lectured there not only on lifting $GL(2)$ to $GL(3)$ but also on Langlands' proof of Artin's Conjecture for tetrahedral representations. Langlands was not there, but he heard about the write-up of my talk, and asked for it before writing up his book on *Base Change for GL(2)*. My lecture was published in the 1977 Springer Notes *Modular Functions of One Variable VI*. Langlands' book on base change – using a trace formula written up in mimeographed notes in 1975 and built on the work of Saito and Shintani – was published in the 1980 *Annals of Mathematics Studies*. Once again, the trace formula proved the "principle of functoriality" for base change (but I cannot say the same for ingredient (ii) of the previous paragraph).

Aside 2 The lift from $GL(2)$ to $GL(3)$ was one of the first nontrivial examples of functoriality proved. The lift to $GL(n)$ (for arbitrary n) seems to be more difficult. All that is known is the cases of n equal to 4 and 5, due to H. Kim and Shahidi in 2002. The proof of these cases is more complicated than $GL(2)$ to $GL(3)$; it uses the results of Kim and Shahidi *plus* earlier works of Piatetski-Shapiro, Cogdell, and many others. The cases of n equal to 3, 4, or 5 are proved using the converse theorem. Although it is known to be true in these cases, Langlands is awaiting a proof using the trace formula; in fact, he's probably right to assume that *every lift to GL(n)* must be true by the trace formula. Such results, suitably generalized, are a small portion of the Langlands Program; still, they imply the generalized Ramanujan and Selberg Conjectures.

Back to Bonn, 1976 Another highlight of this Conference was the appearance of Piatetski-Shapiro, who had just left the Soviet Union that year. He and I had exchanged several letters before July 1976, some about Jacquet–Piatetski–Shalika, some about $L(s, \pi, \mathrm{Sym}(2))$, and some automorphic functions on the metaplectic group. At Bonn, from our very first handshake, we hit it off immediately; in particular, we decided to work together on the metaplectic group.

Piatetski-Shapiro had of course heard about Langlands' work from the early 1960s (especially his theory of Eisenstein series alluded to in his *Group Representations and Automorphic Forms*), and was right up-to-date on the later Langlands Program. I have in mind the following: I had been trying to understand the ideas behind Langlands' 1972 paper "Modular forms and l-adic representations" (*Proceedings of the Antwerp Conference on Modular Forms*, Springer Lecture Notes in Mathematics Vol. 349); not really succeeding, in 1976 I published an expository paper called "Elliptic curves and automorphic representations." It covered Eichler–Shimura theory briefly, and a little bit more of Deligne and Langlands. Piatetski-Shapiro, it turns out, wrote a paper "similar" to Langlands', which appeared just before his in those *Proceedings*.

Some years later, I noticed that Piatetski-Shapiro was the Math reviewer of Langlands' paper! Piatetski-Shapiro wrote (in his review):

> The author considers the relation between the zeta-functions of modular curves and the Jacquet–Langlands zeta-functions for $GL(2)$. The same question was considered in a paper of the reviewer in the *Proceedings*. However, the methods are different. That of the reviewer is close to the classical methods of Eichler and Shimura, while that of the present author is very new and interesting. It is based on the *Selberg trace formula* and it yields the coincidence of the Euler factors not only for the unramified places, but also for all the places that correspond to representations of the principle series. This is a new result.

Once again, the trace formula played its fundamental role.

Looking back, we see that the top people in automorphic forms proper (Langlands, Jacquet, Shalika, Piatetski-Shapiro,...) were working *directly* on problems posed by the Langlands Program. But – *most* of these people were not principally number theorists or algebraic geometers. At that time, number theorists still generally stayed clear of the Langlands Program.

In early 1977, however, things began to change.

The 1977 Summer Research Program of the AMS was devoted to *Automorphic Forms, Representations, and L-Functions*. It was held at Oregon State University, Corvallis, and attracted a much wider group of mathematicians than before. Resulting from "Corvallis" was a wonderful two-volume introduction to the Langlands Program divided into four parts: (1) reductive groups and representations; (2) automorphic forms and representations; (3) automorphic representations and L-functions; and (4) arithmetical algebraic geometry and automorphic L-functions. Langlands gave one of his lectures on Shimura varieties and motives, and several other people covered the *Base Change* theory of Langlands, Saito, and Shintani. I gave a lecture on Howe's dual reductive pairs, and a series of lectures with Jacquet on $GL(2)$ and the trace formula. The most positive thing about that Conference was that it finally brought some number theory people (like Tate, who attended the Conference and had several Ph.D. students using Langlands theory) – and even some algebraic geometry people – smack into the realm of automorphic forms. In other words – slowly but surely – number theory, representation theory, and algebraic geometry, were merging into the domain of Langlands (who, himself, knew all along that they were deeply connected).

Aside 3 The Birth of Langlands–Shahidi Theory Shahidi came to the Institute for Advanced Study in the fall of 1975, right after his doctoral studies under Shalika at Johns Hopkins. What Langlands suggested Shahidi study were the *non*constant (as opposed to the constant) Fourier coefficients of Eisenstein series associated to a pair (G, M) along the lines predicted in a letter to Godement several years before 1975. Working with everything generic,

Shahidi tried his hand with G the exceptional group of type G_2 and M one of its $GL(2)$ Levi subgroups. In 1977 he succeeded: the nonconstant Fourier coefficient involved the L-function of the symmetric *cube* of $GL(2)$. The paper was solicited by Langlands for *Compositio Mathematica* and, as an Editor, Langlands sent it to me to referee. Shahidi called it "Functional equation satisfied by certain L-functions". This was the birth of Langlands–Shahidi theory; it is still going strong today. It was also the birth of two new friendships – between me and between Shahidi, and between Shahidi and Langlands.

In 1979, the Tata Institute of Fundamental Research in Bombay held its first *International Congress on Automorphic Forms, Representation Theory, and Arithmetic*. This was similar to the Bonn Conference of 1976, only smaller. At it, Piatetski-Shapiro and I finished the local–global Shimura correspondence S using the converse theorem for $GL(2)$ and got to explain away some of the mysteries of the theta-function using group representations; as a result, we obtained generalizations of the classical theta results due to Shimura, Deligne, Serre, Stark, and Vigneras of 1973–1977.

Aside 4 After the late 1970s, the theory of metaplectic groups began to take off. The most notable work was done by Jean-Loup Waldspurger. He devoted a deep study of automorphic forms on the cover of $SL(2)$ and automorphic forms on $PGL(2)$. In particular, he discovered an amazing formula relating the Fourier coefficient of a half integral weight form to a central value of an L-function of the corresponding integral weight form; this had an important application to elliptic curves and the congruent numbers problem.

The year 1982–1983 was a Special Year in Lie Group Representations at the University of Maryland, which sponsored five three-week sessions throughout the year, one being "The Langlands Program." Langlands was continuing his study of orbital integrals and transfer factors (with his former student Shelstad), zeta-functions of Shimura varieties (with Harder, Rapoport, and others), and spinoffs that continued through the 1980s. Piatetski-Shapiro and I were now dealing with the group $U(3)$ and the six-dimensional representation of its L-group. One motivation for this comes from the need to relate the poles of the L-function for $U(3)$ to integrals of cusp forms over cycles coming from $U(1, 1)$; the prototype here is the proof of Tate's conjecture for Hilbert modular surfaces due to Langlands, Harder, and Rapaport.

2.4 The Years 1983 until 1988

At the end of summer 1983, our family moved to Israel from Cornell (after a visit to Israel in the spring of 1978 and a Sabbatical year there in 1981–1982). That was also the beginning of Langlands' Special Year on Automorphic Forms at the Institute, which I felt bad about missing. In May 1984, however, Langlands and his wife Charlotte visited Israel.

Figure 2.2 Langlands in Israel (1984). Photo courtesy of Steve Gelbart.

I learned many things about them from this trip. For one thing, they both loved learning about the ins and outs of faraway places, including history. From a first letter before coming, Langlands wrote "I am very much looking forward to the visit to Israel, as is Charlotte, and indeed have been preparing for it for several months now, reading in the Bible and about the history and geography. Since I am starting from scratch, I have a great deal to do."

I also learned that they loved hiking. From a second letter, Langlands wrote "Soudry and Piatetski-Shapiro (who have been in Princeton) have been kind enough to arrange for Charlotte and me to participate in several organized hikes during our stay in Israel. We will need sleeping bags and prefer not to bring our own along, if indeed we possess two. Soudry suggested that it would be no problem to borrow two. Can you help?". Well, we found two, but the Langlands' found the first organized hike too simple, and after that went out on their own!

This next story, however, is another sign of Langlands' love of foreign languages. My wife Mary told me the following: Bob decided that he wanted to know how to read signs that he was seeing in Hebrew, so he asked her for a guide to the Hebrew alphabet, which she gave him, with a page showing what the letters looked like and what sound each letter made. A few days later he told her he was using the guide she had given him to sound out Hebrew words but had a question: "the letter *vav* seems sometimes to be sounded like the English *consonant* 'v', and at other times like the *vowel* sound 'oh'." She thought it was amazing that he had picked this up so quickly, since it had not been mentioned in whatever material she had given him, and it was something that had taken her much longer to grasp. She later turned to me and said "OK, now I know what you mean when you say he's awfully smart!".

In that same year, I wrote the *Bulletin* paper "An elementary introduction to the Langlands Program". Starting from Fermat's work of the integers p mod

4 and ending with Langlands' functoriality, it tried to give a picture of what was known to me of the Langlands philosophy up to then. In 2011, it was republished again with the commentary of *Bulletin* Editor Edward Frenkel.

In 1987, a Conference in honor of the 70th Birthday of Atle Selberg was held in Oslo, Norway, bringing together mathematicians such as Weil, Piatetski-Shapiro, Jacquet, Arthur, and Langlands. My father – also a mathematician – came too, since he had known Selberg well at the Institute since 1947–1948, brought Selberg to Syracuse University the following year, and spent 1951–1952 on a Fulbright Fellowship in Trondheim Norway at Selberg's recommendation; I was only five years old in 1951, but had many memories of Trondheim – some from photographs – and was excited to be in Norway again in 1987.

Right after the Symposium ended, Langlands wrote me the following: "After returning from Oslo, I had lunch in Paris with Coates, and a question came up to which some attention might be given. Which of the two mathematical methods for dealing with the L-functions is most useful for the applications to p-adic L-functions? I have no idea, but the conversation with Coates suggests that the prevailing opinion is that only the technique of zeta-integrals is useful. It would be good to have some concrete evidence for the opposite view." The funny thing was that many years later, neither of us could remember this letter at all! By looking up something else in 2010, I happened upon the 1987 letter by chance. But – in 2010 – it was exactly what Shahidi, Steve Miller, Panchishkin, and I were trying to prove in a paper about p-adic zeta-functions. Langlands' fertile brain was able to predict it years earlier, though p-adic Langlands–Shahidi theory still seems far away.

In the year 1986–1987 a Special Year in Automorphic Forms at the Hebrew University's Institute for Advanced Studies was organized by Piatetski and myself. We had planned a few long-term visitors, like Bernstein, Bump, Howe,

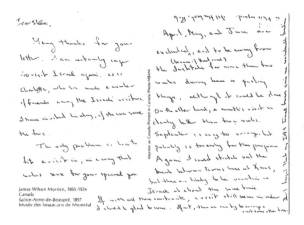

Figure 2.3 Postcard from Langlands.

Rallis, and Sarnak, and also many shorter ones. In the postcard shown here, it seemed difficult to find a good time for Bob to come.

Eventually, we settled on the three weeks around Christmas 1987, and Langlands lectured on "Endoscopy and Transfer" and his *fundamental lemma*.

2.5 From 1988 until Today

Among other things, Wiles's 1995 proof of Fermat's Last Theorem and Ngo Bao Chou's 2009 proof of the *fundamental lemma* are seen as spectacular achievements of a small part of the early Langlands' Program.

Figure 2.4 Bob and I (Quebec, 2002). Photo courtesy of Mary Gelbart.

Figure 2.5 Robert and Charlotte Langlands (2016). Photo courtesy of Steve Gelbart.

Langlands and I still exchange email messages, and enjoy getting together at least once a year. In particular, in 2002, Mary and I spent two days visiting Bob and Charlotte at their summer cottage in northern Quebec. In the evening, when it was completely dark, the four of us agreed to paddle two canoes to the middle of the quiet lake their cottage bordered on. When we got there – all of a sudden – the whole sky lit up with brilliant swirling colors: Mary and I were watching our first *aurora borealis*!

2.6 A Final Word

I feel lucky that my career was shaped by Langlands' Program, and by Langlands himself.

PART II

Langlands as Mentor

3

Langlands and Turkey

Cihan Saçlioğlu

It is an honor to write a memoir on Professor Langlands' association with Turkey. I have stumbled into the honor partly by luck, but mostly thanks to his extraordinary generosity towards this student from his distant past – the 1967–1968 academic year he spent at the Middle East Technical University (METU), Ankara.

It might be helpful to say a few words about the academic and social environment in a city unlikely to be familiar to most readers – *and* the way it was 50 years ago; the past is truly another country. A delightful account of this setting and of professors such as Feza Gürsey, Cahit Arf, and Erdal İnönü who shaped it can be read in the *Hull Reminiscences* (www.hull.ac.uk/php/masrs/reminiscences.html) of the late Professor Ron Shaw, who spent a semester at METU in 1963 four years before Langlands. He had earlier discovered non-Abelian gauge theory in his Cambridge Ph.D. dissertation, independently *of* and nearly simultaneously *with* Yang and Mills He tells about this discovery and his other mathematical and philosophical interests along with hilarious Philip Swallow-like (the main character in David Lodge's campus novel *Changing Places*) anecdotes about his time in Turkey. But the best source about the milieu (for those who can read Turkish!) is Langlands' own "Letter to Turkish readers" (LTR) written in Turkish by him, and included in the Turkish version of Frenkel's *Love and Math* This also explains how he chose to spend a year at METU, Turkey. We'll come back to the letter later.

My story of how I came to meet Langlands begins with my decision to study physics at METU, where Gürsey had set up a graduate-level department of theoretical physics. As students, we became accustomed to seeing luminaries such as Murray Gell-Mann, Sydney Coleman, and other friends of Feza who came for short visits; Geoffrey Chew stayed longer and gave the first-ever course in the world on S-Matrix theory at METU. However, such advanced talks did not penetrate our undergraduate heads and at the end of my first two years, I felt I should be learning more in regular courses. I asked the mathematics department chairman Professor Körezlioğlu what it would take

59

to earn a mathematics degree as well. I had decided I wanted to become a theoretical physicist and assumed that learning more mathematics was bound to be helpful. He drew up a list of additional math courses I should take over the next two years. The first course was Introductory Analysis. I was the only physics student in the class. At the appointed time and classroom, the instructor, a very pleasant young man walked in and introduced himself as Robert Langlands. Little did we know that one year before, this young man had written a now-famous letter to André Weil that changed the course of mathematics. He was going to follow Edmund Landau's *Grundlagen der Analysis,* translating from the German as he went along. Until then, my idea of mathematics had been something along the lines of Thomas' *Calculus and Analytic Geometry*, but here we started with integers and Peano's axioms, introduced rational numbers by dividing integers, and then the irrationals using Dedekind cuts, etc. From the point of view of short-term benefit for an aspiring physicist a more useless course could hardly be imagined; but as a first introduction to rigorous mathematical reasoning, it was an eye (and mind) opener. Langlands' lectures were a model of clarity and order. He would lecture at a perfectly even pace, not too slow, but not too fast for taking notes; everything he uttered was then very neatly written on the blackboard in his characteristic left-leaning lettering; nothing was sketchy or messy. He was very kind and patient with our often dumb and sometimes showoffy questions. The homework assignments were new and challenging. My idea of mathematical difficulty until then consisted of finding clever substitutions to do complicated integrals. Here we were expected to reason our way strictly from the axioms to prove seemingly obvious statements and this was not easy or familiar I soon realized however that unlike me, one classmate, a mathematics major named Mükremin Neşeli from the Aegean town of Muğla, seemed perfectly attuned to this way of thinking, having been led to ideas of continuity and differentiability on his own partly by contemplating a water fountain in a park during his high-school years. He was altogether on another intellectual plane, critically reading Reichenbach, Quine, and Russell on the philosophy of mathematics and Mach on physics; this last impressed him less. He, some others and I got As, but everybody, and especially Langlands, realized that while we were all good students, Mükremin stood out in talent and mathematical maturity. More on this later.

In the spring semester, Langlands offered a course on Differential Geometry, which I of course was not going to miss. I was hoping to learn something about noneuclidean geometry as used in Einstein's theory of gravity, but our textbook only covered curves and surfaces embeddable in three dimensions. Nevertheless, the idea of defining curvature independently of an embedding was again a revelation. Furthermore, through my effusive proselytizing about the importance of mathematics in physics, I had persuaded Yılmaz Akyıldız, an amazingly energetic and talented fellow physics student who came from an engineering program, to take extra math courses also. Using the differential geometry we learned in Langlands' course, plus *very* substantial help from Gürsey, we obtained a spinor version of the Frenet formulae and published it in the *METU Journal*. At the end of the term, as we were working on the final exam for the course, we noticed Langlands was perusing a Turkish newspaper while supervising us. Somehow while teaching his courses, corresponding with Jacquet, working on Class Field Theory from a little known paper of Hasse given to him by Arf, and attending to four young children with his wife Charlotte, he had not only learned enough Turkish to read newspapers, but even prepared and delivered a mathematics talk in Turkish in İzmir (with some linguistic help from Şafak Alpay, another classmate from the analysis course).

At the end of the semester, we heard that Langlands, having been offered a professorship at Yale, was leaving METU. In fact, the offer was made shortly after he had arranged to come to METU, but he asked Yale for a year so he would not break his promise to METU. Professor John Donald, who had come from Berkeley, was leaving too. He had taught us two terms of Linear Algebra using Lang, Hoffman, and Kunze in a crystal-clear way, and as Yılmaz and I happened to be reading Dirac's *Mathematical Principles of Quantum Mechanics* at the time, we were awed to recognize that the same vector space concepts, generalized to an infinite number of dimensions, were at the basis of this fundamental theory in physics. When we asked John Donald whether he could write recommendations in support of our applications to graduate school the next year, he said he would be glad to, but we should make sure to ask Langlands too, as he was "a big professor," while he himself was "a little professor." We did, and, the big professor very kindly accepted. With Gürsey also putting in a few good words for us, Yılmaz was accepted into the graduate physics Ph.D. program at Caltech, and I into the University of Chicago. Langlands was planning to bring Mükremin, who was one year younger than us, to Yale the following year.

That, unfortunately, did not work out. The year was 1968, and radical student movements engulfed campuses. The government's response to it, partly by underhandedly supporting militant rightwing groups, became increasingly uncompromising and violent. Mükremin decided he could not idly stand by in the face of social injustices and the dogmatic responses to them by radical students. He felt he had to take an active part, if only to try to steer others in a more reasonable direction. He was never involved in any violence, except as a victim in mysterious circumstances at the end. Arf, Gürsey, and Langlands had tried to reason with him to stick to mathematics and go to Yale, but to no avail. He definitely had the mathematician's trait, necessary but not always sufficient for achieving nontrivial results, of stubbornly following axioms to their logical, and in his case, fatal end. Langlands tells the story in Turkish in LTR.

It is perhaps of some interest to explain how Langlands maintained his interest in Turkey and the Turkish language. For this we have to jump to 1981, when I was at the Bonn Physikalisches Institut on an Alexander von Humboldt Stiftung fellowship. I heard Langlands was spending a year at Hirzebruch's Max Planck Institute for Mathematics (where I believe he had introduced "The Fundamental Lemma" in the 1970s) a few hundred meters away, so I thought maybe I should go say hello. I was not sure he would remember me, or be eager to refresh the acquaintance. I need not have worried; just like my signing up for the Introductory Analysis course, this turned out to be one of the best moves I ever made. He recognized me right away in spite of my unkempt beard and a few extra pounds, which he mentioned politely as he tried to identify me with the 19-year old he had known. Upon finding out I was married, he invited my wife Ginger and me to dinner at the apartment where he was staying with his wife Charlotte The four of us got together many more times, sometimes with German physics Ph.D. students to whom he was giving lectures (in German, of course) on Statistical Mechanics. This may have been the beginning of his interest in the renormalization group fixed point; it also rekindled his desire to renew his knowledge of Turkish. He invited me to spend the 1985–1986 academic year at the Institute for Advanced Study (IAS), ostensibly as an assistant to answer his questions on quantum field theory, but really to at least temporarily relieve an old student who was finding life difficult in his home country at the time. I spent the IAS year working on generalizing self-duality for Yang–Mills fields to higher dimensions, attending Witten's lectures on String Theory together with Langlands and also studying a deep paper titled *Algebras, lattices and strings* by Peter Goddard and David Olive. It was in this paper that I came across the Montonen–Olive conjecture, which claims a dual relationship between a Yang–Mills theory with coupling constant g and gauge fields in the adjoint representation in the root lattice, and another gauge theory where the fields are solitons of the original gauge fields. The solitons are monopoles sitting in the weight lattice. Their coupling constant is $1/g$; thus the strong coupling regime of one theory, where perturbation theory is useless, is the weak coupling regime of the other. For $N = 4$ supersymmetric Yang–Mills theory in four spacetime dimensions, the two formulations would be expected to become the same.

In 1997, when I was visiting Langlands in Montreal, Drinfeld and others were beginning to develop a theory which was referred to as the geometric Langlands program. Although I knew very little about the original program, I gathered the geometric one was reminiscent of Montonen–Olive duality. I asked Langlands if my impression was correct, after first sketching Montonen and Olive's ideas. I remember his answer was positive, albeit in a noncommittal way, given my very imprecise account. Several years later, Langlands, together with İlhan İkeda (a former student of Shimura) and Ali Altuğ, a Ph.D. student of Langlands and Peter Sarnak, organized a series of lectures in Istanbul to study the relation beween the original Langlands program and its geometric extension. Although he initially expressed some reservations about this extension, he must have later been persuaded about its value, and recently published a number of papers on the subject with Frenkel. Given my inadequate mathematical background, I have only a limited understanding of these papers.

Frenkel later wrote the book *Love and Mathematics* in which he attempts to explain aspects of the Langlands program at a semipopular level. The book was translated to many languages including Turkish, and Langlands contributed the "Letter to Turkish readers" (LTR) in Turkish. From studying Geoffrey Lewis' *Teach Yourself Turkish* in 1967, he has progressed to giving more than one series of mathematics lectures in Turkish in Istanbul and reading advanced Turkish historical or literary texts, many of which he found on his own without my recommendation. In the LTR he explains how he tries to do this in all the countries he visits for a lengthy period. He tries not only to learn languages but to mentally inhabit the places where they arose. In Turkey, we visited together the beautifully situated Assos, where Aristotle spent three years across the island of Lesbos; Nicea, where the concept of the Trinity was forged with some divine inspiration and a little nudging from the Emperor Constantine who was getting impatient with theological disputes; and Troy, which hardly needs an explanatory comment. It is partly his personal way of resisting the homogenization of world culture into a least-common-denominator version of Anglo-American culture, a tendency he abhors. But perhaps it is also a manifestation of his general inclination to look at things and concepts from different viewpoints (in this case, languages), and observing how they are related. I remember him once saying in relation to his work connecting number theory and representations of noncompact groups that it was helpful to keep a number of different things in one's mind at the same time.

His concern for Turkish academic matters has not diminished over the years. When the Turkish government decided in 2011 to change the law governing the choice of new members into the Turkish Academy of Sciences (TÜBA) to allow for government appointees, some of the TÜBA members resigned in protest to form The Science Academy (BA) (a Russian colleague once quipped that Stalin had a number of Soviet Academicians executed, but it had never occurred to him to appoint new members). This was to be independent of the government both in appointments and in funding. When we wrote to Langlands to ask whether he would accept a membership in the new BA, he honored us by quickly replying that of course he would. An important founding member of BA was the late Professor Tosun Terzioğlu, in whose honor Langlands is going to deliver a memorial lecture on September 23, 2018. An interesting detail connecting the locations and personalities is that when Terzioğlu arrived at METU from Germany, he moved into the office Langlands had just vacated. When they met many years later, Langlands asked Terzioğlu if the American-made pencil sharpener he had installed was still working. It was, and had served Terzioğlu well for years.

I cannot resist relating a story he told me about his coming to Turkey on a Ford Foundation sponsored first-class Pan Am flight in 1967. For Charlotte and him, this was their first flight outside of North America, and they were discreetly speculating about their fellow passengers. They were intrigued particularly by a silver-haired Englishman in a bespoke suit. Was he a diplomat, or perhaps an important business executive? The plane landed, everybody proceeded to their respective Ankara destinations, and the Langlandses assumed they would never find out who the mysterious gentleman was – until they ran into him a few days later in the METU Mathematics department! He was Professor F. R. Keogh originally from London, was on leave from the University of Kentucky at the time. He had come to Turkey on the same Ford Foundation grant as Langlands. I am still grateful to Keogh for introducing me to Ahlfors' beautiful book on Complex Analysis in the course he taught on the subject.

4

Reminiscences by a Student of Langlands

Thomas Hales

> We are in a forest whose trees will not fall with a few timid hatchet blows. We have to take up the double-bitted axe and the cross-cut saw, and hope that our muscles are equal to them.
>
> *(– R. P. Langlands)*

Bob Langlands was my thesis advisor at Princeton, from 1984 to 1986. No mathematician has shaped my research career so profoundly as he has. To use Weil's memorable phrase, I offer a few *souvenirs d'apprentissage*.

4.1 Leap to Generality

A stack exchange discussion asks for the largest "leap-to-generality" in mathematical history. Suggestions include the notion of category theory (Eilenberg and MacLane), the rise of abstract algebraic structures, Cantor's set theory, mathematics of the infinite (starting with Archimedes' use of the method of exhaustion), Aristotelian logic, the Turing machine, the foundations of probability (Kolmogorov), and the Langlands program.

Harish-Chandra, Grothendieck, and Kolmogorov were Langlands' early "models for emulation." In Langlands' own words, "not satisfied with partial insights and partial solutions, they [Harish-Chandra and Grothendieck] insisted – not so much in the form of intentions or exhortations as in what they brought to pass – on methods that were adequate to establishing the theories envisaged in their full natural generality."

4.2 Princeton, 1983

By the time I arrived in Princeton as a first-year graduate student in the fall of 1983, I had already acquired interests including Lie theory, representation theory, and the trace formula (thanks to Paul Cohen), p-adic analysis (thanks to J. W. S. Cassels), and modular forms (thanks to John Thompson in the heyday of moonshine).

From my first days at Princeton, I visited the Insititute for Advanced Study (IAS) once or twice a week, attending the Borel–Miličić D-modules seminar, Borel's sixtieth birthday conference, and the Harish-Chandra memorial conference.

Eclipsing everything else that year were the momentous morning and afternoon seminars on the trace formula. It is hard for me to convey how deeply formative those seminars were for me, even if I was not then at a stage to appreciate their full significance. Experts – especially Arthur, Clozel, and Rogawski – encouraged me and taught me the basics.

I first met Bob Langlands in person in January 1984 at a dinner arranged by Helaman Ferguson – somebody that I spent considerable time with during my first year at Princeton. (Helaman's son, Samuel, later became my coauthor on the proof of the Kepler conjecture.) By that spring, Langlands had become my advisor, and I had burrowed my way into the Corvallis conference proceedings. By arriving on scene after the 1977 Corvallis conference, which was still the subject of spirited conversations, I was made to feel I had missed a major event in the history of the Langlands program.

My other main reference that first year was *Les Débuts* (or the purple turtle as we called it), where the the fundamental lemma was first stated. My research problem, broadly stated, was to use Igusa theory to understand the transfer of p-adic orbital integrals between a reductive group and its endoscopic groups.

This remained my primary research interest for 10 years. It has been a great adventure to witness the trajectory of endoscopy over the decades, culminating in Ngô Bao Châu's proof of the fundamental lemma. In research posted to the arXiv in 2018, the fundamental lemma has finally emerged in its natural geometric context, as expressing that some dual abelian varieties have the same p-adic volume.

Already by the time I met him, Langlands had a towering reputation for his mathematical achievements. He once published an unforgettable critical book review with the lead "This is a shallow book on deep matters" that compounded his formidable reputation. I found that he was more mellow in person than his reputation might suggest, and he embodied the Institute's ideal of curiosity driven research. It was heartwarming for me to see Bob in 2018 at the Abel Prize conference in Minneapolis after many years.

4.3 Apprenticeship

As a graduate student, the mathematical facts I learned mattered far less than my apprenticeship as a researcher under Langlands. I arrived with good work habits, a disposition for long calculations, and ambition. Here are a few things my apprenticeship gave me.

4.3.1 Taste

American popular culture failed miserably in conveying great mathematical ideas to me. In my teenage years, my (undeveloped) idea of research mathematics was a confusing amalgamation of general relativity, Thom's catastrophe theory, the Penrose staircase, and stunning continued fraction expansions. I remember wandering through the library stacks and wondering, which of these books matter most?

There is no question that my mathematical taste improved enormously under Langlands. More broadly, lectures by Langlands, Serre, Weil, Borel, Kottwitz, Iwasawa, Thurston, and Witten developed my tastes.

4.3.2 Seclusion

Today, Google Scholar, the arXiv, Wikipedia, and MathOverflow give us nearly instant answers; polymath offers instant collaboration.

Then, there was a widespread belief that serious mathematical research required long periods of intense work in relative seclusion. In Flexner's vision, the Institute "exists as a paradise for scholars who... have won the right to do as they please and who accomplish most when enabled to do so." I did not see Langlands summers, when he went into work-related retreat in Montreal. I developed my own routines of seclusion: research retreats to a family cabin at the Sundance ski resort in Utah, secret study areas, and long runs along the Raritan Canal.

I have never seen Langlands at any mega-conference, and he did not push the chores of professional service. He discouraged rapid publication. He valued mathematical substance and frowned on veneer. Details of proofs mattered. He advised me to pick jobs based on educational merit rather than salary or prestige.

4.3.3 Complexity

There was a communal belief that we were building a monumental edifice that would take many decades to complete. One-hundred-page research papers were the norm (and still are).

As documented in Wikipedia's list of long mathematical proofs, it is no coincidence that some of the longest papers in mathematics are in this field: Langlands (Eisenstein series), Arthur (trace formula), Waldspurger (stable trace formula), Lafforgue (Langlands conjecture for the general linear group over function fields); or in neighboring fields in papers by Grothendieck, Hironaka, Harish-Chandra, Cartan, and Deligne.

As his student, I learned how to hold onto a problem that might take years to solve. I learned how to build evidence to support a hunch, how to follow a lead, and how to bury a fruitless idea and move on. These skills have transferred to my other large-scale research projects.

Researchers in the Langlands program were expected to rapidly assimilate many research fields. See Knapp's nine-page reading list "Prerequisites for the Langlands Program," which expands to thousands of pages of readings in algebraic geometry, Lie theory and algebraic groups, representation theory, algebraic number theory, and modular forms.

Yet book lists misguide us. Library scholarship is not mathematical research, and Langlands himself gave me just a few required readings. When we met, it was entirely focused on what I was able to calculate or figure out, and on his suggestions for figuring things out better.

4.3.4 Exoticism

Langlands has described his recurrent dream of escaping – in T. E. Lawrence style – "into the life and language of some exotic land and beginning anew," leading to his expeditions to Ankara. When I was his student, he had just completed his French lectures on stabilization of the trace formula and was collaborating with Rapoport on their German masterpiece "Shimuravarietäten und Gerben."

Langlands's exoticism is inseparable from his mathematical oeuvre. To me, it explains how he went straight from a Yale thesis on PDEs and analytic semigroups, to teaching a graduate course at Princeton on class-field theory ("I still knew almost nothing about the subject, had only two weeks to prepare, was very young, and scared stiff"). With minimal ado, he would jump into an exotic field, then reconnect it with the ever-expanding Langlands program. His new beginnings are memorable, such as when he wrote that if his proofs seem clumsy, it was because he "has not cocycled before and has only minimum control of his vehicle."

4.3.5 Examples and Generalization

Langlands made many detailed studies of special cases, to shed light on the general theory. There was base change for $GL(2)$, Jacquet–Langlands for $GL(2)$, Labesse–Langlands for $SL(2)$, Igusa theory and endoscopy for $SL(3)$, representations of abelian algebraic groups, and a partial stable trace formula for $SL(n)$.

There were several axes of generalization. What works for $SL(2)$ (or even $GL(1)$) should work for all reductive groups. What works for the field of rational numbers should work for all global fields. What works for one local field should work for all. There should be a parallel between local and global theories, related by local–global principles.

Methods were always under assessment: could they encompass the general case?

4.3.6 Freedom

Langlands' interests were already shifting to percolation theory when I was his student. I was to be the last of his PhD students in the Langlands program for decades.

The winter of the Langlands program lasted several years, starting with Langlands' switch to percolation theory, and continuing until Wiles' announcement of Fermat's Last Theorem. Although significant activity continued during the winter years, insiders and outsiders alike had become increasingly disenchanted by the glacial pace towards solutions to the central problems of the program.

At the time, Langlands would sometimes baffle his audience and speak on percolation theory to an audience clearly expecting automorphic representations. As his student, I have claimed the same freedom to pursue my mathematical interests wherever they lead, however baffling. On one level, I have left the Langlands program behind. But on another level, I have remained a true student of Langlands by claiming this freedom.

5

Graduate School with Langlands

S. Ali Altuğ

I was extremely fortunate to have Robert Langlands and Peter Sarnak as my advisors in graduate school. I got to see both of their perspectives and approaches first-hand. It was a unique opportunity that has shaped my view of mathematics and research.

When I was asked to contribute to the current volume with a short recollection of my time in graduate school with Langlands, I initially thought of writing more about my nonmathematical interactions with him. After all, I've known him for a considerable amount of time, and there is already plenty written about his mathematics. When I started writing, however, a profusion of memories made it difficult to organize these thoughts in a way that would do them justice. Consequently, the end result happened to revolve more around our mathematical interactions. I hope it still conveys the magnitude of his impact on my apprenticeship.

I believe it is appropriate to begin with a disclosure. Unsurprisingly, both Langlands and Sarnak had tremendous influence on the way I approach mathematical research. Although they each have their unique styles, they also have many aspects in common, and it turns out that it is quite difficult for me to attribute individual traits. What comes next is about my interactions with Langlands, but many of the lessons learned are indeed from a combination of both Langlands and Sarnak. **I met Langlands for the first time out of a sheer coincidence** in the Fall of 2004 in Turkey. He was that year's speaker for the annual Arf Lectures[1] at the Middle East Technical University in Ankara, where I was an undergraduate at the time. Although Langlands' connections to Turkey are no secret, as a junior in college, I embarrassingly had no idea who he was, and I certainly did not know anything about the celebrated Langlands Program. All I knew was that he was a mathematician from the IAS at Princeton; I was interested in number theory (or rather, what I thought

---

[1] These are annual lectures in the memory of the Turkish number theorist Cahit Arf, as in the Hasse–Arf theorem.

was number theory); and I was planning on applying to graduate schools soon. So I thought it couldn't hurt to attend his lectures.

Indeed, looking at the announcement for Langlands' talks, it was not easy to tell what he was going to talk about, let alone that he was a number theorist – the two talks were titled: *Conformal Field Theory and the Mathematician* and *Descartes ile Fermat*. The first one turned out to be about percolation theory, and the second was about a historical perspective on a problem of Pappus on conic sections. Nevertheless, one thing was quite clear: that these were no ordinary talks. What made the talks even more extraordinary was that the second one was actually in Turkish. It is not an everyday event to see a foreign mathematician come to Turkey and give a talk in Turkish!

After a true tail event,[2] I crossed paths with Langlands at the posttalk reception. He was keen to talk to students (seemingly more so than anyone else) and hearing about their mathematical interests. It was fascinating, at least to me at the time, to see how much he enjoyed speaking the language.[3] For reasons that are a mystery to me to this day, he appeared to be quite interested in hearing about what I would like to do in graduate school, and told me to keep in touch. This short and very much serendipitous introduction was pivotal in my life and career.

Graduate school By the time I arrived at Princeton in 2006, I knew a little bit more about automorphic forms and had at least heard of the Langlands Program, but was still at the crawling stage and rather far from walking. In an attempt to figure out what I should focus on, I started talking to both Langlands and Sarnak. They were both quite welcoming; honestly, much more than I anticipated. Indeed, it was only much later I learned that Langlands may not have had the reputation of being particularly approachable. In my experience, although he is definitely tough, he is also kind and supportive, albeit sometimes in his own way.

After talking to them for nearly two years, my nascent research projects finally started materializing. Sarnak suggested several problems, lying between automorphic forms and analytic number theory, which eventually evolved into joint work with J. Tsimerman on half-integral weight Ramanujan conjecture. This project took much longer than either of us had initially anticipated, and turned out to be the perfect introduction to research (if you think it will take x, in reality it will likely take somewhere in $(3x, \infty)$). I also remember thinking that the whole project had ended up being overly technical. In retrospect, I might have judged too soon; I should have waited until I started working with the trace formula.

---

[2] The full story involves many people and would take this whole article to recount.
[3] I don't think Turkish is unique to that end. He clearly enjoys foreign languages.

At the other end, Langlands had a separate project in mind. During one of our first meetings, he told me to read his recent (at the time) paper "Beyond endoscopy." The paper described an approach towards the functoriality conjectures, using a combination of the trace formula with analytic number theory, and my task would be to investigate it for $GL(2)$.[4] I spent pretty much all of the second half of graduate school, as well as the early days of my postdoc, working exclusively on beyond endoscopy.

The problem, following the trend of functoriality and trace formula related matters, was technical with many moving parts. Incidentally, I trudged through many failed attempts. Around the same time, Langlands was also thinking about problems of a similar nature (his interests shifted towards the geometric theory shortly after), and as a result we got to have many exchanges during this period.

We would meet frequently in his office at the IAS, almost always about an hour before tea time. Meetings would start with a chat, in Turkish, about essentially anything – he was often interested in hearing about Turkey: friends, family, mathematics, etc. I suppose he also considered these chats as opportunities to practice his Turkish. Next, the conversation would move into math and simultaneously switch to English.[5]

Then came tea time, a short but necessary break, and then back to his office for more math. I quickly learned the first lesson after the first tea time: *no cups on his table without a coaster!* And he always made sure this was enforced.

These meetings were remarkably inspirational. Not because immediate technical problems got resolved or anything of that nature – such a thing rarely happens in an advisor meeting anyway. But through these meetings I got to observe the way he approached research and, maybe even more importantly, developed an appreciation for how research works:[6] Psychology is a huge component, and disappointment is par for the course. Indeed one is stuck more often than one isn't – it is how one responds to these challenges that matters the most.

I recall once walking into his office after a particularly brutal series of failed attempts on a part of the problem. I must have had such a dismal face, as he immediately asked what was wrong. The problem was, of course, that I

---

[4] More precisely, for $GL(2)$ and the *standard representation*.

[5] Frankly, the switch was mainly because of me. If it were up to Langlands, he would have had no problem going on about the trace formula in Turkish.

This indeed ended up happening in 2009 when we were preparing to give a lecture series at Galatasray University in İstanbul (with K.İ. İkeda and C. Saçlıoğlu) on functoriality and related topics. Langlands insisted that the lectures be in Turkish. This may sound straightforward, given we were going to be in Turkey; however, at that time automorphic forms vocabulary was in its infancy (we often needed to invent terms). At some point, I mentioned to him that it would be easier if we did the lectures in English. His response was: *If I can do it, you can do it!* Needless to say, the lectures were all delivered in Turkish.

[6] This process is of course quite standard for any researcher.

was stuck. He sat me down and told me about the times he was working on Eisenstein series. He told me that there was a period when he would go through the proof every single day – a very complicated induction, which at the end culminated in a book over 300 pages in length (Springer LNM 544), only to discover that something was not working. He would then isolate the problem, find a counterexample for the issue, and then go through the entire induction with slightly different conditions. This apparently went on for an excruciating amount of time until, well, until everything finally worked out. It was moments like these that made it easier to bear the tribulations of doing research.[7]

Impact I believe Langlands' biggest impact on me has been through his fortitude. It has been an invaluable experience to see him tackle problems, not just on his home turf, as one might say, but sometimes also in somewhat distant areas.

This is especially easy to relate to for a graduate student, as one frequently finds oneself working on problems in areas where one doesn't have much prior experience. A combination of lack of patience and frustration may, and often does, lead to getting lost in peripheral issues as well as intricacies of technical machinery. All this makes it easy to forget the fundamental questions and principles. Watching Langlands' approach has been a constant reminder to not forget the basic problems. His extraordinary tenacity, in mathematical research and beyond, is something that I will always look up to.

We stayed in touch after graduate school and I try to visit him whenever I am around Princeton. He still likes to hear and talk about, and even visit, Turkey. Indeed, recently we had the pleasure of hosting him in İstanbul.[8] I particularly enjoy our conversations around history and (Turkish) literature, even though I am often reminded of my own ignorance. Although our mathematical interests have been evolving in somewhat different directions, I still, and will continue to, learn from him.

---

[7] Perhaps it's true that misery loves company.

[8] Although İstanbul in 2018 is markedly different from the first time he visited in 1967, I think he still enjoyed the visit quite a bit, especially the Beşiktaş–Üsküddar ferry ride.

6

My Unforgettable Early Years at the Institute

Enstitüde Unutulmaz Erken Yıllarım

Dinakar Ramakrishnan

'And what was it like,' I asked him, 'meeting Eliot?'
'When he looked at you,' he said, 'it was like
standing on a quay, watching the prow of the Queen Mary
come towards you, very slowly.'

– from "Stern" by Seamus Heaney
in memory of Ted Hughes, about the time he met T. S. Eliot

It was a fortunate stroke of serendipity for me to have been at the Institute for Advanced Study in Princeton, twice during the 1980s, first as a postdoctoral member in 1982–1983, and later as a Sloan Fellow in the Fall of 1986. I had the privilege of getting to know Robert Langlands at that time, and, needless to say, he has had a larger than life influence on me. It wasn't like two ships passing in the night, but more like a rowboat feeling the waves of an oncoming ship, with the reverberations lasting long.

Langlands and I did not have many conversations, but each time we did, he would make a Zen-like remark which took me a long time, at times months (or even years), to comprehend. Once or twice it even looked like he was commenting not on the question I posed, but on a tangential one; however, after much reflection, it became apparent that what he had said had an interesting bearing on what I had been wondering about, and it always provided a new take, at least to me, on the matter. Most importantly, to a beginner in the field, as I was then, he was generous to a fault, always willing, whenever asked, to explain the subtle aspects of his own work.

I first heard of Langlands when I was trying to think about what I should try to work on for my doctoral thesis. It was during 1976–1977, and I was in the second year of my graduate studies at Columbia, learning about Number theory (using the books of Lang, Davenport, and Artin and Tate) and Algebraic Groups (using Borel) as my two special topics in which I was going to have an oral exam, later in May. In the Spring of 1977, I went to visit my friend Sheldon Kamienny, who had just begun his graduate work at Harvard; I had moved to mathematics after getting a master's in electrical engineering, when Sheldon was an undergraduate mathematics major. During that visit to Harvard, I was fortunate to meet John Tate, who exuded genial wisdom. I mentioned to him

74

that I was trying to come to a decision on what to work on, when he said, "Isn't Hervé Jacquet at Columbia? Lots of people in number theory want to understand what Langlands is proposing, and if as a graduate student you can learn the basics of infinite-dimensional representations from Jacquet, it will give you an advantage and propel you forward," or something to that effect! Tate was very gracious and quite open to new directions, which might not have been the prevailing view of all the number theorists then. The revolutionary idea that noncommutative harmonic analysis on Lie groups should have a say on the decomposition of primes in nonabelian extensions of the rational field was too foreign, and was not so easily accepted! Spurred by Tate's comments, I tried to learn a bit about the sweeping ideas of Langlands, and took a look at the forbiddingly long masterpiece *Automorphic Forms on GL(2)* by Jacquet and Langlands, which resisted my initial forlorn attempts to comprehend. After my orals I asked Jacquet if he would agree to guide me, and he graciously said yes, a bit after seeing how I progressed; it was helpful to first read portions of Godement's *Notes on Jacquet–Langlands* and Gelbart's *Automorphic Forms on Adèle Groups*. I slowly got drawn into this large area of mathematics which has come to be known as the Langlands program, and was also helped by Diana Shelstad, who was at Columbia then. I learned from Jacquet, a wonderful advisor, a lot about Whittaker functions, global and local, and I worked on the spectral decomposition of the L^2 Gelfand–Graev representation of $GL(n)$ over any local field, archimedean or not, with a sliver of it applying to general reductive groups G. While doing that (starting with some notes of Jacquet in the $GL(2)$ case), I tried to understand in my own way, for inspiration and analogy, a little bit of the monumental work of Langlands on the spectral decomposition of $L^2(\Gamma \backslash G)$. I learned about his philosophy of cusp forms being the fundamental building blocks for automorphic forms; I did notice the astonishing difficulty of Langlands' formidable chapter 7, which I could not wrap my head around, let alone decipher or grasp in any real sense. In my (easier) situation, the "cusp forms" were just the generic discrete series representations, and a description of the continuous spectrum involved making the measure explicit in terms of ratios of gamma factors.

I had been to the Institute before that postdoctoral year. In the Fall of 1975, when I had just started at Columbia, I took a train from New York to Princeton and attended a lecture of André Weil there, a follow up to his lectures the previous year on Eisenstein and Kronecker, which was being turned into a lovely monograph. I was tongue tied in front of so many luminaries, the "*who's who*" of number theory, even all of mathematics, who were there. Nevertheless, I learned a lot from the (accessible) lecture as well as from talking to C. S. Seshadri and Madhav Nori, whom I met for the first time. I think Langlands was already there, but I didn't meet him then, nor any other permanent member. At least I got to know the Institute grounds by walking in the woods and around the beautiful pond, thinking about Weil's 1949 paper, "Number of solutions of equations in finite fields," which I was carrying with me, dealing with diagonal varieties, due to my interest then on Gauss and Jacobi sums. I was to learn later that, more importantly, that paper of Weil had concluded with an amazing set of conjectures for all varieties over finite fields, the most difficult of which, the Riemann hypothesis in that setting, had in 1972–1973 been resolved in a monumental work of Deligne. Equally, I had no idea then that, in 1974, Langlands had posed a far reaching conjecture of his own in De Kalb, IL, on the points modulo p of modular varieties bearing the name of Shimura in "Some contemporary problems with origins in the Jugendtraum, Mathematical Developments arising from Hilbert Problems," relating them to automorphic forms and their L-functions. I am mentioning all this to give an idea of how exciting, and equally forbidding, the Institute was in those days.

Before arriving at the Institute, I was a Dickson Instructor at the University of Chicago during 1980–1982, due to some happy happenstance. In my last year as a student at Columbia I took a break from writing and finalizing my thesis, and managed to derive a new result on the ubiquitous Dilogarithm function and its connections to the Heisenberg group. It got Spencer Bloch interested, due to the connection to the tame symbol for algebraic curves, leading to a job at Chicago. I was also partly helped by the fact that the chair there, Paul Sally, was one of only two people in the world (besides me and my advisor) who found my thesis interesting! Paul was a wonderfully hilarious personality, with a pirate's eye patch, who gave me very good comments on a paper I was writing based on my thesis, and introduced me to his students Allen Moy, Dan Bump, and Sol Friedberg. However, he was so busy being chair that we had very little time to discuss some new mathematics projects. So I became involved much more with Spencer's group, working on questions arising from algebraic cycles and K-theory, and in the process got to know many including Richard Swan, V. Srinivas, and Chad Schoen.

Coming back to representation theory, the second person who showed an interest in my thesis topic was Harish-Chandra, thanks to Paul's prodding me to send him a copy of the dissertation. Surprisingly, a few months before I arrived at the IAS, I received a handwritten letter from Harish-Chandra saying that he had completely solved my thesis problem for all reductive groups G, using his powerful theory of wave packets and the Schwartz space. I was filled with admiration since my ad hoc methods for $GL(n)$ would never do the trick in general. I was also relieved in a way to have to find other problems. It is sad that Harish-Chandra never got around to writing out the details of his work on this topic, which would have been a major contribution. However, there is a sketch of his method in Volume IV of his *Collected Papers*, edited by V. S. Varadarajan.

When I arrived at the Institute in September 1982, I went up to Langlands' office in the main building, which used to be Einstein's office in the old days, I think, and said hello. Then I moved my books, etc., to my office in Building C (which has now changed and is no longer with mathematics anymore). I found that my office (on the second floor) was just two doors down from Harish-Chandra's, and downstairs near the entrance was Armand Borel's office. I don't think there was any postdoctoral member who worked anywhere as hard as Harish-Chandra or Borel. I was quite embarrassed, but I was human and had other interests like English literature and music. I was attending four seminars a week; one of them was the algebra seminar at the University, to which I rode my bicycle. (I saw Langlands riding his bike too everywhere.) But I think Borel was attending six seminars, while giving talks at two of them! I went to his seminar as well and learned a bit about the L^2 cohomology of arithmetic groups Γ, which I had been introduced to earlier by Avner Ash, while he was a Ritt Assistant Professor at Columbia during my graduate student days there. At Princeton University, I got to know Goro Shimura , Nick Katz, and Andrew Wiles, and I went to a few of their lectures; they were inspiring. I had met Andrew before, during one of my visits to Harvard in 1977, when he had just arrived as a Benjamin Pierce Lecturer; he was very friendly and told me something *clairvoyant* in retrospect, namely that the key to understanding the arithmetic of elliptic curves and their generalizations lay in the Selmer group! He is my contemporary, but I consider him among my teachers, along with Jacquet, Bloch, Langlands, and Shalika – in chronological order.

Before discussing the mathematical insights I received from Langlands during those two years, let me briefly mention some occurrences, partly of a personal nature. In 1982–1983, I was shocked one day to find André Weil walk up to me in the reading room and ask me if I knew Sanskrit. He seemed pleased when I said "Yes, I have learned some." He asked for my favorite work in it, and I said *Shakuntala*, to which Weil nodded in approval and said Kalidasa, the author of the play, was gifted. I never discussed mathematics with Weil, but attended his historical lectures, which I loved. At the Institute there were many other postdoctoral fellows, and I became close to them, like the analytic number theorists Brian Conrey, Amit Ghosh, and Dan Goldston. One day Amit introduced me to Atle Selberg, who was extremely nice to me and answered my question as to what made him think of the powerful technique now known as the Rankin–Selberg method in automorphic forms. It was illuminating as I had asked earlier Robert Rankin the same question while at a conference in Britain, and I now realized that they had completely different interests and motivations in coming up with that method – independently. There was some ongoing activity in the field of Automorphic Forms at the Institute, which led to my meeting Jim Cogdell, Laurent Clozel, Gerard Laumon, and Jean-Luc Brylinski. That year I got married to Pat Carlton, and the reception at my IAS flat was attended by Charlotte Langlands, as well as by Lily and Harish-Chandra. That year I also got to know Don Blasius very well, whom I had met during the previous Fall during a brief stay at Princeton, but who was now a Ritt Assistant Professor at Columbia.

Let me now recount two specific things I learned from Langlands during those two years in the 1980s that shaped my research. The first was his explanation of the notion of *distinction* for cuspidal automorphic representations π of $GL(2)$ over a quadratic field F (over the rational field Q), relative to the subgroup $GL(2)/Q$, and the relationship to the Asai L-function, which played a key role in his work with Harder and Rapoport on the Tate conjecture for Hilbert modular surfaces X over Q. I used this indirectly in my work on the Beilinson conjecture for periods of integrals on X arising from K_1 of X, in modern terms from a suitable motivic cohomology group of X. I needed to tackle a problem in the nondistinguished case, and the integral representation arising from their work was crucial, as well as its geometric meaning, relating the residue to the nontriviality of the π-component of the diagonal cycle. It led to a *Comptes Rendus* Note and a preprint. Later, in the summer and early Fall of 1986, I proved with Kumar Murty (whom I met that year), the full Tate conjecture for Hilbert modular surfaces, which came down to establishing it over nonabelian extensions. The Tate classes over abelian fields had been shown by Harder, Langlands, and Rapoport to be accounted for by the Hirzebruch–Zagier cycles, which are Hecke translates of embedded modular curves, and they are all defined over cyclotomic fields. Kumar and I did not have any algebraic cycles to play with that were defined over nonabelian fields, so we decided to show that every Tate class gave rise to a Hodge class, which came down to exhibiting a period relation (in our perspective), allowing us to appeal to the Hodge conjecture for divisors, known by Lefschetz. Coincidentally, almost to the day when we finished the project, or perhaps a day earlier, there appeared another proof, by Christophe Klingenberg, a student of Harder, and his method was to restrict to $SL(2)$ and make use of endoscopy. In a sense his approach was more in line with the Langlands philosophy. Anyhow, it motivated me to learn more about the stable trace formula, especially for $SL(2)$, and this is the second insight of Langlands that impacted my work, albeit some years later. I learned of the problem of multiplicity one for $SL(2)$, and came to know his approach to prove it using the adjoint transfer of automorphic forms from $SL(2)$ to $PGL(3)$, which was being carried out by Y. Z. Flicker. Somehow, I wanted to find an alternate way to establish this multiplicity one result for $SL(2)$, and later I found such a path, by making use of a different functoriality. I needed the conjectured automorphy of the Rankin–Selberg product, sometimes called the automorphic tensor product, of two cusp forms π, π' on $GL(2)$, again predicted by Langlands to exist. I had to show that if they had the same adjoint squares, whose automorphy in $GL(3)$ was known by Gelbart and Jacquet, then π and π' are twist equivalent, as required by the seminal work of Labesse and Langlands, which I read carefully in 1986. So I needed to prove the automorphy of $\pi \times \pi'$, which I managed to do after some years, and one of the tools was the strengthened converse theorem for $GL(4)$ due to Cogdell and Piatetskii-Shapiro, along with a lemma coming out of earlier work with Blasius, and some local identities for triple product L-functions, as well as some bounds for the growth of Eisenstein series.

One thing Langlands impressed upon me in 1986 was that while instances of his general conjecture are useful to isolate and prove, such as the reciprocity conjecture relating the (appropriate) representations of a connected reductive group G to certain morphisms from the Weil (or the conjectural Langlands) group to the dual group $^L G$, the entire picture of functoriality must be viewed all together, like a complex pattern, to understand better the interconnections between different aspects of functoriality. Many years later, I found an example of what he was talking about in a joint work of mine with Dipendra Prasad, which I would now like to explain. The local Langlands conjecture (LLC), a theorem for $GL(n)$ with different proofs (due independently to Harris and Taylor, Henniart, and recently Scholze) asserts over a p-adic field F that irreducible n-dimensional representations σ of the Weil group correspond to supercuspidal representations π' of $GL(n, F)$, which in turn correspond, by the (extended) Jacquet–Langlands correspondence, to certain irreducible representations π of D^*, with D being the central division algebra over F of dimension n^2 and invariant $1/n$. One sees easily that σ is selfdual if and only if π is selfdual. However, if we ask whether σ and π are simultaneously orthogonal, i.e. leaves invariant a symmetric bilinear form, then the answer is not predicted by LLC. Prasad and I could show that surprisingly, when n is even, σ is symplectic if and only if π is orthogonal! The way we did it was to use the functoriality from the relevant classical groups to $GL(n)$. So in effect, it is still as predicted by Langlands, but one has to use LLC in conjunction with this functoriality.

I also learned from Langlands, either then or later, that one way to understand the conjectural automorphic Galois group, called the Langlands group by Jim Arthur, is to think of its essential consequence predicting the existence, for each cuspidal automorphic representation π of $GL(n)$, a reductive subgroup $H(\pi)$ of $GL(n, C)$ not contained in any Levi subgroup; this is what happens when π is cohomological and corresponds to an n-dimensional Galois representation ρ with the Zariski closure of its image being (the p-adic avatar of) $H(\pi)$. The beauty is that this reductive subgroup is supposed to exist even for nonalgebraic cusp forms!

In the Fall of 1986 I got to meet Gerd Faltings at the University, and also managed to attend a course taught by Andrew Wiles at the University on Hilbert modular forms, and there I met three exceptional graduate students, namely Michael Larsen, Richard Taylor, and Fred Diamond. Even then it was clear Richard was going to be a major player.

One day that year Langlands made an offhand remark that I filed away in my mind and stored for later rumination. There had been some casual discussion of the Riemann zeta function, and Langlands wondered if something could be made of it occurring as a factor of L-functions of appropriate Eisensteinian forms on $GL(n)$. I said nothing, but remembered then that Selberg had conjectured that every Dirichlet series in his class admitting a pole must be divisible by the Riemann zeta function. Many years later I came to think about Siegel zeros, which are spurious real zeros just to the left of $s = 1$, and wished I could say something about such zeros of automorphic L-functions of $GL(n)$. Spurred by a comment of Goldfeld at a conference, Jeff Hoffstein and I managed to show that the L-functions of cusp forms on $GL(2)$ had no such zeros. More importantly, we showed that for any $n > 1$, if we assumed Langlands' functoriality, say just for the Rankin–Selberg product $GL(k) \times GL(m) \to GL(km)$ (for suitable k, m), then no cuspidal L-function $L(s)$ of $GL(n)$ could have a Siegel zero; it was deduced by exhibiting an automorphic L-function $D(s)$ of positive type which was divisible by a higher power of $L(s)$ than the order of pole (of $D(s)$). Thus functoriality does in an indirect way have something to do with zeros inside the critical strip, albeit just inside. Curiously this line of argument fails for $n = 1$, the key (in fact only) unresolved case being that of a quadratic Dirichlet character. It certainly seems like $GL(1)$ is harder than $GL(n)$ in this regard, because there is not enough room, perhaps in the same way low-dimensional topology is harder than in higher dimensions.

Something that impressed me about Langlands was his desire and ability to master several foreign languages, like the fact that he wrote his paper with Harder and Rapoport in German; my own German improved by studying that paper minutely, which incidentally exposed me to that beautiful word "*ausgezeichnet*" (distinguished)! When I got to know that he was fluent in Turkish, I became more than a bit curious, as his visitor Cihan Saçlıoğlu told me he thought many names of dishes in Indian restaurants seemed to be cognates of Turkish words. This led me to study a bit of the etymology of words in Hindustani, which is really a continuous spectrum of languages between Hindi at one end and Urdu at the other, with people in different regions of India speaking at a different point in the "moduli space." Indeed many words have Turkish origins; even the name of the language Urdu means "army" in Turkish, but it also means "camp" in Farsi, so we can't be sure where it came from; but we do know that Urdu evolved when the Mughal army recruits from Tajik and Uzbek lands to the north (in Central Asia) began mixing expressions of Turkish, Farsi, and Sanskritic origins; some Arabic words got thrown in as well. Like children do in many of those lands, I also got exposed to the stories of Nasruddin Hodja in India, especially the comics made in Bombay (now Mumbai). The word Hod (or Khod) is of Farsi origin meaning "God," and Hodja is a cleric. The Albanian name Hoxha, pronounced exactly as Hodja (though routinely mispronounced in Europe), means the same, apparently, and it presumably came from Turkey. Anyhow all this led me to learn a little bit of some Turkish; the subtitle of this *recollection article* is in it and means the same as the English title.

Langlands also shared with me some of his unpublished notes, one on the representations of abelian algebraic groups, another giving his thoughts on the theta correspondence, and yet another on his factorization of the Artin root number. The very first one, which extended class field theory from the multiplicative group to arbitrary tori, was later published in the Olga Taussky memorial volume, in a special issue of the *Pacific Journal of Mathematics* in 1989. The last one never got in print, perhaps because, soon after his result was known, Deligne found a short global argument. It is still very valuable to understand Langlands' approach, which could be helpful in other contexts.

I quickly learned that Langlands' vision was so broad, and all encompassing, that I could find problems which probed and combined my three areas of interest, namely representation theory, analytic theory of L-functions and arithmetic geometry. He also said once: "Think conceptually in the most general framework, but then prove something nontrivial in some case!". He also disliked routine extensions and incremental advances, and approved of taking bold and courageous steps. I have always treasured his advice and have passed on to my students some of the wisdom he imparted.

One of the works of Langlands I am enamored of is his old paper on the volumes of fundamental domains of arithmetic groups Γ. Expressing the volume as the integral over the fundamental domain $'\Omega$ of the constant function, which is orthogonal to the cuspidal spectrum, and using his spectral decomposition of this orthogonal part in $L^2(\Gamma \backslash G)$ via Eisenstein series, he expressed the volume of $'\Omega$ as ratios of products of suitable L-values, generalizing old formulae of Siegel. I think this method of his is a tour de force. I was originally thinking of writing an exposition of it for the second part of this volume, but then I found out (through Clozel) that Michael Rapoport had already written such an exposition, which can be found on his website, and I have nothing more to contribute to it.

I would be remiss if I do not acknowledge the hospitality shown by both Bob and Charlotte Langlands over the years. Charlotte, among her many talents, is a gifted sculptor, and it is always a pleasure to view her astounding works. The bust she made of Weil is at the Institute, and the one she made of Harish-Chandra is at the HRI (Harish-Chandra Research Institute) in Allahabad, India.

It is very rare indeed that a perspicacious person such as Langlands enthralls us with his insights, making fundamental contributions in an ancient and challenging subject. Therefore we are grateful to be so fortunate. Many future generations will continue to be influenced by his thinking and predictions. In one sense, a lot has been proven and clarified due to the works of several people, in particular in the field of Shimura varieties, L-functions and their relationship to Galois representations, and also for geometric analogues where Q is replaced by C. However, from another perspective, only the surface has been scratched and many basic questions still remain unanswered – even for $GL(2)$ over Q, which bodes well for the future.

It is not in the stars to hold our destiny but in ourselves.

(William Shakespeare)

PART III

Langlands as Friend

7

My Reminiscences of Bob Langlands
at the University of British Columbia

Alan Goodacre

In this chapter, I would like to say a few words about my recollections of Bob Langlands when we were at the University of British Columbia (UBC).

These recollections are from our third year at UBC when were both in "Honors Physics and Mathematics."

Bob and I were still in some classes together in fourth year but he had switched from "Honors Physics and Mathematics" to "Honors Mathematics" so there was less overlap between us. Also he and Charlotte had gotten married before our fourth-year classes started. Incidentally, she was crowned "Totem Queen" during the 1954–1955 year – Bob's second year at UBC.

As most people know, Bob did his MSc in Mathematics at UBC and then went to Yale for his PhD. I did my MA in Physics at UBC and then went to work at the Dominion Observatory in Ottawa. A decade later I took educational leave to do my PhD in Geophysics at the University of Durham in England.

We hadn't remained in touch until 1983, when I happened to read a book about the Institute for Advanced Studies. In one of the chapters the author was describing how the mathematicians were filing into a lecture and I saw the name Langlands. I figured this must be Bob so I looked him up in a *Who's Who* and sure enough it was he. So I wrote to Bob and it turned out that he was lecturing at the University of Quebec in Montreal for the Summer. My wife, Lise, and I drove to Montreal and had a fine visit with Bob and Charlotte, catching up with news about our children, etc.

Fast forward a couple of decades and Bob was wondering what I was doing, so Charlotte found me via the Internet. Most of my stuff is under A. K. Goodacre rather than the name Alan Goodacre, which has been taken over by some chap in England!

By this time I had retired and Bob and Charlotte's daughters were in the Ottawa area, so we managed to get together once a year or so depending upon Bob and Charlotte's busy schedule. I mention their visits in recent years because I want to mention the bumble bee. We had been sitting in the back yard one Summer day and it was time for me to take Bob and Charlotte back to Thomasin's place. A big bumble bee had gotten into the screened-in gazebo

85

and wouldn't come out even though the door was open. I finally closed the door, but as we walked towards our car Bob turned back and propped open the door of the gazebo. He didn't want the bumble bee to be trapped. This hasn't much to do with Mathematics but tells you what kind of a person Bob is.

When Lise and I got back home the bumble bee was still in the gazebo with its still open door, but I finally got him to climb on to a broom handle and put him outside where he flew away. I made sure to email Bob to let him know the bumble bee was free.

Now, here are some of my specific recollections of Bob Langlands from our time at UBC.

I first met Bob in 1955 when we were both enrolled in the third-year "Honors Physics and Mathematics" course at UBC. I expect I would have met Bob before then but I had spent my first two years of university at Victoria College, now University of Victoria. The required courses in third year included three Mathematics courses: "Differential Calculus," "Integral Calculus and Differential Equations," and "Algebra and Geometry."

As I recall, we were supposed to have tests each week in rotation. The first test in "Differential Calculus" was set by Professor Leimanis, the second test a week later in "Integral Calculus and Differential Equations" by Professor Goodspeed, but for some reason the tests were then discontinued. I had initially found it a bit of a leap from second-year Mathematics at Victoria College to third-year Honors at UBC, but I passed the first test satisfactorily. When Professor Goodspeed started to hand out the results of the second test he mentioned that Bob had originally gotten 85 percent but he had scaled Bob's mark up to 100 percent and everyone else by the same factor. Except that when he came to me he said I was lucky because I had a larger scaling factor than the rest of the class. He had scaled my 3 percent mark up to 4 percent! After Professor Goodspeed had finished handing out the test papers Bob raised his hand and said he thought that the answer marked wrong was actually correct. Professor Goodspeed went over to Bob, looked at Bob's paper and agreed. Professor Goodspeed then said he would go to the registrar and raise Bob's mark to 118 percent! Whether Professor Goodspeed was joking or not, I remember the incident as though it happened yesterday.

The three "mathematicians" in the Honors courses were Bob, Klaus Hoechsmann, and Roger Purvis. I recall that they would discuss various mathematical concepts together. For example, one day Bob and Klaus were discussing Hausdorff Spaces. I remember the name Hausdorff because it is German, as is Klaus' family name. I thought that Roger, unlike Klaus, was somewhat shy. As the school year progressed I could clearly see that Mathematics was Bob's "forté."

Bob and I were partners in the laboratory section of the third-year "Physical Optics" course. I think we made a good team. We tended to split up the duties, with my taking the experimental part and Bob's taking the theoretical part.

I recall in particular a delicate experiment where I set up a diffraction grating and Bob used a Cornu spiral in describing the results. If I had a mathematical question Bob would answer it in an easy-to-understand way.

Professor Jennings taught "Algebra and Geometry," the third of the three required Mathematics courses. In his last lecture at the end of the school year he finished up a bit early, so in the remaining time Bob, Klaus, and Roger gathered around him discussing topics that they found interesting. I was standing back a bit listening to their conversation and remember that Fermat's Last Theorem came up. Professor Jennings said something to the effect that Fermat's Last Theorem might not be a good choice for a thesis topic if one wanted to graduate in a finite length of time! It is rather inspiring to note that some of Bob's subsequent work played an important role in the proof of Fermat's Last Theorem, and that Bob was to become "Nulli Secundus" in his chosen field.

8

Robert P. Langlands: l'homme derrière le mathématicien

Claude Levesque

Extended English Abstract *The goal of this chapter is to give some tidbits about the man behind the mathematician Robert P. Langlands, well known for the program that bears his name. The paper starts with the story of a honorary doctorate given to him by Laval University on the occasion of the Canadian Mathematical Society summer meeting in 2002. He prepared a speech to be delivered in French during the ceremony and asked me to correct the mistakes (he was not as fluent in French as he is now). I gave him a report that was longer than the document he previously handed me, and in which I gave detailed explanations for each correction I suggested to make. This pleased him.*

A few weeks later, I dared to ask him to deliver during the CMS conference a lecture in mathematics in English. I was not aware that he has strict rules concerning the choice of a language for a lecture. He wrote me back that if he had to speak in English, he prefered to refuse his honorary doctorate and to not attend the CMS meeting. Of course, he immediately got a message that he was welcome to speak in French.

We had, later, many interesting exchanges concerning his two papers written in French and published respectively in the old ANNALES DES SCIENCES MATHÉMATIQUES DU QUÉBEC *and in the new* ANNALES MATHÉMATIQUES DU QUÉBEC. *I guess he liked my long reports with detailed explanations. By the way, those two papers played a key role for attracting important papers, in particular in Number Theory. Indeed, he was and has always been very interested in developing mathematics in Canada and particularly in Quebec.*

Professor Langlands obtained many prestigious prizes in mathematics, the Shaw Prize, the Wolf Prize, and the Abel Prize, to name but a few. Sometimes those prizes came with large amounts of money, part of which he used to make important gifts for the development of mathematics, but no details are given here out of respect for his private life and his modesty.

Professor Langlands is fluent in four languages: English, French, German, and Turkish. For him, a serious mathematician should be able to read mathematical papers in English, in French, in German... and eventually in Latin in order to be able to read the work of Gauss in the original language. Although he wonders about his ability, I can testify that Professor Langlands is very good at speaking in French.

It happened that Langlands, when he was young, made a window frame at the store of his father, and the frame had the wrong dimensions. So one wonders whether or not he had some abilities for doing some work with his hands. Let me cross the t's by stating that Robert Langlands is a competent handyperson who built, with his sons, two cottages on the bank of Lake Gustave (in the north of Montreal).

Professor Langlands loved to spend quiet summers at his cottage with his wife, who is a professional marble sculptor. He likes to do some sports like walking, skating during winter, and riding a bicycle during summer.

Right now, he enjoys having a peaceful environment, trying to avoid trips and enjoying all the pleasant opportunities of life, though he is not very optimistic about the future of the world.

Résumé *Le but de cet article est de décrire quelques aspects de l'homme derrière le mathématicien Robert P. Langlands, bien connu pour le programme qui porte son nom. Cet article commence avec l'histoire d'un doctorat honoris causa que l'Université Laval lui a décerné en 2002 lors d'une réunion de la Société mathématique du Canada. À cette occasion, le professeur Langlands a préparé un discours de remerciements en français et il m'a demandé d'en corriger les fautes, ce qui ne me semblait pas une mince tâche. Suite à ma requête de donner une conférence mathématique en anglais, il a mentionné que s'il devait le faire en anglais, il préférait renoncer à son doctorat honorifique. Il a des règles précises et des standards quant au choix des langues pour ses exposés. Il a reçu plusieurs doctorats d'honneur. Aussi de nombreux grands prix, dont en particulier le prix Shaw, le prix Wolf et le prix Abel. Avec une partie de l'argent reçu avec ses prix, il a fait des dons importants pour le développement des mathématiques, mais par respect pour sa modestie je ne donne pas de détails sur ce sujet. C'est en particulier grâce à lui que les* ANNALES MATHÉMATIQUES DU QUÉBEC *ont atteint un haut niveau de qualité. C'est un homme humble, modeste, un grand amant de la nature et un bon bricoleur habile de ses mains. Il est plein de gratitude envers son épouse. C'est aussi un sportif, mais il n'aime pas la compétition. Il recherche maintenant la paix de l'esprit, mais affirme qu'il n'est pas très optimiste pour l'avenir du monde.*

8.1 Introduction: Un doctorat honorifique

C'est un peu un concours de circonstances, mais me voilà au sein d'un groupe de personnes qu'on dit proches du professeur Robert P. Langlands et à qui on a demandé de raconter quelques faits reliés à ce très grand mathématicien qui a marqué le domaine des mathématiques en énonçant des conjectures incroyablement profondes et en écrivant des articles sur un programme énormément ambitieux qui porte son nom, le cas $n = 1$ étant, ni plus ni moins, la théorie (abélienne) du corps de classes.

Sur cet homme, que dire qui ait du sens? D'un côté, il y a mon cerveau qui me suggère d'être neutre et bref et d'énoncer des faits sans périphrase. De l'autre, il y a mon coeur qui me dit de ne pas être froid, mais d'être chaleureux et de tout mettre en valeur. J'ai décidé d'écouter mon cerveau... qui m'a dicté de suivre mon coeur..., quoique, pour citer une diaphore de Pascal, *"le coeur a ses raisons que la raison ne connaît point"*.

En 2002 se tenait à l'Université Laval la réunion d'été de la Société mathématique du Canada (SMC). J'ai alors pensé que c'était une belle occasion pour décerner un doctorat *honoris causa* à un mathématicien important. Pour moi, le candidat idéal (lire une valeur sûre) que je voulais proposer était le professeur Robert Langlands. En moins d'une heure, grâce au moteur de recherche google, je monte un dossier sur le candidat et j'envoie ledit dossier au doyen de la faculté des sciences et de génie de l'Université Laval. Le lendemain, le doyen vient me voir à mon bureau pour me dire que j'avais un dossier gagnant.

Le professeur Langlands a alors préparé un texte en français qu'il planifiait de lire lors de la remise dudit doctorat d'honneur. Il ne maîtrisait pas encore très bien la langue française et il me demanda si j'accepterais de lire au préalable son document de 4 ou 5 pages et d'y corriger les fautes. C'était tout un honneur pour moi, quoique cela représentait tout un défi car, d'un point de vue psychologique, je ne pouvais certainement pas dresser sans explication une liste bébête de fautes à corriger.

Je me demandais bien comment un cul-terreux comme moi pourrait procéder pour dire à quelqu'un, que je considérais comme un demi-dieu, qu'il y avait des fautes de français. Bien que je me voyais vraiment indigne de délier les courroies de ses sandales, j'ai pris mon courage à deux mains et je lui ai remis un document deux fois plus long que le sien avec ma liste de changements à apporter et surtout avec des explications fort détaillées expliquant pourquoi chaque changement était justifié. Cela lui a bien plu, semble-t-il, car en quelque sorte je suis temporairement devenu l'un de ses professeurs de français (lire conseillers), comme vous le verrez plus tard dans cet article.

8.2 Il ne plie pas l'échine

Comme organisateur de cette réunion d'été de la SMC, je me disais que je devais privilégier l'usage de la langue anglaise pour les exposés mathématiques, vu que les participants étaient pour la plupart unilingues anglophones. J'ai alors suggéré au professeur Langlands de donner son exposé mathématique en anglais. C'était méconnaître l'homme intransigeant qu'il est lorsque le sujet des langues est sur le tapis. C'est un homme de principes et il a des règles strictes que nous décrirons à la section 8.

Ma requête a donc provoqué son ire: une sainte colère, bien sûr. (Cela lui arrivait de temps en temps, paraît-il, lorsqu'il était plus jeune, mais évidemment c'était toujours pour de bonnes causes, pas par caprice.) Il m'envoie alors un message clair et sans équivoque me disant que s'il est obligé de donner son exposé en anglais, il renonce tout de suite à son doctorat honorifique et annule sa participation au congrès de la SMC. *"Ah!"*, me suis-je dit, *"C'est dur la vie académique!"* Il n'en fallait pas plus pour lui écrire immédiatement que je retirais ma requête et qu'il était invité à parler en français. J'ajoutai qu'il fut un temps où j'étais un ardent défenseur de la langue française au Québec et qu'en vieillissant je m'étais ramolli, mais qu'il me donnait le goût de reprendre le bâton du pèlerin et de repartir en croisade. Ceci semble l'avoir rassuré parce que, sur-le-champ, il m'envoya un message électronique indiquant qu'il serait présent à cette réunion de la SMC.

8.3 Un homme généreux avec de nombreux prix

Il a reçu au fil des ans de nombreuses récompenses. Mentionnons tout d'abord ses doctorats d'honneur. À l'époque, je lui ai demandé combien de doctorats *honoris causa* il avait reçus dans sa vie. Il a mentionné qu'il en avait perdu le compte. D'aucuns pourraient voir dans cette réponse un signe de vantardise, mais ce n'est pas un bon décodage. Selon moi, le fait d'accepter ces honneurs est plus pour faire plaisir à ses hôtes que pour se faire plaisir à lui-même. Quand on y pense sérieusement, qui a gagné le plus avec ce doctorat d'honneur décerné par l'Université Laval? Est-ce lui ou est-ce l'Université Laval? Je vous laisse deviner ma réponse.

Au cours de sa carrière, le professeur Langlands a reçu de nombreux prix: le prix de l'Académie Nationale des sciences (États-Unis) décerné pour la première fois en mathématiques, le prix Cole de l'AMS, le prix Steele de l'AMS, le prix Humboldt, le prix Jeffery-Williams de la SMC, le prix Nemmers, le prix Wolf, le prix Shaw, la Grande Médaille de l'Académie des Sciences (France) et tout récemment le prix Abel. Ceci m'amène à mentionner une de ses grandes qualités: il est très généreux. Certains de ces prix venaient avec de gros montants d'argent; il a utilisé une partie de cet argent pour faire des dons importants pour le développement des mathématiques; je suis bien à l'aise de le mentionner, mais pour respecter sa modestie et sa vie privée, je ne veux pas donner plus de détails.

Il connaît tout le respect que j'ai pour tous ces prix, mais il avoue être *"parfois prêt à hausser les épaules"*. Quoique ce soit une opinion personnelle, j'ai l'impression que pour lui, les grands prix (et même la médaille Fields) ne servent pas très bien la cause du progrès des mathématiques, mais contribuent plus à créer de la compétition.

Il s'est fait un plaisir d'aider de jeunes mathématiciens. Par exemple, il y a une jeune Chinoise qui était intéressée par l'étude de son programme et il lui a envoyé un message indiquant en long et en large comment procéder.

Il y a près de dix ans, les ANNALES DES SCIENCES MATHÉMATIQUES DU QUÉBEC sont devenues les ANNALES MATHÉMATIQUES DU QUÉBEC. Cette revue mathématique est maintenant un journal d'un très bon niveau grâce à un excellent article conjoint de Edward Frenkel, Robert Langlands et Ngô Bảo Châu [F-L-N] et aussi grâce à un autre article très important de Robert Langlands [L1]. Nous, les Québécois, nous leur devons, à tous les trois, une fière chandelle. Le professeur Langlands lui-même a insisté sur le fait que c'était important d'aider cette revue québécoise. Il a raison; la qualité attire et génère la qualité. Vraiment, le professeur Langlands a toujours eu à coeur le développement des mathématiques au Canada, et particulièrement au Québec.

8.4 Un homme humble

Quand il a reçu la Grande Médaille de l'Académie des Sciences (France), il s'est demandé pendant son discours si c'était lui qui avait conçu le programme visionnaire qui porte son nom ou si ce n'était pas plutôt son frère jumeau. Pour apprécier cette belle figure de style, il faut savoir que Robert Langlands n'a pas de frère jumeau. Il était un peu comme un champion olympique qui 30 ans plus tard se demande devant un miroir si c'est vraiment lui qui a obtenu le record du monde.

Soit dit en passant, pour ses articles conjoints, il tient mordicus à ne pas mentionner explicitement qui a fait quoi. Redit autrement, il ne cherche pas à se mettre en valeur ou en évidence.

Il affirme d'ailleurs qu'il se sent *"bien parmi les modestes de ce monde"*. Il n'a d'ailleurs jamais caché qu'il venait d'un milieu modeste. Je pense qu'en fait, il en est fier. Son père était charpentier et avait un magasin dans lequel il vendait des matériaux de construction et fabriquait des cadres de portes et de fenêtres. Son grand-père était bûcheron. Son beau-père a aussi été bûcheron avant de devenir peintre en bâtiment.

Jeune adolescent, il voulait cesser d'aller à l'école, décrocher comme on dit au Québec. Son but était de traverser le Canada sur le pouce avec un ami. Quand il a eu 15 ans, sa mère a réussi à le convaincre de faire une autre année d'étude. Pendant cette année, il a eu un enseignant qui a reconnu le talent du jeune Robert et qui a pris une heure complète de son cours pour le convaincre qu'il devait aller à l'université.

À 16, 17 ans, durant ses vacances d'été, Robert Langlands était camionneur. C'était un petit camion d'une tonne et demie, pour être précis. Il chargeait à la main des sacs de ciment, du bois d'oeuvre ou d'autres matériaux pour aller en faire la livraison dans un village voisin.

L'histoire veut qu'il ait fabriqué un cadre de fenêtre qui était de la mauvaise grandeur. Imaginez le drame: c'était pour son futur beau-père qui n'avait alors *"aucune idée de l'avenir infortuné qui attendait sa fille"*. Cette antiphrase provient de Langlands lui-même et me fait conclure qu'il aime parfois jongler et jouer avec les mots.

8.5 Un amant de la nature

Avec l'aide de ses fils, il a bâti dans le nord de Montréal un chalet sur le bord du Lac Gustave, près de la ville de Morin Heights. *"Dans les pays d'en haut"*, se plaît-il à dire, pour employer une expression de nos ancêtres québécois. En fait, il en a bâti un deuxième qui servait pour les invités. Il affirme n'avoir joué qu'un rôle de bricoleur compétent et que tous les membres de sa famille (épouse, enfants, petits-enfants) sont plus doués que lui pour effectuer des travaux manuels. L'un de ses fils est d'ailleurs entrepreneur dans le domaine de la construction de maisons.

Il y a quelques années, il a invité à son chalet quelques mathématiciens dont moi-même. Il m'a montré les rénovations qu'il a faites à ses chalets. Étant moi-même bricoleur à mes heures, j'ai pu constater *de visu* qu'il est tout le contraire d'un nul pour le bricolage: les cadres de fenêtres m'ont semblé parfaits...

Après un délicieux repas préparé par son épouse, il a suggéré de faire du canot. Tous, sauf moi, se sont désistés (probablement par crainte de chavirer) et j'ai eu le privilège de faire une longue randonnée en canot avec lui. Nous nous sommes arrêtés au bout du lac pour admirer les plongeons d'un huard: une merveilleuse communion avec la nature! Il était d'ailleurs très attaché à un couple de huards qu'il a vu au lac Gustave pendant plusieurs années consécutives. Ce couple de huards a eu une portée d'oisillons seulement une année et cette infertilité semblait le chagriner.

8.6 Une merveilleuse épouse

Ce serait un crime de lèse-majesté de ne pas mentionner qu'il est très attaché à son épouse Charlotte Lorraine Cheverie, qu'il admire beaucoup, en particulier, pour son esprit d'abnégation. C'est une sculpteure professionnelle sur marbre et son nom d'artiste est Charlotte Langlands. Ses merveilleuses oeuvres d'art me font saliver. Pour le plus grand plaisir de son mari, elle a en particulier sculpté la tête d'André Weil... et celle de son époux.

C'est ici le bon moment de mentionner que la nouvelle passion du professeur Langlands est la généalogie, en particulier, les ancêtres de son épouse. Il a pu vérifier que lui et son épouse sont de vieille souche canadienne. Il se plaît à dire que la généalogie *"donne souvent un côté personnel à l'histoire"*.

8.7 Un sportif

Le professeur Langlands aime pratiquer des sports, mais jamais sous forme de compétition. Bref, que ce soit pour les mathématiques ou pour le sport, je le répète, pas de compétition! Il aime la marche. Avec son épouse, il patine l'hiver, en particulier lorsqu'il est à Montréal. L'été, c'est le vélo. Avec un vieux vélo, pour ne pas dire une épave. Il a beaucoup adoré faire du vélo avec son grand ami Helmut Koch. Sur la toile, on trouve un reportage de la cérémonie de la remise du prix Abel et on le montre en train de faire du vélo. Vous pourrez vérifier qu'il n'est pas assis sur un beau Cervelo P5, mais plutôt sur une vieille bécane, ce qui rend ce petit film encore plus attachant.

8.8 Les langues

La langue maternelle du professeur Langlands est l'anglais, mais il a appris le français, le turc, l'allemand. Il parle donc couramment quatre langues. Il a débuté l'apprentissage du russe et lors de son séjour en Inde, il a commencé à apprendre le hindi. Il m'a confié qu'il voulait aussi apprendre l'italien; il a lu quelques volumes en italien, mais a rapidement oublié ce qu'il a appris. Au Québec et en France, il fera ses exposés en français; en Turquie, ce sera en turc; en Allemagne c'est en allemand. Il s'en fait un devoir et un plaisir. Pour lui, c'est un choix personnel qu'il défend férocement. D'aucuns pourraient penser qu'il est entêté, pour ne pas dire têtu, mais personnellement je pense qu'il a, ce que l'on apelle, le courage de ses idées et qu'il accepte d'en payer le prix.

Il a écrit qu'il aime baragouiner les langues. L'emploi de ce dernier verbe témoigne de l'étendue de son vocabulaire et met en évidence sa modestie, car il fait plus que baragouiner les langues: il en parle plusieurs couramment. Du moins, je le pense quoique je ne connaisse rien du turc et très peu de l'allemand.

À un moment donné, il a écrit: *"Je commence à comprendre que je n'écrirai jamais le français sans erreur"*. Ceci dénote son souci de perfection, mais je tiens à mettre les points sur les *i* : le professeur Langlands maîtrise très bien la langue de Molière.

Au cours des sept dernières années, il s'est concentré sur la rédaction d'un article mathématique en russe. Il m'a affirmé qu'il n'est pas suffisamment à l'aise pour parler en russe, mais qu'il le lit assez couramment.

Selon le professeur Langlands, c'est un devoir de tout mathématicien
sérieux de pouvoir lire les articles mathématiques en anglais, en français, en
allemand; aussi en latin, afin de pouvoir lire les écrits de Gauss dans la langue
originale. Pour lui, c'est un scandale, un déshonneur, que dis-je, une honte que
l'anglais devienne une langue universelle en mathématiques. Quand je lui ai
mentionné que l'anglais pourrait désormais jouer le rôle que le latin jouait il
y a deux cents ans, il s'est énergiquement opposé en mentionnant (sans ire)
que le latin est une langue morte, alors que l'anglais n'est pas une langue
morte, puis il a ajouté des commentaires pleins de sens, mais qui ne m'ont
pas complètement convaincu, du moins, pour le moment...

Si vous lisez ses écrits, vous verrez qu'il aime faire de longues phrases, de
très longues phrases, autant en français qu'en anglais. Même si parfois je dois
les relire et en faire une petite analyse grammaticale, cela me fascine de voir
qu'il maîtrise bien cet art d'impliquer plusieurs thèmes dans une même phrase.
Le premier paragraphe de l'introduction de cet article se veut d'ailleurs un clin
d'oeil à ses longues phrases, et j'accepte à l'avance de passer pour un épigone.

8.9 La langue française

Pour ses deux longs articles en français, respectivement dans les ANNALES DES
SCIENCES MATHÉMATIQUES DU QUÉBEC [F-L-N] et dans les ANNALES
MATHÉMATIQUES DU QUÉBEC [L1], nous avons fréquemment échangé nos
opinions. Ce fut enrichissant et pour moi et aussi pour lui, je présume, parce
que fréquemment nous avions tous les deux raison, même si nous semblions
avoir des opinions différentes. Pour vérifier quelques-unes de mes affirmations,
il consultait parfois ses documents (dictionnaires et volumes de grammaire),
mais voyez-y un souci de perfection et non une volonté de prouver qu'il avait
raison. Vous ne serez sûrement pas surpris d'apprendre qu'à quelques reprises,
c'est lui qui a corrigé des fautes dans certaines de mes remarques, ce qui m'a
à chaque fois fait plaisir, car cela témoignait du fait qu'il prenait très à coeur
la rédaction de ses écrits en français.

Voici quelques exemples de nos échanges sur la langue française:

- La conjonction de subordination "*quoique*" commande une élision
  seulement devant les pronoms, ce que j'ai appris grâce à lui. Il n'y a
  cependant pas unanimité chez les grammairiens.
- Doit-on mettre "*difficile de faire*" ou bien "*difficile à faire*"? (J'ai
  mentionné que cela dépend s'il y a un complément direct ou non.)
- Doit-on mettre "*continuer de*" ou bien "*continuer à*"?
- Doit-on mettre "*s'attendre de*" ou bien "*s'attendre à*"?
- Doit-on mettre le subjonctif présent ou l'indicatif présent: "*Supposons que
  n soit un entier pair*" ou bien "*Supposons que n est un entier pair*"?

(J'ai indiqué que les deux façons sont bonnes et que selon moi le subjonctif
a une belle saveur de voeu pieux.)

_ Est-ce que la locution adverbiale "*À toutes fins utiles*" s'écrit au pluriel ou
au singulier?

_ Est-ce que la locution adverbiale "*À portée de mains*" s'écrit au pluriel ou
au singulier?

Parfois le professeur Langlands a des arguments des plus convaincants.
Lorsque je lui ai mentionné de ne pas utiliser le verbe "*bâcler*" pour ses
travaux, je lui disais que la connotation dudit mot était plutôt négative,
mais il m'a cité "*L'amour bien bâclé*" de l'auteur-poète-chanteur québécois
Gilles Vigneault..., comme si c'était faisable de BIEN bâcler un travail...
Assurément à son corps défendant, il m'a cloué le bec et j'en fus bouche bée.

J'ai été ému d'apprendre qu'il avait lu le roman "*Maria Chapdelaine*" de
l'écrivain français Louis Hémon.[1] C'est clair que le professeur Langlands a
une très grande culture. Il a lu des volumes importants en anglais, en français,
en turc, en allemand...

Lors d'un repas à Québec, j'ai assisté à une sorte de débat entre lui et le
célèbre avocat québécois Jacques Larochelle portant sur plusieurs sujets, dont
André Weil et sa soeur Simone (parfois c'était sur Simone Weil et son frère
André). À tour de rôle, on se relançait avec des détails historiques. Un beau
duel amical entre deux érudits!

8.10 Sur la beauté des théories mathématiques

Le professeur Langlands s'intéresse à l'histoire des mathématiques, mais dans
les faits beaucoup plus à, selon moi, l'évolution des idées en mathématiques.
En 1999–2000, il a donné au IAS une série très intéressante de 16 exposés
intitulée "*The practice of mathematics*" (voir [L2]), portant sur l'évolution
des mathématiques au fil des siècles. Il a débuté ses exposés en affirmant
ne prendre pour acquis que la bonne volonté des membres de l'audience,
son but étant d'expliquer aux non-mathématiciens du grand public le "quoi"
et le "pourquoi" des mathématiques ("*the what and why*"). Il ajouta que
s'il se donnait à lui-même ces exposés, ce serait parfois difficile. Il avoua
même devoir faire face à son ignorance: "*I am brought face to face with
my ignorance*". Ce n'est pas faire preuve d'une grande originalité que de
mentionner que j'aimerais posséder son ignorance, mais je tiens à mentionner

---

[1] Je remercie Rob Harron qui m'a appris que Louis Hémon est français, bien qu'il soit souvent
considéré comme un auteur québécois. Ce roman décrit merveilleusement bien la vie des colons
du Québec d'il y a 100 ans.

que son affirmation n'est pas de la fausse humilité, mais reflète vraiment sa personnalité: se consacrer coeur, corps et âme à l'apprentissage et au développement des mathématiques.

En 2010, le professeur Langlands donna à l'Université de Notre Dame un exposé intitulé " *Is there beauty in mathematical theories?*" C'est un très beau survol de l'évolution des mathématiques au fil des siècles (voir [L3]). Aussi surprenant que cela puisse paraître, le professeur Langlands se dit incapable de définir les notions de beauté et d'esthétisme, tout particulièrement en mathématiques. C'est vrai que la notion de beauté est relative et subjective, mais beaucoup de membres de la communauté mathématique sont d'accord pour dire que le programme de Langlands est la réalisation parfaite de la beauté en mathématiques.

J'en arrive à croire que le professeur Langlands est aussi un brin philosophe. Il affirme que les deux qualités frappantes des concepts mathématiques sont qu'ils ont simultanément la possibilité de se développer par eux-mêmes et d'avoir une validité permanente: *"Two striking qualities of mathematical concepts are that they are simultaneously pregnant with possibilities for their own development, and are of permanent validity"*. Il y a dans son exposé de belles pensées profondes. Il rappelle que dans le passé on a fait jouer à Dieu et au diable des rôles importants en mathématiques. C'est fascinant!

Il cite un passage d'une lettre de Jacobi à Legendre: *"Fourier avait l'opinion que le but pricipal des mathématiques était l'utilité publique et l'explication des phénomìes naturels; mais un philosophe comme lui aurait dû savoir que le but unique de la science, c'est l'honneur de l'esprit hmain, et que sous ce titre, une question de nombres vaut autant qu'une question du système du monde"*. Je pense que c'est bon de rappeler fréquemment à la société l'importance de l'HONNEUR DE L'ESPRIT HUMAIN, que ce soit pour la culture, les arts, la peinture, l'histoire, la poésie… et les mathématiques.

8.11 Conclusion

Il y a beaucoup à dire sur une personne aussi célèbre que le professeur Robert P. Langlands, un Canadien dont nous sommes tous fiers, et en particulier un Québécois dont je suis très fier. En fait, je le considère plutôt comme un citoyen du monde. Il est à une étape de sa vie où ce qui prime pour lui, c'est la paix de l'esprit. Depuis quelques années, il semble ne plus aimer voyager. Il mentionne: *"En effet, je trouve que j'ai vu assez du monde. Il m'intéresse toujours, mais je trouve que je l'aime mieux en lisant qu'en voyageant. Le monde contemporain et ses habitants me découragent."*

En particulier, il n'aime plus prendre l'avion. Dans les aéroports, il n'aime pas se soumettre aux questions des douaniers et des préposés car, affirme-t-il, il perd un peu plus du peu de liberté qu'il lui reste. Il essaie de jouer de tous les bons moments de la vie, bien que sans être un pessimiste de nature, il n'est pas très optimiste pour l'avenir du monde. " *C'est triste mais les maux d'aujourd'hui semblent ne pas être résolubles. Nous sommes entourés par la folie, …*"

Rappelons que cet important programme de Langlands est né d'une lettre de 17 pages qu'il a écrite et envoyée à André Weil en 1967 et à qui il mentionnait:

«*If you are willing to read it as pure speculation, I would appreciate that.*

If not – I am sure you have a waste basket handy.»

Ces deux phrases décrivent merveilleusement bien la personnalité du professeur Robert P. Langlands.

Bien que je sois un néophite en la matière, je me permets de mentionner que je considère le programme qui porte son nom comme

la Voie, la Vérité, la Vie.

C'est presque blasphématoire d'utiliser les mots du grand prophète Jésus, mais je ne donne à ces mots aucun sens religieux ou spirituel. Seulement un sens mathématique:

- *la Voie*, car c'est un chemin (un lien) qui unifie deux mondes en mathématiques, l'un algébrique et l'autre analytique; pour être plus précis, le lien est entre la théorie des nombres algébriques et l'analyse harmonique;
- *la Vérité*, car vraiment, mais vraiment, il y a un pont entre ces deux mondes, quoique ce pont soit plus virtuel que concret pour le moment;
- *la Vie*, car ce pont apporte à ces deux mondes de la lumière et un surplus de vie.

En conclusion, derrière ce mathématicien, oui, il y a un homme, mais derrière ce mathématicien il y a aussi un prophète, non pas un prophète inspiré par la divinité, mais un prophète qui a été inspiré par des éclairs de génie… et non par son frère jumeau…

8.12 Épilogue

Le 12 avril 2015, le professeur Robert P. Langlands m'écrivait:

«*Je me suis dit que j'ai un travail à faire - construire une théorie géométrique convenable des formes automorphes. Dans la théorie que les émigrés russes essaient de construire, il manque une théorie spectrale (donnée par des séries d'Eisenstein dans la théorie arithmétique). Les Russes veulent utiliser la théorie des faisceaux dans la forme créée par Grothendieck, mais à mon avis il faut plutôt une théorie analytique dans laquelle la géométrie différentielle joue un rôle majeur. Je crois savoir quoi faire – plus ou moins – mais il y a beaucoup à apprendre sur la géométrie différentielle et sur la théorie spectrale générale des opérateurs dans les espaces hilbertiens pour en arriver à construire une théorie convenable, ou bien il faut trouver quelque part une littérature encore inconnue pour moi.»*

En [L4], on trouve son article [*iztvestiya.pdf*] écrit en russe dans lequel il indique comment il conçoit cette construction d'une *"théorie géométrique convenable des formes automorphes"*; c'est peut-être son testament mathématique. On trouve aussi en [L4] ses commentaires [*comments-onprevious.pdf*] sur cet article en russe, de même que des précisions sur sa conception de la théorie géométrique et de la théorie arithmétique.

Je vous ai parlé de l'homme derrière le mathématicien, mais la relation est réflexive car derrière l'homme il y aura toujours le mathématicien. Bien qu'il ait des idées très précises sur la théorie des formes automorphes et qu'il a sûrement des conseils à donner aux jeunes qu veulent poursuivre son programme, cet homme jeune de coeur et d'esprit, né le 6 octobre 1936, pense qu'après avoir consacré 65 ans aux mathématiques, il est temps de passer à des choses différentes. Vous pouvez être assurés que les projets ne manquent pas.

Bibliographie

[F-L-N] Edward Frenkel, Robert Langlands, Ngô Bảo Châu, Formule des traces et fonctorialité: le début d'un programme, *Ann. Sci. Math. Québec* **34** (2010), no. 2, 199–243.

[L1] Robert Langlands, Singularités et transfert, *Ann. Math. Québec* **37**, no. 2 (2013), 173–253.

[L2] Robert Langlands, *The practice of mathematics*, www.math.ias.edu/practice

[L3] Robert Langlands, *Is there beauty in mathematical theories*, Lecture in 2010 at University of Notre Dame, published by University of Notre Dame and by *Bhāvanā* **3**, No. 3. http://publications.ias.edu/sites/default/files

[L4] Robert Langlands, *The work of Robert Langlands*, publications.ias .edu, section 12, The geometric theory: [iztvestiya.pdf] et [comments-on-previous.pdf]

9

Un homme de culture et de nature

Claude Pichet

Die Grenzen meiner Sprache bedeuten die Grenzen meiner Welt

(Wittgenstein)

Abstract. *Robert Langlands has made a major contribution to mathematics mostly through the creation of the program that bears his name. He has received numerous awards for his work in mathematics but the Fields prize eluded him, which is unfortunate. But there is more to him than mathematics. This brief testimonial tries to show the other aspects of his personality with respect to culture and nature. He has a love of the humanities and of languages. He reads, speaks and writes in several languages: English, French, German, Turkish, and Russian. He also has a love of the outdoors and I have had the chance of making several canoe and hiking expeditions with him.*

9.1 Introduction

Ce texte est rédigé en français. Il y a aujourd'hui une hégémonie de la langue anglaise dans les travaux scientifiques qui a l'avantage principal de permettre la dissémination des travaux scientifiques à l'ensemble de la planète mais qui a comme désavantage d'occulter les particularités associées à chaque langue et culture. Cette uniformisation et cette perte de diversité amènent probablement une diminution et un appauvrissement des productions intellectuelles dans tous les domaines. Robert Langlands est un partisan convaincu de la diversité et de la richesse portées par les langues et l'a d'ailleurs démontré par ses travaux. Il est un des seuls mathématiciens contemporains qui a rédigé ses travaux dans plusieurs langues. Il a rédigé des articles scientifiques en anglais, français, allemand, turc et russe. Son article le plus récent qui est, selon ses dires, son chant du cygne mathématique a d'ailleurs été conçu et écrit en russe. Dans un livre qui veut expliquer ses travaux et présenter sa personnalité, il est donc normal que les langues qu'il a utilisées dans ses travaux soient représentées. On peut simplement déplorer que les exigences de l'édition n'aient pu permettre

Figure 9.1 Robert Langlands en Bretagne, juin 2018. Crédit photo: Claude Pichet.

que l'allemand, le russe et le turc n'apparaissent. La citation de Wittgenstein mise en exergue de ce texte s'applique bien à lui.

J'ai rencontré Robert Langlands en 1986. Il était venu donner un exposé à l'Université de Montréal sur des travaux de Harish-Chandra. J'ai aussi assisté à une série de conférences qu'il a données sur la physique statistique, sujet auquel il commençait à s'intéresser. À la fin de la série de conférences, les participants lui ont donné quelques livres choisis parmi les classiques de la littérature québécoise. Je discutai ensuite avec lui de ces choix. Ce fut le début de notre amitié.

Ses talents mathématiques sont connus et célébrés mais il existe d'autres dimensions à sa personnalité qui méritent d'être connues et reconnues. J'aimerais ici témoigner de ses talents mathématiques mais surtout de son intérêt pour la culture, plus particulièrement la littérature, et de son amour de la nature et des expéditions faites à pied et en canot. Mais revenons tout d'abord sur sa carrière mathématique et sur un événement qui ne s'est pas produit.

9.2 D'abord, le mathématicien

L'obtention du prix Abel en 2018 démontre les qualités mathématiques de Robert Langlands. Ce prix couronne sa carrière mathématique qui s'étale sur 60 années. Il a aussi obtenu plusieurs autres prix et récompenses (Prix Wolf, Steele, Shaw, etc...) mais la médaille Fields, qui récompense les mathématiciens de moins de 40 ans, lui a échappé alors que les astres étaient alignés pour qu'il l'obtienne en 1972 ce qui aurait eu pour conséquence un retentissement majeur pour lui et la communauté mathématique canadienne. La genèse de ce qu'il est convenu de nommer le "Programme de Langlands" date de la décennie entre 1960 et 1970 et l'architecture de ses travaux était bien établie avant 1972. Les grandes lignes de ce programme sont esquissées dans la lettre qu'il a envoyée à André Weil en 1966, lettre qui est d'ailleurs la source de la légende autour de ses travaux. On parle souvent d'occasion manquée lorsque une action qui aurait pu se réaliser ne s'est pas accompli. On peut dire que le le Congrès International des mathématiciens qui se tenait en 1972 à Vancouver au Canada en fut une. Il aurait été tout à fait approprié qu'il obtienne la médaille Fields lors de ce congrès car il est canadien, originaire de la région de Vancouver, avait moins de 40 ans et avait effectué des travaux mathématiques importants. Un bémol cependant. La réputation de la médaille Fields s'est faite sur un malentendu et une fausse conception du travail des mathématiciens basée sur la supériorité de la jeunesse et qui laisse croire que les travaux importants sont toujours effectuées avant l'âge de 40 ans. Il est vrai que les personnes jeunes ont plus d'énergie que les personnes âgées mais la somme de connaissances accumulées par les mathématiciens plus âgés leur permet aussi de continuer à contribuer au monde mathématique.

Les travaux de Robert Langlands sont difficiles à comprendre car ils utilisent plusieurs concepts qui ne font pas partie des concepts généralement enseignés aux apprentis mathématiciens. Je pense ici aux adèles et à l'analyse qui leur est associée. Comme il l'a souligné, tout son travail peut se caractériser comme une nouvelle façon de faire de la théorie des nombres. Il a utilisé les notions et outils de la théorie des groupes développés d'abord par Évariste Galois et Sophus Lie, ce qui lui a permis de faire des avancées spectaculaires dans le domaine de la compréhension de la nature des nombres. Bien que ses travaux n'aient probablement pas la même envergure que ceux de Newton et Leibniz qui ont mené à la création du calcul différentiel et intégral inventé au dix-septième siècle et qu'il a fallu près de 300 ans pour comprendre et simplifier suffisamment pourqu'il puisse être enseigné au niveau collégial, les travaux de Robert Langlands méritent d'être simplifiés pour être mieux compris et il faut donc espérer que les générations futures de mathématiciens s'attelleront à vulgariser au sens noble du terme les travaux de Langlands et que le temps qui passe permettra de décider si ces travaux mériteront la considération qu'ils obtiennent maintenant auprès des experts du domaine.

Robert Langlands est le successeur et l'héritier des mathématiciens alle-mands des dix-neuvième et vingtième siècles, à savoir Gauss, Kronecker, Dedekind, Hilbert, Frobenius, Hecke, Artin et autres. Ses travaux ne sont pas des pics isolés dans l'univers mathématique mais s'inscrivent dans une continuité historique. Les autres chapitres de ce livre présentent les travaux de Robert Langlands en détail et avec une compétence largement supérieure à la mienne mais j'aimerais, malgré tout, indiquer en quoi sa contribution est reliée aux travaux de Gauss et de ses successeurs. Gauss a découvert et prouvé en 1801 la loi de la réciprocité quadratique. Cette loi nous permet de déterminer la nature des entiers premiers de l'anneau des entiers associé à $\mathbb{Q}(\sqrt{d})$ Cette loi a ensuite été généralisée pour les cas cubique et biquadratique. Ensuite, l'école allemande menée par Emil Artin, Erich Hecke et David Hilbert a proposé une version plus étendue de cette loi dans la première moitié du vingtième siècle pour le cas des extensions obtenues en ajoutant à un corps les racines d'une équation dont le groupe de Galois est commutatif. Cette version porte le nom de théorie des corps de classe. Ce que Langlands a fait est de proposer un programme pour étendre cette théorie aux cas où le groupe de Galois n'est pas commutatif.

Le lecteur courageux peut s'attaquer à l'édifice patiemment construit par Robert Langlands et ses collaborateurs. Robert Langlands a d'ailleurs rédigé quelques textes qui expliquent de façon relativement élémentaire la nature de ses travaux en théorie des nombres. Je vous invite à lire aussi les introductions de ses textes mathématiques car on y retrouve des accents littéraires que l'on ne rencontre pas souvent dans des articles de recherche ce qui m'amène à considérer l'aspect littéraire de sa personnalité.

9.3 Un homme de culture

Tous lui reconnaissent un intellect hors du commun mais ce ne sont pas tous qui connaissent son intérêt pour le monde à l'extérieur des mathématiques. Je veux simplement témoigner de son intérêt pour la culture classique, les langues et les activités associées à la nature, à savoir la randonnée et le canotage. Tout d'abord, parlons de son intérêt pour les langues et la littérature. Il parle et lit l'anglais, l'allemand, le français, le russe et le turc. Comme il a séjourné assez longtemps en Allemagne, en Turquie et en Russie, il a eu la politesse d'en apprendre les langues. Mais il a aussi une connaissance de l'hindi et de l'arménien. Il a séjourné en Inde, patrie de Harish Chandra, qui a eu une grande importance dans sa carrière et auquel il rend souvent hommage dans ses écrits, et en a profité pour apprendre le hindi. D'ailleurs, une partie de la correspondance relative à ce séjour a été rédigée dans cette langue. Lors de ses séjours en Turquie, il a visité des cimetières dans lesquels les inscriptions sur les pierres tombales étaient rédigées en arménien et il a appris à déchiffrer ces inscriptions. Il m'a d'ailleurs expliqué qu'il existe plusieurs variantes de l'arménien et que l'arménien utilisé en Arménie est différent de celui de la diaspora arménienne.

C'est une chose de connaître et parler les langues, c'en est une autre de lire la littérature écrite dans ces langues. Les œuvres littéraires qui deviennent des classiques permettent de comprendre les sociétés dans lesquelles les auteurs vivent même si les sujets traités dans les œuvres n'ont pas de lien direct avec la vie courante. Pensons à Asimov qui est un écrivain américain dont les œuvres se déroulent dans des univers imaginaires qui en disent beaucoup sur la société américaine. Lire Proust, Thomas Mann, Tolstoï, Roquebrune et Tampinar permet d'élargir les horizons et de comprendre les sociétés dont sont issus les auteurs. J'ai eu la chance de discuter avec lui de plusieurs de ces oeuvres et il m'a recommandé la lecture de certaines. Il m'a fait découvrir les mémoires de jeunesse de Gorki et la poésie de Yeats. Bien que je sois du Québec, il m'a permis de découvrir certains auteurs québecois dont Robert de Roquebrune, que, à ma courte honte, je ne connaissais que de nom. Comme je l'ai déjà indiqué, cet aspect se retrouve dans ses textes mathématiques dont les préfaces particulièrement témoignent de son intérêt pour la chose littéraire. Il fut une époque qui s'est terminée vers 1900 où il était possible à des individus d'avoir une vision encyclopédique du monde, pensons à Diderot. Cette période est terminée mais certains essaient de croire que cette vision est encore d'actualité. Robert Langlands est de ce nombre.

Il existe chez lui un attachement au Canada qui se manifeste en partie par son intérêt pour la langue française qui est une langue officielle au Canada. Il a donné une partie du prix Abel pour des projets reliés à l'utilisation de la langue française dans les mathématiques, car comme je l'ai mentionné, il s'inquiète de l'utilisation presque exclusive de la langue anglaise dans les travaux scientifiques. La richesse de la culture d'une langue se mesure sans doute au nombre de locuteurs de cette langue mais aussi à la capacité qu'elle a d'être utilisée dans le cadre des activités normales d'une société et le travail mathématique en est une.

9.4 Un homme de la nature

Robert Langlands est né au Canada et a toujours conservé sa citoyenneté canadienne. Il s'est marié jeune, au début de la vingtaine avec Charlotte Cheverie, qu'il a connue à l'école secondaire et avec laquelle il est toujours marié. Ils ont eu quatre enfants et ont maintenant des petits-enfants et des arrières-petits-enfants. Trois de ses quatre enfants résident au Canada et le quatrième demeure dans l'État du Vermont aux États-Unis à quelques kilomètres de la frontière canadienne. Il a longtemps passé l'été avec sa famille dans un chalet sans électricité dans les Laurentides au nord de Montréal.

Robert Langlands est conscient de sa valeur intellectuelle mais reste modeste et mène une vie simple et quelque peu frugale. Il a passé son adolescence à aider son père dans son travail d'entrepeneur en construction et il a donc l'habitude des travaux manuels ardus et épuisants. De plus, pour réussir dans son domaine au niveau auquel il s'est hissé, il faut être muni d'une détermination et d'une intensité peu communes. Son charisme particulier est l'expression de sa ténacité et du sentiment de devoir qui s'exprime dans

Figure 9.2 Robert Langlands et Claude Pichet sur le sentier des douaniers en Bretagne, juin 2018. Crédit photo: Claude Pichet.

l'aphorisme "Fais ce que dois". Ces caractéristiques qui peuvent faire croire à de la brusquerie lorsqu'elles se manifestent dans un contexte professionnel et lors de colloques deviennent un atout dans la pratique de la marche et du canotage. Son caractère intense peut se manifester sans problèmes dans ces activités car elles sont essentiellement solitaires et permettent le dépassement.

C'est peut-être un poncif de présenter les Canadiens comme des descendants des coureurs des bois, mais il y a une part de vérité dans cette image. Les vastes espaces canadiens ne sont pas une légende et existent en effet et il est possible de parcourir lacs et rivières en canot dans des parcs nationaux. Ces expéditions font rêver et elles nous confortent dans l'image du Canadien coureur des bois, cependant il pleut, il fait froid, il faut transporter les canots et préparer les repas dans la nature ce qui n'est pas de tout repos. J'ai eu le plaisir d'effectuer de nombreuses randonnées avec lui et, beau temps mauvais temps, il marche, canote et ne se plaint jamais. Un compagnon idéal!

La géographie des Laurentides, chaîne de montagnes au nord de Montréal est particulière: des montagnes assez peu élevées avec un relief doux et de très nombreux lacs. La région est relativement peu peuplée et il y a de nombreux parcs nationaux et réserves fauniques. Certains lacs sont très grands avec de nombreuses baies et îles qui font qu'il est facile de s'y perdre. Je me permets de nommer les principaux endroits où nous avons canoté: Parc de la Vérendrye, Réserve faunique Papineau-Labelle, lac Taureau et Réserve faunique Mastigouche. Voici deux anecdotes qui décrivent bien Robert Langlands: un jour sur le lac Taureau, il m'expliqua qu'il venait de faire la critique d'un livre qu'il n'avait pas du tout apprécié et dans le feu de son argumentation, il s'emporta et fit presque chavirer notre esquif, un autre soir, au parc de la Vérendrye, je le retrouvai en pleine nuit couché sur une roche plate car il voulait contempler les aurores boréales.

La randonnée est une activité qui convient bien aux intellectuels. Cheminer en surveillant le sentier permet de décompresser et de réfléchir. Il aime marcher et profite de ses séjours à Montréal pour parcourir les rues de son quartier et visiter les cimetières situés sur le Mont-Royal. Souvenons-nous que les cimetières sont des dépositaires du passé. Nous avons aussi effectué quelques longues randonnées, plus de 700km, et bien que ce texte ne soit pas un récit de voyage, j'aimerais souligner celles qui me rappellent les meilleurs souvenirs: le sentier de Stevenson dans les Cévennes, le sentier des douaniers en Cornouailles et la traversée du parc de la Gaspésie.

9.5 Conclusion

J'aimerais terminer mon témoignage en citant la dernière phrase de l'autobiographie de Jean-Paul Sartre qui s'applique bien à Robert Langlands: Tout un homme, fait de tous les hommes et qui les vaut tous et que vaut n'importe qui.

PART IV

Surveys of the Langlands Program –
Early Years

10

An Introduction to Langlands Functoriality

James Arthur

The Principle of Functoriality has long been regarded as the centre of the Langlands Program. More recently, it has had to share the spotlight with Reciprocity, Langlands' conjecture that relates automorphic representations with motives from algebraic geometry. However, the two principles are closely related, and in any case, Reciprocity came at the end of the decade that followed the years 1960–1967 that are the focus of this volume.

Functoriality famously had its roots in the 17-page letter that Langlands gave to André Weil in 1967 [L2]. He wrote some of the details shortly afterwards in the article he dedicated to Salomon Bochner [L3]. It represented a very different direction for Langlands after his monumental volume [L1] on Eisenstein series, which was largely analysis. Langlands credits Bochner, an analyst himself, with directing him towards number theory, especially, I believe, class field theory and its long-sought nonabelian generalization.

There are several ways to introduce functoriality to a general reader. One is as a series of identities (reciprocity laws) that relate families of conjugacy classes

$$c = \{c_p : p \nmid N\}$$

in different complex groups. As objects with complex coordinates, but parameterized by prime numbers, these families are easy to imagine as gateways to higher arithmetic. There are already a number of introductions to functoriality from this point of view. (See for example [A, Section 4].) We shall take a slightly different approach here, one that is closer to the way Langlands originally presented functoriality[1] in [L3]. We shall describe it as a fundamental property of L-functions, and especially the arithmetic L-functions introduced and studied by Emil Artin. The reader can refer also to the chapter [S] by Shahidi in this volume, which is an introduction to Langlands' L-functions that is largely complementary to what is contained here.

---

[1] The name *functoriality* was introduced only later, along with other terms such as *automorphic representation*

This chapter is intended to be a short historical introduction to functoriality. I have not taken up a well-founded suggestion of Matthew Sunohara to break the chapter into sections, which I have formulated as:

(1) Introduction
(2) From Euler's product to Artin reciprocity to Tate's thesis
(3) Artin L-functions and Godement–Jacquet L-functions
(4) On Langlands' seven questions
(5) Four applications.

This would have added clarity, but I have preferred to keep the narrative as informal as possible. I would like to thank Matthew for this suggestion and for other thoughtful comments, most of which I have adopted.

L-functions have a long history, with roots in both analytic and algebraic number theory. For the former, one can look back to Euler. He introduced the infinite series

$$\zeta(s) = \sum_{n=1}^{\infty} \frac{1}{n^s}$$

for real numbers $s > 1$, proved that it had what is now called an Euler product

$$\zeta(s) = \prod_{p}(1 - p^{-s})^{-1}, \tag{10.1}$$

and studied its behaviour near $s = 1$. For algebraic number theory, one thinks of Gauss and his famous law of quadratic reciprocity. This formula anticipated what we look for even today in the study of number fields, and can be used in this connection to define the coefficients of the first L-functions. These in some sense represent an early model for the general automorphic L-function of Langlands.

Recall that a Dirichlet series is an infinite series of the form

$$\sum_{n=1}^{\infty} a_n n^{-s},$$

for complex coefficients a_n and a complex variable s. If the coefficients satisfy a bound

$$|a_n| \leq C n^{\alpha}, \quad n \in \mathbb{N},$$

for a positive number α, the series converges in the right half-plane $\mathrm{Re}(s) > \alpha + 1$. The original model is of course Riemann's extension

$$\zeta(s) = \sum_{n=1}^{\infty} n^{-s}, \quad \mathrm{Re}(s) > 1,$$

of Euler's series. It converges to an analytic function of s in the right half-plane $\mathrm{Re}(s) > 1$. It also has analytic continuation to a meromorphic function of $s \in \mathbb{C}$, whose only singularity is a simple pole at $s = 1$, and which satisfies a functional equation relating its values at s and $1 - s$. In addition, the Riemann zeta function has an Euler product. By the fundamental theorem of arithmetic, it can be represented as a product

$$\zeta(s) = \prod_p (1 - p^{-s})^{-1} = \prod_p \left(\sum_{k=0}^{\infty} (p^k)^{-s} \right)$$

of Dirichlet series attached to prime numbers.

An *L-function* is a Dirichlet series with supplementary properties. There seems to be no universal agreement as to the definition, but let us say that an *L*-function is a Dirichlet series that converges in some right half-plane, and that has an Euler product of the general form[2]

$$L^{\infty}(s) = \prod_p \left(1 + \sum_{k=1}^{\infty} c_{p,k} p^{-ks} \right),$$

for complex numbers $c_{p,k}$. We will not insist on analytic continuation and functional equation, simply because this has not been established for many of the *L*-functions that arise naturally, even though it is widely expected to hold.

Much of the history of nineteenth-century number theory concerns *L*-functions, explicitly or implicitly. Not surprisingly, it was Dirichlet who introduced the first *L*-functions after the original Euler product. These were the series

$$L(s, \chi) = \sum_{n=1}^{\infty} \chi(n) n^{-s} = \prod_p (1 - \chi(p) p^{-s})^{-1}$$

attached to Dirichlet characters χ on $(\mathbb{Z}/N\mathbb{Z})^{\times}$, for any positive integer N. It is understood that $\chi(n)$ depends only on the congruence class of n modulo N, and that $\chi(n) = 0$ if n has any prime factors that divide N. The case that $N = 1$ is of course Euler's product (10.1). Dirichlet studied these objects as functions of a positive real variable s. The series converges absolutely if $s > 1$, but Dirichlet showed that if $\chi \neq 1$, the series converges conditionally for $0 < s \leq 1$ and that $L(1, \chi) \neq 0$. He used this in 1837 to prove that there are infinitely many primes in any arithmetic series

$$\{a + nd : n > 0\}, \quad (a, d) = 1.$$

---

[2] The superscript ∞ follows the conventions of Langlands. It is used to indicate that the *L*-function is incomplete, in the sense that the product is missing an archimedean factor that would greatly simplify an expected functional equation if it were included.

Twenty years later, Riemann studied Euler's series as a function of a complex variable s. He proved that it has meromorphic continuation to the complex plane, as we have already noted, and that the product

$$L(s,1) = \pi^{-s/2}\Gamma(s/2)\zeta(s)$$

satisfies the functional equation

$$L(s,1) = L(1-s,1).$$

He also showed that $L(s,1)$ is entire, apart from a simple pole at $s = 1$. Finally, he introduced his hypothesis that the only zeros of $L(s,1)$ lie on the line $\text{Re}(s) = 1/2$. This would imply a very strong asymptotic estimate for the number

$$\pi(x) = |\{p \leq x\}|$$

of primes less than or equal to a large number x.

The Riemann hypothesis is of course completely open today. What is less widely discussed is a common belief that the analogue of the Riemann hypothesis holds for all arithmetic L-functions, apart from certain obvious exceptions. In particular, it is thought to hold for the general Langlands L-functions $L(s,\pi,\rho)$ that are the topic of this article. This would imply equally strong asymptotic estimates for the arithmetic data that go into the coefficients of these Dirichlet series.

Later nineteenth-century contributions to the developing theory of L-functions include Kummer's 1851 generalization of Dirichlet L-functions to cyclotomic fields, Dedekind's 1893 generalization of the Riemann zeta function to arbitrary number fields, and Weber's 1897 generalization of Dirichlet L-functions to arbitrary fields. Weber then used the Dedekind zeta function to prove what was called the first inequality (later demoted to the second inequality!), a fundamental early step in the development of abelian class field theory. About this time, the turn of the century, Hilbert was laying the foundations of what would become the modern outline of class field theory. He worked with the Hilbert class field of a given number field F, the maximal *unramified* abelian extension of F, rather than the maximal abelian extension. However, his framework offered a new perspective, and anticipated what would be used in the general ramified case. He also rewrote Gauss' Law of quadratic reciprocity as a product formula for the Hilbert symbol. This made quadratic reciprocity the foundation of the simplest case of class field theory (with its corresponding Dirichlet series), the quadratic extensions of the field $F = \mathbb{Q}$.

We can now go on to Artin L-functions, which we are treating as a foundation for the L-functions introduced by Langlands. For a short but systematic account of the history of class field theory, the reader can consult [Con]. The article [Cog] is an interesting informal introduction to Artin L-functions.

Suppose that F is a number field, and that K/F is a finite Galois extension. Recall that almost all prime ideals $\mathfrak{p}$ in F are unramified in K, and that for any such $\mathfrak{p}$, the Frobenius class $\mathrm{Frob}_\mathfrak{p} = \Phi_\mathfrak{p}$ is a canonical conjugacy class in the Galois group

$$\Gamma_{K/F} = \mathrm{Gal}(K/F).$$

We thus obtain a family

$$\{\Phi_\mathfrak{p} : \mathfrak{p} \notin S\}$$

of conjugacy classes, parameterized by the prime ideals $\mathfrak{p}$ outside some chosen finite set that includes all the ramified primes. This is a fundamental datum attached to K/F, which is given entirely in terms of $\mathfrak{p}$. Recall also that $\mathfrak{p}$ is said to *split completely* in K if $\Phi_\mathfrak{p}$ is the identity element 1 in $\Gamma_{K/F}$. It is then a well-known fact that the map

$$\{K/F\} \longrightarrow \mathrm{Spl}(K/F), \tag{10.2}$$

from the set of finite Galois extensions K/F of F to the set of families $\mathrm{Spl}(K/F) = \mathrm{Spl}^S(K/F)$ of primes $\mathfrak{p} \notin S$ that split completely in K, is injective.[3] Therefore the map parameterizes the finite Galois extensions K/F in terms the data $\mathrm{Spl}(K/F)$.

Suppose for example that $F = \mathbb{Q}$. We can represent K as the splitting field of an irreducible monic polynomial $f(x) \in \mathbb{Z}[x]$ of degree n. There is then an embedding of the Galois group $\Gamma_{K/\mathbb{Q}}$ into the symmetric group S_n, which is canonical up to conjugacy. The conjugacy classes in S_n correspond to partitions of n. In particular, if $\Gamma_{K/\mathbb{Q}}$ equals S_n, which is what happens generically, we can identify the various Frobenius classes for $K/\mathbb{Q}$ with partitions of n. In fact, it follows from the basic theory that the Frobenius class of an unramified prime p is the partition defined by the irreducible factors of $f(x)$ modulo p. This gives a concrete realization of a deep phenomenon. In particular, even without the restriction on $\Gamma_{K/F}$, $\mathrm{Spl}(K/F)$ is the set of primes p such that $f(x)$ breaks into linear factors modulo p.

Emil Artin used the families of conjugacy classes $\Phi_\mathfrak{p}$ to construct the L-functions that bear his name. For the coefficients, he had to attach complex parameters to the conjugacy classes in $\Gamma_{E/F}$. His idea was to take not just the Galois extension K of F, but also a finite-dimensional complex representation

$$r : \Gamma_{K/F} \longrightarrow \mathrm{GL}(n, \mathbb{C})$$

---

[3] The choice of S is immaterial. One could take $\mathrm{Spl}(K/F)$ to be the set of equivalence classes of families $\mathrm{Spl}^S(K/F)$, in which $\mathrm{Spl}^S(K/F)$ is equivalent to $\mathrm{Spl}^{S'}(K/F)$ if the intersection of the two sets has finite complement in each one. The mapping (10.2) then remains injective with this interpretation of the right-hand side.

of its Galois group. The conjugacy classes in $\Gamma_{K/F}$ would then be mapped to semisimple conjugacy classes in $GL(n,\mathbb{C})$ (of finite order), which could then be parameterized by their characteristic polynomials. The local Artin L-function at an unramified prime $\mathfrak{p} \notin S$ of F is defined in these terms as

$$L_{\mathfrak{p}}(s,r) = \det(1 - (N\mathfrak{p})^{-s}r(\Phi_{\mathfrak{p}}))^{-1}.$$

The (unramified) global Artin L-function for r is then the Euler product

$$L^S(s,r) = \prod_{\mathfrak{p} \notin S} \det(1 - (N\mathfrak{p})^{-s}r(\Phi_{\mathfrak{p}}))^{-1}.$$

If we identify $\mathfrak{p}$ with its normalized valuation v, as is convenient, we can also write

$$L^S(s,r) = \prod_{v \notin S} L_v(s,r) = \prod_{v \notin S} \det(1 - q_v^{-s}r(\Phi_v))^{-1}, \qquad (10.3)$$

where q_v is the order of the residue class field of $\mathfrak{p}$, and S is now understood to include the finite set S_∞ of normalized archimedean valuations of F as well as the nonarchimedean valuations at which r ramifies.

An L-function is not just a way to package arithmetic data. It should also lead ultimately to fundamental asymptotic properties of these data. A necessary condition for this would be that the L-function have meromorphic continuation to a suitable function on the complex plane. Artin conjectured that for any r, $L^S(s,r)$ could be completed with a suitable contribution $L_S(s,r)$ from the places in S so that the resulting product

$$L(s,r) = L_S(s,r)L^S(s,r) \qquad (10.4)$$

has meromorphic continuation, and satisfies a functional equation

$$L(s,r) = \varepsilon(s,r)L(1-s,r^\vee), \qquad (10.5)$$

for the contragredient representation

$$r^\vee(\sigma) = {}^t r(\sigma^{-1}), \quad \sigma \in \Gamma_{K/F},$$

and a monomial

$$\varepsilon(s,r) = ab^s, \quad a \in \mathbb{C}^\times, b \in \mathbb{C}. \qquad (10.6)$$

He conjectured further that for irreducible r, $L(s,r)$ is entire unless r is the trivial one-dimensional representation (in which case his L-function is just the completed Riemann zeta function

$$L(s,1) = L_\infty(s,1)\zeta(s) = \pi^{-s/2}\Gamma(s/2)\zeta(s),$$

which is entire apart from a simple pole at $s = 1$). It is the last statement that is the deepest. It is known today simply as *the* Artin conjecture.

Artin proved the earlier assertions of his conjecture in a way that became part of the motivation for Langlands' principle of functoriality. The heart of what he established was the case that K/F is abelian. One might imagine that there would be a direct proof of the Artin conjecture in this case. However, that is not the way the mathematical world was put together. Artin gave a decidedly indirect proof that even today seems extraordinary. He showed that every abelian Artin L-function was a Hecke L-function, the class of concrete L-functions that arose from analysis, and for which Hecke had been able to establish the desired analytic properties. This was class field theory. Artin studied what was known, and extended it to what was required for his purposes. The result was the Artin Reciprocity Law, which we state in adèlic terms as follows.

Artin Reciprocity Law *Suppose that K/F is an abelian extension of the number field F. Then there is a canonical isomorphism*

$$\theta_{K/F} : C_F/N_{K/F}(C_K) \xrightarrow{\sim} \Gamma_{K/F}. \tag{10.7}$$

We recall here that the *adèle* ring of F is a topological direct limit

$$\mathbb{A}_F = \varinjlim_S \left(\prod_{v \in S} F_v \times \prod_{v \notin S} \mathcal{O}_S \right),$$

where S ranges over finite sets of valuations of F that contain the set S_∞ of archimedean valuations. Then $\mathbb{A}_F$ is a locally compact ring (commutative, with identity 1), which contains the diagonal image of F as a discrete, cocompact subring. It was introduced by Artin and Whaples in 1945, following the earlier introduction of its group of units, the *idèle* group

$$I_F = \mathbb{A}_F^\times = \mathrm{GL}(1, \mathbb{A}_F),$$

by Chevalley in 1940. The *idèle class group* is the quotient

$$C_F = I_F/F^\times,$$

while its quotient

$$C_F/N_{K/F}(C_K) = I_F/F^\times N_{K/F}(I_K) \tag{10.8}$$

is the domain of the Artin map $\theta_{K/F}$ in (10.7). The norm map $N_{K/F} : C_K \to C_F$ is built in the obvious way from the usual norm maps between local and global fields.

The definition of the Artin map is built on local class field theory, which asserts that there is a canonical isomorphism

$$\theta_{K_w/F_v} : F_v^\times/N_{K_w/F_v}(K_w^\times) \longrightarrow \Gamma_{K_w/F_v},$$

for any completions F_v of F and K_w of K over F_v. Since $\Gamma_{K/F}$ is abelian, this provides a canonical embedding of Γ_{K_w/F_v} into $\Gamma_{K/F}$ that depends only on F_v (and not the choice of the field K_w over F_v). The product over v of these embeddings then gives a well-defined mapping

$$\theta_{K/F} : I_F/N_{K/F}(I_K) \longrightarrow \Gamma_{K/F}.$$

The deepest property, the one that makes this global mapping a "reciprocity law," is the fact that its kernel equals the image of the subgroup $F^\times$ of I_F. Therefore $\theta_{K/F}$ descends to a mapping on the domain (10.8) of (10.7). This seems to have been a point that late-nineteenth-century number theorists struggled with. It was eventually clarified (at least for unramified extensions K/F) by Hilbert.

We have formulated the Artin Reciprocity Law in this detail so as to serve as a foundation for Langlands' nonabelian generalization. Our description does rely on local class field theory, which was an important advance in its own right. However, one could set up the Artin map $\theta_{K/F}$ in a more elementary, if less elegant, way by restricting the factors θ_{K_w/F_v} of $\theta_{K/F}$ to the unramified places of K/F, where they can be defined in terms of Frobenius elements in $\Gamma_{K/F}$. The adèlic formulation itself could be replaced by the more concrete (but also more cumbersome) classical description in terms of "moduli" and "conductors." For further information, a reader might consult the Wikipedia articles "Artin reciprocity law" and "Symbols (number theory)" from June 3, 2020. The second of these is like a one-page history of class field theory, in the form of a list of symbols for the evolving reciprocity maps, from the *Legendre symbol* of quadratic reciprocity to the *Hilbert symbol* for Kummer extensions, and then finally, to the *Artin symbol* for arbitrary abelian extensions.

An abelian Artin L-function of degree 1 is defined by a character ξ on the abelian Galois group $\Gamma_{K/F}$ on the right-hand side of (10.7). A character χ on the domain at the left in (10.7) defines a Hecke L-function. The two definitions can be seen to match under the isomorphism $\theta_{K/F}$, thereby giving an identity

$$L(s,\xi) = \prod_v L_v(s,\xi_v) = \prod_v L_v(s,\chi_v) = L(s,\chi) \qquad (10.9)$$

of the two kinds of abelian L-functions. Hecke used harmonic analysis to show that his abelian L-functions satisfied analogues of all the conditions conjectured by Artin. Therefore the abelian L-functions of Artin also satisfy the assertions of his conjecture. It was in its form of a correspondence $\xi \to \chi$ of abelian global characters, and the resulting identity (10.9) of L-functions, that Langlands generalized Artin reciprocity.

Artin used his reciprocity law and the Hecke L-functions $L(s,\chi)$ it provided to prove some of the assertions of his general conjecture. The idea was to decompose a general representation r of $\Gamma_{K/F}$ into a virtual linear combination

$$r = \sum_i a_i \, \mathrm{Ind}_{\Gamma_i}^{\Gamma}(\xi_i), \quad \Gamma = \Gamma_{K/F},$$

for one-dimensional representations ξ_i of cyclic subgroups Γ_i of $\Gamma = \Gamma_{K/F}$. Artin proved that this could be done for rational numbers a_i. Standard properties of L-functions then provided a corresponding product decomposition

$$L(s,r) = \prod_i L(s,\xi_i)^{a_i}$$

of $L(s,r)$ into abelian L-functions over cyclic extensions K/F_i. Artin was then able to use this to establish the analytic continuation and functional equation of $L(s,r)$. (See [Cog, p. 10] for further remarks on this, including its relation to the later Brauer induction theorem.) What the decomposition did *not* give was Artin's conjectural assertion that $L(s,r)$ is entire. The problem is the contribution to the product of the negative numbers a_i, from which the zeros of $L(s,\xi_i)$ could contribute poles to $L(s,r)$. This crosses into the domain of the Riemann hypothesis and its analogues for Hecke L-functions. The phenomenon is certainly interesting but, by itself at least, does not offer any help with the last assertion of the conjecture, the assertion now known as the Artin conjecture.

Hecke studied the L-functions $L(s,\chi)$ for what he called Grössencharaktere (now known simply *Hecke characters*). They amount to characters on the full idèle class group C_F, not just those that descend to $C_F/N_{K/F}(C_K)$, for a general number field F, although the idèles were not introduced until 20 years later. Hecke's L-functions represent a major generalization of the Dirichlet L-functions and the Dedekind zeta functions, and indeed, all of the extensions of these functions that had previously been studied. To establish their analytic properties Hecke relied on the classical Mellin transform and classical Poisson summation. In this, he was following Riemann, but with arguments that were by necessity considerably more sophisticated.

Thirty years later, Tate's 1950 Ph.D. thesis [T1] gave a different way of looking at both Hecke's proofs and his results. Tate had the advantage of being able to work with the idèles $I_F = \mathbb{A}_F^\times$ that had been introduced 10 years earlier by Chevalley, and in terms of which we stated the Artin reciprocity law. The heart of his proof was the application of the Poisson summation formula for the discrete subgroup F of the adèle ring $\mathbb{A}_F$. The simplicity of Tate's arguments led to an important refinement of the functional equation

$$L(s,\chi) = \varepsilon(s,\chi)L(1-s,\overline{\chi})$$

for the given L-function

$$L(s,\chi) = L_S(s,\chi)L^S(s,\chi).$$

This was a decomposition of the global monomial in the equation into a product

$$\varepsilon(s,\chi) = \prod_{v\in S} \varepsilon(s,\chi_v,\psi_v)$$

of local monomials

$$\varepsilon(s, \chi_v, \psi_v) = \varepsilon(\chi_v, \psi_v) q_v^{-n_v(s-\frac{1}{2})},$$

where q_v equals the residual degree of F_v if v is nonarchimedean, and equals 1 if v is archimedean, and ψ is a nontrivial additive character on the quotient $\mathbb{A}_F/F$.

We are working towards the 1968 preprint [L3] of Langlands, and the seven questions it posed, in our attempt to understand the origins of functoriality. The logical next step in our exposition here is the generalization of Tate's thesis from GL(1) to GL(n) by Jacquet and Godement [GJ]. It was not published until 1972, well after [L3], but its future existence was clearly part of Langlands' thinking. He was in regular communication with Godement throughout the 1960s, and he mentions the extension of Tate's thesis in his paper [L3] as an essential premise for his conjectures.

Suppose for a moment that G is any reductive group over a number field F, and that π is an automorphic representation of G (which is to say, of the locally compact group $G(\mathbb{A}_F)$). This term was not in use at the time of [L3], as we observed in footnote 1. Its formal definition did not come until the Corvallis conference [BJ], [L5] 10 years later. Langlands simply referred to π as an irreducible representation of $G(\mathbb{A}_F)$ that "occurs in" $L^2(G(F)\backslash G(\mathbb{A}_F))$, a description that gives a good idea of the concept, even if it is also somewhat more restrictive than what became the general definition. Langlands also took for granted that an automorphic representation π has a unique (restricted) tensor product decomposition

$$\pi = \widetilde{\bigotimes_v} \pi_v, \quad \pi_v \in \Pi(G_v),$$

into irreducible representations π_v of the local components $G_v = G(F_v)$ of $G(\mathbb{A}_F)$, almost all of which are unramified. The formal proof of this by Flath also came 10 years later at the Corvallis conference [F].

We recall that an irreducible representation π_v of G_v is *unramified* if G_v is quasisplit over F_v and split over some unramified extension E_v of F_v, and if the restriction of π_v to a suitable (hyperspecial) maximal compact subgroup K_v of $G_v = G(F_v)$ contains the trivial one-dimensional representation of K_v. Langlands introduced this notion in [L3] (again without the name). He then observed that there was a bijective correspondence $\pi_v \rightarrow c(\pi_v)$ from the unramified representations π_v of G_v to the semisimple conjugacy classes c_v in the L-group[4]

$$^L G_v = \hat{G}_v \rtimes \Gamma_{E_v/F_v} \longleftrightarrow {}^L G = \hat{G} \rtimes \Gamma_{E/F}, \quad \hat{G}_v = \hat{G},$$

---

[4] Langlands had earlier introduced the fundamental notion he called the *associate group* for G, and that Borel later named the L-group in [B1]. The notation here is due to Kottwitz [K].

whose image in Γ_{E_v/F_v} projects onto the Frobenius class Φ_v. This is a consequence of the classification of complex-valued homomorphisms on the Hecke algebra $C_c(K_v\backslash G_v/K_v)$ (under convolution) or equivalently, the description of the unramified principal series for G_v. The automorphic representation π of G thus gives rise to a family

$$c(\pi) = \{c_v(\pi) = c(\pi_v) : v \notin S\}$$

of semisimple conjugacy classes in ^LG, where $S \supset S_\infty$ is again a finite set of valuations of F outside of which π_v is unramified.

Returning to the volume of Godement–Jacquet, we take G equal to $GL(n)$ over F. An automorphic representation π of G now gives a family

$$c(\pi) = \{c_v(\pi) = c(\pi_v) : v \notin S\}$$

of semisimple conjugacy classes in the complex group $^LG = \hat{G} = GL(n,\mathbb{C})$. A very special case of Langlands' general definitions (which we will come to presently) is then the associated family

$$L_v(s,\pi) = L(s,\pi_v) = \det(1 - q_v^{-s}c_v(\pi))^{-1}, \quad v \notin S,$$

of unramified local L-functions, and the unramified global Euler product

$$L^S(s,\pi) = \prod_{v \notin S} L_v(s,\pi), \qquad (10.3')$$

which converges for $\mathrm{Re}(s)$ in some right half-plane.

The main theorem of [GJ] applies to the basic case that π is cuspidal. It asserts that $L^S(s,\pi)$ can be expanded by a finite product

$$L_S(s,\pi) = \prod_{v \in S} L_v(s,\pi) = \prod_{v \in S} L(s,\pi_v)$$

of ramified local L-functions so that the resulting completion

$$L(s,\pi) = L_S(s,\pi)L^S(s,\pi) \qquad (10.4')$$

has meromorphic continuation, and satisfies a functional equation

$$L(s,\pi) = \varepsilon(s,\pi)L(1-s,\pi^\vee), \qquad (10.5')$$

for the contragredient representation $\pi^\vee = {}^t\pi(x^{-1})$, and a finite product

$$\varepsilon(s,\pi) = \prod_{v \in S} \varepsilon(s,\pi_v,\psi_v) \qquad (10.6')$$

of local monomials

$$\varepsilon(s,\pi_v,\psi_v) = \varepsilon(\pi_v,\psi_v)q_v^{-n_v(s-\frac{1}{2})}$$

that depend on the local components ψ_v of a nontrivial additive character ψ on $\mathbb{A}_F/F$. Moreover, $L(s,\pi)$ is entire unless $n = 1$ and $\pi(x) = |x|^u$ for some $u \in \mathbb{C}$. This theorem is the natural generalization of the theorem of Tate for $n = 1$, itself a refinement of the fundamental results of Hecke. The restriction to cuspidal π is not a serious impediment. With techniques from Langlands' theory of Eisenstein series [L1], the theorem can be extended to arbitrary automorphic representations π of G.

With the Galois L-functions $L(s, r)$ of degree n and the automorphic L-functions $L(s, \pi)$ for $\mathrm{GL}(n)$, our exposition has acquired a certain symmetry. The Galois representation r can in fact be made independent of the finite Galois extension K/F, simply by taking it to be a continuous, complex representation of the absolute Galois group $\Gamma_F = \mathrm{Gal}(\overline{F}/F)$. For it would then automatically factor through a finite quotient $\Gamma_{K/F}$. The other discrepancy between the two theories is more significant. It is the lack of a local factorization for Artin's Galois ε-factor $\varepsilon(s, r)$ that would match the canonical factorization (10.6′) of the automorphic ε-factor $\varepsilon(s, \pi)$ attached to ψ_v. For Langlands, this was a serious deficiency, given the local classification he had in mind for Question 6 of [L3], which we will come to presently. He worked hard in the late 1960s to establish a local construction of Artin ε-factors $\varepsilon(s, r_v, \psi_v)$. He eventually succeeded, but did not include all of the details in his long treatise [L4]. Soon afterwards, Deligne was able to find a simpler global solution of the problem [T2].

The upshot is that the two theories are completely parallel. Taken together, they very much resemble what had been established in the abelian case of $\mathrm{GL}(1)$. Given the Artin Reciprocity Law, a reader might well wonder whether every Artin L-function of a degree n representation r of Γ_F is a Godement–Jacquet L-function of an automorphic representation π of $\mathrm{GL}(n)$. That is, whether there is an injective correspondence $r \to \pi$ such that

$$L(s, r) = \prod_v L_v(s, r) = \prod_v L_v(s, \pi) = L(s, \pi). \qquad (10.9′)$$

We would then have a reciprocity law that amounted to nonabelian class field theory. If so, would it then be the final word on the subject?

There are three points to consider in regard to the last question. One would be the uncomfortable prospect of having to prove such a broad nonabelian reciprocity law, given the historical difficulty in establishing just the abelian theory. Nonabelian class field theory, whatever form it might take, was obviously going to be very deep. It would be reassuring to think that the problem at least had some further structure. A second point concerns this last possibility. Suppose that r' is an irreducible Galois representation of degree n', and that ρ' is an irreducible n-dimensional representation of $\mathrm{GL}(n', \mathbb{C})$. The composition

$$r : \Gamma_F \xrightarrow{r'} \mathrm{GL}(n', \mathbb{C}) \xrightarrow{\rho'} \mathrm{GL}(n, \mathbb{C})$$

is then a Galois representation (frequently irreducible) of degree n. The Frobenius classes that define the Artin L-functions $L^S(s, r')$ and

$$L^S(s, r) = L^S(s, \rho' \circ r')$$

satisfy the obvious relation

$$r(\Phi_v) = (\rho' \circ r')(\Phi_v), \quad v \notin S.$$

How could this be reflected in the corresponding automorphic representations? Finally, the work of Harish-Chandra has taught us that representations should be studied uniformly for all groups. If some interesting phenomenon is discovered in one group, or one family of groups such as $\{\mathrm{GL}(n)\}$, it should be investigated for all groups. What are the implications of this for automorphic L-functions?

These considerations were undoubtedly part of the thinking of Langlands that led up to the Principle of Functoriality. However, perhaps the most decisive hints were in his theory of Eisenstein series. They came from the L-functions that he discovered in the global intertwining operators $M(w, \lambda)$ from his functional equations for Eisenstein series. Thus informed by his general results on Eisenstein series, as well as his study of Artin L-functions and abelian class field theory, and perhaps above all, his earlier study of the work of Harish-Chandra, Langlands put his ideas together in the letter to Weil and the paper [L3]. It was clear to him that the theory should indeed encompass the automorphic representations π of an arbitrary reductive group G over F. He actually took F to be any global field, but we shall continue to assume that it is a number field.

At the beginning of [L3], Langlands introduced the L-group. This was a sweeping new idea in its own right. He then defined the semisimple conjugacy classes $c_v(\pi) = c(\pi_v)$ in $^L G$ attached to the unramified constituents of an automorphic representation π, which of course also gave the family $c(\pi) = \{c_v(\pi)\}$ we have described above. But he wanted also to attach L-functions to these objects. This was not immediately clear, since unramified L-functions had always been defined as characteristic polynomials, and the L-group $^L G$ that contains the conjugacy classes from π does not usually come with a general linear group. Langlands' solution was simple and elegant. It was to attach another datum to π, a finite-dimensional representation

$$\rho : {}^L G \longrightarrow \mathrm{GL}(n, \mathbb{C})$$

of the L-group. The unramified local and global L-functions could then be defined as

$$L_v(s, \pi, \rho) = L(s, \pi_v, \rho_v) = \det(1 - q_v^{-s} \rho_v(c_v(\pi)))^{-1}$$

and

$$L^S(s,\pi,\rho) = \prod_{v \notin S} L_v(s,\pi,\rho). \tag{10.3''}$$

Langlands formulated his ideas as a series of questions. The first was designed to frame the entire discussion in terms of L-functions. It asked whether the unramified L-functions $L^S(s,\pi,\rho)$ above have the same analytic properties as in the special case of Godement–Jacquet.

Question 1 *Given G/F, π and ρ as above, is it possible to define local L-functions*

$$L_v(s,\pi,\rho) = L(s,\pi_v,\rho_v)$$

and epsilon factors

$$\varepsilon_v(s,\pi,\rho,\psi) = \varepsilon(s,\pi_v,\rho_v,\psi_v) = \epsilon(\pi_v,\rho_v,\psi_v)q_v^{-n_v(s-\frac{1}{2})}$$

at the ramified (and archimedean) places $v \in S$ so that if

$$L_S(s,\pi,\rho) = \prod_{v \in S} L_v(s,\pi,\rho)$$

and

$$\varepsilon(s,\pi,\rho) = \prod_{v \in S} \varepsilon_v(s,\pi,\rho,\psi), \tag{10.6''}$$

then the completed global L-function

$$L(s,\pi,\rho) = L_S(s,\pi,\rho)L^S(s,\pi,\rho) \tag{10.4''}$$

has meromorphic continuation to the complex plane with only finitely many poles, and satisfies the functional equation

$$L(s,\pi,\rho) = \varepsilon(s,\pi,\rho)L(1-s,\pi,\rho^\vee), \tag{10.5''}$$

for $\rho^\vee = {}^t\rho(g^{-1})$?

Langlands then alluded to the case that G equals $\mathrm{GL}(n)$ and ρ is the standard n-dimensional representation of $\hat{G} = \mathrm{GL}(n,\mathbb{C})$. In this case, $L(s,\pi,\rho) = L(s,\pi)$ is the Godement–Jacquet L-function. Referring to ongoing work of Godement, Langlands offered the expectation that the assertions of Question 1 would be answered affirmatively in this case. The other special case of immediate interest was for G equal to the trivial group $\{1\}$. This of course forces the automorphic representation to be trivial, but $\rho = r$ can still be an arbitrary complex representation of the L-group

$$^LG = \{1\} \rtimes \Gamma_F,$$

or in other words, a continuous representation of the absolute Galois group Γ_F. In this case, $L(s,\pi,\rho) = L(s,r)$ is an arbitrary Artin L-function. By introducing the further L-functions $L(s,\pi,\rho)$, with the plausible hope that they too have the desired analytic properties, Langlands does indeed impose further structure on the general problem of relating $L(s,r)$ to $L(s,\pi)$. This becomes more vivid as we go along.

After stating Question 1, Langlands wrote, "The idea that led Artin to the general [abelian] reciprocity law suggests that we try to answer [Question 1] in general by answering a further series of questions." It would be very interesting to trace through the details of Artin's proof with Langlands' questions as a guide, but I have not done so. The remaining six questions are divided into three pairs, each consisting of a local and a global version of a question. Question 2 and 3 concern how the L-functions behave under inner twists. It is Question 4 and 5 that introduce local and global functoriality, our main topic. Questions 6 and 7 represent a generalization of parts of functoriality, with the Weil group W_F in place of the Galois group Γ_F.

Recall that an arbitrary reductive group G over F can be obtained uniquely from a quasisplit group G^* (a group that contains a Borel subgroup B^* over F) by twisting the Galois action on G^* by inner automorphisms. The L-group $^LG^*$ of G^* is then equal to that of G. Questions 2 and 3 ask whether the automorphic representation theory of G is similar to that of G^*. More precisely, is there a correspondence (binary relation) $\pi_v \to \pi_v^*$ of representations over each localization F_v such that $L(s,\pi_v,\rho_v)$ equals $L(s,\pi_v^*,\rho_v)$? Then if $\pi = \tilde{\bigotimes}_v \pi_v$ is automorphic, is $\pi^* = \tilde{\bigotimes}_v \pi_v^*$ also automorphic, thereby giving an identity $L(s,\pi,\rho) = L(s,\pi^*,\rho)$ of automorphic L-functions for different groups? The answers to these questions are turning out to be interesting and subtle. The representation theory of inner twists is now treated as part of a different theory, Langlands' conjectural theory of endoscopy, which began to evolve in the 1970s. This means that for these questions on L-functions and functoriality, one usually takes G to be quasisplit over F.

For Questions 4 and 5 on functoriality, we take G' and G to be two quasisplit groups over the number field F, related by an L-homomorphism

$$\rho' : {}^LG' \longrightarrow {}^LG$$

between their L-groups. Question 4 asks whether there is a local correspondence $\pi_v' \to \pi_v$ between the irreducible representations π_v' and π_v of $G'(F_v)$ and $G(F_v)$ such that

$$L(s,\pi_v,\rho_v) = L(s,\pi_v',\rho_v \circ \rho_v')$$

and

$$\varepsilon(s,\pi_v,\rho_v,\psi_v) = \varepsilon(s,\pi_v',\rho_v \circ \rho_v',\psi_v)$$

for every complex finite-dimensional representation ρ_v of $^L G_v$ and every nontrivial additive character ψ_v on F_v. This is **local functoriality.** Question 5 then asks if $\pi' = \tilde{\bigotimes}_v \pi'_v$ is automorphic for G', and $\pi'_v \to \pi_v$ for every v, whether $\pi = \tilde{\bigotimes}_v \pi_v$ is automorphic for G. This is **global functoriality.** It implies the identity

$$L(s, \pi, \rho) = L(s, \pi', \rho \circ \rho')$$

of global L-functions for every complex, finite-dimensional representation ρ of $^L G$. At first glance, it might in fact seem like a harmless assertion. This perhaps accounts for how long it took to be accepted by the mathematical community for what it was, a revolutionary change in our understanding of number theory.

The last two questions concern the special case of functoriality in which $G' = \{1\}$. Then ρ' is an L-homomorphism from the Galois group to the dual group $^L G$ of the given quasisplit group. For these questions, Langlands replaced the Galois groups Γ_{F_v} and Γ_F by the local and global Weil groups W_{F_v} and W_F. My understanding is that he learned of these objects in his discussions with Weil, and that he was very happy to discover that they would become a natural part of his theory. Weil had introduced his groups[5] in 1951, as objects that behaved very much like Galois groups. In particular, he was able to attach L-functions (local or global) to finite-dimensional representations ϕ of the relevant Weil group, thereby providing an important generalization of Artin L-functions.

Question 6 asks whether there is a correspondence $\phi_v \to \pi_v$, which takes L-homomorphisms $\phi_v : W_{F_v} \to {}^L G_v$ to irreducible representations π_v of $G(F_v)$, such that

$$L(s, \pi_v, \rho_v) = L(s, \rho_v \circ \phi_v)$$

and[6]

$$\varepsilon(s, \pi_v, \rho_v, \psi_v) = \varepsilon(s, \rho_v \circ \phi_v, \psi_v),$$

for every complex finite-dimensional representation ρ_v of $^L G_v$, and every nontrivial additive character ψ_v of F_v. Question 7 then asks if $\phi : W_F \to {}^L G$

---

[5] We recall that the Weil group (over F_v or F) is a locally compact group, with a canonical mapping into the corresponding Galois group (over F_v or F), whose image is dense. The pullback of the mapping then gives an injection $r \to \phi$ from Galois representations to Weil group representations. The Weil L-function for the image of r of course then coincides with the Artin L-function of r. (See [T2].)

[6] Langlands was anticipating the results of [L4] here, which included the existence of local ε-factors for Weil groups.

is an L-homomorphism, and $\pi = \tilde{\bigotimes}_v \pi_v$ for local images $\phi_v \to \pi_v$ of the correspondence, whether π is automorphic for G. This would imply the identity

$$L(s, \pi, \rho) = L(s, \rho \circ \phi)$$

of global L-functions attached to complex, finite-dimensional representations ρ of $^L G$. The local Question 6 has turned out to be particularly important. Langlands later wrote Π_{ϕ_v} for the set of images π_v of a given ϕ_v under the correspondence $\phi_v \to \pi_v$. The local *Langlands classification*, or *local Langlands correspondence*, is the conjecture that the L-packets Π_{ϕ_v} are finite, disjoint sets, whose union over ϕ_v is the set of all[7] irreducible representations of $G(F_v)$. This is now known for quasisplit classical groups, and for all real groups, but otherwise remains largely open. It is now treated as part of Langlands theory of endoscopy.

Let us add a couple more remarks to our description of the questions of [L3]. We have tried to motivate the Principle of Functoriality according to the presentation of Langlands, as a natural outgrowth of the theory of L-functions. More narrowly, we could think of it simply as an attempt to understand Artin L-functions, and to prove the Artin conjecture that $L(s, r)$ is entire. We think back to the simple question we raised on the possible correspondence $r \to \pi$ from Galois representations of degree n to automorphic representations of GL(n) (with its associated matching (10.9$'$) of L-functions). The three points we raised then are clearly accounted for in the greatly expanded theory encompassed by Langlands' seven questions. For a start, the automorphic L-functions $L(s, \pi, \rho)$ of Question 1 are attached to automorphic representations π of a general group G, not just GL(n). Secondly, the seven questions reveal a vast, previously hidden, structure that surrounds the original two L-functions $L(s, r)$ and $L(s, \pi)$. And finally, the automorphic interpretation of the Artin L-function $L(s, r) = L(s, r', \rho')$ attached to an n-dimensional representation ρ' of GL$(n', \mathbb{C})$ is just the existence of an automorphic representation π of GL(n) attached to the given automorphic representation π' of GL(n') such that $L(s, \pi) = L(s, \pi', \rho')$. This is functoriality itself, or rather the special case of it for general linear groups.

As we have noted at the beginning, one can also motivate functoriality simply as a set of reciprocity laws among concrete arithmetic data. Recall that an automorphic representation π of a reductive group G over F comes with a family

---

[7] This is correctly stated here only if the local field F_v is archimedean. If F_v is nonarchimedean, one must replace W_{F_v} with the larger group $W_{F_v} \times \mathrm{SU}(2)$ in order to account for the Steinberg representation π_v of $G(F_v)$, and more generally, what are called the special representations.

$$c(\pi) = \{c_v(\pi) = c(\pi_v) : v \notin S\}$$

of semisimple conjugacy classes in $^L G$. Suppose that G', π', G and ρ' are as in the statement of functoriality from Langlands' Questions 4 and 5. Then functoriality asserts the existence of an automorphic representation π of G such that

$$c(\pi) = \rho'(c(\pi')).$$

In other words, for each v outside a finite set S, the conjugacy class in $^L G$ that contains $\rho'(c(\pi_v))$ equals $c(\pi_v)$. These data should among other things govern the fundamental structure of arithmetic algebraic varieties, and their motivic components. It seems truly remarkable that they should satisfy such concrete relations.

We conclude by recalling the four fundamental applications of functoriality sketched by Langlands at the end of his paper [L3].

(i) *Artin L-functions and nonabelian class field theory* We have already commented on this, but it is worth repeating, since it is by any measure what mathematicians have been searching for ever since Artin. It is the case of functoriality with $G' = \{1\}$, $\rho' = r$ an n-dimensional representation of the L-group group $^L G' = \Gamma_F$, and $\rho = \mathrm{St}_n$ the standard representation of $\mathrm{GL}(n, \mathbb{C})$. The assertion of functoriality is that there is an automorphic representation π of $\mathrm{GL}(n)$ such that

$$L(s, r) = L(s, \pi).$$

This is the original desired identity (10.9') that we have just been discussing. It characterizes the arithmetic data that classify Galois extensions of F in analytic terms. It also tells us any irreducible Artin L-function is a cuspidal Godement–Jacquet L-function for $\mathrm{GL}(n)$, and hence entire.

(ii) *Analytic continuation and functional equation* This is a generalization of (i) to an arbitrary automorphic L-function $L(s, \pi, \rho)$, attached to an automorphic representation π of G and an N-dimensional representation ρ of $^L G$. Functoriality asserts that there is an automorphic representation π_N of $G_N = \mathrm{GL}(N)$ such that $L(s, \pi, \rho_N \circ \rho)$ equals $L(s, \pi_N, \rho_N)$, for any complex representaiton ρ_N of $^L G_N$. If we take ρ_N to be the standard representation St_N of G_N, the assertion becomes

$$L(s, \pi, \rho) = L(s, \pi_N).$$

In other words, any automorphic L-function is a Godement–Jacquet L-function. It therefore has meromorphic continuation and functional equation, with only finitely many poles.

(iii) *Generalized Ramanujan conjecture* The generalized Ramanujan conjecture asserts that a cuspidal automorphic representation $\pi = \tilde{\bigotimes}_v \pi_v$ of GL(n) is *tempered*. This means that the character

$$f_v \longrightarrow \mathrm{tr}(\pi(f_v)), \quad f_v \in C_c^\infty(\mathrm{GL}(n, F_v)),$$

of each local constituent π_v of π is tempered, in the sense that it extends to a continuous linear form on the Schwartz space $\mathcal{C}(\mathrm{GL}(n, F_v))$ on $\mathrm{GL}(n, F_v)$ defined by Harish-Chandra. We recall that the classical Ramanujan conjecture applies to the case $n = 2$, and π comes from the cusp form of weight 12 and level 1. It was proved by Deligne, who established more generally (for $n = 2$) that the conjecture holds if π is attached to any holomorphic cusp form. The case that π comes from a Maass form remains an important open problem. Langlands observed that functoriality, combined with expected properties of the correspondence $\pi' \to \pi$, would imply the generalized Ramanujan conjecture for GL(n). His representation theoretic argument is strikingly similar to Deligne's geometric proof.

(iv) *Sato–Tate conjecture* The Sato–Tate conjecture for the distribution of the numbers $N_p(E)$ of solutions (mod p) of an elliptic curve E over $\mathbb{Q}$ has a general analogue for automorphic representations. Suppose for example that π is a cuspidal automorphic representation of GL(n). The generalized Ramanujan conjecture of (iii) asserts that the conjugacy classes, represented by diagonal S_n-orbits

$$c_p(\pi) = S_n \cdot \begin{pmatrix} c_{p,1}(\pi) & & 0 \\ & \ddots & \\ 0 & & c_{p,n}(\pi) \end{pmatrix},$$

have eigenvalues of absolute value 1. The generalized Sato–Tate conjecture describes their distribution in the maximal compact torus $U(1)^n$ of the dual group $\mathrm{GL}(n, \mathbb{C})$. If π is *primitive* (a notion that requires functoriality even to define), the distribution of these classes should be given by the weight function in the Weyl integration formula for the unitary group $U(n)$. Langlands sketched a rough argument for establishing such a result from general functoriality. Clozel, Harris, Shepherd-Barron, and Taylor followed this argument in their proof of the original Sato–Tate conjecture, but using base change for GL(n) and deformation results in place of functoriality.

I would like to express my gratitude to Robert Langlands, for his friendship and encouragement over the many years since I first met him in 1968, and also

Figure 10.1 Jim Arthur and Bob Langlands, courtesy of the Simons Foundation.

for what he has given to everyone in his beautiful and profound mathematical contributions. They offer inspiration for all of us in these troubled times when we are most in need of it.

References

[A] J. Arthur, *The work of Robert Langlands*, to appear in *The Abel Prize 2018–2022*, Helge Holden, Ragni Piene (editors).

[B1] A. Borel, *Formes automorphes et séries de Dirichlet (d'après R. P. Langlands)*, Séminaire Bourbaki, no. 466, 1975.

[BJ] A. Borel and H. Jacquet, Automorphic forms and automorphic representations, in *Automorphic Forms, Representations and L-functions*, *Proc. Sympos. Pure Math.* vol. 33, Part 1, Amer. Math. Soc. 1979, 189–202.

[Cog] J. Cogdell, *On Artin L-functions*, https://people.math.osu.edu/cogdell.1/artin-www.pdf

[Con] K. Conrad, *History of class field theory*, https://kconrad.math.uconn.edu/blurbs/gradnumthy/cfthistory.pdf

[F] D. Flath, Decomposition of representations into tensor products, in Automorphic Forms, Representations and L-functions, *Proc. Sympos. Pure Math.* vol. 33, Part 1, Amer. Math. Soc. 1979, 179–184.

[GJ] R. Godement and H. Jacquet, *Zeta Functions of Simple Algebras*, Lecture Notes in Math. 260, Springer, New York, 1972.

[K] R. Kottwitz, Stable trace formula: cuspidal tempered terms, *Duke Math. J.* **51** (1984), 611–650.

[L1] R. P. Langlands, *On the Functional Equations Satisfied by Eisenstein Series*, Lecture Notes in Math. 544, Springer, New York, 1976 (from original manuscript completed in 1965).

[L2]   , *Letter to A. Weil*, 1967, https://publications.ias.edu/letter-to-Weil

[L3]   , *Problems in the Theory of Automorphic Forms*, Lecture Notes in Math. 170, Springer, New York, 1970, 18–61 (from original manuscript completed in 1968).

[L4]   , *On the Functional Equation of the Artin L-functions*, 1970, https://publications.ias.edu/rpl/paper/61

[L5]   , On the notion of an automorphic representation. A supplement to the preceding paper, in *Automorphic forms, Representations and L-functions, Proc. Sympos. Pure Math.* vol. 33, Part 1, Amer. Math. Soc. 1979, 203–207.

[S] F. Shahidi, Automorphic L-Functions, in this volume.

[T1] J. Tate, Fourier Analysis in Number Fields and Hecke's Zeta-Functions (Thesis 1950), in *Algebraic Number Theory*, Thompson Book Company, 1967, 305–347.

[T2]   , Number theoretic background, in *Automorphic Forms, Representations and L-functions, Proc. Sympos. Pure Math.* vol. 33, Part 2, Amer. Math. Soc. 1979, 3–26.

11

In the Beginning: Langlands' Doctoral Thesis

Derek W. Robinson

11.1 Introduction

Bob Langlands' mathematical research career effectively began with his 1960 PhD thesis at Yale. It was a remarkable beginning to a remarkable career, but a beginning which largely went unnoticed. It was remarkable as Bob wrote the thesis, with no direct guidance, during the first year of graduate studies; in his own words "it all happened in a hurry." This did have the serendipitous outcome that during the second year of graduate work he was free to let his interests wander in different directions. In particular his attention focused on Selberg's work on spectral theory of Lie groups, a direction of research which led, within a few years, to the famous Langlands program. The thesis was also remarkable as the major part was never published in full. The thesis consisted of two chapters. The first chapter, approximately a third of the thesis, resolved a problem of Hille [Hil50] in the arcane area of Lie semigroups. This material was subsequently published in the *Canadian Journal of Mathematics* [Lan60a]. The second longer and more interesting chapter only surfaced as a short announcement in the *Proceedings of the National Academy of Sciences* [Lan60b]. This announcement is less than two pages in length and provides an excellent illustration of Polonius' aphorism that 'Brevity is the soul of wit.' Unfortunately the brief account failed to give an intelligible explanation of the detailed results. At least it was well beyond my wit and ken when I first tried to understand it in 1986. The second chapter developed a theory of general-order elliptic operators affiliated with a continuous representation of a Lie group. This theory was then used to resolve a problem inherent in the work of Harish-Chandra [HC53] concerning the analytic structure of the representations. The problem was very topical in the late 1950s and its solution was a remarkable achievement for a first year graduate student without guidance.

The Delphic nature of the *Academy* announcement meant that the principal results of the thesis passed with little attention. Much later, in the mid 1980s, I was interested in the integration of representations of Lie algebras and this led to an investigation of the regularity properties of Lie group representations.

At this point I noticed a reference to Bob's *Academy* note in a paper of Roe Goodman [Goo71]. The results stated in the note were clearly of relevance to my interests at the time and I searched the literature for further evidence of the thesis. Frustrated by my failure to find any other published trace of the work I eventually wrote a letter to Bob asking if he still had a copy of his thesis. Much to my surprise I found a copy in my mailbox a couple of weeks later. Now, almost 60 years later, the thesis is widely available. A photocopy of the original, which was typed by Bob's wife Charlotte in 1959, has been posted on the Princeton website http://publications.ias.edu/rpl/section/3. The thesis was subsequently retyped at Yale sometime after Summer 1968 and distributed to a few people. A revised and corrected version of the second typescript, prepared by myself in collaboration with Anthony Pulido, is also on the Princeton website. An alternative more-detailed presentation of the material in the second part of the thesis can also be found in Chapters I and III of my book *Elliptic Operators and Lie Groups* [Rob91]. This latter presentation provides an extended description that largely follows the reasoning of the thesis. But both the original and my alternative version are quite complicated and difficult to follow. Helgason's remark in his review of Bob's 1963 paper on automorphic forms "the proof involves many interesting ideas and techniques, which, however, do not ... emerge from the overly condensed exposition with the clarity they deserve" could well apply. Consequently the current intention is to attempt to give a different, more pedestrian, description which explains the principal difficulties and results of the thesis. Although this description adopts the general strategy of the original and uses the same techniques the reasoning follows a rather different route.

The story of the differential and analytic structure of representations of Lie groups started in 1947 with a key observation of Gårding [Går47]. He remarked that every continuous Banach space representation of a Lie group determines a representation of the associated Lie algebra. The image of an element of the Banach space under the action of the representation is a function on the group and the representatives of the Lie algebra can be viewed as first-order differential operators with smooth coefficients acting on these functions. Gårding established by a regularization technique that the corresponding subspace of infinitely often differentiable functions is dense. But in 1953 Harish-Chandra [HC53] observed that the Gårding subspace was not very satisfactory for many purposes and proposed restricting consideration to the subspace of the Gårding space formed by the functions with convergent Taylor series. He also introduced the terminology well-behaved functions, or well-behaved vectors. It was not, however, apparent that the subspace spanned by such functions would automatically be dense in the representation space. Nevertheless, Harish-Chandra did prove the density for certain representations of a specific class of groups. His proof relied heavily on the structure theory of Lie groups. Subsequently, in 1958, Cartier and Dixmier [CD58] introduced the alternative terminology analytic functions and established the density property for a wider class of groups and representations. In particular, they established that the analytic functions are dense for every unitary representation of an arbitrary Lie group. This was the first indication that the density property of the analytic functions might be a universal property, valid for general representations of arbitrary Lie groups. This universality was the final conclusion, Theorem 10, of Bob's thesis. Strangely enough the result is not explicitly stated in the *Academy* announcement although it is a direct inference of the final sentence.

The problem of the density of the analytic functions also attracted the attention of Ed Nelson, who was a postdoctoral fellow at the Institute of Advanced Studies (IAS) in Princeton. Nelson [Nel59] established the density property for general Lie groups and representations independently of Langlands although they were aware of each other's interests. As Bob explained

> I do not think that Nelson ever saw my thesis. What happened . . . is that
> Lennie Gross, who was then an instructor at Yale, invited two graduate students
> John Frampton and myself to drive with him to the IAS, where he was to visit
> friends from his graduate student years in Chicago, among them Ed Nelson and
> Paul Cohen. So it came that Nelson and I discovered a common interest. As
> I have mentioned before, it was that conversation, which I would guess went into
> some detail, that provided Ed and myself with whatever we knew of each
> other's work.

The only other remaining clue to the overlap of their works is a reference to a forthcoming paper of Nelson in the introduction to Bob's thesis together with a reference to a related paper of Nelson and Stinespring [NS59]. In addition there is a reference to Nelson's published paper [Nel59] in the *Academy* announcement. The road trip to Princeton later had an unforeseen consequence with long term implications. As Bob wrote recently "I also believe that my invitation to come to Princeton as an instructor then came to me without any further action on my part, thus I made no application. I do not know any longer from whom the invitation came, but it was clear that the recommendation had come from Ed." As far as I am aware Nelson never made any specific reference to Bob's thesis or the announcement in the *Proceedings of the Academy* in any of his publications.

The reference to Bob's *Academy* note in Roe Goodmans's paper [Goo71] was a result of Roe spending the academic year 1968–1969 at the IAS on leave from MIT. By this time Bob had moved back to Yale and Nelson's interest had moved on to Constructive Quantum Field Theory. Fortunately Roe became aware of Bob's thesis during this period and even obtained a copy. He wrote that it was "perhaps at the Institute or at the historic old Fine Hall library" although he subsequently admitted that "perhaps it was at MIT." There is now no trace of the thesis in the Princeton library records so the origin of the copy is obscure. It was possibly a copy of the second retyped version. In any case

Roe wrote that he "read it in detail" and was probably the first and last reader until Bob sent me a copy in 1986.

The general strategy of Langlands and Nelson was very similar but differed significantly in detail. Both used techniques of semigroup theory and parabolic partial differential equations. The common idea was to construct elliptic operators as polynomials in the representatives of the Lie algebra and to argue that these operators generated continuous analytic semigroups which mapped the representation space into the subspace of analytic functions. The density property was then a direct consequence of the semigroup continuity. The implementation of this strategy differed between the two authors. Nelson only considered operators in the form of Laplacians, sum of squares of the Lie algebra representatives, but Langlands analyzed polynomials of all orders with complex coefficients. We will not attempt to describe Nelson's arguments albeit to say that they relied in part on probability theory and in part on results of Eidelman [Eid56] on analyticity of solutions of parabolic differential equations together with some observations of Gårding on the decrease properties of these solutions. A complete, simplified, version of Nelson's proof was subsequently given by Gårding [Går60]. Although we will not discuss Nelson's work it should be noted that it dealt with a broad range of topics involving analytic elements. In particular it became well known in the mathematical physics community for its results on single operator theory, the intracies involved in the addition of unbounded operators, etc. I became acquainted with it in the early 1960s some 25 years before I encountered Langlands' thesis.

The arguments in the thesis proceeded in four major steps, the first three of which corresponded to the theorems in the *Academy* note. First, it was necessary to prove that the (closure of the) elliptic operators did indeed generate analytic semigroups. Since Bob was considering operators of general order this was a very complicated technical problem whose resolution was given by Theorems 7 and 8 of the thesis and stated as Theorems 1 and 2 of [Lan60b]. Secondly, as a consequence of approaching the problem in such generality, it was straightforward to deduce that the semigroups mapped into the Gårding subspace of infinitely often differentiable functions. Next his analysis established that the action of the semigroups was determined by a universal integrable kernel. This was Theorem 9 of [Lan60a] and Theorem 3 of [Lan60b]. Finally it was necessary to prove that the semigroups in fact map into the subspace of analytic functions. This was Theorem 10 in the thesis but only appeared as a passing remark at the end of the *Academy* paper. It was at this last stage that the proof depended on the theory of parabolic partial differential equations with analytic coefficients. In fact Langlands cites the work of Eidelman [Eid56], which was also used by Nelson. Since Eidelman's paper is in Russian this was a barrier in 1986 to my comprehension of the proof. This difficulty was compounded by Bob's observation in his thesis that

"The facts we need from this paper are not explicitly stated as theorems and the proofs are not given in complete detail. However, since the proofs are quite complicated . . . we prefer not to perform the calculations in detail here." Fortunately this final stage of the proof can be completed by a quite different argument. The Eidelman results on the analyticity of solutions of parabolic equations were expressed in terms of complex variables but one can instead use real variable arguments. This approach was given in detail in Chapter II of [Rob91] and we will give a streamlined version in the sequel.

Since our aim is to be as elementary as possible we will take a different approach to both Langlands and Nelson. We will work in Langlands' framework with general-order elliptic operators and first give an elementary proof that the density property of the analytic functions holds for all representations of the multidimensional Euclidean group of translations. This result is essentially a straightforward exercise in Fourier analysis. Secondly we transfer the conclusions for the Euclidean group to all continuous representations of a general Lie group by using an alternative version of the parametrix arguments developed by Langlands in combination with some relatively simple functional analytic arguments. In the course of our argument we establish that the elliptic operators do generate holomorphic semigroups whose action is given by integrable kernels. In fact we also deduce that these kernels satisfy Gaussian-style bounds. All these conclusions are reached by variations of the arguments of Langlands' thesis supplemented by other results developed in the 1950s, the heyday of semigroup theory. We conclude with an overview of other lines of investigation which developed from the thesis work.

11.2 The Euclidean Group

Our discussion of the density of analytic functions starts with an examination of the representations of the Euclidean group $\mathbf{R}^d$ of translations. Since this group is commutative there are no complications of structure theory but it nevertheless remains to establish that the density property is universal for all representations of the group. In the $\mathbf{R}^d$-case we argue that the universality follows once one has demonstrated the density of the analytic functions for the unitary representation of translations on $L_2(\mathbf{R}^d)$. Thus the proof of the density is reduced to understanding a relatively simple unitary representation and then lifting, or transferring, the result to a general representation. A similar strategy works in the general case, as we demonstrate in Section 11.3.

We begin by recalling some well-known properties of the unitary representation of $\mathbf{R}^d$ by translations on $L_2(\mathbf{R}^d)$, the Hilbert space of square integrable functions with respect to Lebesgue measure and with norm $\| \cdot \|_2$. The group representation T is given by the family of operators defined by $(T(y)\varphi)(x) = \varphi(x - y)$ for all $x, y \in \mathbf{R}^d$ and $\varphi \in L_2(\mathbf{R}^d)$. If $x_1, \ldots, x_d$ is a basis of $\mathbf{R}^d$ then the generators of the one-parameter subgroups $t \in \mathbf{R} \mapsto T(tx_k)$ of translations in the coordinate directions are given by $-\partial_k$ where $\partial_k = \partial/\partial x_k$. The partial derivatives are the representatives of the Lie algebra. Next adopt the multi-index notation $x^\alpha = x_{k_1} \ldots x_{k_n}$, $\partial^\alpha = \partial_{k_1} \ldots \partial_{k_n}$, etc., where $\alpha = (k_1, \ldots, k_n)$ and the $k_j \in \{1, \ldots, d\}$. Further denote the length of α by $|\alpha| = n$.

The differential structure of the L_2-representation is described by a well-known family of Sobolev spaces. The subspace $L_{2;n}(\mathbf{R}^d)$ of differential functions of order n is the common domain $\bigcap_{\{\alpha:|\alpha|=n\}} D(\partial^\alpha)$ of all the nth-order differential operators and the subspace of infinitely often differentiable functions is given by $L_{2;\infty}(\mathbf{R}^d) = \bigcap_{n \geq 0} L_{2;n}(\mathbf{R}^d)$. It follows by a standard argument with an approximate identity that $L_{2;\infty}(\mathbf{R}^d)$ is dense in $L_2(\mathbf{R}^d)$. Finally the function $\varphi \in L_{2;\infty}(\mathbf{R}^d)$ is defined to be an analytic function for translations if $\sum_{k \geq 1} (s^k/k!) N_k(\varphi) < \infty$ for some $s > 0$ and an entire analytic function if the sum converges for all $s > 0$. Here N_k is the seminorm on $L_{2;k}(\mathbf{R}^d)$ defined by $N_k(\varphi) = \sup_{\{\alpha:|\alpha|=k\}} \|\partial^\alpha \varphi\|_2$. This is the real analytic definition originally considered by Harish-Chandra [HC53]. There is, of course, an equivalent complex analytic definition in terms of extensions of the functions $y \in \mathbf{R}^d \mapsto T(y)\varphi$ to strips in $\mathbf{C}^d$ but we will not need to consider such extensions. We will, however, need to norm the subspaces $L_{2;k}(\mathbf{R}^d)$ and it is convenient to set $\|\varphi\|_k = \sup_{\{0 \leq l \leq k\}} N_l(\varphi)$.

First, following Langlands, we introduce the mth-order partial differential operators $H = \sum_{\{\alpha:|\alpha| \leq m\}} c_\alpha (-\partial)^\alpha$ with coefficients $c_\alpha \in \mathbf{C}$ where m is an even integer. Then H is defined to be strongly elliptic if there is a $\mu > 0$ such that

$$\mathrm{Re}\left((-1)^{m/2} \sum_{\{\alpha:|\alpha|=m\}} c_\alpha \xi^\alpha\right) \geq \mu \, |\xi|^m \tag{11.1}$$

for all $\xi \in \mathbf{R}^d$. The largest value of μ is called the ellipticity constant of H. Thus if $h(\xi) = \sum_{\{\alpha:|\alpha| \leq m\}} c_\alpha (i\xi)^\alpha$ then there are $\lambda \in (0, \mu]$ and $\omega \geq 0$ such that $\mathrm{Re}\, h(\xi) \geq \lambda \, |\xi|^m - \omega$ for all $\xi \in \mathbf{R}^d$. Note that the strong ellipticity condition only involves the real part of the principal coefficients, i.e. those with $|\alpha| = m$. Moreover, it is easily established that the condition is independent of the choice of coordinate basis. A nonsingular transformation of the basis does not affect the validity of the condition. Although Langlands mainly considers strongly elliptic operators he does in part examine properties of elliptic operators whose coefficients satisfy the weaker condition

$$\left| \sum_{\{\alpha:|\alpha|=m\}} c_\alpha \xi^\alpha \right| \geq \mu \, |\xi|^m \tag{11.2}$$

for all $\xi \in \mathbf{R}^d$. For brevity we will, however, concentrate on the strongly elliptic case.

Secondly, let $\widetilde{\varphi} \in L_2(\mathbf{R}^d)$ denote the Fourier transform of $\varphi \in L_2(\mathbf{R}^d)$, i.e.

$$\widetilde{\varphi}(\xi) = (2\pi)^{-d/2} \int_{\mathbf{R}^d} dx \, e^{-ix.\xi} \, \varphi(x).$$

Then $\widetilde{(-\partial_k\varphi)}(\xi) = (i\xi_k) \, \widetilde{\varphi}(\xi)$ and $(\widetilde{H\varphi})(\xi) = h(\xi) \, \widetilde{\varphi}(\xi)$, i.e. the elliptic differential operators act as multiplication operators on the Fourier space with multiplier h. In particular H is a closed operator on the subspace of functions $\varphi \in L_2(\mathbf{R}^d)$ such that $h\widetilde{\varphi} \in L_2(\mathbf{R}^d)$. Therefore H generates a semigroup S whose action is given by

$$(S_t\varphi)(x) = (2\pi)^{-d/2} \int_{\mathbf{R}^d} d\xi \, e^{ix.\xi} \, e^{-th(\xi)} \, \widetilde{\varphi}(\xi) = (K_t * \varphi)(x)$$

for all $t \geq 0$ where $K_t(x) = (2\pi)^{-d/2} \int_{\mathbf{R}^d} d\xi \, e^{ix.\xi} \, e^{-th(\xi)}$. The semigroup property follows from the action by multiplication on the Fourier space and this also ensures that the kernel is a convolution semigroup, i.e. $K_{s+t}(x) = \int_{\mathbf{R}^d} dz \, K_s(z) \, K_t(x-z)$.

Several key properties of S follow immediately from the Fourier transform definition together with Plancherel's theorem, i.e. the identities $\|\varphi\|_2 = \|\widetilde{\varphi}\|_2$. First one verifies easily that S is strongly continuous, i.e. $\lim_{t\to 0} \|(S_t - I)\varphi\|_2^2 = 0$. In addition the S_t satisfy the operator bounds $\|S_t\| \leq \exp(\omega t)$ with ω the constant in the ellipticity bound on $\operatorname{Re} h$. Explicitly one has

$$\|S_t\varphi\|_2^2 = \int_{\mathbf{R}^d} d\xi \, e^{-2t \operatorname{Re} h(\xi)} \, |\widetilde{\varphi}(\xi)|^2 \leq e^{2\omega t} \, \|\varphi\|_2^2.$$

More interestingly, one calculates that the semigroup S maps $L_2(\mathbf{R}^d)$ into the subspace $L_{2;\infty}(\mathbf{R}^d)$. For example.

$$\|\partial^\alpha S_t\varphi\|_2 \leq \left(\int_{\mathbf{R}^d} d\xi \, |\xi|^{2|\alpha|} e^{-2t \operatorname{Re} h(\xi)} \, |\widetilde{\varphi}(\xi)|^2 \right)^{1/2}$$

$$\leq t^{-k/m} \left(\int_{\mathbf{R}^d} d\xi \, (t|\xi|^m)^{2|\alpha|/m} e^{-2(\lambda t|\xi|^m - \omega t)} \, |\widetilde{\varphi}(\xi)|^2 \right)^{1/2}$$

$$\leq C_k \, t^{-k/m} \, e^{\omega t} \, \|\varphi\|_2$$

for all α with $|\alpha| = k$ where $C_k > 0$. The first step uses the ellipticity bound on $h(\xi)$ and the second follows from an estimate $(t|\xi|^m)^{2|\alpha|/m} \leq c_\lambda \exp(2\lambda t|\xi|^m)$. Consequently S_t maps $L_2(\mathbf{R}^d)$ into $L_{2;k}(\mathbf{R}^d)$ for each $k \geq 1$ and hence into $L_{2;\infty}(\mathbf{R}^d)$. Unfortunately this argument does not give a good control on the constants C_k. Nevertheless one can control the growth property by an iterative argument starting from the bounds on the first derivatives $\|\partial_j S_t\varphi\|_2$.

It follows from the foregoing that there is a $C_1 > 0$ such that

$$N_1(S_t\varphi) = \sup_{j\in\{1,\dots,d\}} \|\partial_j S_t\varphi\|_2 \leq C_1 \, t^{-1/m} \, e^{\omega t} \, \|\varphi\|_2 \qquad (11.3)$$

for all $\varphi \in L_2(\mathbf{R}^d)$. But then

$$N_k(S_t\varphi) \leq \sup_{j_1,\dots,j_k} \|(\partial_{j_1} S_{t/k}) \dots (\partial_{j_k} S_{t/k})\varphi\|_2 \leq C_1^k (t/k)^{-k/m} e^{\omega t} \|\varphi\|_2$$

for all $k \geq 1$ and $\varphi \in L_2(\mathbf{R}^d)$. Hence, by Stirling's formula, there are $a, b > 0$ such that

$$N_k(S_t\varphi) \leq a\, b^k\, t^{-k/m}\, (k!)^{1/m}\, e^{\omega t}\, \|\varphi\|_2 \tag{11.4}$$

for all $k \geq 1$, $t > 0$ and $\varphi \in L_2(\mathbf{R}^d)$. Therefore $S_t\varphi$ is an entire analytic function for translations. But $\lim_{n\to\infty} \|S_t\varphi - \varphi\|_2 = 0$. So the entire analytic functions are dense in $L_2(\mathbf{R}^d)$ and we have proved more than we set out to do.

After this initial skirmish with the L_2-representation of $\mathbf{R}^d$ we next explain how one can transfer the density result for the analytic functions to a general continuous Banach space representation. It is here that estimates on the semigroup kernel K_t are of importance. In addition there are two new elements entering the arguments, the continuity and the boundedness properties of the representation. Let χ be a Banach space and U a continuous representation of $\mathbf{R}^d$ by bounded operators $U(x)$, $x \in \mathbf{R}^d$, on χ. There are two types of continuity of interest, strong continuity $\|(U(x) - I)\varphi\| \to 0$ as $|x| \to 0$, and weak* continuity. But a basic result of Yosida establishes that strong continuity is equivalent to weak continuity, i.e. equivalent to the conditions

$$\lim_{|x|\to 0} (f, U(x)\varphi) = (f, \varphi) \tag{11.5}$$

for all $\varphi \in \chi$ and $f \in \chi*$, the dual of χ. Alternatively, if χ is the dual of a Banach space χ_*, the predual of χ, then U is weak* continuous if $f \circ U \in \chi_*$ for all $f \in \chi_*$ and (11.5) is valid for all $\varphi \in \chi$ and $f \in \chi_*$. Thus both types of continuity can be handled similarly. It also follows from the group property and either form of continuity that there are $M \geq 1$ and $\rho \geq 0$ such that one has bounds

$$\|U(x)\| \leq M\, e^{\rho|x|} \tag{11.6}$$

for all $x \in \mathbf{R}^d$. Here $\|\cdot\|$ indicates the standard operator norm associated with the Banach space.

Now to emulate the earlier L_2-arguments for the Banach space representation it is necessary to define an analogue of the semigroup S. One direct way of doing this is by noting that on $L_2(\mathbf{R}^d)$ the semigroup satisfies

$$(S_t\varphi)(x) = \int_{\mathbf{R}^d} dy\, K_t(y)\, \varphi(x - y) = \int_{\mathbf{R}^d} dy\, K_t(y)\, (T(y)\varphi)(x).$$

Thus, formally at least, one has the operator representation

$$S_t = \int_{\mathbf{R}^d} dy\, K_t(y)\, T(y) = T(K_t) \tag{11.7}$$

where the integral is in the weak sense. Therefore the correct analogue of S in the Banach space setting should be the semigroup

$$S_t^{(U)} = \int_{\mathbf{R}^d} dy \, K_t(y) \, U(y) = U(K_t) \tag{11.8}$$

where the integral is in the weak or weak* sense. This is an observation that we will use in the subsequent discussion of general Lie groups. But in order to make sense of either of the relations (11.7) or (11.8) one needs control on the growth properties of the kernel K_t. In fact the kernel and its derivatives satisfy Gaussian-type bounds.

Proposition 11.2.1 *There exist $b > 0$ and $\omega \geq 0$, and for each multi-index α an $a_\alpha > 0$, such that*

$$|(\partial^\alpha K_t)(x)| \leq a_\alpha \, t^{-(d+|\alpha|)/m} \, e^{\omega t} \, e^{-b(|x|^m/t)^{1/(m-1)}}$$

for all $x \in \mathbf{R}^d$ and all $t > 0$.

Proof First consider the case that $\alpha = 0$. Then by contour integration one deduces that

$$|K_t(x)| = (2\pi)^{-d/2} \left| \int_{\mathbf{R}^d} d\xi \, e^{ix.(\xi+i\eta)} \, e^{-th(\xi+i\eta)} \right|$$

$$\leq (2\pi)^{-d/2} \int_{\mathbf{R}^d} d\xi \, e^{-x.\eta} \, e^{-t \, \mathrm{Re} \, h(\xi+i\eta)}$$

for all $\eta \in \mathbf{R}^d$. But then there are $\lambda > 0$ and $\sigma, \omega \geq 0$ such that

$$\mathrm{Re} \, h(\xi + i\eta) \geq \lambda \, |\xi|^m - \sigma \, |\eta|^m - \omega.$$

Therefore $|K_t(x)| \leq a \, e^{\omega t} \, e^{-\eta.x} \, e^{\sigma t |\eta|^m}$ and the required bound follows by minimizing with respect to η.

Secondly, if $\alpha \neq 0$ then the derivatives introduce additional multipliers $(i\xi)^\alpha$ on the Fourier transform. But for each $\varepsilon > 0$ there is a $k_{\alpha,\varepsilon} > 0$ such that

$$|(\xi + i\eta)^\alpha| \leq (t|\xi + i\eta|^m)^{|\alpha|/m} \, t^{-|\alpha|/m} \leq k_{\alpha,\varepsilon} \, t^{-|\alpha|/m} \, e^{\varepsilon t (|\xi|^m + |\eta|^m)}.$$

Then, if ε is sufficiently small, the estimates for the derivatives follow as above.   □

One also has analogous bounds on the functions $x \to x^\beta (\partial^\alpha K_t)(x)$.

Corollary 11.2.2 *There exist $b > 0$ and $\omega \geq 0$, and for each pair of multi-indices α, β an $a_{\alpha,\beta} > 0$, such that*

$$|x^\beta (\partial^\alpha K_t)(x)| \leq a_{\alpha,\beta} \, t^{-(d+|\alpha|-|\beta|)/m} \, e^{\omega t} \, e^{-b(|x|^m/t)^{1/(m-1)}}$$

for all $x \in \mathbf{R}^d$ and all $t > 0$.

Proof The statement follows by remarking that for each $\varepsilon > 0$ there is an $l_{\beta,\varepsilon} > 0$ such that

$$|x^\beta| \le |x|^{|\beta|} = (|x|^m/t)^{|\beta|/m} \, t^{|\beta|/m} \le l_{\beta,\varepsilon} \, e^{\varepsilon(|x|^m/t)^{1/(m-1)}} \, t^{|\beta|/m}.$$

Thus if ε is sufficiently small the bounds follow from those of Proposition 11.2.1 with a slightly smaller value of b. $\qquad\square$

The bounds of the corollary will be applied in the sequel to differential operators whose effective order is lower than the nominal order. If $D_n = \sum_{\{\alpha; |\alpha| \le n\}} c_\alpha \, \partial^\alpha$ is an nth-order partial differential operator with smooth coefficients supported in a compact neighborhood of the origin and $(\partial^\beta c_\alpha)(0) = 0$ for all β with $|\beta| < (|\alpha| - k) \vee 0$ then D_n is defined to have effective order k. Corollary 11.2.2 then gives the bounds

$$|(D_n K_t)(x)| \le a_{\alpha,\beta} \, t^{-(d+k)/m} \, e^{\omega t} \, e^{-b(|x|^m/t)^{1/(m-1)}}$$

for all $x \in \mathbf{R}^d$ and all $t > 0$, i.e. the order of the t singularity is governed by the effective order k rather than the real order n.

The pointwise bounds on the kernel and its derivatives allow one to deduce various weighted bounds. For example, the bounds of Proposition 11.2.1 give

$$\sup_{x \in \mathbf{R}^d} e^{\rho|x|}|(\partial^\alpha K_t)(x)| \le a_\alpha \, t^{-(d+|\alpha|)/m} \, e^{\omega t} \sup_{x \in \mathbf{R}^d} \left(e^{\rho|x|} e^{-b(|x|^m/t)^{1/(m-1)}} \right)$$

$$\le a_\alpha \, t^{-(d+|\alpha|)/m} \, e^{\omega'(1+\rho^m)t}$$

with $\omega' > 0$. Weighted L_1-bounds are also valid.

Proposition 11.2.3 *There is an $\omega > 0$ and for each multi-index α an $a_\alpha > 0$, such that*

$$\int_{\mathbf{R}^d} dx \, |(\partial^\alpha K_t)(x)| \, e^{\rho|x|} \le a_\alpha \, t^{-|\alpha|/m} \, e^{\omega(1+\rho^m)t}$$

for all $\rho, t > 0$.

These bounds follow straightforwardly from the estimates of Proposition 11.2.1.

The weighted estimates of Proposition 11.2.3 combined with the continuity bounds (11.6) imply

$$\left| \int_{\mathbf{R}^d} dy \, K_t(y) \, (f, U(y)\varphi) \right| \le M \int_{\mathbf{R}^d} dy \, |K_t(y)| \, e^{\rho|y|} \, \|f\| \cdot \|\varphi\|$$

$$\le a \, M \, e^{\omega(1+\rho^m)t} \, \|f\| \cdot \|\varphi\|$$

for all $f \in \chi^*$, or χ_*, and $\varphi \in \chi$. Thus $S_t^{(U)} = U(K_t)$, formally given by (11.8), is indeed well-defined as a bounded operator on χ, for each $t > 0$, and one has bounds

$$\|S_t^{(U)}\| \le a\,M\,e^{\omega(1+\rho^m)t}$$

for all $t > 0$. Then since the K_t form a convolution semigroup it follows straightforwardly that the $S_t^{(U)}$ form a continuous semigroup with the type of continuity dictated by the continuity of the representation, either strong or weak*. But one can also identify the generator of $S^{(U)}$.

First let $X_1, \ldots, X_d$ denote the generators of the one-parameter semi-groups $t \in \mathbf{R} \mapsto U(tx_k)$ in the coordinate directions $x_1, \ldots, x_d$, e.g. $X_k = \lim_{t \to 0}(U(tx_k) - I)/t$. Then set $\chi_n = \bigcap_{\{\alpha:|\alpha|=n\}} D(X^\alpha)$ and $\chi_\infty = \bigcap_{n \ge 1} \chi_n$. These subspaces correspond in an obvious way to the nth-order differentiable functions and the infinitely often differentiable functions respectively. Sec-ondly, note that

$$X_k S_t^{(U)} \varphi = \lim_{t \to 0} \int_{\mathbf{R}^d} dy\, K_t(y)\,(U(y + tx_k) - I)\varphi/t$$

$$= \lim_{t \to 0} \int_{\mathbf{R}^d} dy\,(K_t(y - tx_k) - K_t(y))\,U(y)\varphi/t$$

$$= \int_{\mathbf{R}^d} dy\,(-\partial_k K_t)(y)\,U(y)\varphi.$$

Then by iteration

$$X^\alpha S_t^{(U)} \varphi = U((-\partial)^\alpha K_t)\varphi = \prod_{j \in \alpha} U(-\partial_j K_{t/|\alpha|})\varphi \qquad (11.9)$$

for all $\varphi \in \chi$ and all α. The right-hand side is well-defined by another application of the bounds of Proposition 11.2.3. In fact the proposition leads to bounds

$$\|X^\alpha S_t^{(U)} \varphi\| \le a_\alpha\,M\,t^{-|\alpha|/m}\,e^{\omega(1+\rho^m)t}\,\|\varphi\| \qquad (11.10)$$

for all α. Therefore one concludes that $S_t^{(U)}\chi \subseteq \chi_\infty$. Thirdly, define the (entire) analytic functions in χ, corresponding to the representation U, to be the subspace of $\varphi \in \chi_\infty$ for which $\sum_{k \ge 1}(s^k/k!)N_k(\varphi) < \infty$ for (for all) some $s > 0$ where N_k is now the seminorm on χ_k defined by $N_k(\varphi) = \sup_{\{\alpha:|\alpha|=k\}}\|X^\alpha\varphi\|$ and the corresponding norms are again defined by $\|\varphi\|_k = \sup_{\{0 \le l \le k\}} N_l(\varphi)$.

The principal structural result following from this discussion is the density of the entire analytic functions.

Proposition 11.2.4 *The operators* $S_t^{(U)} = U(K_t)$ *are well-defined as strong, or weak*, integrals on* χ. *They form a strongly, or weakly*, continuous holomorphic semigroup whose generator is the strong, or weak*, closure of the operator* $H^{(U)} = \sum_{\{\alpha:|\alpha|\le m\}} c_\alpha X^\alpha$ *on* χ_m. *Moreover, there are* $a, b > 0$ *and* $\omega \ge 0$ *such that*

$$N_k(S_t^{(U)}\varphi) \le a\, b^k\, t^{-k/m}\, (k!)^{1/m}\, e^{\omega t}\, \|\varphi\| \qquad (11.11)$$

for all $k \ge 1$, $t > 0$ and $\varphi \in \chi$. Hence $S_t^{(U)}$ maps χ into the subspace of entire analytic functions for U. Therefore the latter subspace is strongly, or weakly*, dense in χ.

Proof We have already discussed the definition of the S_t as a continuous semigroup. Next the bounds on the seminorm start by iterating the bounds

$$N_1(S_t\varphi) \le C_1\, t^{-1/m}\, e^{\omega(1+\rho^m)t}\, \|\varphi\|,$$

which follow for all $\varphi \in \chi$ from the estimates (11.10) with $|\alpha| = 1$. The bounds (11.11) then follow from the factorization $X^\alpha S_t^{(U)}\varphi = X_{j_1} S_{t/n}^{(U)} \ldots X_{j_n} S_{t/n}^{(U)}\varphi$ if $\alpha = \{j_1, \ldots, j_n\}$ in direct analogy with the argument for translations on $L_2(\mathbf{R}^d)$. In fact the bounds (11.11) are the direct generalization of the L_2-bounds (11.4). The mapping property and the density property are then an immediate consequence, as before.

It remains to identify the generator of the semigroup. But $S_t^{(U)}\chi \subseteq \chi_m = D(H_{(U)})$ by the mapping property. In particular $S_t^{(U)}D(H_{(U)}) \subseteq D(H_{(U)})$. Therefore $D(H_{(U)})$ is a core of the semigroup generator, i.e. the generator is the closure of $H_{(U)}$ with respect to the norm $\|\cdot\|_m$. $\square$

The proposition establishes the density of the analytic functions for all the continuous representations of $\mathbf{R}^d$ and there are two features of the analysis that persist in the subsequent discussion of general Lie groups. First we have shown that the analytic properties of all continuous representations U can be inferred from those of left translations T by replacing the semigroup $S_t = T(K_t)$ by the transferred operator $S_t^{(U)} = U(K_t)$. This transference technique carries over to the general situation. Then properties governed by the semigroups can be analyzed by considering the universal semigroup kernel on the $L_p(\mathbf{R}^d)$-spaces. Chapter II of Bob Langlands' thesis was based on these tactics although the presentation was somewhat different. Our approach in Section 11.3 puts a different emphasis on the semigroup kernel and its properties.

Before proceeding to general groups we sketch a class of examples which illustrate the diversity of representations of $\mathbf{R}^d$ and the breadth of application of Proposition 11.2.4. Let $\mathcal{L}(\mathfrak{A})$ denote the space of all bounded operators on the Hilbert space $L_2(\mathbf{R}^d)$ equipped with the usual operator norm. This space is in fact an algebra equipped with an adjoint operator corresponding to the adjoint operation on the operators. It is a von Neumann algebra, or a W^*-algebra depending on choice of terminology. It clearly contains all bounded multiplication operators but also all bounded functions of the partial derivatives ∂_k. The latter operators typically act by convolution. Now there is an action τ of $\mathbf{R}^d$ on the algebra given by $A \in \mathcal{L}(\mathfrak{A}), x \in \mathbf{R}^d \mapsto \tau_x(A) =$

$L(x)AL(x)^{-1} \in \mathcal{L}(\mathfrak{A})$. The τ form a group of *-automorphisms of $\mathcal{L}(\mathfrak{A})$ that is weakly* continuous. Alternatively there are a great variety of *-subalgebras that are closed in respect to the operator norm, C^*-algebras, which are invariant under τ. On these algebras the automorphisms are weakly (strongly) continuous. All the foregoing results apply to this range of examples. My interest in the differential structure of representations of Lie groups was initially sparked by such examples and my involvement in the application of operator algebras to quantum field theory and quantum statistical mechanics (see [BR87] [BR97]).

11.3 General Groups

The preceding analysis of representations of $\mathbf{R}^d$ gives a simple illustration of the principal results in the second part of Bob Langlands' thesis. In addition it provides a starting point and a strategy for the analysis of the continuous representations of a general Lie group G. We have shown that the analytic structure of the representations of $\mathbf{R}^d$ can be inferred from the study of the representation of the group as left translations on the L_p-spaces and we next explain how this approach can be modified and expanded to understand the representations of the general group. The resulting strategy is essentially the same as in the thesis but our tactics are somewhat different. Both the original proofs and the following arguments use three basic techniques, the exponential map, the parametrix method, and a transference technique. We assume the reader is conversant with the definition and the standard properties of the exponential map and we do not dwell on the details. In contrast we will elaborate on the formulation and structure of the parametrix method. There is little to add about the transference techique as it is applied exactly as in the $\mathbf{R}^d$-case.

Section 11.3.1 is devoted to the parametrix method as a precursor to the applications in the following two subsections. In the first of these (Section 11.3.2) we use a parametrix to establish that each strongly elliptic operator affiliated with the group representation generates a holomorphic semigroup determined by an integral kernel satisfying Gaussian-type bounds. The proof starts from the estimates for the Euclidean group discussed in the previous section. Then in Section 11.3.3 we analyze the connection between the semigroups and the analytic elements of the representation. Throughout we use real analytic arguments and avoid complex function theory.

11.3.1 The Parametrix Method

Let G be a connected d-dimensional Lie group with a corresponding Lie algebra $\mathfrak{g}$. The main framework of our analysis is the representation L of the group by left translations on the space $L_2(G\,;dg)$ of functions over G that are square integrable with respect to the Haar measure dg. Explicitly

$$(L(g)\varphi)(h) = \varphi(g^{-1}h),$$

for all $\varphi \in L_2(G\,;dg)$ and $g,h \in G$. It is, however, convenient to consider in addition the representation L on the associated spaces $L_p(G\,;dg)$. The representation L is usually referred to as the left regular representation although it is the direct analogue of translations in the Euclidean case. Note that for $p \in [1,\infty)$ the representation is strongly continuous and for $p = \infty$ it is weakly* continuous. All subsequent topological properties, and in particular continuity and density properties, are understood in the strong topology if $p \in [1,\infty)$, or the weak* topology if $p = \infty$.

The first step in the analysis is to note that the Lie group is a manifold that is locally diffeomorphic to $\mathbf{R}^d$ under the usual exponential map. Therefore the semigroup kernel constructed in the previous section for a given strongly elliptic operator on $L_2(\mathbf{R}^d)$ can be used as a local approximation for the kernel of the corresponding elliptic operator on $L_2(G\,;dg)$. Subsequently, starting from the local approximation, one can construct iteratively a family of functions on G that formally corresponds to the semigroup kernel in the general representation. The iterative method which lies at the heart of this approach is Langlands' version of Levi's parametrix method [Lev07] dating back to 1907. Despite its early origins the method was not well known in the 1950s and Bob no doubt learnt of it from Felix Browder's Yale lectures, which he observed to be "of a singularly instructive nature." I certainly learnt of it from reading Bob's thesis 28 years later and my initial understanding was enhanced by reading Avner Friedman's book [Fri64] on parabolic differential equations. The method can be formulated as a technique for solving parabolic and elliptic differential equations starting from local approximations obtained by fixing the coefficients at a given point. In semigroup theory the two approaches give approximations to the semigroup and resolvent, respectively. These approximations are analogous to the well-known "time-dependent" and "time-independent" methods of perturbation theory although one does not have a perturbation in any conventional sense. In the Australian vernacular it is a Claytons[1] perturbation theory. We rely on the parabolic parametrix method. It has two remarkable features. First it leads to a global solution starting from the initial local approximation. Technically this arises because the terms in the series expansion are given by convolution of terms localized in a fixed compact region. Therefore larger distances only arise in higher-order terms. Secondly, the expansion has extremely good convergent properties. The solution of the

---

[1] In the late 1970s when I first came to Australia there was a vigorous government campaign against drink driving. Concurrently a nonalcoholic drink with a color resembling whisky and appropriately bottled was heavily promoted under the brand name Claytons "as the drink you have when you're not having a drink." Since then the prefix Claytons has been commonly used to indicate an ersatz product lacking the vital ingredient.

parabolic equation is a function over $\mathbf{R}_+ \times G$ with the variable $t \in \mathbf{R}_+$ interpretable as the time parameter. The series expansion is in powers of t and is uniformly convergent over G for all $t > 0$. In many time-dependent problems the perturbation series are only convergent for small times but in the current context one obtains convergence for all times as a consequence of the Gaussian bounds on the local Euclidean approximant.

The local approximation procedure starts with the exponential map. Recall that if $a \in \mathfrak{g}$ then $\exp(a) \in G$ is defined by $\exp(a) = \gamma(1)$, where $\gamma : \mathbf{R} \mapsto G$ is the unique one-parameter subgroup of G whose tangent vector at the identity e of G is equal to a. The map is a diffeomorphism of a neighborhood of the origin in $\mathfrak{g}$ to a neighborhood of $e \in G$. In the case of a matrix Lie group $\exp(X)$ coincides with the usual definition of the exponential of the matrix X by a power series expansion. In the physics literature it is commonplace to use the notation e^X for the map since it shares many of the basic features of the standard exponential. Now let $a_1, \ldots, a_d$ be a vector space basis of the Lie algebra $\mathfrak{g}$ of G. Then $t \mapsto \exp(-t a_j)$ is a one-parameter subgroup of G and the corresponding left translations $t \mapsto L(\exp(-t a_j))$ form a continuous one-parameter group on each of the spaces $L_p(G ; dg)$. Let A_j denote the generator of this group. For example, if $G = \mathbf{R}^d$ then $A_j = -\partial_j$. Now we consider mth-order operators

$$ H = \sum_{\alpha;\ |\alpha| \le m} c_\alpha A^\alpha $$

with $c_\alpha \in \mathbf{C}$ and m an even integer. The domain of H in $L_p(G ; dg)$ is the subspace $L_{p;m}(G) = \bigcap_{|\alpha| \le m} D(A^\alpha)$ of m-times left-differentiable functions. It is not difficult to establish that $L_{p;m}(G ; dg)$ is dense in $L_p(G ; dg)$. Hence H is densely defined. Then the adjoint H^* of H is densely defined. Hence H is closable and, for simplicity, we retain the notation H for the closure. Moreover, the subspace $L_{p;\infty}(G ; dg) = \bigcap_{m \ge 0} L_{p;m}(G ; dg)$ of C^∞-functions is a core for each H. The operators H are the direct analogue of those examined in the previous section for the Euclidean group and the definition of strong ellipticity (11.1) and the definition of the ellipticity constant are unchanged. Again the strong ellipticity condition is independent of the choice of basis of $\mathfrak{g}$ and also of the lack of commutativity. If, for example, one replaces $A_i A_j$ by $A_j A_i$ in the definition of H one effectively introduces a modification $A_j A_i - A_i A_j$ which is linear in the A by the structure relations of $\mathfrak{g}$. Therefore reordering only changes the lower-order terms, those with $|\alpha| < m$, and does not affect the principal terms, those with $|\alpha| = m$. It also follows that the product $H_1 H_2$ of two strongly elliptic operators with real coefficients H_1 and H_2 of orders m_1 and m_2, respectively, is a strongly elliptic operator of order $m_1 m_2$. The principal coefficients of the product operator are products of the principal coefficients of the component operators.

Our immediate aim is to establish that each (closed) strongly elliptic H generates a continuous semigroup S on $L_2(G\,;dg)$ whose action is given by a kernel K satisfying Gaussian-type bounds. As mentioned above we approach this problem by first constructing a family of functions K by a local approximation with the kernel of the analogous operator on the Euclidean group, corresponding formally to the semigroup kernel. Then we verify that the family does indeed have the correct properties for a semigroup kernel and that H is the generator of the semigroup.

The motivation for the construction is the observation that the kernel K, if it exists, should be a solution of the parabolic equation

$$(\partial_t + H)K_t = 0$$

for $t > 0$ with the initial condition $K_t \rightarrow \delta$ as $t \rightarrow 0$. Alternatively if one defines $K_t = 0$ for $t \leq 0$ then $(t,g) \mapsto K_t(g)$ from $\mathbf{R} \times G$ into $\mathbf{C}$ should be the fundamental solution for the heat operator $\partial_t + H$, i.e., one should have

$$((\partial_t + H)K_t)(g) = \delta(t)\,\delta(g) \tag{11.12}$$

for all $t \in \mathbf{R}$ and $g \in G$. Now the parametrix method expresses K as a "perturbation" expansion in the "time" variable t.

Let $\Omega \subset G$ be an open relatively compact neighborhood of the identity $e \in G$ and B_0 an open ball in $\mathfrak{g}$ centred at the origin such that $\exp|_{B_0} : B_0 \rightarrow \Omega$ is an analytic diffeomorphism. Set $a_x = \sum_{i=1}^d x_i a_i$, for $x \in \mathbf{R}^d$, and $B = \{x \in \mathbf{R}^d \,:\, a_x \in B_0\}$. Then for $\varphi \colon \Omega \rightarrow \mathbf{C}$ define $\hat{\varphi} \colon B_0 \rightarrow \mathbf{C}$ by $\hat{\varphi}(x) = \varphi(\exp(a_x))$. If Ω is small enough the image of Haar measure under this map is absolutely continuous with respect to Lebesgue measure. In particular, there exists a positive C^∞-function σ on B, bounded from below by a strictly positive constant, such that all derivatives are bounded on B and such that

$$\int_\Omega dg\,\varphi(g) = \int_B dx\,\sigma(x)\,\hat{\varphi}(x) \tag{11.13}$$

for all $\varphi \in L_1(\Omega\,;dg)$. We normalize the Haar measure dg such that $\sigma(0) = 1$ and choose the modulus on $\mathfrak{g}$ such that $|a_x| = |x|$ for all $x \in B$.

The key feature of the exponential map is the existence of C^∞-vector fields $X_1,\ldots,X_d$ on B, i.e. first-order partial differential operators with coefficients in $C_c^\infty(B)$, with the property

$$(X_k\hat{\varphi})(x) = (\widehat{A_k\varphi})(x) = (A_k\varphi)(\exp(a_x)) \tag{11.14}$$

for all $\varphi \in C_c^\infty(\Omega)$, where the $A_1,\ldots,A_d$ are generators of left translations. Thus the X_k are representatives on $L_2(\Omega\,;dg)$ of the Lie algebra $\mathfrak{g}$. Moreover,

$$X_k\hat{\varphi} = -\partial_k\hat{\varphi} + Y_k\hat{\varphi} \tag{11.15}$$

for $\varphi \in C_c^\infty(\Omega)$ where the Y_k are again C^∞-vector fields. But the crucial feature is that the Y_k have effective order zero as defined in Section 11.2.

Explicitly, $Y_k = \sum_{l=1}^{d} c_{kl}\partial_l$ with coefficients $c_{kl} \in C_c^\infty(B)$ which have a first-order zero at the origin. This property is a consequence of the Baker–Campbell–Hausdorff formula. This asserts that if Ω is sufficiently small and $\exp(a), \exp(b)$ are both in Ω then there is a $c(a,b) \in \mathfrak{g}$ such that $\exp(-a)\exp(b) = \exp(c(a,b))$ and

$$c(a,b) = -a + b + R(a,b)$$

where the remainder $R(a,b)$ is a sum of multicommutators of a and b with each commutator containing at least one a and one b. Now $X_k\hat{\varphi}$ is the transform of the limit as $s \to 0$ of the expression

$$s^{-1}\Big((L(\exp(sa_k))\varphi)(\exp(a_x)) - \varphi(\exp(a_x))\Big)$$

$$= s^{-1}\Big(\varphi(\exp(c(sa_k,a_x))) - \varphi(\exp(a_x))\Big).$$

The identity (11.15) follows immediately. The term $-\partial_k$ originates from the leading term $-sa_k$ in the expression for $c(sa_k,a_x)$. The Y_k, however, stem from the term linear in s which occurs in the remainder $R(sa_k,a_x)$. This term is of the form $\sum_{l=1}^{d} c_{kl}a_l$ and, importantly, the coefficients c_{kl} have a first-order zero. The latter property is a consequence of the structure relations of the Lie algebra since the remainder is a sum of multicommutators, each of which contains at least one a_x. Therefore it follows from the identity (11.15) that

$$\widehat{H\varphi} = \widehat{H}_0\hat{\varphi} + \widehat{H}_1\hat{\varphi}, \tag{11.16}$$

where $\widehat{H}_0 = \sum_\alpha c_\alpha(-\partial)^\alpha$ is the operator with constant coefficients corresponding to H on $\mathbf{R}^d$ and $\widehat{H}_1$ is an operator of effective order at most $m - 1$. In particular, the coefficients of $\widehat{H}_1$ have a zero at the origin. This local representation of H on $\mathbf{R}^d$ is the starting point for constructing the semigroup corresponding to H on G.

Let $\widetilde{K}_t$ denote the kernel associated with $\widehat{H}_0$ on $\mathbf{R}^d$ but with $\widetilde{K}_t = 0$ if $t \le 0$. Further, let $\chi \in C_c^\infty(\Omega)$ with $0 \le \chi \le 1$ and $\chi = 1$ in a neighborhood of the identity. Then define $K_t^{(0)}$ by setting $\widehat{K}_t^{(0)} = \hat{\chi}\,\widetilde{K}_t$ on B. It follows immediately from (11.16) that

$$((\partial_t + H)K_t^{(0)})\widehat{\ }(x) = ((\partial_t + \widehat{H}_0)(\hat{\chi}\,\widetilde{K}_t))(x) + (\widehat{H}_1(\hat{\chi}\,\widetilde{K}_t))(x)$$

$$= \delta(t)\,\delta(x) + \widehat{M}_t(x),$$

where $\widehat{M}_t = \widehat{D}\widetilde{K}_t$ with $\widehat{D}$ a partial differential operator of the form $\sum_k \hat{\chi}_k(x)\,M^{(k)}$, the sum is finite, the $\hat{\chi}_k \in C_c^\infty(B)$ and the $M^{(k)}$ are operators of effective order at most $m - 1$. Therefore the corresponding functions $K_t^{(0)}$ and M_t on G have compact support and satisfy the heat equation

$$((\partial_t + H)K_t^{(0)})(g) = \delta(t)\,\delta(g) + M_t(g) \tag{11.17}$$

for all $g \in G$ and $t > 0$. But it follows from the heat equations (11.12) and (11.17) that

$$K_t - K_t^{(0)} = -\int_0^t ds\, K_{t-s} * M_s \qquad (11.18)$$

for all $t \in \mathbf{R}$ where $*$ denotes the usual convolution product on G. Thus defining the convolution product $\hat{*}$ on $\mathbf{R} \times G$ by

$$(\varphi \,\hat{*}\, \psi)_t(g) = \int_0^t ds\, (\varphi_{t-s} * \psi_s)(g) = \int_0^t ds \int_G dh\, \varphi_{t-s}(h)\psi_s(h^{-1}g),$$

one has

$$K_t = K_t^{(0)} - (K \,\hat{*}\, M)_t \qquad (11.19)$$

for all $t \in \mathbf{R}$. Then by iteration this latter equation gives

$$K_t = K_t^{(0)} - (K^{(0)} \hat{*} M)_t + (K^{(0)} \hat{*} M \hat{*} M)_t - \cdots \qquad (11.20)$$

This is the parabolic parametrix expansion alluded to above. It represents the solution of the parabolic equation (11.12) and is the analogue of the series expansion encountered in "time-dependent" perturbation theory. It is the principal tool we use in Section 11.3.2 to construct the kernel $K = \{K_t\}_{t>0}$ and demonstrate that the family K_t forms a convolution semigroup. Although we do not need the elliptic or "time-independent" version of the expansion we note that it can be defined by Laplace transformation from the parabolic version.

One can in principle define functions L_λ for sufficiently large λ by

$$L_\lambda = \int_0^\infty dt\, e^{-\lambda t}\, K_t.$$

Formally L_λ is the solution of the elliptic equation $((\lambda I + H)L_\lambda)(g) = \delta(g)$. Then, assuming the L_λ are well-defined, the parametrix identity (11.19) immediately gives the relations

$$L_\lambda = L_\lambda^{(0)} - L_\lambda * N_\lambda,$$

where $L_\lambda^{(0)}$ and N_λ are the Laplace transforms of $K_t^{(0)}$ and M_t, respectively. Now by iteration one obtains the elliptic version of the parametrix expansion

$$L_\lambda = L_\lambda^{(0)} - L_\lambda^{(0)} * N_\lambda + L_\lambda^{(0)} * N_\lambda * N_\lambda - \cdots$$

Note that it follows from the definition of the M_t that $N_\lambda = DL_\lambda^{(0)}$ where D is a partial differential operator of effective order at most $m - 1$. The unifying feature of the two versions of the parametrix method is that they both lead to inverses of partial differential operators. The K_t are inverses of the operator $(\partial_t + H)$ on $\mathbf{R}_+ \times G$ and the L_λ are inverses of the operator $(\lambda I + H)$ on G.

Next we turn to the problem of proving that the parametrix relation (11.19) and the expansion (11.20) are well-defined and the K_t form a convolution semigroup.

11.3.2 Kernels and Semigroups

The initial step in constructing the semigroup kernel corresponding to H is to prove that the expansion (11.20) determines a unique bounded integrable function K_t for all $t > 0$. Subsequently we derive more-detailed boundedness and smoothness property and establish that the family of functions form a convolution semigroup.

It follows from the definition of $\widehat{K}_t^{(0)}$ and $\widehat{M}_t$ that there are $a, b > 0$ and $\omega \geq 0$ such that

$$|\widehat{K}_t^{(0)}(x)| \leq a\, t^{-d/m} e^{\omega t} e^{-b(|x|^m/t)^{1/(m-1)}} \tag{11.21}$$

and

$$|\widehat{M}_t(x)| \leq a\, t^{-(d+m-1)/m} e^{\omega t} e^{-b(|x|^m/t)^{1/(m-1)}} \tag{11.22}$$

for all $x \in \mathbf{R}^d$ and all $t > 0$. These bounds follow from Proposition 11.2.1 and Corollary 11.2.2 if $x \in B$ but then are obviously true for all $x \in \mathbf{R}^d$ since both functions have support in B. Next we convert these bounds into bounds on $K_t^{(0)}$ and M_t on G.

One can associate with G and the left-invariant Haar measure dg a modulus $g \in G \mapsto |g|$ as the shortest length measured by dg of the absolutely continuous paths from g to e. Then the modulus is locally equivalent to the modulus on $\mathfrak{g}$ by the exponential map. In particular there is a $c > 0$ such that

$$c^{-1}|a| \leq |\exp(a)| \leq c|a|$$

for all $a \in B_0$. Therefore, by the choice of modulus on $\mathfrak{g}$, one has

$$c^{-1}|x| \leq |\exp(a_x)| \leq c|x|$$

for all $x \in B$. In particular $|x| \geq c^{-1}|\exp(a_x)|$ for all $x \in B$. Since $\widehat{K}_t^{(0)}$ and $\widehat{M}_t$ have support in B the bounds (11.21) and (11.22) immediately translate into bounds on $K_t^{(0)}$ and M_t. Explicitly there are $a, b > 0$ and $\omega \geq 0$ such that

$$|K_t^{(0)}(g)| \leq a\, t^{-d/m} e^{\omega t} e^{-b c^{-1}(|g|^m/t)^{1/(m-1)}} \tag{11.23}$$

and

$$|M_t(g)| \leq a\, t^{-(d+m-1)/m} e^{\omega t} e^{-b c^{-1}(|g|^m/t)^{1/(m-1)}} \tag{11.24}$$

for all $g \in G$ and all $t > 0$. These estimates immediately yield the basic existence result for the K_t.

Proposition 11.3.1 *Define $K_t^{(n)}$ recursively by $K_t^{(n)} = -(K^{(n-1)} * M)_t$ where $K^{(0)}$ and M are defined as above. It follows that the series $\sum_{n \geq 0} K_t^{(n)}$ is $L_p(G; dg)$-convergent to a limit $K_t \in L_p(G; dg)$ for all $p \in [1, \infty]$ and $t > 0$. The function $t > 0 \mapsto K_t$ is continuous and satisfies the heat equation (11.12).*

Proof It suffices to prove that the series is L_1-, and L_∞-, convergent because the L_p-convergence is then an immediate consequence. The L_1-convergence is particularly easy because the estimates (11.23) and (11.24) imply that

$$\|K_t^{(0)}\|_1 \leq a\, e^{\omega t} \qquad \text{and} \qquad \|M_t\|_1 \leq a\, t^{-(m-1)/m} e^{\omega t} \qquad (11.25)$$

for suitable $a > 0$ and $\omega \geq 0$. For example, one can establish analogous estimates for the $L_1(\mathbf{R}^d; \sigma dx)$ norms of $\widehat{K}^{(0)}$ and $\widehat{M}$ from (11.21) and (11.22) and these translate into the $L_1(G; dg)$ bounds by (11.13). But the recursion inequalities

$$\|K_t^{(n)}\|_1 \leq \int_0^t ds\, \|K_{t-s}^{(n-1)}\|_1 \|M_s\|_1 \qquad (11.26)$$

follow from the definition of the $K_t^{(n)}$. Then, arguing by induction, one establishes bounds

$$\|K_t^{(n)}\|_1 \leq a\, b^n\, (t^n/n!)^{1/m} e^{\omega t}$$

for all $n \geq 0$ and all $t > 0$. In particular the series is L_1-convergent for all $t > 0$.

The L_∞-convergence is slightly more complicated. It relies on the L_1-bounds (11.25) together with the analogous L_∞-bounds

$$\|K_t^{(0)}\|_\infty \leq a\, t^{-d/m} e^{\omega t} \qquad \text{and} \qquad \|M_t\|_\infty \leq a\, t^{-(m-1)/m}\, t^{-d/m} e^{\omega t}. \quad (11.27)$$

Now one has the recursion relations

$$\|K_{t-s}^{(n-1)} * M_s\|_\infty \leq \int_G dh |K_{t-s}^{(n-1)}(h)|\, |M_s(h^{-1}g)| \leq \|K_{t-s}^{(n-1)}\|_1 \|M_s\|_\infty$$

but these do not immediately give useful bounds on $\|(K^{(n-1)} \hat{*} M)_t\|_\infty$ because the bounds (11.27) on $s > 0 \mapsto \|M_s\|_\infty$ are not integrable at $s = 0$. One does, however, have the alternative bounds

$$\|K_{t-s}^{(n-1)} * M_s\|_\infty \leq \int_0^t ds\, \|K_{t-s}^{(n-1)}\|_\infty \sup_{g \in G} \int_G dh\, |M_s(h^{-1}g)|.$$

But now the problem is that the left-invariant measure dh is not necessarily right-invariant. Nevertheless,

$$\int_G dh\, |M_s(hg)| = \Delta(g)^{-1} \int_G dh\, |M_s(h)|,$$

where Δ is the modular function. Since M_s has support in a compact s-independent set and Δ is locally bounded one then concludes that there is a $\gamma \geq 1$ such that

$$\|K_t^{(n)}\|_\infty = \|(K^{(n-1)}\hat{*}M)_t\|_\infty \leq \gamma \int_0^t ds \, \|K_{t-s}^{(n-1)}\|_\infty \|M_s\|_1.$$

Then combination of these observations readily gives the recursive inequalities

$$\|K_t^{(n)}\|_\infty \leq \gamma \int_0^t ds \left(\|K_{t-s}^{(n-1)}\|_\infty \|M_s\|_1 \right) \wedge \left(\|K_{t-s}^{(n-1)}\|_1 \|M_s\|_\infty \right).$$

$$(11.28)$$

These inequalities lead to finite bounds on each $\|K_t^{(n)}\|_\infty$ since $s \mapsto \|K_{t-s}^{(0)}\|_\infty$ is integrable at $s = 0$ and $s \mapsto \|K_{t-s}^{(0)}\|_1$ is integrable at $s = t$. Another induction argument indeed establishes bounds

$$\|K_t^{(n)}\|_\infty \leq a \, b^n \, (t^n/n!)^{1/m} \, t^{-d/m} e^{\omega t}$$

for suitable a, b, ω, uniformly for all $t > 0$ and $n \geq 0$. Hence one obtains uniform convergence of the series for K_t.

Secondly, similar estimates allow one to verify that $t \mapsto K_t$ is continuous and satisfies the heat equation (11.12). For example, the heat equation is established from a term by term calculation with the expansion (11.20). Explicitly, one has

$$(\partial_t + H)(K_t^{(0)} * \varphi) = \varphi + M_t * \varphi$$

$$-(\partial_t + H)((K^{(0)} \hat{*} M)_t * \varphi) = -M_t * \varphi + (M \hat{*} M)_t * \varphi, \quad \text{etc.,}$$

by use of the heat equation (11.17). Addition of the terms to nth order gives cancellations leaving a single term composed of convolutions of $(n+1)$-factors M_t with φ. This remainder converges to zero as $n \to \infty$ by estimates of the foregoing type. $\qquad\square$

The proof of Proposition 11.3.1 immediately leads to bounds

$$\|K_t\|_1 \leq a \, e^{\omega t} \quad \text{and} \quad \|K_t\|_\infty \leq a \, t^{-d/m} \, e^{\omega t}$$

for all $t > 0$. But a slight elaboration of the proof yields much stronger results.

Corollary 11.3.2 *Let U_ρ denote the operator of multiplication by the function $e^{\rho|g|}$ where $\rho \geq 0$. Then there are $a > 0$ and $\omega \geq 0$ such that*

$$\|U_\rho K_t\|_1 \leq a \, e^{\omega(1+\rho^m)t} \quad \text{and} \quad \|U_\rho K_t\|_\infty \leq a \, t^{-d/m} \, e^{\omega(1+\rho^m)t}$$

$$(11.29)$$

for all $\rho, t > 0$. Hence there are $a, b > 0$ and $\omega \geq 0$ such that

$$|K_t(g)| \leq a\, t^{-d/m}\, e^{\omega t}\, e^{-b(|g|^m/t)^{1/(m-1)}} \qquad (11.30)$$

for all $g \in G$ and $t > 0$.

Proof The proof is a simple repetition of the preceding arguments applied to the weighted functions $U_\rho K_t^{(n)}$. Now

$$|U_\rho K_t^{(n)}| \leq \left(|U_\rho K^{(n-1)}| \,\hat{*}\, |U_\rho M| \right)_t$$

as a consequence of the recursive definition of $K^{(n)}$ and the triangle inequality for the modulus $|g|$. Therefore the estimates (11.26) and (11.28) are valid with $K^{(n)}$ and M replaced by $U_\rho K^{(n)}$ and $U_\rho M$. But to make recursive estimates one needs to replace the bounds (11.25) and (11.27) by bounds on $U_\rho K^{(0)}$ and $U_\rho M$. It follows, however, from Proposition 11.2.3, and the discussion preceding it, that the introduction of the weight U_ρ merely introduces an additional factor $e^{\omega \rho^m t}$ to the $K^{(0)}$ and M bounds, e.g. one has $\|U_\rho K_t^{(0)}\|_1 \leq a\, e^{\omega(1+\rho^m)t}$. Therefore the induction argument now gives bounds (11.29). But the second of these bounds also gives

$$|K_t(g)| \leq a\, t^{-d/m}\, e^{\omega t} \inf_{\rho \geq 0} e^{-\rho|g|}\, e^{\omega \rho^m t}$$

which immediately yields (11.30).   □

Slight elaborations of these arguments establish differentiability properties of the K_t.

Corollary 11.3.3 *The functions K_t are in $L_{p;m}(G;dg)$ for all $p \in [1,\infty]$. Moreover, there are $a > 0$ and $\omega \geq 0$ such that*

$$\|U_\rho A^\alpha K_t\|_1 \leq a\, t^{-|\alpha|/m}\, e^{\omega(1+\rho^m)t} \qquad and$$

$$\|U_\rho A^\alpha K_t\|_\infty \leq a\, t^{-(d+|\alpha|)/m}\, e^{\omega(1+\rho^m)t}$$

for all $t > 0$, $\rho \geq 0$ and α with $|\alpha| \leq m$.

The L_1-estimates correspond to the bounds of Proposition 11.2.3 in the Euclidean case and the L_∞-estimates lead to Gaussian-type bounds

$$|(A^\alpha K_t)(g)| \leq a\, t^{-(d+|\alpha|)/m}\, e^{\omega t}\, e^{-b(|g|^m/t)^{1/(m-1)}} \qquad (11.31)$$

analogous to those of Proposition 11.2.1 by the remark at the end of the previous proof.

Proof of Corollary 11.3.3   First note that it suffices to consider the case $\rho = 0$ since the factor U_ρ can be added by the argument used to prove Corollary

11.3.2. Secondly, observe that $K_t^{(0)} \in D(A^\alpha)$ for all α. Consequently $K_t^{(n)} \in D(A^\alpha)$ for all n and α. Moreover,

$$A^\alpha K_t^{(n)} = -((A^\alpha K^{(n-1)}) \,\hat{*}\, M)_t.$$

Therefore the recursion bounds (11.26) and (11.28) are valid with $K^{(n)}$ replaced by $A^\alpha K^{(n)}$. Thirdly, the $\mathbf{R}^d$-bounds of Proposition 11.2.1 give estimates

$$\|A^\alpha K_t^{(0)}\|_1 \le a_\alpha \, t^{-|\alpha|/m} e^{\omega t} \quad \text{and} \quad \|A^\alpha K_t^{(0)}\|_\infty \le a_\alpha \, t^{-(d+|\alpha|)/m} e^{\omega t}$$

for suitable $a_\alpha > 0$ and $\omega \ge 0$. The derivatives A^α introduce the additional factors $t^{-|\alpha|/m}$. But $s \mapsto \|K_{t-s}^{(0)}\|_1$ is integrable at $s = t$ if and only if $|\alpha| \le m - 1$. If the latter condition is satisfied then the sum of the sequence $A^\alpha K_t^{(n)}$ is L_1-convergent by the proof of Proposition 11.3.1 to a limit $K_{\alpha;t}$ satisfying bounds $\|K_{\alpha;t}\|_1 \le a \, t^{-|\alpha|/m} e^{\omega t}$. Moreover, the sum of the $A^\alpha K_t^{(n)}$ is uniformly convergent to $K_{\alpha;t}$ with bounds $\|K_{\alpha;t}\|_\infty \le a \, e^{\omega t} \, t^{-(d+|\alpha|)/m}$. Next one must prove that K_t is in $L_{\infty;m-1}(G\,;dg)$ and the sum of the series is indeed equal to $A^\alpha K_t$. But left translations are weak* continuous on $L_\infty(G\,;dg)$. Therefore $\varphi \in D(A_j)$ if and only if $\sup_{s\in(0,1]} s^{-1} \|(I - L(sa_j))\varphi\|_\infty < \infty$. Since one has $s^{-1}\|(I-L(sa_j))K_t^{(n)}\|_\infty \le \|A_j K_t^{(n)}\|_\infty$ it immediately follows by the convergence argument that $K_t \in D(A_j)$ and $K_{\{j\};t} = A_j K_t$. Higher-order derivatives are treated by iterating this argument.

Finally one must deal with the case $|\alpha| = m$. This is achieved by splitting the integral over s into two components, the integral over $[0, t/2]$, which causes no problem, and the integral over $\langle t/2, t]$. Then if $\alpha = \{k\}\cup\alpha'$ with $|\alpha'| = m-1$ one has

$$(A^\alpha K_{t-s}^{(n-1)} * M_s)(g) = -\int_G dh\, (A^{\alpha'} K_{t-s}^{(n-1)})(h)(A_k L(h)M_s)(g).$$

But

$$(A_k L(h)M_s)(g) = L(h)\frac{d}{du}(L(h^{-1})\exp(u\,a_k)L(h)M_s)(g)\Big|_{u=0}$$

$$= L(h)\frac{d}{du}(\exp(u\,x_k(h).a)M_s)(g)\Big|_{u=0},$$

where $|x_k|_\infty \le \sigma e^{\tau|h|}$ with $\sigma \ge 1$ and $\tau \ge 0$. Therefore

$$\|A_k L(h)M_s\|_\infty \le \sigma \, e^{\tau|h|} \sup_{1\le j\le d} \|A_j M_s\|_\infty$$

and

$$\int_{t/2}^t ds\, \|A^\alpha K_{t-s}^{(n-1)} * M_s)\|_\infty \le \sigma \int_{t/2}^t ds\, \|U_\tau A^{\alpha'} K_{t-s}^{(n-1)}\|_\infty \sup_{1\le j\le d} \|A_j M_s\|_\infty.$$

Thus the integral is finite. Using this decomposition gives an alternative recursion inequality, which allows one to argue as before.   □

The foregoing argument can be iterated to deduce that $K_t \in L_{p;\infty}(G;dg)$. It is a rather tedious method of deducing that the kernel is infinitely often differentiable and it does not give good control over the growth of the bounds as $|\alpha|$ increases. An alternative argument can be constructed using arguments of elliptic regularity.

The expansion used to construct the kernel K has a notable localization feature. The zero-order approximant $K^{(0)}$ and the remainder function M are supported by Ω. Then since $K^{(n)}$ involves a convolution of $K^{(0)}$ and n copies of M it must be supported by Ω^n. Thus the larger distance behaviour of the kernel is only affected by the higher-order terms.

Proposition 11.3.1 establishes that the K_t satisfy the heat equation (11.12) and the expectation is they are the kernel of a holomorphic semigroup $S_t = L(K_t)$ on the $L_p(G;dg)$-spaces. Our next aim is to explain this property on the Hilbert space $L_2(G;dg)$. First since $\|K_t\|_1 \leq a\,e^{\omega t}$ by the earlier estimates it follows that the S_t are bounded operators on each of the L_p-spaces with operator norms satisfying $\|S_t\|_{p\to p} \leq a\,e^{\omega t}$. Moreover, one has bounds

$$\|S_s - S_t\|_{p\to p} \leq a\,\|K_s - K_t\|_1$$

uniformly for s,t in bounded intervals of $\langle 0,\infty\rangle$. Therefore the continuity of the K_t implies that the S_t are uniformly continuous on bounded intervals. This is a characteristic of holomorphic semigroups. Moreover, it follows from the parametrix construction and the bounds on the $K_t^{(n)}$ that

$$\|S_t\varphi - L(K_t^{(0)})\varphi\|_p \leq a\,t\,\|\varphi\|_p$$

for all $\varphi \in L_p(G;dg)$. Since $L(K_t^{(0)})\varphi \to \varphi$ as $t \to 0$ it follows that S_t converges to the identity as $t \to 0$. Hence we set $S_0 = I$. Now the main problem is to prove that the S_t have the semigroup property. But

$$(S_s S_t - S_{s+t})\varphi = L(K_s * K_t - K_{s+t})\varphi,$$

so it is equivalent to prove that the K_t form a convolution semigroup. This can, however, be achieved by L_2-arguments.

Proposition 11.3.4 *The operators $S_t = L(K_t)$ form a holomorphic semigroup on $L_2(G;dg)$ whose generator is (the L_2-closure of) H.*

The semigroup property is established by first examining the real part of H which is both strongly elliptic and symmetric on $L_2(G;dg)$.

Lemma 11.3.5 *Each closed symmetric strongly elliptic operator H on $L_2(G;dg)$ is self-adjoint and lower semibounded.*

The semiboundedness is the important feature for the proof of the proposition. Its validity was conjectured at the end Section 6 of Ed Nelson's 1959 article [Nel59] on analytic vectors. Roe Goodman raised it again in Section 3 of his 1971 article [Goo71] on regularity properties of Lie group representations. Neither author appreciated that it was a straightforward corollary of Langlands' parametrix arguments. This was not observed until 1989. It was then pointed out in a short note I wrote with the late Ola Bratteli, Fred Goodman, and Palle Jørgensen [BGJR89]. The main aim of the latter note was to establish a Lie group version of the classic Gårding inequality. This will be discussed further in Section 11.4.

The proof of the semiboundedness is based on resolvent arguments.

Proof of Lemma 11.3.5   Since the kernel K_t corresponding to H satisfies $\|K_t\|_1 \leq a\, e^{\omega t}$ the Laplace transforms

$$L_\lambda = \int_0^\infty dt\, e^{-\lambda t}\, K_t$$

are well-defined for $\lambda > \omega$ and $\|L_\lambda\|_1 \leq a\,(\lambda - \omega)^{-1}$.

Moreover the operators R_λ defined by $R_\lambda \varphi = L_\lambda * \varphi$ are bounded on $L_2(G\,;dg)$ with $\|R_\lambda \varphi\|_2 \leq a\,(\lambda-\omega)^{-1}\|\varphi\|_2$. But K_t satisfies the heat equation (11.12) by Proposition 11.3.1. Therefore

$$(\lambda I + H)R_\lambda \varphi = \int_0^\infty dt\, e^{-\lambda t}((\lambda I + H)K_t) * \varphi$$

$$= \varphi + \int_0^\infty dt\, e^{-\lambda t}((\lambda - \partial_t)K_t) * \varphi = \varphi$$

for all $\varphi \in C_c^\infty(G)$ and then by continuity $(\lambda I + H)R_\lambda \varphi = \varphi$ for all $\varphi \in L_2(G\,;dg)$. Therefore the range of $(\lambda I + H)$ is equal to $L_2(G\,;dg)$. This range condition together with the symmetry of H implies that H is self-adjoint.

Next one has $(\psi, \varphi) = ((\lambda I + H)R_\lambda \psi, \varphi) = (\psi, R_\lambda^*(\lambda I + H)\varphi)$ for all $\varphi, \psi \in D(H)$ and $\lambda > \omega$. Therefore $\varphi = R_\lambda^*(\lambda I + H)\varphi$ and

$$\|\varphi\|_2 = a\,(\lambda - \omega)^{-1}\|(\lambda I + H)\varphi\|_2$$

for all $\varphi \in D(H)$. Then it follows by spectral theory that the self-adjoint operator H is lower semibounded.   □

Now we are prepared to prove the proposition.

Proof of Proposition 11.3.4   First let $H^\dagger$ be the formal adjoint of H, i.e. the strongly elliptic operator with coefficients $c^\dagger(\alpha) = (-1)^{|\alpha|}c(\alpha_*)$ where $\alpha_* = (i_n, \ldots, \alpha_1)$ when $\alpha = (i_1, \ldots, i_n)$. Then the real part $H_R = (H + H^\dagger)/2$ of H is a symmetric strongly elliptic operator. Therefore Lemma 11.3.5 holds with H replaced by H_R. In particular the closure of H_R is lower semibounded. Therefore there is a $\nu \geq 0$ such that

$$((H\varphi, \varphi) + (\varphi, H\varphi))/2 \geq -\nu \|\varphi\|_2^2$$

for all $\varphi \in D(H)$. Next observe that if $\varphi_t \in D(H)$ satisfies the Cauchy equation

$$\frac{d}{dt}\varphi_t + H\varphi_t = 0 \qquad (11.32)$$

for all $t > 0$ then

$$\frac{d}{dt}\|\varphi_t\|_2^2 = -(H\varphi_t, \varphi_t) - (\varphi_t, H\varphi_t) \leq 2\nu \|\varphi_t\|_2^2.$$

Therefore $t \mapsto e^{-\nu t}\|\varphi_t\|_2$ is a decreasing function. Now suppose $\varphi_t^{(1)}$ and $\varphi_t^{(2)}$ both satisfy (11.32) and $\varphi_t^{(1)} \to \varphi$, $\varphi_t^{(2)} \to \varphi$ as $t \to 0$. Then $\varphi_t^{(1)} - \varphi_t^{(2)}$ also satisfies the equation but $\varphi_t^{(1)} - \varphi_t^{(2)} \to 0$ as $t \to 0$. Hence, as a consequence of the foregoing decrease property, $\varphi_t^{(1)} = \varphi_t^{(2)}$, i.e. the solution of (11.32) is uniquely determined by the initial data $\varphi = \varphi_0$.

Now $S_t L_2(G\,;dg) \subseteq L_{2;m}(G\,;dg) \subseteq D(H)$ by Corollary 11.3.3. Therefore $\varphi_t = S_{t+s}\varphi$, with $s > 0$, satisfies (11.32) with initial data $\varphi_0 = S_s\varphi$ for all $\varphi \in L_2(G\,;dg)$. Moreover, $\varphi_t = S_t S_s \varphi$ satisfies the equation with the same initial data. Hence

$$(S_{t+s} - S_t S_s)\varphi = 0$$

for all $\varphi \in L_2(G\,;dg)$. This establishes that S is a semigroup. But the generator H_S of S is an extension of H on $L_{2;m}(G\,;dg)$. Then since $S_t L_{2;m}(G\,;dg) \subseteq L_{2;m}(G\,;dg)$ it follows that $L_{2;m}(G\,;dg)$ is a core of H_S. Therefore $H_S = H$.

Finally, the semigroup S is holomorphic if and only if $S_t L_2(G\,;dg) \subseteq D(H)$ and, in addition, one has bounds $\|H S_t\|_{2 \to 2} \leq a\, t^{-1}$ on the L_2-operator norm for all $t \in \langle 0, 1]$. But both these properties are a consquence of Corollary 11.3.3. In particular

$$H S_t\varphi = \int_G dg\,(HK_t) * \varphi$$

for all $\varphi \in L_2(G\,;dg)$ since $K_t \in L_{2;m}(G\,;dg)$. Then the L_1-estimate of the corollary gives

$$\|H S_t\varphi\|_2 \leq a \sup_{|\alpha| \leq m} \|A^\alpha S_t\varphi\|_2 \leq a'\, t^{-1}\, e^{\omega t} \|\varphi\|_2$$

for all $t > 0$. Hence S is holomorphic. $\square$

Finally, consider a general continuous representation U of G on a Banach space χ. One can again associate with a fixed basis $a_1, \ldots, a_d$ of the

Lie algebra $\mathfrak{g}$ the generators $A_{U;j}$ of the one-parameter groups $t \in \mathbf{R} \mapsto U(\exp(-ta_j))$ acting on χ and introduce the corresponding subspaces $\chi_n = \bigcap_{\{\alpha:|\alpha|=n\}} D((A_U)^\alpha)$. The seminorms N_k are again defined on the χ_k by

$$N_k(\xi) = \sup_{\{\alpha:|\alpha|=k\}} \|(A_U)^\alpha \xi\|$$

and the norms $\|\cdot\|_k$ are defined by $\|\xi\|_k = \sup_{\{0 \le l \le k\}} N_l(\xi)$. Now the strongly elliptic operator H is the closure of the operator defined on χ_m in terms of the c_α and the A_U. It again follows from the continuity of U that there exist $M \ge 1$ and $\rho \ge 0$ such that one has bounds

$$\|U(g)\| \le M e^{\rho|g|}$$

for all $g \in G$. But the kernel K_t associated with H acting on $L_1(G;dg)$ satisfies the weighted bounds of Corollary 11.3.2. Therefore the conclusions of Proposition 11.3.4 for left translations L of the group on $L_2(G;dg)$ transfer to the representation U on χ.

Theorem 11.3.6 *The operators $\{S_t^{(U)}\}_{t>0}$ defined by*

$$S_t^{(U)} = U(K_t) = \int_G dg \, K_t(g) \, U(g)$$

form a holomorphic semigroup $S^{(U)}$ on the Banach space χ such that $S_t^{(U)} \chi \subseteq \chi_\infty$ for all $t > 0$. The generator of the semigroup is the closure of H on χ_m.

Proof The $S^{(U)}$ are well-defined as bounded operators on χ by the bounds of Corollary 11.3.2. Then since S is a semigroup for the left regular representation L of G on $L_2(G;dg)$, by Proposition 11.3.4, the K_t must form a convolution semigroup, i.e. $K_{s+t} = K_s * K_t$ for all $s,t > 0$. Therefore $S^{(U)}$ is a continuous semigroup on χ. The remaining statements follow from the properties of the kernel K once one notes that

$$U(\exp(sa))S_t^{(U)}\xi = \int_G dg \, K_t(g) U(\exp(sa)g)\xi$$

$$= \int_G dg \, K_t(\exp(-sa)g))U(g)\xi$$

for $a \in \mathfrak{g}$, all small s and all $\xi \in \chi$. Thus $A_U S_t^{(U)}\xi = U(AK_t)\xi$ for all $\xi \in \chi$, where the A_U are generators in the representation U and the A are the left regular representatives. Then by iteration $(A_U)^\alpha S_t^{(U)} = U(A^\alpha K_t)\xi$ for all α. Note that K_t is m-times left differentiable, again by Corollary 11.3.3, but in fact it is infinitely often differentiable by the remark following the proof of the corollary. Hence $S_t^{(U)}\chi \subseteq \chi_m \subseteq D(H)$. Since χ_m is $S^{(U)}$-invariant it is a core of the semigroup generator. Thus the generator must be H. Finally the holomorphy of $S^{(U)}$ follows by the reasoning in the proof of Proposition 11.3.4. $\square$

One also has the following extension of Lemma 11.3.5.

Corollary 11.3.7 *Assume χ is a Hilbert space, U a unitary representation and the strongly elliptic operator H corresponding to U is symmetric. Then H is self-adjoint and lower semibounded with a bound which is independent of the choice of unitary representation.*

Proof Let K_t be the kernel corresponding to H and set $S_t^{(U)} = U(K_t)$. It follows by the proof of Lemma 11.3.5 that the generator of $S^{(U)}$ is equal to H and is self-adjoint. But $\|U(g)\| = 1$ for all $g \in G$ by unitarity where $\|\cdot\|$ the Hilbert space operator norm. Then $\|S_t^{(U)}\| \leq \|K_t\|_1 \leq a\,e^{\omega t}$ for all $t > 0$. Hence the generator of $S^{(U)}$ is self-adjoint and lower semibounded by spectral theory. The bound does not depend on U. It is equal to the bound on $L_2(G;dg)$. $\square$

Another consequence of Proposition 11.3.4 is that the parametrix function L_λ is the kernel of the resolvent of H on $L_2(G;dg)$. Explicitly

$$L_\lambda * \varphi = \int_0^\infty dt\, e^{-\lambda t} K_t * \varphi = \int_0^\infty dt\, S_t \varphi = (\lambda I + H)^{-1}\varphi$$

for all $\varphi \in L_2(G;dg)$ and $\lambda > \omega$. Then it follows by a similar calculation that $U(L_\lambda)$ is the kernel of the generator of the semigroup $S^{(U)}$ on χ. Thus the functions K_t and L_λ determine the action of the semigroup and resolvent for all continuous representations of the group.

Theorem 11.3.6 encompasses two of the principal results of Langlands' 1960 thesis, Theorems 8 and 9. Elaborations of these theorems are given by Theorem I.5.1 and Theorem III.2.1 in my 1991 book [Rob91]. The common features of these earlier descriptions and the current presentation are the exponential map and the parametrix technique but the proofs differ in some features. For the interested reader we make a few remarks on the relations between the three versions.

In our current notation, Theorem 8 of the thesis established that the closure of H in the continuous representation U generates a holomorphic semigroup $S^{(U)}$. Then Theorem 9 showed that the action of $S^{(U)}$ is given by a family of measures which form a convolution semigroup. Subsequently, after the statement of Theorem 10, it was shown that these measures are absolutely continuous with respect to Haar measure, i.e. the $S^{(U)}$ has a kernel $K^{(U)}$ in the current terminology. Theorem 10 derives the basic complex analytic properties of the semigroup. We have bypassed Theorem 7 of the thesis, which gives an identification of the adjoint of H in the representation U. In fact the identification is valid for the wider class of elliptic operators with coefficients satisfying (11.2). Strong ellipticity is not necessary. The proof, however, requires the parametrix expansion for the resolvent and a version of elliptic regularity for operators with continuous coefficients. This approach

also leads to the alternative proof that $K_t \in L_{p;\infty}(G;dg)$ alluded to after
the proof of Corollary 11.3.3. The original proof of the generator property in
Theorem 8 of the thesis utilized Theorem 7 and, of course, required strong
ellipticity. We have avoided these arguments for the sake of brevity. Bob cites
Theorem 12.8.1 of Hille and Phillips' book [HP57] in the derivation of the
generator property although this appears to require some poetic license. Further
details can be found in Sections I.4 and I.5 of [Rob91]. We note that the
identification given by the missing Theorem 7 is a direct corollary of Theorem
11.3.6 if H is strongly elliptic.

The main distinction between the developments of the thesis and the current
exposition is largely in emphasis. The primary focus of the thesis is the
semigroup $S^{(U)}$ and the kernel $K^{(U)}$ is a secondary artefact. In the preceding
analysis, however, the kernel plays the principal role and the semigroup is
constructed from the kernel. The current presentation is based on three papers
I coauthored [BGJR89] [Rob93] [ER96] some 30–35 years after Bob wrote
his thesis. Nevertheless, the additional arguments are all from the late 1950s.
For example, the uniqueness criteria for the Cauchy problem are discussed
at length in Chapter III of Hille and Phillips' 1957 book [HP57] and the
characterization of holomorphy of S by a norm estimate on HS_t was a 1958
result of Yosida [Yos58]. Finally, the characterization of cores of semigroup
generators by semigroup invariance is essentially contained in Nelson's 1959
paper [Nel59].

We will return to the discussion of other properties that can be deduced
by variations of these arguments in Section 11.4. But next we turn to the
explanation of the last result, Theorem 10, of Langlands' thesis.

11.3.3 Differential and Analytic Structure

Theorem 10 of the thesis established the result highlighted in the introduction,
the density of the analytic elements $\chi_a(U)$ of a general continuous representation U of a Lie group. Now

$$\chi_a(U) = \left\{ \xi \in \chi_\infty : \sum_{k\geq 1}(s^k/k!)N_k(\xi) < \infty \text{ for some } s > 0 \right\},$$

where the seminorms are defined in terms of the representatives A_U of the Lie
algebra. The strategy of the thesis was to prove that $\chi_a(U)$ contains the dense
subspace $\chi_a(H)$ of analytic elements of each strongly elliptic operator H, i.e.
the subspace

$$\chi_a(H) = \left\{ \xi \in \chi_\infty : \sum_{k\geq 1}(s^k/k!)\|H^k\xi\| < \infty \text{ for some } s > 0 \right\}.$$

The density of the latter subspace follows because H generates the holomorphic semigroup $S^{(U)}$, by Theorem 11.3.6. In particular it follows that

$$\chi_a(H) = \bigcap_{t>0} S_t^{(U)} \chi.$$

The density of $\chi_a(H)$ is then an immediate consequence since $S_t^{(U)} \to I$ as $t \to 0$. We give a different proof of this result based on arguments of a 1988 paper [BGJR88] whose principal focus was the integrability of representations of Lie algebras. This is the original problem which attracted my attention to the Lie group theory and led me to search out Bob's thesis. The proof uses real analytic arguments which replace the application of Eidelman's results on the analyticity of the solutions of parabolic equations cited in the thesis. This application is, by Bob's own admission, quoted in the Introduction, somewhat nebulous. The alternative proof, which we next outline, is given in greater detail in Chapter II of [Rob91]. It is a basically a straightforward exercise in functional analysis.

Theorem 11.3.8 *There exist $a, b > 0$ and $\omega \geq 0$ such that*

$$\| S_t^{(U)} \xi \|_n \leq a\, b^n\, n!\, t^{-n/m}\, e^{\omega t}\, \|\xi\| \tag{11.33}$$

for all $\xi \in \chi$, $n \in \{0, 1, 2 \ldots\}$ and $t > 0$. Therefore $S_t^{(U)} \chi \subseteq \chi_a(U)$ for all $t > 0$. Consequently $\chi_a(H) \subseteq \chi_a(U)$ and $\chi_a(U)$ is dense in χ.

Proof Throughout the proof we omit the index U. This should cause no confusion as all calculations are within the representation. We also use the convention that a, b, ω, etc., are n-independent constants whose values might vary line by line. Any n-dependence is explicitly noted by suffices.

First observe that $S_t^{(U)} \chi \subseteq \chi_\infty$ for all $t > 0$ by Theorem 11.3.6. Therefore it is sufficient to derive the estimates (11.33) for all $\xi \in \chi_\infty$. In particular, if $\lambda \in \langle 0, 1 \rangle$ then $\| S_t \xi \|_n = \| S_{(1-\lambda)t} (S_{\lambda t} \xi) \|_n$ with $S_{\lambda t} \xi \in \chi_\infty$ satisfying bounds $\| S_{\lambda t} \xi \| \leq M\, e^{\omega \lambda t} \|\xi\|$. Therefore the bounds (11.33) for general $\xi \in \chi$ follow from those for $\xi \in \chi_\infty$, albeit with increased values of a, b and ω.

Secondly, it suffices to establish (11.33) for all $t \in \langle 0, t_0]$ for some $t_0 > 0$. This follows since $\| S_t \xi \|_n \leq (a\, b^n\, n!\,) \| S_{t-t_0} \xi \|$ for $t > t_0$ as a consequence of the bounds for $t \leq t_0$. But $\| S_{t-t_0} \xi \| \leq M\, e^{\omega(t-t_0)} \|\xi\| \leq M\, t^{-n/m}\, e^{\omega t}\, \|\xi\|$ for $t > t_0$. Therefore one obtains the bounds (11.33) for all $t > 0$, again with increased values of a, b and ω. Hence the proof is now reduced to considering $\xi \in \chi_\infty$ and small $t > 0$. The starting point is the following observation for small n.

Lemma 11.3.9 *There is an $a > 0$ such that $\| S_t \xi \|_n \leq a\, t^{-n/m}\, \|\xi\|$ for all $\xi \in \chi$, $t \in \langle 0, 1]$ and $n = 0, 1, \ldots, m - 1$.*

Proof It follows from the L_1-estimates of Corollary 11.3.3 that

$$\|S_t\xi\|_n = \sup_{\{\alpha:|\alpha|\leq n\}} \|A^\alpha S_t\xi\| = \sup_{\{\alpha:|\alpha|\leq n\}} \|U(A^\alpha K_t)\xi\|$$

$$\leq M \sup_{\{\alpha:|\alpha|\leq n\}} \|U_\rho A^\alpha K_t\|_1 \|\xi\| \leq a\, t^{-n/m}\|\xi\|$$

by the (M,ρ)-continuity estimates for the representation. $\square$

If, however, $\xi \in \chi_\infty$ then $\|S_t\xi\|_n$ is not singular at $t = 0$.

Lemma 11.3.10 *There is a $b > 0$ such that $\|S_t\xi\|_n \leq a\,(\|H\xi\| + \|\xi\|)$ for all $\xi \in \chi_\infty$, $t \in \langle 0, 1]$ and $n = 0, 1, \ldots, m - 1$.*

Proof Since one has

$$\|A^\alpha(\lambda I + H)^{-1}\xi\| \leq \int_0^\infty ds\, e^{-\lambda s}\|A^\alpha S_s\xi\|$$

$$\leq \int_0^\infty ds\, e^{-\lambda s}\|U(A^\alpha K_s)\xi\|$$

$$\leq a \int_0^\infty ds\, e^{-(\lambda-\omega)s} s^{-n/m}\|\xi\|$$

the statement of the lemma follows by replacing ξ by $(\lambda I + H)S_t\xi$ and choosing λ sufficiently large. $\square$

The rest of the proof consists of "bootstrapping" the small n-estimates of Lemma 11.3.9 into universal estimates with the correct quantitative behaviour for small $t > 0$. This is achieved by recursive arguments.

Assume one has bounds

$$\|S_t\xi\|_n \leq c_n\, t^{-n/m}\|\xi\| \tag{11.34}$$

with $c_n > 0$ for all $\xi \in \chi_\infty$, $t \in \langle 0, 1]$ and all $n \leq k(m - 1)$ for some $k \geq 1$. It follows from the lemma this assumption is indeed valid for $k = 1$, with $c_1 = \cdots = c_{m-1} = a$. Next we argue that similar bounds are valid for all $n \leq (k + 1)(m - 1)$ and we also obtain estimates on the corresponding c_n. A crucial part of the argument is the observation that the commutators $(\mathrm{ad}\, A^\alpha)(A^\beta) = A^\alpha A^\beta - A^\beta A^\alpha$ have the property

$$\|(\mathrm{ad}\, A^\alpha)(A^\beta)\xi\| \leq a_{\alpha,\beta}\|\xi\|_{\alpha+\beta-1}$$

for all α, β and all $\xi \in \chi_\infty$ as a result of the structure relations of $\mathfrak{g}$. In particular, the commutator gives a unit reduction to the apparent operator order.

Let $A^\alpha = A^{\alpha_0} A^{\alpha_1}$ with $|\alpha| = n \geq m$ and $|\alpha_0| = m - 1$ so $|\alpha_1| = n - m + 1$. Then

$$A^\alpha S_t \xi = A^{\alpha_0} S_s (A^{\alpha_1} S_{t-s} \xi) + A^{\alpha_0} (\text{ad } A^{\alpha_1})(S_s) S_{t-s} \xi$$

$$= A^{\alpha_0} S_s (A^{\alpha_1} S_{t-s} \xi) - \int_0^s du \, A^{\alpha_0} S_u (\text{ad } A^{\alpha_1})(H) S_{t-u} \xi$$

for all $s \in \langle 0, t \rangle$ and $\xi \in \chi_\infty$. Since H is a polynomial in A^β with $|\beta| \leq m$ one then deduces from (11.34) that

$$\| S_t \xi \|_n \leq a \, c_{n-m+1} \, s^{-(m-1)/m} (t-s)^{-(n-m+1)} \| \xi \|$$

$$+ b \, n \int_0^s dr \, r^{-(m-1)/m} \| S_{t-r} \xi \|_n,$$

where $a, b > 0$ are independent of n. Now it remains to solve these inequalities.

First set $s = \mu t$ with $\mu \in \langle 0, 1 \rangle$ and $r = ut$. Then

$$\| S_t \xi \|_n \leq a \, c_{n-m+1} \, t^{-n/m} \mu^{-(m-1)/m} (1 - \mu)^{-(n-m+1)} \| \xi \|$$

$$+ b \, n \, t^{1/m} \int_0^\mu du \, u^{-(m-1)/m} \| S_{t(1-u)} \xi \|_n.$$

Second, choose $\mu = n^{-m}$. Then $\mu^{-(m-1)/m} = n^{m-1}$ and $(1 - \mu)^{-(n-m+1)} \leq b_m$ where $b_m = \sup_{n \geq m} (1 - n^{-m})^{-n/m} < \infty$. Therefore

$$\| S_t \xi \|_n \leq a \, n^{m-1} \, c_{n-m+1} \, t^{-n/m} \| \xi \|$$

$$+ b \, n \, t^{1/m} \int_0^{n^{-m}} du \, u^{-(m-1)/m} \| S_{t(1-u)} \xi \|_n.$$

But iteration of this inequality p-times gives

$$\| S_t \xi \|_n \leq a \, n^{m-1} \, c_{n-m+1} \, t^{-n/m} \| \xi \| \sum_{l=0}^{p-1} (c \, t^{1/m})^l + R_{p,n}(t)$$

with $c = bm$ independent of n and

$$R_{p,n}(t) = (b \, n \, t^{1/m})^p \int_0^{n^{-m}} du_1 \, u_1^{-(m-1)/m}$$

$$\cdots \int_0^{n^{-m}} du_p \, u_p^{-(m-1)/m} \| S_{t(1-u_1)\ldots(1-u_p)} \xi \|_n.$$

But if $t < t_0$ with $c \, t_0 < 1$ then in the limit $p \to \infty$ one has

$$\| S_t \xi \|_n \leq a \, n^{m-1} \, c_{n-m+1} \, t^{-n/m} \| \xi \| + \limsup_{p \to \infty} R_{p,n}(t)$$

for all $t < t_0$ and $\xi \in \chi_\infty$. Next we argue that the limit of the remainder term is zero.

We may assume $t \le t_0 < 1$ and $n \ge 1$. Therefore one immediately has bounds

$$R_{p,n}(t) \le (b\, t_0)^p \left(n \int_0^{n^{-m}} du\, u^{-(m-1)/m} \right)^p \sup_{t \in \langle 0,1]} \|S_t \xi\|_n$$

$$= (c\, t_0)^p \sup_{t \in \langle 0,1]} \|S_t \xi\|_n.$$

Now $n \le (k+1)(m-1) < (k+1)m$. Then, by a relatively straightforward extension of Lemma 11.3.10, one obtains bounds $\sup_{t \in \langle 0,1]} \|S_t \xi\|_n \le a_k (\|H^{k+1}\xi\| + \|\xi\|)$, e.g. if the principal coefficients of H are real then H^{k+1} is strongly elliptic and the lemma is valid with H replaced by H^{k+1} and m replaced by $(k+1)m$. Hence

$$R_{p,n}(t) \le a_k (c\, t_0)^p (\|H^{k+1}\xi\| + \|\xi\|)$$

for all $\xi \in \chi_\infty$. Thus $R_{p,n}(t) \to 0$ as $p \to \infty$ for all $t \in \langle 0, t_0 \rangle$ if $c\, t_0 < 1$. Combining these observations one concludes that

$$\|S_t \xi\|_n \le a\, n^{m-1}\, c_{n-m+1}\, t^{-n/m} \|\xi\|$$

for all $t < t_0$, with t_0 sufficient small, and all $\xi \in \chi_\infty$. Moreover, by induction, these estimates are valid for all $n \ge m - 1$.

Next if $k(m-1) \ge n > (k-1)(m-1)$ then by iteration

$$c_n \le a^{k-1} \left(\prod_{l=0}^{k-2} (n - l(m-1))^{m-1} \right) c_{n-(k-1)(m-1)} \le a^{k-1} n^n c_{m-1}$$

because the product has $k - 1$-factors each bounded by n^{m-1}. But $k - 1 < n/(m-1)$ and $n^n \le e^n n!$. Thus the bounds take the form $a\, b^n\, n!$ for all $n \ge 1$. Finally, since $N_n(\xi) \le \|\xi\|_n$ for all $\xi \in \chi_\infty$ one has

$$\sum_{k \ge 1} (s^k/k!)\, N_k(S_t \xi) \le \sum_{k \ge 1} (s^k/k!)\, \|S_t \xi\|_k \le a \sum_{k \ge 1} (bst^{-1/m})^k e^{\omega t} \|\xi\|$$

for all $\xi \in \chi$ and $bs < t^{1/m}$. Therefore $S_t \chi \subseteq \chi_a(U)$ for all $t > 0$. $\quad\square$

Theorem 11.3.8 achieves the aim set out in the introduction, it establishes the density of the analytic elements for all continuous representations of a general Lie group. This property was the final conclusion, Theorem 10, of Langlands' 1960 thesis. It is also the final conclusion of our explanation of the results of the thesis. But we are not finished. To conclude we give a summary of consequences of the thesis results which have been subsequently established.

Most of these results date from the Gaussian revolution in semigroup theory which started slowly in 1967 with Aronson's paper [Aro67] on bounds on solutions of parabolic equations and which peaked in the 1980s.

11.4 Consequences

The early results of Langlands on the differential and analytic structure of Lie groups have developed in two different, but related, frameworks. First, there has been considerable progress in the framework of strongly elliptic operators described above. Secondly, the theory has been generalized to a broader class of subelliptic operators. The latter operators are defined as polynomials in an algebraic basis of the Lie algebra $\mathfrak{g}$, i.e., a linearly independent subset of $a_1, \ldots a_{d_1} \in \mathfrak{g}$ whose Lie algebra spans $\mathfrak{g}$. We will briefly describe some of the key features of the developments in the strongly elliptic setting and then comment on the more complicated subelliptic situation. We have seen in the foregoing that despite appearances the strongly elliptic theory remains largely commutative. The subelliptic theory, in contrast, contains a genuine noncommutative element.

The first topic of our discussion is extensions of the classical Gårding inequality to unitary representations of Lie groups. These inequalities characterize the notion of strong ellipticity and provide a basis for the definition of subellipticity for operators of general order.

Proposition 11.4.1 *Let U be a unitary representation of G on a Hilbert space χ and H a strongly elliptic operator with ellipticity constant μ. Then for each $\lambda \in \langle 0, \mu \rangle$ there is a representation independent $\nu \geq 0$ such that*

$$\mathrm{Re}(\varphi, H\varphi) \geq \lambda N_{m/2}(\varphi)^2 - \nu \|\varphi\|^2 \qquad (11.35)$$

for all $\varphi \in \chi_m$.

Proof Since H is strongly elliptic with ellipticity constant μ it follows that the real part of $H - \lambda A^{\alpha*} A^\alpha$, with $|\alpha| = m/2$, is a symmetric strongly elliptic operator with ellipticity constant $\mu - \lambda$. Therefore it follows from Corollary 11.3.7 that $\mathrm{Re}\, H - \lambda A^{\alpha*} A^\alpha$ is essentially self-adjoint and lower semibounded by $-\nu_\alpha I$ where $\nu_\alpha = \inf_{t>0} \log \|K_t^{(\alpha)}\|_1$ and $K_t^{(\alpha)}$ is the kernel corresponding to the operator $\mathrm{Re}\, H - \lambda A^{\alpha*} A^\alpha$. Hence

$$\mathrm{Re}(\varphi, H\varphi) \geq \lambda \|A^\alpha \varphi\|^2 - \nu_\alpha \|\varphi\|^2$$

for all $\varphi \in \chi_m$. But ν_α is independent of the particular unitary representation. Therefore (11.35) follows by taking the supremum over the α. □

The inequality (11.35) is a Lie group version of the classic Gårding inequality for strongly elliptic divergence form operators with bounded continuous coefficients on $L_2(\mathbf{R}^d)$. But there are other possible formulations. For example,

$$\text{Re}(\varphi, H\varphi) \geq \lambda \, (\varphi, \Delta^{m/2}\varphi) - \nu \, \|\varphi\|^2 \qquad (11.36)$$

for all $\varphi \in \chi_m$ with Δ the Laplacian corresponding to the basis a_i in the representation U. The proof follows as before since $H - \lambda \, \Delta^{m/2}$ is strongly elliptic for all $\lambda \in \langle 0, \mu \rangle$.

Proposition 11.4.1 establishes that strong ellipticity of H implies the Gårding inequalities (11.35) for any unitary representation of G. Moreover, since the the strong ellipticity condition (11.1) is just a restriction on the coefficients c_α of the operator it also implies the Gårding inequalities for the unitary representation of $\mathbf{R}^d$ by left translations on $L_2(\mathbf{R}^d)$. Conversely assume that (11.35) is valid without the strong ellipticity restriction (11.1). Then choose $\varphi \in C_c^\infty(\Omega)$ with $\hat{\varphi}(x) = e^{i\eta \cdot x}\chi(x)$ where χ is a C^∞-function supported in a ball of radius r centred at the origin and we have again used the exponential map and the notation of Section 11.3.1. Evaluating (11.35) with φ for large η and small r then yields the strong ellipticity condition. These arguments are summarized by the following.

Corollary 11.4.2 *The strong ellipticity condition* (11.1) *is equivalent to the Gårding inequality* (11.35) *in a unitary representation of G, or equivalent to* (11.35) *for left translations on $L_2(\mathbf{R}^d)$.*

A similar conclusion follows with the Gårding inequality (11.35) replaced by the alternative formulation (11.36). In particular, both forms of the Gårding inequality are equivalent.

The second topic of discussion is a regularity property for all unitary representations of G that follows from the Gårding inequality and is of significance for a more-detailed understanding of the analytic structure of the group representations.

Proposition 11.4.3 *Adopt the assumptions of Proposition 11.4.1. Then there is a representation independent $a > 0$ such that*

$$N_m(\varphi) \leq a \, (\|H\varphi\| + \|\varphi\|) \qquad (11.37)$$

for all $\varphi \in \chi_m$.

Proof Let $H_1 = H^\dagger H$. Then H_1 is a strongly elliptic operator with ellipticity constant $\mu_1 \geq \mu^2$. Then for each $\lambda \in \langle 0, \mu \rangle$ there is a $\nu \geq 0$ such that

$$\|H\varphi\|^2 = \text{Re}(\varphi, H_1\varphi) \geq \lambda^2 \, N_m(\varphi)^2 - \nu^2 \, \|\varphi\|^2$$

for all $\varphi \in \chi_m$ by (11.35) applied to H_1. The value of ν is independent of the particular unitary representation. Thus

$$N_m(\varphi)^2 \leq \lambda^{-2}\|H\varphi\|^2 + \nu^2\|\varphi\|^2 \leq (\lambda^{-1}\|H\varphi\| + \nu\|\varphi\|)^2$$

for all $\varphi \in \chi_m$. Now set $a = \lambda^{-1} \vee \nu$.   □

The regularity property (11.37) was first established for Laplacians and unitary representations by Nelson [Nel59], Section 6, by an algebraic calculation. (A simplified version of Nelson's result is given by [Rob88a], Lemma 1.7.) It should be emphasized that the property is not valid for all representations. For example, if one considers the representation of $\mathbf{R}^d$ by left translations on $L_p(\mathbf{R}^d)$ and sets $H = \Delta$, the standard Laplacian operator, then Calderón [Cal61] has shown that (11.37) is valid for all $p \in \langle 1, \infty\rangle$. Nevertheless it fails if $p = 1$ or $p = \infty$. There are locally integrable functions ξ such that $\Delta\xi$ is also locally integrable but the mixed derivatives $\partial_i\partial_j\xi$ are not. This pathology has a long history going back at least to Petrini's 1908 paper [Pet08]. It was, however, still a topical problem in the 1960s, and the L_1 and L_∞ counterexamples can be found in [Orn62] and [LM64], respectively. The situation is even more complicated. The Euclidean group is also represented by left translations on the spaces $C(\mathbf{R}^d)$ and, more generally, $C^k(\mathbf{R}^d)$. But (11.37) fails on $C^k(\mathbf{R}^d)$ for all integers $k \geq 1$.

The third topic we address is a characterization of the analytic elements for general group representations following suggestions of Roe Goodman [Goo69a] [Goo69b]. This characterization is in terms of fractional powers of the strongly elliptic operators H and its proof depends on the regularity property although the conclusion is independent of the property. We next give a brief description of this result.

The earlier discussion of the analytic elements compared two series with general terms $N_k(\xi)/k!$ and $\|H^k\xi\|/k!$, respectively. Convergence of the first series characterized the analytic elements $\chi_a(U)$ of the representation U and convergence of the second characterized the analytic elements $\chi_a(H)$ of the strongly elliptic operator H. The arguments of Section 11.3 established that $\chi_a(H) \subseteq \chi_a(U)$ and consequently $\chi_a(U)$ is dense in the representation space U. But $N_k(\xi)$ involves k-derivatives whilst $\|H^k\xi\|$ involves km-derivatives. As Goodman pointed out, it is more appropriate to compare the series with terms $N_k(\xi)/k!$ and $\|H^k\xi\|/(km)!$. The latter series is, however, related to the series characterizing the analytic elements of the fractional power $H^{1/m}$ of H. The general theory of fractional powers of semigroup generators was developed in the late 1950s and a summary of the basic properties can be found in Chapter IX of Yosida's book [Yos80] on functional analysis or Chapter 1 of Triebel's book on interpolation theory [Tri78]. For current purposes it suffices to know that if H generates a uniformly bounded semigroup then the fractional powers H^γ with $\gamma \in \langle 0, 1\rangle$ are well-defined and generate uniformly bounded holomorphic semigroups. But by the arguments of Section 11.3 each strongly elliptic operator H corresponding to a group representation generates a continuous semigroup S satisfying operator bounds $\|S_t\| \le a\, e^{\nu t}$ for some $\nu \ge 0$ and all $t > 0$. Therefore the uniform boundedness property can be arranged by replacing H with $H + \nu I$. This replacement does not change the space of analytic elements of H. Hence in the following discussion we will assume that $H^{1/m}$ is well-defined and satisfies the standard properties of fractional powers, e.g. $(H^{1/m})^k = (H^k)^{1/m} = H^{k/m}$. Then the analytic elements $\chi_a(H^{1/m})$ of $H^{1/m}$ are defined as the $\xi \in \chi_\infty$ such that

$$\chi_a(H^{1/m}) = \{\xi \in \chi_\infty : \sum_{k \ge 1} \|H^{k/m}\xi\|/k! < \infty\}.$$

In fact one does not need to consider fractional powers to analyze this subspace.

Lemma 11.4.4 *The following conditions are equivalent:*

I $\xi \in \chi_a(H^{1/m})$,

II $\sum_{k \ge 1} \|H^k \xi\|/(km)! < \infty$.

Proof I$\Rightarrow$II   If the series defining $\chi_a(H^{1/m})$ is finite then Condition II is evident.

II$\Rightarrow$I   Since $H^{1/m}$ generates a continuous semigroup there is a $C > 0$ such that

$$\|H^{l/m}\xi\| \le C\,(\|H\xi\| + \|\xi\|)$$

for all $\xi \in D(H)$ and all $l \in \{1, \dots, m-1\}$. Therefore

$$\sum_{n \ge 1} \|H^{k/m}\xi\|/k! \le m\,C \sum_{k \ge 0} \Big(\|H^{k+1}\xi\| + \|H^k\xi\|\Big)\Big/(km)! < \infty$$

for all ξ satisfying Condition II. $\square$

The final conclusion of Goodman's observations on fractional powers is a complete characterization of the analytic elements for an arbitrary group representation.

Theorem 11.4.5 *If H is a strongly elliptic operator associated with the Banach space representation (χ, U) then $\chi_a(U) = \chi_a(H^{1/m})$.*

This result was established in several stages.

First Goodman [Goo69a] established the characterization for all unitary representations and H a Laplacian by a modification of Nelson's theory of operator dominance [Nel59].

Secondly, Goodman's paper also had a brief but intriguing appendix contributed by Nelson that gave an elegant argument indicating a similar characterization was valid for all Banach space representations of the group satisfying the regularity condition (11.37). The conclusion of Nelson's suggestion was stated for Laplacians in Corollary A.1 of [Goo69a] although Nelson remarked that his arguments applied to higher orders.

Thirdly, I extended the Goodman–Nelson result to general representations by a longish detour through the interpolation spaces between the C^k-subspaces χ_k of the representation space. My first foray in this direction [Rob88b] was again for Laplacians. I was aware of Langlands' thesis at that time but was not ambitious enough to extend the interpolation arguments to higher order operators. This final step was described in Chapter II of my book [Rob91]. The reason behind the use of the interpolation arguments was quite simple. The analytic properties were not affected by transferring to the interpolation spaces but the regularity properties were improved.

The conclusion of the first two stages are summarized by the following proposition.

Proposition 11.4.6 (Goodman–Nelson) *Let H be a strongly elliptic operator associated with the Banach space representation (χ, U) satisfying $N_m(\xi) \leq a \left(\|H\xi\| + \|\xi\| \right)$ for some $a > 0$ and all $\xi \in \chi_m$. It follows that $\chi_a(U) = \chi_a(H^{1/m})$.*

Note that the proposition applies to unitary representations because unitarity implies the regularity assumption by Proposition 11.4.3. Goodman's proof for unitary representations was based on Nelson's original theory of operator domination [Nel59] but the argument advanced by Nelson for representations satisfying the regularity assumption was an extension of this theory. The inclusion $\chi_a(U) \subseteq \chi_a(H^{1/m})$ is quite elementary and does not depend on domination theory. For example, if $\xi \in \chi_\infty$ then there is a $C > 0$ such that $\|H^k\xi\| \leq C^k \|\xi\|_{km}$ for all $k \geq 1$. Therefore if $\xi \in \chi_a(U)$ it follows from Lemma 11.4.4 that $\xi \in \chi_a(H^{1/m})$. The proof of the converse inclusion is, however, more delicate. Nelson's method was based on a recursive argument involving the structure relations of the Lie algebra somewhat similar to the reasoning used in Section 11.3.3. The argument depends critically on the regularity condition. This allows one to estimate products A^β with $|\beta| = m$ by a single H, e.g. if $A^\alpha = A^\beta A^\gamma$ with $|\beta| = m$ then $\|A^\alpha\xi\| \leq C \left(\|HA^\gamma\xi\| + \|A^\gamma\xi\| \right)$. Then one can commute the factor H to the right of the A^γ and the additional commutator term is a sum of products A^δ with $|\delta| \leq |\alpha| - 1$, i.e. it is a lower-order correction. This is the start of the recursive argument. In the simplest case, $G = \mathbf{R}^d$, all the A commute and the regularity condition $\|\xi\|_m \leq C \left(\|H\xi\| + \|\xi\| \right)$ iterates in this manner to give $\|\xi\|_{km} \leq (2C)^k (\|H^k\xi\| + \|\xi\|)$. Hence if $\xi \in \chi_a(H^{1/m})$ then $\sum_{k \geq 1} \|\xi\|_{km}/(km)! < \infty$ by Lemma 11.4.4 and this suffices to deduce that $\xi \in \chi_a(U)$. Details of the general case are more complicated since one has to control the lower-order terms that arise from the lack of commutation. Details are given in Chapter II of [Rob91]. We will not persevere with the argument but instead explain how to deduce Theorem 11.4.5 from Proposition 11.4.6.

First, it is not surprising that the regularity condition is not necessary. The Goodman–Nelson arguments essentially use this condition to make a term by term comparison of the exponential series characterizing $\chi_a(H^{1/m})$ with the series characterizing $\chi_a(U)$. But such a comparison is clearly stronger than necessary for the inclusion $\chi_a(H^{1/m}) \subseteq \chi_a(U)$. I was aware of this problem by the early 1970s but only realized how to solve it some 10–15 years later. In the meantime, my interests were directed to quite different topics. My idea in the mid 1980s was to exploit the theory of interpolation spaces and reduce the problem to a similar problem for an auxiliary representation on a Banach space intermediate to the C^k-subspaces χ_k. In fact it suffices to consider a space intermediate to χ and χ_1.

The second observation is that each of the subspaces χ_k is invariant under the representation U. Hence $U_k = U|_{\chi_k}$ is a representation of G on χ_k. Moreover the C^l-subspaces $\chi_{k;l}$ of U_k are equal to the C^{k+l}-subspaces of U, i.e. $\chi_{k;l} = \chi_{k+l}$. Next by a standard procedure of real interpolation (see, for example, [Rob91] Section II.4.1) one can introduce a family of Banach spaces χ_γ, with $\gamma \in \langle 0,1 \rangle$, such that each space is invariant under U and $\chi_1 \subseteq \chi_\gamma \subseteq \chi$. Let U_γ denote the corresponding representations. The embeddings are continuous and one has bounds $c \|\xi\| \le \|\xi\|_\gamma \le C \|\xi\|_1$ for some $c, C > 0$ and all $\xi \in \chi_1$. Consequently, the C^k-subspaces of the representation U_γ satisfy $c \|\xi\|_k \le \|\xi\|_{\gamma;k} \le C \|\xi\|_{k+1}$ for all $k \ge 1$ and all $\xi \in \chi_\infty$. Therefore $\chi_a(U) = \chi_a(U_\gamma)$. Although there is still a term by term comparison of the two relevant series there is a slippage of one term in the comparison which does not affect the conclusion.

Now consider the comparison of the powers of H. For simplicity we use H as a common notation for the operators associated with each of the representations U, U_γ and U_1. Then one has $c \|H^k\xi\| \le \|H^k\xi\|_\gamma \le C \|H^k\xi\|_1$ for all $\xi \in \chi_\infty$. But since $\|\xi\|_1 \le a (\|H\xi\| + \|\xi\|)$ it follows that $\|H^k\xi\|_1 \le a (\|H^{k+1}\xi\| + \|H^k\xi\|)$. Hence the series $\|H^k\xi\|/(km)!$ and $\|H^k\xi\|_\gamma/(km)!$ are simultaneously convergent. Therefore $\chi_a(H^{1/m}) = \chi_{\gamma;a}(H^{1/m})$. Combining these conclusions one obtains the reduction result.

Lemma 11.4.7 $\chi_a(U) = \chi_a(H^{1/m})$ *if and only if* $\chi_{\gamma;a}(U_\gamma) = \chi_{\gamma;a}(H^{1/m})$ *for some* $\gamma \in \langle 0,1 \rangle$.

It might appear that this manipulation with the interpolation spaces has achieved very little. The problem for the representation (χ, U) has been identified with the analogous problem for the representation (χ_γ, U_γ). But the redeeming feature, the magic of the interpolation argument, is that the latter representation satisfies the regularity condition necessary for the Goodman–Nelson result, Proposition 11.4.6. Explicitly, there is an $a_\gamma > 0$ such that

$$\|\xi\|_{\gamma;m} \leq a_\gamma \left(\|H\xi\|_\gamma + \|\xi\|_\gamma\right) \tag{11.38}$$

for all $\xi \in \chi_{\gamma;m}$. Therefore $\chi_{\gamma;a}(U_\gamma) = \chi_{\gamma;a}(H^{1/m})$ by Proposition 11.4.6 and consequently $\chi_a(U) = \chi_a(H^{1/m})$ by Lemma 11.4.7. Thus the statement of Theorem 11.4.5 is established. The only problem remaining is to explain why the regularity property (11.38) is valid for the intermediate representations (χ_γ, U_γ) even if it is not valid for the representations (χ, U) and (χ_1, U_1). This is a convoluted story.

Interpolation has a long history starting with the work of Riesz in 1926. But in the late 1950s there was an explosion of interest in the subject motivated by problems of partial differential operators, approximation theory and singular integration. Many of the new developments concerned the classical spaces of functions over $\mathbf{R}^d$ but there were also new ideas on abstract methods of interpolation. One of the main motivations for the construction of new function spaces was indeed the regularity condition (11.38) (see [Tri83] pages 38–40). This led to the construction of various families of spaces satisfying the regularity condition intermediate to the C^k-spaces $L_{p;k}(\mathbf{R}^d)$ associated with left translations on the L_p-spaces. So the proof of Theorem 11.4.5 for the $\mathbf{R}^d$-theory could have been completed by appealing to the results described, for example, in the books of Triebel [Tri78] [Tri83]. Unfortunately, there was no equivalent theory for representations of Lie groups, although Peetre gave some partial results in [Pee70]. Nevertheless, the methods required to describe the general situation were available. In particular, there was a detailed description of interpolation methods for semigroups acting on abstract Banach spaces in the book by Butzer and Berens [BB67] on approximation theory. This theory was largely based on ideas of Peetre on methods of real interpolation between general Banach spaces. We now sketch its application to the representation (χ, U) of the Lie group G.

First, if χ_1 is the C^1-subspace of the representation space χ then $\chi_{\gamma, p} = (\chi, \chi_1)_{\gamma, p}$ is defined as the space of $\xi \in \chi$ such that the seminorm $N_{\gamma, p}(\xi) = (\int_0^\infty dt\, t^{-1}\, (t^{-\gamma}\kappa_t(\xi))^p)^{1/p}$ is finite where $\kappa_t(\xi) = \inf_{\xi_1 \in \chi_1}(\|\xi - \xi_1\| + t\|\xi_1\|_1)$ and $p \in [1, \infty)$. The interpolation function κ_t gauges the relevant importance of the representation space χ and the C^1-subspace χ_1. Consequently γ gives a measure of the smoothness of ξ with the choice of p giving an extra gradation. If $\xi \in \chi_1$ then $\kappa_t(\xi)$ tends to zero as $t \to 0$. Secondly, if S is the continuous semigroup generated by the strongly elliptic operator H associated with (χ, U) then $\chi_{\gamma, p}^S$ is defined as the subspace of $\xi \in \chi$ for which the seminorm $N_{\gamma, p}^S(\xi) = (\int_0^\infty dt\, t^{-1}\, (t^{-\gamma}\|(I - S_t)\xi\|)^p)^{1/p}$ is finite. Since the semigroup S is holomorphic it also follows that $\chi_{\gamma, p}^S$ is the subspace of ξ for which the seminorm $(\int_0^\infty dt\, t^{-1}\, (t^{1-\gamma}\|HS_t\xi\|)^p)^{1/p}$ is finite. In these definitions the γ and p measure the smoothing properties of the semigroup

S for small t. The only apparent group connection between $\chi_{\gamma,p}$ and $\chi_{\gamma,p}^{S}$ is the first space involves the C^{1}-subspace χ_{1} of the representation space whilst the second space depends indirectly on the representation U through the strongly elliptic operator H. Nevertheless the two intermediate spaces both give a measure of smoothness and the striking conclusion is that they are equal, modulo a slight change of parameter. Specifically, $\chi_{\gamma,p} = \chi_{(\gamma/m),p}^{S}$ with equivalence of the natural norms. In fact there is even a third chacterization of these spaces directly involving the representation U. The space $\chi_{\gamma,p}$ consists of the $\xi \in \chi$ for which the seminorm $\int_{\mathcal{O}} dg\, |g|^{-d}(|g|^{-\gamma}\|(I - U(g)\xi\|)^{p}$ is finite where $\mathcal{O}$ is an open neighborhood of the identity in G. These results are given by Proposition II.4.3 and Theorem II.6.1 of [Rob91] although they were well known for $G = \mathbf{R}^{d}$ much earlier.

Once one has the identification $\chi_{\gamma,p} = \chi_{(\gamma/m),p}^{S}$ it is relatively straightforward to deduce the regularity property (11.38) for the intermediate spaces $\chi_{\gamma} = \chi_{\gamma,p}$. One key observation is that $\xi_{1} = S_{t}\xi \in \chi_{1}$ for each $\xi \in \chi$. Therefore the decomposition $\xi = (\xi - \xi_{1}) + \xi_{1}$ takes the form $\xi = (I - S_{t})\xi + S_{t}\xi$ and allows one to estimate $\kappa_{t}(\xi)$ in terms of $\|(I - S_{t})\xi\|$ and $\|S_{t}\xi\|_{1}$. The details are given in Theorem II.4.5 of [Rob91] but the ideas are just borrowed from the $\mathbf{R}^{d}$-theory developed in the 1960s that can be found in [BB67] or [Tri78] among many other places.

This completes our discussion of the characterization of analytic elements, Theorem 11.4.5, and our summary of the developments concerning the higher-order strongly elliptic operators introduced by Langlands. The most striking aspect of these results is their universal nature, e.g. $\chi_{a}(U) = \chi_{a}(H^{1/m})$ for all the suitably normalized mth-order operators H independent of the group structure. The conclusions are basically locally and are independent of the Lie algebraic details. One can obtain more detailed global results by specializing to second-order operators such as Laplacians or to restricted classes of groups. But the global analysis requires the introduction of quite different techniques, e.g. generalized Nash inequalities [Rob91] or Harnack inequalities [VSCC92]. The conclusions are also sensitive to the large scale geometry of the group. A detailed analysis of these properties for groups of polynomial growth can be found in [DER03].

We conclude with a brief discussion of a slightly different topic, general-order subelliptic operators.

The subelliptic theory is formulated in a similar manner to the strongly elliptic theory but the vector space basis $a_{1}, \ldots, a_{d}$ of the Lie algebra $\mathfrak{g}$ is replaced by an algebraic subbasis $a_{1}, \ldots, a_{d_{1}}$, i.e. a linearly independent set of elements which generate $\mathfrak{g}$ algebraically. Then one can define differential operators as polynomials of the representatives A_{k} of the a_{k} in the subbasis. The properties of second-order subelliptic operators, so-called "sums of squares," have been extensively studied since Hörmander's fundamental 1967 paper [Hör67]. If, however, one tries to develop the structure of higher order subelliptic operators following the outlines of the strongly elliptic theory one immediately encounters several new obstacles.

The first substantial obstacle is to find a replacement for the strong ellipticity condition (11.1). This is an $\mathbf{R}^d$-condition on the coefficients which does not reflect the restraints imposed by the subellipticity condition. Secondly, it is not at all clear that there is an alternative version of the parametrix arguments. This problem is related to the previous difficulty. The parametrix expansion for operators associated with the group G is the analogue of perturbation theory with the unperturbed system given by operators corresponding to $\mathbf{R}^d$. Fortunately both these obstacles can be avoided by a rather different common approach. The basic idea is to introduce a "simpler" group G_0, related to G but with a streamlined algebraic structure dictated by the subelliptic basis, as a replacement for $\mathbf{R}^d$.

First, however, define the subelliptic distance $|g|_1$ as the shortest length of the absolutely continuous paths from g to e following the directions of the algebraic subbasis. It is not evident that one can find connecting paths of this type for each $g \in G$ but this is a result of Caratheodory's early research into thermodynamics [Car09]. Moreover, it follows that a ball of radius δ measured with respect to this distance behaves as δ^D as $\delta \to 0$, where D is an integer, the local subelliptic dimension. It can be calculated as follows. Let $\mathfrak{g}_1$ denote the linear span of the algebraic basis $a_1, \ldots, a_{d_1}$ and $\mathfrak{g}_j$ the span of the algebraic basis together with the corresponding multiple commutators of order less than or equal to j. Then $\mathfrak{g}_1 \subset \mathfrak{g}_2 \subset \cdots \subset \mathfrak{g}_r = \mathfrak{g}$, where r is an integer, the rank of the algebraic basis. Next set $\mathfrak{g}'_1 = \mathfrak{g}_1$, and $\mathfrak{g}'_j$ the vector space complement of $\mathfrak{g}_{j-1}$ in $\mathfrak{g}_j$. This yields the direct sum decomposition $\mathfrak{g} = \mathfrak{g}'_1 \oplus \mathfrak{g}'_2 \oplus \cdots \oplus \mathfrak{g}'_r$ of the Lie algebra. Then D is given by $D = \sum_{j=1}^{r} j \, (\dim \mathfrak{g}'_j)$.

Secondly, the group G_0 is is defined by a contraction procedure. Define γ as the family of maps of $\mathfrak{g}$ into $\mathfrak{g}$ such that $\gamma_t(a) = t^k a$ for all $a = \mathfrak{g}'_k$ and $t > 0$. Then $\mathfrak{g}_0$ is defined as the vector space $\mathfrak{g}$ equipped with the Lie bracket

$$[a,b]_0 = \lim_{t \to 0} \gamma_t^{-1}([\gamma_t(a), \gamma_t(b)]).$$

It follows that $\mathfrak{g}_0$ is a nilpotent Lie algebra and $a_1, \ldots, a_{d_1}$ is an algebraic basis of $\mathfrak{g}_0$ of rank r. Moreover, the dilations γ_t are automorphisms of $\mathfrak{g}_0$. Then G_0 is defined as the connected, simply connected, Lie group with Lie algebra $\mathfrak{g}_0$. It is this group which acts as the local approximation to G in the subelliptic theory. The simplifying feature of G_0 is the existence of the dilations γ_t, which allow scaling arguments to extend local properties globally. Note that if $a_1, \ldots, a_{d_1}$ is a vector space basis of $\mathfrak{g}$ then $\mathfrak{g}_1 = \mathfrak{g}$ and $\gamma_t(a) = ta$ for all $a \in \mathfrak{g}$. Therefore $\gamma_t^{-1}([\gamma_t(a), \gamma_t(b)]) = t \, [a, b] \to 0$ as $t \to 0$ and $\mathfrak{g}_0$ is abelian. Thus $G_0 = \mathbf{R}^d$ in conformity with the earlier strongly elliptic case.

Thirdly, the notion of subellipticity of the operator $H = \sum_{\alpha:|\alpha|\leq m} c_\alpha A^\alpha$, where the multi-indices α only involve the indices $\{1,\ldots,d_1\}$ of the subbasis, is defined in a manner that simulates the definition of strong ellipticity. Since Corollary 11.4.2 establishes that strong ellipticity is equivalent to the Gärding inequality (11.35) for left translations on $L_2(\mathbf{R}^d)$ we define the operator H to be subelliptic on G if an analogous inequality is satisfied on $C_c(G_0)$. Explicitly, H is subelliptic on G if

$$\mathrm{Re}(\varphi, H\varphi) \geq \lambda\, N'_{m/2}(\varphi)^2 - \nu\, \|\varphi\|^2 \qquad (11.39)$$

for all $\varphi \in C_c^\infty(G_0)$ where N'_k is the seminorm given by restricting the supremum in the earlier definition of N_k to multi-indices in the subelliptic directions $a_1,\ldots,a_{d_1}$. Therefore H is subelliptic on G if and only if it is subelliptic on G_0. This definition removes the first obstacle cited above.

Next, as a preliminary to developing a parametrix formalism for the subelliptic operators in general representations of G, one must first analyze the operators in the left regular representation of G_0 on $L_2(G_0)$. In the strongly elliptic case with $G_0 = \mathbf{R}^d$ this was accomplished in Section 11.2 largely by techniques of Fourier analysis. In the subcoercive case the situation is more complicated. It is, however, facilitated by the nilpotent structure and the homogeneity properties with respect to dilations. In particular one establishes that each closed subelliptic operator H generates a continuous semigroup S on the L_p-spaces over G_0 with a kernel satisfying mth-order Gaussian bounds similar to those given by (11.31) but with $|g|$ replaced by $|g|_1$ and d replaced by D. Subsequently the properties of H for the nilpotent group G_0 are extended to the corresponding operator on the group G by the parametrix arguments. The reasoning is not substantially different. Although the conclusions for the semigroup structure in the subelliptic case are directly analogous to those of the strongly elliptic case their implications for the differential and analytic structure are considerably weaker. One striking difference is the failure of the Goodman characterization of the analytic functions in terms of fractional powers of the elliptic operators.

If the semigroup S generated by the mth-order subelliptic operator H is uniformly bounded one can define $H^{1/m}$ as before and the characterization of the subspace of analytic elements $\chi_a(H^{1/m})$ given by Lemma 11.4.4 is still valid. Hence $\chi'_a(U) \subseteq \chi_a(H^{1/m})$, where $\chi'_a(U)$ is the subspace of analytic elements of the representation U defined with the subelliptic seminorms N'_k. It is not, however, true that $\chi'_a(U) = \chi_a(H^{1/m})$ even for second-order operators and unitary representations. Example 8.7 of [ER94a] gives a counterexample based on the left regular representation of the group of rotations on $\mathbf{R}^3$ and the standard Laplacian. Nevertheless many regularity results have been established in the subelliptic case. Details can be found in [ER94b] [ER95] and [ERS97].

Finally we note that there is a third class of elliptic operators that can be analyzed by Langlands' methods, weighted strongly elliptic operators.

Subellipticity is based on the idea that there is a certain subset of preferred directions. In the weighted theory all directions are allowed but some have greater weight, or preference, than others. An extensive analysis of this class of operators, along the foregoing lines, can be found in [ER94c]. In particular there is an analogous theory of holomorphic semigroups generated by weighted operators. Again this leads to a good understanding of the corresponding differential structure and the structure of the weighted analytic elements. For example the characterization of the analytic elements in terms of the corresponding elements of fractional powers remains valid. Nevertheless there are significant differences introduced by the weighting. In conclusion the structural properties of the analytic elements in the broader context of subelliptic operators or weighted elliptic operators still pose intriguing open problems 60 years after Langlands' thesis work.

References

[Aro67] ARONSON, D. G., Bounds for the fundamental solution of a parabolic equation. *Bull. Amer. Math. Soc.* **73** (1967), 890–896.

[BB67] BUTZER, P. L., and BERENS, H., *Semi-groups of Operators and Approximation*. Die Grundlehren der mathematischen Wissenschaften 145. Springer-Verlag, Berlin, 1967.

[BGJR88] BRATTELI, O., GOODMAN, F. M., JØRGENSEN, P. E. T., and ROBINSON, D. W., The heat semigroup and integrability of Lie algebras. *J. Funct. Anal.* **79** (1988), 351–397.

[BGJR89] BRATTELI, O., GOODMAN, F. M., JØRGENSEN, P. E. T., and ROBINSON, D. W., Unitary representations of Lie groups and Gårding's inequality. *Proc. Amer. Math. Soc.* **107** (1989), 627–632.

[BR87] BRATTELI, O., and ROBINSON, D. W., *Operator Algebras and Quantum Statistical Mechanics*, vol. 1. Second edition. Springer-Verlag, New York, 1987.

[BR97] , *Operator Algebras and Quantum Statistical Mechanics*, vol. 2. Second edition. Springer-Verlag, New York, 1997.

[Cal61] CALDERÓN, A.-P., Lebesgue spaces of differentiable functions and distributions. In *Partial Differential Equations*, vol. 4 of Proc. Sympos. Pure Math., 33–49. Amer. Math., Soc., Providence, 1961.

[Car09] CARATHEODORY, C., Untersuchungen über die Grundlagen der Thermodynamic. *Math. Ann.* **67** (1909), 355–386.

[CD58] CARTIER, P., and DIXMIER, J., Vecteurs analytiques dans les représentations de groupes de Lie. *Amer. J. Math.* **80** (1958), 131–145.

[DER03] DUNGEY, N., ELST, A. F. M. TER, and ROBINSON, D. W., *Analysis on Lie Groups with Polynomial Growth*, vol. 214 of Progress in Mathematics. Birkhäuser Boston Inc., Boston, 2003.

174 Derek W. Robinson

[Eid56] EIDELMAN, S., On the fundamental solutions of parabolic systems. *Mat. Sbornik N.S.* **38(80)** (1956), 51–92.

[ER94a] ELST, A. F. M. TER, and ROBINSON, D. W., Subelliptic operators on Lie groups: regularity. *J. Austr. Math. Soc. (Series A)* **57** (1994), 179–229.

[ER94b] , Subcoercivity and subelliptic operators on Lie groups I: Free nilpotent groups. *Potential Anal.* **3** (1994), 283–337.

[ER94c] , Weighted strongly elliptic operators on Lie groups. *J. Funct. Anal.* **125** (1994), 548–603.

[ER95] , Subcoercivity and subelliptic operators on Lie groups II: The general case. *Potential Anal.* **4** (1995), 205–243.

[ER96] , Elliptic operators on Lie groups. *Acta Appl. Math.* **44** (1996), 133–150.

[ERS97] ELST, A. F. M. TER, ROBINSON, D. W., and SIKORA, A., Heat kernels and Riesz transforms on nilpotent Lie groups. *Coll. Math.* **74** (1997), 191–218.

[Fri64] FRIEDMAN, A., *Partial Differential Equations of Parabolic Type*. Prentice-Hall, Inc., Englewood Cliffs, N.J., 1964.

[Går47] GÅRDING, L., Note on continuous representations of Lie groups. *Proc. Nat. Acad. Sci. U.S.A.* **33** (1947), 331–332.

[Går60] , Vecteurs analytiques dans les représentations des groupes de Lie. *Bull. Soc. Math. France* **88** (1960), 73–93.

[Goo69a] GOODMAN, R., Analytic domination by fractional powers of a positive operator. *J. Funct. Anal.* **3** (1969), 246–264.

[Goo69b] , Analytic and entire vectors for representations of Lie groups. *Trans. Amer. Math. Soc.* **143** (1969), 55–76.

[Goo71] , Some regularity theorems for operators in an enveloping algebra. *J. Diff. Eq.* **10** (1971), 448–470.

[HC53] HARISH-CHANDRA, Representations of a semisimple Lie group on a Banach space. I. *Trans. Amer. Math. Soc.* **75** (1953), 185–243.

[Hil50] HILLE, E., Lie theory of semi-groups of linear transformations. *Bull. Amer. Math. Soc.* **56** (1950), 89–114.

[Hör67] HÖRMANDER, L., Hypoelliptic second order differential equations. *Acta Math.* **119** (1967), 147–171.

[HP57] HILLE, E., and PHILLIPS, R. S., *Functional Analysis and Semigroups*, vol. 31 of American Mathematical Society Colloquium Publications. American Mathematical Society, Providence, Rhode Island, 1957.

[Lan60a] LANGLANDS, R. P., On Lie semi-groups. *Canad. J. Math.* **12** (1960), 686–693.

[Lan60b] , Some holomorphic semi-groups. *Proc. Nat. Acad. Sci. U.S.A.* **46** (1960), 361–363.

[Lev07] LEVI, E. E., Sulle equazioni lineari totalmente ellittiche alle derivate parziali. *Rend. del Circ. Mat. Palermo* **24** (1907), 275–317.

[LM64] LEEUW, K. DE, and MIRKIL, H., A priori estimates for differential operators in L_∞ norm. *Illinois J. Math.* **8** (1964), 112–124.

[Nel59] NELSON, E., Analytic vectors. *Ann. Math.* **70** (1959), 572–615.

[NS59] NELSON, E., and STINESPRING, W. F., Representation of elliptic operators in an enveloping algebra. *Amer. J. Math.* **81** (1959), 547–560.

[Orn62] ORNSTEIN, D., A non-inequality for differential operators in the L_1 norm. *Arch. Rational Mech. Anal.* **11** (1962), 40–49.

[Pee70] PEETRE, J., Non-commutative interpolation. *Matematiche (Catania)* **25** (1970), 159–173.

[Pet08] PETRINI, H., Les dérivées premières et secondes du potentiel. *Acta Math.* **31** (1908), 127–332.

[Rob88a] ROBINSON, D. W., The differential and integral structure of representations of Lie groups. *J. Operator Theory* **19** (1988), 95–128.

[Rob88b] , Lie groups and Lipschitz spaces. *Duke Math. J.* **57** (1988), 357–395.

[Rob91] , *Elliptic Operators and Lie Groups*. Oxford Mathematical Monographs. Oxford University Press, Oxford, 1991.

[Rob93] , Strongly elliptic and subelliptic operators on Lie groups. In *Quantum and non-commutative analysis (Kyoto, 1992)*, vol. 16 of Math. Phys. Stud., 435–453. Kluwer Academic Publishers, Dordrecht, 1993.

[Tri78] TRIEBEL, H., *Interpolation Theory, Function Spaces, Differential Operators*. North-Holland, Amsterdam, 1978.

[Tri83] , *Theory of function spaces*. Birkhäuser Verlag, Basel, 1983.

[VSCC92] VAROPOULOS, N. T., SALOFF-COSTE, L., and COULHON, T., *Analysis and Geometry on Groups*. Cambridge Tracts in Mathematics 100. Cambridge University Press, Cambridge, 1992.

[Yos58] YOSIDA, K., On the differentiability of semigroups of linear operators. *Proc. Japan Acad.* **34** (1958), 337–340.

[Yos80] , *Functional Analysis*. Sixth edition, Grundlehren der mathematischen Wissenschaften 123. Springer-Verlag, New York, 1980.

12

The Langlands Spectral Decomposition

Jean-Pierre Labesse

Abstract. We review the standard definitions for basic objects in automorphic theory and then give an overview of Langlands' fundamental results established in [14]. We try to explain ideas behind the proof when reasonably simple, following mainly the surveys [16] and [1]. We emphasize the role of the truncation operator, which appears for the first time, but in some guise, in [16]. In the last sections we explain the formal aspects of the spectral decomposition for the space of K-invariant functions on $GL(2)$ and $GL(3)$ being otherwise rather sloppy on analytic questions.

We assume the reader is familiar with basic representation theory, linear algebraic groups and adèles. We must apologize for copying, most of the time, parts of [16] and [1]. We have given in greater detail only some elementary arguments that were maybe a bit too sketchy in these references. On the other hand, we have made no attempt to be more explicit than these surveys on the most difficult part of the proof.

12.1 Introduction

12.1.1 Langlands' Fundamental Paper

Consider a connected reductive group $\mathbf{G}$ defined over $\mathbb{Q}$ and Γ an arithmetic subgroup of the connected reductive Lie group $G = \mathbf{G}(\mathbb{R})^0$. The spectral decomposition of the right regular representation of G on the Hilbert space $\mathcal{L}$ of square integrable function on the quotient space $\Gamma \backslash G$:

$$\mathcal{L} = L^2(\Gamma \backslash G)$$

is one of the two most important results of the long paper written by Langlands in the early 1960s and completed in 1964 under the title:

On the functional equations satisfied by Eisenstein series.

The second main result is given in the title. Both rely on the analytical continuation of Eisenstein series $E(x; \Phi, \lambda)$ and of intertwining operators $M(s, \lambda)$ that show up in their functional equations. A mimeographed version of this long and difficult manuscript was circulated but remained unpublished until 1976 when it appeared, together with a Preface and four appendices added, as Springer Lecture Notes 544 [14].

This almost explicit description (see 12.1.5 below) of the spectral decomposition obtained by Langlands is a basic tool for the study of automorphic forms, in particular to establish the Trace Formula which, in turn, is one of the most powerful tools toward this study. In fact, in the early 1960s, Langlands already had it in mind to work on a generalization of Selberg's Trace Formula to all reductive groups. Moreover, instances of intertwining operators $M(s, \lambda)$ were soon recognized as expressible in terms of a new family of L-functions [17] and this, in turn, was at the origin of the Functoriality Conjectures [18].

12.1.2 Classical versus Adèlic Framework

The original monograph is written in the classical (nonadèlic) language and deals with discrete subgroups Γ of G satisfying technical assumptions that are automatically satisfied when Γ is an arithmetic subgroup and a fortiori when Γ is a congruence subgroup. General arithmetic subgroups are important in geometry while congruence subgroups and Hecke operators play a central role in number theory. The consideration of Hecke operators leads naturally to work with projective limits of coverings defined by congruence subgroups and this essentially amounts to work with $\mathbf{G}(\mathbb{Q}) \backslash \mathbf{G}(\mathbb{A})$ but thus we bypass more general arithmetic subgroups. It turns out to be simpler to deal with adèlic quotients. Moreover it is also necessary to use adèlic language for the formulation of Langlands functoriality principles. We observe that the appendices II, III and IV added in [14] by Langlands in the 1970s to the original monograph are written in adèlic language. From now on, we restrict ourselves to the adèlic setting, nevertheless we shall use a set of notation similar to what is used in the classical setting.

12.1.3 Remarks on the Strategy

An abstract notion tells us that the spectral decomposition of $\mathcal{L}$ can be described in terms of generalized vectors $E(\bullet; \pi)$ in the sense of Gelfand: any $\varphi \in \mathcal{L}$ can be written at least in a formal way

$$\varphi(x) = \int_{\pi \in \Pi} \hat{\varphi}(\pi) E(x; \pi) \, d\mu(\pi) \qquad \text{with} \qquad \hat{\varphi}(\pi) = \, <\varphi, E(\bullet; \pi)>_{\mathcal{L}},$$

where Π parameterizes a set of data attached to unitary representations of G. Langlands shows that these generalized vectors can be constructed via the meromorphic continuation of the so-called Eisenstein series. When $\mathbf{G} = GL(1)$ over $\mathbb{Q}$, Eisenstein series are elementary objects, namely they are characters of the form

$$E(x; \Phi, \lambda) = |x|^\lambda \Phi(x)$$

with $\lambda \in i\mathbb{R}$ and Φ a Dirichlet character. In such a case the spectral decomposition is nothing but an instance of Pontryagin theory (i.e. Fourier inversion for locally compact abelian groups). In general, Eisenstein series are of the form (see below for notation)

$$E(x; \Phi, \lambda) = \sum_{\gamma \in \Gamma_P \backslash \Gamma} e^{<\lambda+\rho_P, H_P(\gamma x)>} \Phi(\gamma x),$$

where λ is a parameter in the complexification of a finite-dimensional real vector space and Φ is an automorphic form on a parabolic subgroup. These series converge in some domain but may not converge for the purely imaginary values of λ we are interested in; then one needs to establish their meromorphic continuation.

For Eisenstein series on groups of $\mathbb{Q}$-rank-one, a proof of the meromorphic continuation has been obtained by Selberg (see [23] for a brief account). There are various independent approaches. New insights were necessary to deal with the general case and Langlands' proof is quite involved. He proves rather directly, in chapter 6 of [14], the meromorphic continuation and the functional equations satisfied by intertwining operators and Eisenstein series when Φ is cuspidal by a reduction to the rank-one case. But, in general, when Φ is only assumed to belong to the discrete spectrum of a parabolic subgroup, the meromorphic continuation is obtained at the same time as the full spectral decomposition in the very difficult chapter 7. In fact, this chapter is famous for being *almost impenetrable*, as said by Langlands himself in his Preface to the Springer Lecture Notes.

12.1.4 Further References

A brief survey of [14] is to be found in Langlands' article [16] for the AMS Conference in Boulder in 1965. Another survey, due to J. Arthur [1], was written for the AMS Conference in Corvallis in 1977. The easiest part of the proof, i.e. the results of the first six chapters of [14], may also be found in Harish-Chandra's Lecture Notes [9]. The "Paraphrase" by Mœglin and Waldspurger [20] gives a complete and detailed proof in the adèlic language for groups over arbitrary global fields (i.e. number fields or function fields), also valid for metaplectic groups.

12.1.5 Recent Progress

The spectral decomposition is explicitly given by Langlands in terms of two black boxes. The first one is the cuspidal spectrum of Levi subgroups. The only information given is that the cuspidal spectrum is discrete with finite multiplicity. Little progress has been made and the conjectures describing this spectrum in term of a dual object are still mainly out of reach. The second black box is the residual spectrum arising from poles of Eisenstein series, i.e. the discrete but noncuspidal spectrum. For $GL(n)$ with $n \geq 5$ the combinatorics of the residues are already so involved that the explicit description of the residue spectrum, conjectured by Jacquet, was only achieved by Mœglin and Waldspurger in [19]. For classical groups, recent progress by Mœglin, Arthur and others has been made, but for arbitrary reductive groups the goal is still out of reach. A result, due to Franke [7], shows that all automorphic forms are finite linear combinations of Eisenstein series, up to maybe taking residues and derivatives. A different proof for the meromorphic continuation of Eisenstein series was announced by Bernstein a long time ago and is the subject of a recent paper [6].

Acknowledgements

I am grateful to Dinakar Ramakrishnan and to Jean-Loup Waldspurger for useful suggestions and criticisms.

12.2 Basic Objects and Main Theorems

12.2.1 Notation

If **P** is a connected linear algebraic group over $\mathbb{Q}$ we write P for $\mathbf{P}(\mathbb{A})$ where $\mathbb{A}$ is the ring of adèles of $\mathbb{Q}$, P_∞ for the real Lie group $\mathbf{P}(\mathbb{R})$, P_f for the group $\mathbf{P}(\mathbb{A}_f)$ of points over the finite adèles $\mathbb{A}_f$, and Γ_P for the group of rational points $\mathbf{P}(\mathbb{Q})$. Let $\mathfrak{X}(P)$ be the group of rational characters of P and consider

$$\mathfrak{a}_P = \mathrm{Hom}(\mathfrak{X}(P), \mathbb{R}).$$

This is a real vector space whose dimension is denoted a_P. We denote by

$$H_P : P \to \mathfrak{a}_P$$

the map induced by

$$x \mapsto (\chi \in \mathfrak{X}(P) \mapsto \log(|\chi(x)|))$$

and by P^1 its kernel. Since we are dealing with a number field the map H_P is surjective (this would not be true for function fields) and there exists a section of the surjective map H_P with values in $\mathfrak{A}_P$, the connected component of the group of real points of the maximal $\mathbb{Q}$-split torus A_P in the center of a Levi subgroup M_P in P:

$$\mathfrak{A}_P = A_P(\mathbb{R})^0 \subset M_P.$$

In other words H_P induces an isomorphism

$$\mathfrak{A}_P \to \mathfrak{a}_P.$$

The evaluation of a linear form $\lambda \in \mathfrak{a}_P^* \otimes \mathbb{C}$ on a vector $H \in \mathfrak{a}_P$ will be denoted $\langle \lambda, H \rangle$.

12.2.2 Parabolic Subgroups and Iwasawa Decomposition

Now consider a connected reductive group $\mathbf{G}$ defined over $\mathbb{Q}$. According to our convention we put $G = \mathbf{G}(\mathbb{A})$ and $\Gamma_G = \mathbf{G}(\mathbb{Q})$ or even simply Γ if no confusion may arise. As in the classical setting, Γ is a discrete subgroup in G and the quotient $\Gamma \backslash G$ is of finite volume when $\mathbf{G}$ is semisimple.

We choose a minimal rational parabolic subgroup P_0 in G and a Levi decomposition

$$P_0 = M_0 N_0.$$

Consider a standard (rational) parabolic subgroup P with Levi decomposition $P = M_P N_P$ where $M_0 \subset M_P$. We denote by Δ_P the set of nonzero elements in the projection on $(\mathfrak{a}_P/\mathfrak{a}_G)^*$ of the set Δ_{P_0} of simple roots. We denote by ρ the half sum of positive roots, and by ρ_P its projection on $(\mathfrak{a}_P/\mathfrak{a}_G)^*$. We shall, from time to time, use the bilinear scalar product on $(\mathfrak{a}_P/\mathfrak{a}_G)^* \otimes \mathbb{C}$ deduced from the Killing form and we shall denote by $\langle \lambda, \mu \rangle$ its value on the couple (λ, μ). This notation implicitly implies that we identify $\mathfrak{a}_P/\mathfrak{a}_G$ with its dual via this bilinear form.

We choose a "good" maximal compact subgroup K of G, i.e. of the form $K = K_\infty \times \prod_p K_p$ where K_∞ is a maximal compact subgroup in G_∞, and K_p is a special in $G_p = \mathbf{G}(\mathbb{Q}_p)$ for all prime p and hyperspecial for almost all p (see [24] for these notions). For example, when $\mathbf{G} = GL(n)$ we take $K_p = GL(n, \mathbb{Z}_p)$. Then we have the Iwasawa decomposition

$$G = PK = N_P M_P K.$$

The homomorphism $H_P : P \to \mathfrak{a}_P$ is trivial on N_P and on $K \cap P$. It extends to a function from G onto $\mathfrak{a}_P$, again denoted H_P, satisfying

$$H_P(pk) = H_P(p) \quad \text{for} \quad p \in P \quad \text{and} \quad k \in K.$$

This allows us to view functions on $\mathfrak{a}_P$ as functions on G. The Weyl chamber in $\mathfrak{a}_{P_0}/\mathfrak{a}_G$ is the cone defined by the inequalities

$$\langle \alpha, H \rangle > 0 \quad \text{for} \quad \alpha \in \Delta_{P_0}.$$

Fundamental weights define another cone whose characteristic function is denoted $\hat{\tau}_{P_0}$. More generally, one defines in a similar way characteristic functions $\hat{\tau}_P$ of cones in $\mathfrak{a}_P/\mathfrak{a}_G$ for any parabolic subgroup using $\check{\Delta}_P$ the basis dual to the basis of coroots (see [1] or [13]). They will appear below in the definition of the truncation operator (Section 12.2.4).

12.2.3 The Right Regular Representation

In the following we shall consider only functions invariant under $\mathfrak{A}_G$. Endowed with a G-invariant measure, the quotient

$$X_G = \mathfrak{A}_G \Gamma \backslash G \simeq \Gamma \backslash G^1$$

is of finite volume [3]. Let $\mathcal{L}$ be the Hilbert space of square integrable functions on X_G:

$$\mathcal{L} = L^2(X_G).$$

We want to understand the spectral decomposition $\mathcal{L}$ under of the right regular representation R of G in $\mathcal{L}$, i.e. the map $G \times \mathcal{L} \to \mathcal{L}$ defined by

$$(x, \varphi) \mapsto R(x)\varphi \quad \text{where} \quad (R(x)\varphi)(y) = \varphi(yx)$$
$$\text{for} \quad x \in G, \quad y \in X_G \quad \text{and} \quad \varphi \in \mathcal{L}$$

where, by abuse of notation, we use the same letter for an element of G and its image in X_G. Consider now a smooth compactly supported function f on G. An operator $R(f)$ is defined by integration of f against the right regular representation:

$$(R(f)\varphi)(y) = \int_G f(x)\varphi(yx)\,dx.$$

Here dx is some Haar measure on G. The spectral decomposition of such operators, which is intimately related to the spectral decomposition of $\mathcal{L}$, is a tool and a goal of automorphic theory. The main concern for the spectral decomposition is to understand the discrete spectrum

$$\mathcal{L}_{\text{disc}} := L^2_{\text{disc}}(X_G),$$

which is the Hilbert direct sum of irreducible subspaces in $\mathcal{L}$. The trivial representation 1_G is an obvious but already quite interesting constituent. The way it appears in the spectral decomposition via residues of Eisenstein series leads to the proof of Weil's conjecture for Tamagawa numbers (see 12.3.5).

12.2.4 Constant Terms and Truncation Operators

Two operations play an essential role in the theory of Eisenstein series and of the trace formula as well. The simpler one is the computation of the constant term along a parabolic subgroup P: $\varphi \mapsto \varphi_P$. The second one is the truncation operator $\varphi \mapsto \Lambda^T \varphi$. The definition of the truncation operator in general is due to J. Arthur [2]. Its properties, recalled below, rely on reduction theory and combinatorial arguments already present in [16]. In fact, generalizing an operation used in Selberg's approach for $\mathbb{Q}$-rank-one groups in [23], Langlands constructs, in sections 8 and 9 of [16], variants of Eisenstein series attached to cusp forms, denoted $E''(\bullet, \Phi, \lambda)$, that are nothing but $\Lambda^T E(\bullet, \Phi, \lambda)$, and he computes the scalar product of two such functions. He thus gets a formula, recalled in Theorem 12.3.2.1, which is an explicit and rather simple expression in term of intertwining operators. This is a key tool for the analytic continuation of Eisenstein series: it plays a role similar to the Maass–Selberg relations used by Harish-Chandra's in chapter IV of [9]. Although not fully recognized, this implicitly uses the full power of truncation operators and (G, M)-families,[1] two techniques that play an essential role in establishing the Trace Formula, a fact of which Langlands was already aware in the early 1960s: he says it explicitly in section 7 p. 243 of [16].

These operations are elementary when dealing with modular forms. Consider a modular form f of even weight k for $GL(2, \mathbb{Z})$ whose q-expansion in the upper half plane $\mathfrak{H}$ is of the form

$$f(z) = \sum_0^\infty a_n q^n \qquad \text{with} \qquad q = e^{2\pi i z}.$$

Its constant term is a_0 and its truncated avatar is the function in the fundamental domain:

$$\mathfrak{D} = \{z \in \mathbb{C} \mid |\text{Re}(z)| \leq 1/2 \ \text{and} \ |z| \geq 1\}$$

equal to

$$f(z) \ \text{if} \ \text{Im}(z) \leq e^T \text{and} \ f(z) - a_0 \ \text{if} \ \text{Im}(z) > e^T \ \text{for} \ z \in \mathfrak{D}$$

when T is a positive real number. The modular form f corresponds to an automorphic form φ on

$$Y = GL(2, \mathbb{Q}) \backslash GL(2, \mathbb{A}) / GL(2, \overline{\mathbb{Z}}) \simeq GL(2, \mathbb{Z}) \backslash GL(2, \mathbb{R})$$

defined by

$$\varphi(g) = J(g, i)^k f(g.i) \ \text{for} \ g = \begin{pmatrix} a & b \\ c & d \end{pmatrix} \in GL(2, \mathbb{R})$$

with

$$g.i = \frac{ai + b}{ci + d} = z \in \mathbb{C} - \mathbb{R} \ \text{and} \ J(g, \tau)^k = (\det(g))^{k/2}(c\tau + d)^{-k}.$$

---

[1] This is a combinatorial technique implicit in Langlands' papers and used extensively by Arthur in his work on the Trace Formula. See section 1.10 in [13] for a synthetic account.

The constant term φ_P is given by $J(g,i)^k a_0$ and $\Lambda^T \varphi$ is the function on Y such that

$$\Lambda^T \varphi(g) = \varphi(g) \quad \text{if } \mathrm{Im}(g.i) \le e^T \text{ and}$$
$$\Lambda^T \varphi(g) = (f(g.i) - a_0) J(g,i)^k \quad \text{if } \mathrm{Im}(g.i) > e^T$$

when T is a positive real number and $g.i = z \in \mathfrak{D}$.

Now return to the general case. Taking the constant term along a parabolic subgroup P is the operation that transforms a locally integrable function φ on X_G into a function φ_P on

$$X_{P,G} = \mathfrak{A}_G \Gamma_P N_P \backslash G$$

defined by the integral

$$\varphi_P(x) = \int_{\Gamma_P \cap N_P \backslash N_P} \varphi(nx)\, dn$$

where dn gives measure 1 to $\Gamma_P \cap N_P \backslash N_P$. The following formal computation plays a fundamental role:

Lemma 12.2.4.1 *Let φ be a function on X_G and ϕ a function on $X_{P,G}$. Denote by E_ϕ the function on X_G defined by*

$$E_\phi(x) = \sum_{\Gamma_P \backslash \Gamma_G} \phi(\gamma x)$$

provided the sum is convergent. Then, if integrals are convergent,

$$\langle E_\phi, \varphi \rangle_{X_G} = \int_{X_G} E_\phi(x) \overline{\varphi(x)}\, dx = \int_{\mathfrak{A}_G \Gamma_P \backslash G} \phi(x) \overline{\varphi(x)}\, dx$$
$$= \int_{X_{P,G}} \phi(x) \overline{\varphi_P(x)}\, dx.$$

The truncated function $\Lambda^T \varphi$ is an alternate sum indexed by standard parabolic subgroups of series over $\Gamma_P \backslash \Gamma_G$ of terms that are products of characteristic functions $\hat{\tau}_P$ of cones in $\mathfrak{a}_P$ translated by some parameter $T \in \mathfrak{a}_{P_0}$ (or rather its projection on $\mathfrak{a}_P / \mathfrak{a}_G$) times constant terms φ_P:

$$\Lambda^T \varphi(x) = \sum_{P \supset P_0} (-1)^{a_P - a_G} \sum_{\gamma \in \Gamma_P \backslash \Gamma_G} \hat{\tau}_P(H_P(\gamma x) - T) \varphi_P(\gamma x).$$

The series are trivially convergent since reduction theory shows that, given a compact set Ω, then for $x \in \Omega$ there is only a finite number of $\gamma \in \Gamma_P \backslash \Gamma_G$ such that the expression $\hat{\tau}_P(H_P(\gamma x) - T)$ does not vanish, and moreover it will vanish identically if P is a proper subgroup and if T is far enough from the walls of the Weyl chamber.

Proposition 12.2.4.2 *(i) Given a compact set $\Omega \subset G$ then, provided T is far enough from the walls of the Weyl chamber, one has for any locally integrable function φ:*

$$\mathbf{\Lambda}^T \varphi(x) = \varphi(x) \qquad \text{for all} \qquad x \in \Omega.$$

(ii) $\mathbf{\Lambda}^T$ induces a self-adjoint idempotent operator on the Hilbert space $\mathcal{L}$ (i.e. an orthogonal projector)

$$\mathbf{\Lambda}^T = (\mathbf{\Lambda}^T)^* = (\mathbf{\Lambda}^T)^2.$$

(iii) $\mathbf{\Lambda}^T$ transforms functions of uniform moderate growth into rapidly decreasing functions. In particular, given a smooth compactly supported function f on G the compositum

$$\mathbf{\Lambda}^T \circ R(f)$$

is an operator of Hilbert–Schmidt type.

Proof Assertion (i) follows immediately from the above remarks. For a proof of assertion (ii) and (iii) we refer to refer to chapters 4 and 5 of [13].[2]   □

12.2.5 Automorphic Forms

Automorphic forms are functions φ on X_G that are smooth, K-finite, of uniform moderate growth and annihilated by an ideal $\mathfrak{J}$ of finite codimension in the center $\mathfrak{z}(\mathfrak{g}_\infty)$ of the enveloping algebra of the Lie algebra $\mathfrak{g}_\infty$ of G_∞:

$$\varphi \star \mathfrak{c} = 0 \qquad \text{for all} \qquad \mathfrak{c} \in \mathfrak{J}.$$

Here $\star$ denotes the convolution product and $\mathfrak{c}$ is viewed as a distribution supported at the origin in G. The K-finiteness may be expressed by asking that

$$\varphi \star e = \varphi,$$

where e is an idempotent defined by a measure supported on K associated to a finite dimensional representations of K. For a more detailed definition we refer to [4], [5], [9], or [20]. An important property is that the space of automorphic forms φ annihilated by any given ideal $\mathfrak{J}$ and such that $\varphi \star e = \varphi$ for any given e, is a finite-dimensional vector space. This implies (see [9] for example) that, given an automorphic form φ, there exists a smooth K-finite function f compactly supported on X_G such that

$$\varphi \star f = \varphi.$$

---

[2] The reader should be warned that the proofs given in [2] have to be slightly corrected when dealing with arbitrary reductive groups since some arguments may not apply for nonsplit groups.

Since automorphic forms are of uniform moderate growth, the truncation operator transforms any automorphic form φ into a rapidly decreasing function $\Lambda^T \varphi$ and, in particular, one has $\Lambda^T \varphi \in \mathcal{L}$.

12.2.6 Cusp Forms

An automorphic form φ is said to be cuspidal (or a cusp form) if, for any parabolic subgroup $P \neq G$, the constant terms φ_P vanish identically. It suffices to check this for standard parabolic subgroups. A key observation is that the truncation operator acts by the identity on a cusp form φ:

$$\Lambda^T \varphi = \varphi.$$

Then, assertion (iii) of Proposition 12.2.4.2 implies that cusp forms are rapidly decreasing on X_G and in particular are square-integrable. They generate in $\mathcal{L}$ the cuspidal spectrum

$$\mathcal{L}_{\text{cusp}} := L^2_{\text{cusp}}(X_G).$$

Consider a smooth compactly supported function f on G. The restriction of $R(f)$ to $\mathcal{L}_{\text{cusp}}$ is a trace class operator; this follows from three observations: (i) the factorization theorem of Dixmier–Malliavin tells us that f can be written as a finite sum of convolution products:

$$f = f_1 \star f_2 + \cdots + f_{2r-1} \star f_{2r}$$

for some integer r,
(ii) $\Lambda^T R(f_i) R(f_j) \Lambda^T$ being a product of two Hilbert–Schmidt operators is of trace class,
(iii) $\Lambda^T R(f) \Lambda^T \varphi = R(f)\varphi$ when φ is cuspidal.

This implies that the cuspidal spectrum $\mathcal{L}_{\text{cusp}}$ decomposes as a discrete sum with finite multiplicities of irreducible representation and hence is a subspace of the discrete spectrum $\mathcal{L}_{\text{disc}}$. This result was first obtained by Gelfand and Piatetskii-Shapiro. Unless X_G is compact, $\mathcal{L}_{\text{cusp}}$ is only a strict subspace of $\mathcal{L}_{\text{disc}}$: in fact, whenever X_G is not compact, the trivial representation is discrete but not cuspidal.

12.2.7 Automorphic Forms on Parabolic Subgroups

Consider a standard parabolic subgroup P and the quotient space

$$X_P = \mathfrak{A}_P \Gamma_P N_P \backslash G.$$

Let us denote by

$$\mathcal{L}_P = L^2(X_P)$$

the Hilbert space generated by functions that are square integrable on X_P for the right G-invariant measure:

$$\langle \Phi, \Phi \rangle_{\mathcal{L}_P} = \int_K \int_{\Gamma_P N_P \backslash P^1} \Phi(pk)\overline{\Phi(pk)}\, dp\, dk.$$

The Haar measure dk is normalized so that vol $(K) = 1$ and dp is a right P^1-invariant measure. The space X_P is of finite volume. Observe that $\mathcal{L}_G = \mathcal{L}$.

We say that a function Φ on X_P is automorphic on P if it is K-finite and if, for all $x \in G$, the functions $m \mapsto \Phi(mx)$ for $m \in M_P$ are automorphic forms on the Levi subgroup of P. We say that an automorphic form Φ is cuspidal on P if, moreover, the functions $m \mapsto \Phi(mx)$ are cusp forms.

12.2.8 Representations $I_{P,\lambda}$ and Intertwining Operators

Given Φ on X_P and $\lambda \in (\mathfrak{a}_P/\mathfrak{a}_G)^* \otimes \mathbb{C}$ we introduce

$$\Phi_\lambda(x) = e^{\langle \lambda + \rho_P, H_P(x) \rangle} \Phi(x)$$

and define a representation $I_{P,\lambda}$ by

$$(I_{P,\lambda}(y)\Phi)(x) = e^{\langle \lambda + \rho_P, H_P(xy) - H_P(x) \rangle} \Phi(xy)$$

or, equivalently,

$$(I_{P,\lambda}(y)\Phi)_\lambda(x) = \Phi_\lambda(xy).$$

Given two standard parabolic subgroups $\mathbf{P}$ and $\mathbf{Q}$ we denote by $W(\mathfrak{a}_P, \mathfrak{a}_Q)$ the set of elements of minimal length in the Weyl group W^G of $\mathbf{G}$ such that $s(\mathfrak{a}_P) = \mathfrak{a}_Q$. Recall that two standard parabolic subgroups P and Q are said to be associated if $W(\mathfrak{a}_P, \mathfrak{a}_Q)$ is nonempty. We denote by $W(\mathfrak{a}_P, Q)$ the set of elements of minimal length such that $s(\mathfrak{a}_P) \supset \mathfrak{a}_Q$. Let w denote an element in Γ representing $s \in W(\mathfrak{a}_P, \mathfrak{a}_Q)$ and consider the function Ψ on X_Q defined by the integral, when convergent:

$$\Psi(x) = e^{-\langle s\lambda + \rho_K, H_Q(x) \rangle} \int_{N_w} e^{\langle \lambda + \rho_P, H_P(w^{-1}nx) \rangle} \Phi(w^{-1}nx)\, dn,$$

where $N_w = wN_P w^{-1} \cap N_Q \backslash N_Q$ or, equivalently,

$$\Psi_{s\lambda}(x) = \int_{N_w} \Phi_\lambda(w^{-1}nx)\, dn.$$

One defines an operator $M(s, \lambda)$, which intertwines $I_{P,\lambda}$ and $I_{Q,s\lambda}$ by putting

$$\Psi = M(s, \lambda)\Phi .$$

These operators are products of local analogues:

$$M(s,\lambda) = M_\infty(s,\lambda) \times \prod_p M_p(s,\lambda),$$

where the product is over prime numbers.

If $\mathbf{G} = GL(2)$ and $\Phi(x) \equiv 1$, using this product decomposition it is easy to compute $M(s,\lambda)\Phi$ when s is the nontrivial element in the Weyl group: if $\lambda = \sigma\rho = \frac{1}{2}\sigma\alpha$, where α the positive root and $\sigma \in \mathbb{C}$ with $\mathrm{Re}(\sigma) > 1$, one finds

$$M(s,\lambda)\Phi = m(s,\lambda)\Phi \ \text{ with }\ m(s,\lambda) = m_\infty(s,\lambda)\prod_p m_p(s,\lambda) = \frac{L(\sigma)}{L(1+\sigma)},$$

where L is the complete Riemann Zeta function

$$L(\sigma) = L_\infty(\sigma)\prod_p L(\sigma) \ \text{ with }\ L_\infty(\sigma) = \pi^{-\sigma/2}\Gamma(\sigma/2) \ \text{ and}$$

$$L_p(\sigma) = \frac{1}{1-p^{-\sigma}}.$$

Similar explicit formulas apply for local intertwining operators acting on automorphic forms that are right-invariant under K_p when G_p is quasisplit and K_p is hyperspecial. For example, according to Proposition 3.2 in [12], if G_p is the quasisplit form of $SU(3)$, the special unitary group in three variables attached to the unramified quadratic extension of $\mathbb{Q}_p$ at a prime $p \neq 2$, the local factor $m_p(s,\lambda)$ is

$$m_p(s,\lambda) = \frac{(1 - p^{-2(\sigma+1)})(1 + p^{-2\sigma-1})}{(1 - p^{-2\sigma})(1 + p^{-2\sigma})}$$

when $\lambda = \sigma\rho$. For more examples, but in arbitrary rank, see Section 12.3.5 and 12.4.1 below and reference [12].

12.2.9 Eisenstein Series

Series of the form

$$E_k(z,s) = \sum_{(c,d)} \frac{\mathrm{Im}(z)^{s-k/2}}{(cz+d)^k|cz+d|^{2s-k}}$$

with $z \in \mathfrak{H}$ (the upper half plane) and $\mathrm{Re}(s)$ large enough, were studied by Hecke, Maass, and Selberg. When $2s = k$ they appear in the theory of elliptic modular forms while, when $k = 0$, these series (or rather their analytic continuation) were used by Selberg to describe the spectral decomposition of $L^2(SL(2,\mathbb{Z})\backslash\mathfrak{H})$. In what follows we shall deal with generalizations of these forms.

Consider $\lambda \in \mathfrak{a}_P^* \otimes \mathbb{C}$ and a function Φ on X_P. One defines, when convergent, an Eisenstein series by

$$E(x; \Phi, \lambda) = \sum_{\gamma \in \Gamma_P \backslash \Gamma} \Phi_\lambda(\gamma x) = \sum_{\gamma \in \Gamma_P \backslash \Gamma} e^{<\lambda + \rho_P, H_P(\gamma x)>} \Phi(\gamma x).$$

This is a function on X_G. The following lemma is an immediate consequence of Lemma 12.2.4.1.

Lemma 12.2.9.1 *Eisenstein series are orthogonal to cusp forms.*

In fact, Eisenstein series will allow us to construct the orthogonal supplement to the space of cusp forms in $\mathcal{L}$. Eisenstein series intertwine representations $I_{P,\lambda}$ and the right regular representations for functions on X_G: if f is a smooth compactly supported function on $\mathfrak{A}_G \backslash G$ one has

$$E(x; I_{P,\lambda}(f)\Phi, \lambda) = E(x; \Phi, \lambda) \star f.$$

When Φ is a cusp form on P, the series defining $E(x; \Phi, \lambda)$ converge for $\mathrm{Re}(\lambda) \in C_P + \rho_P$, where C_P denotes the projection of the Weyl chamber on $\mathfrak{a}_P^*$. This follows from the rapid decay of cusp forms on X_P and of the convergence of a similar series but where $\Phi = 1$ identically. For such λ and Φ the intertwining operator

$$\Phi \mapsto M(s, \lambda)\Phi$$

is also given by a convergent integral. The interplay between Eisenstein series and intertwining operators shows up when computing their constant terms.

Lemma 12.2.9.2 *Consider an Eisenstein series defined by Φ cuspidal on P. The constant term $E_Q(x; \Phi, \lambda)$ along any parabolic subgroup Q vanishes unless there is an s in the Weyl group such that $s(\mathfrak{a}_P) \supset \mathfrak{a}_Q$. In this case one has*

$$E_Q(x; \Phi, \lambda) = \sum_{s \in W(\mathfrak{a}_P, Q)} E^Q(x; M(s, \lambda)\Phi, s\lambda),$$

where $E^Q(x; M(s, \lambda)\Phi, s\lambda)$ is the Eisenstein series on Q defined by $M(s, \lambda)\Phi$. When Q is associate to P this can be written

$$E_Q(x; \Phi, \lambda) = \sum_{s \in W(\mathfrak{a}_P, \mathfrak{a}_Q)} e^{<s(\lambda) + \rho_Q, H_Q(x)>} M(s, \lambda)\Phi(x).$$

12.2.10 Meromorphic Continuation and Spectral Decomposition

Before giving some hints toward the proof we state the main theorems of chapter 7 (reformulated in adèlic language in appendix II) of [14].

Theorem 12.2.10.1 *Assume that Φ is automorphic on P and that $m \mapsto \Phi(mx)$ belongs to the discrete spectrum[3] of M_P for any x. Consider $\lambda \in \mathfrak{a}_P^* \otimes \mathbb{C}$.*
(i) The series defining the Eisenstein series $E(x; \Phi, \lambda)$ and the integral defining the intertwining operators $M(s, \lambda)$ are convergent when $\mathrm{Re}(\lambda)$ belongs to some translate of the Weyl chamber.
(ii) They have a meromorphic continuation on the whole space $\mathfrak{a}_P^* \otimes \mathbb{C}$.
(iii) The operators $M(s, \lambda)$ satisfy functional equations: for $s \in W(\mathfrak{a}_P, \mathfrak{a}_Q)$ and $t \in W(\mathfrak{a}_Q, \mathfrak{a}_R)$

$$M(st, \lambda) = M(s, t\lambda) M(t, \lambda).$$

(iv) Moreover

$$M(s, -\overline{\lambda})^* = M(s, \lambda)^{-1}.$$

In particular, $M(s, \lambda)$ is unitary when λ is purely imaginary.
(v) The automorphic forms

$$x \mapsto E(x; \Phi, \lambda)$$

obtained by meromorphic continuation of Eisenstein series satisfy the functional equations:

$$E(x; \Phi, \lambda) = E(x; M(s, \lambda)\Phi, s\lambda).$$

(vi) Eisenstein series are analytic when λ is purely imaginary.

Denote by $\mathcal{H}_P^0$ the space of square integrable automorphic forms on X_P and by $\mathcal{H}_P$ its closure. The proof of the analytic continuation of general Eisenstein series is entangled with the proof of the following theorem.

Theorem 12.2.10.2 *The discrete spectrum $\mathcal{L}_{\mathrm{disc}}$ is generated by automorphic forms that appear as residues of Eisenstein series constructed from automorphic forms in the cuspidal spectrum of Levi subgroups. In particular $\mathcal{L}_{\mathrm{disc}} = \mathcal{H}_G$.*

Consider functions F_P on Λ_P with values in $\mathcal{H}_P$ such that, for some φ smooth, K-finite and compactly supported on X_G and for any $\Phi \in \mathcal{H}_P^0$

$$\langle F_P(\lambda), \Phi \rangle_{\mathcal{L}_P} = \hat{\varphi}(\Phi, \lambda) = \langle \varphi, E(\bullet; \Phi, \lambda) \rangle_{\mathcal{L}} = \int_{X_G} \varphi(x) \overline{E(x; \Phi, \lambda)} \, dx.$$

The functional equations satisfied by Eisenstein series implies that

$$\hat{\varphi}(\Phi, \lambda) = \hat{\varphi}(s\lambda, M(s, \lambda)\Phi),$$

which is equivalent to

$$\langle F_P(\lambda), \Phi \rangle_{\mathcal{L}_P} = \langle F_Q(s\lambda), M(s, \lambda)\Phi \rangle_{\mathcal{L}_Q},$$

---

[3] Thanks to Franke's results [7] this hypothesis can be removed: automorphic is enough.

where Q is the standard parabolic subgroup with Levi subgroup $\mathbf{M}_Q = s(\mathbf{M}_P)$. This tells us that functions F_P satisfy functional equations:

$$F_Q(s\lambda) = M(s, \lambda) F_P(\lambda).$$

Let $\mathcal{A}$ be the set of associate classes of standard parabolic subgroups and $\mathcal{A}'$ a set of representatives of these classes. Let $w(P)$ be the order of $W(\mathfrak{a}_P, \mathfrak{a}_P)$ and $n(\mathfrak{a}_P)$ the number of chambers in $(\mathfrak{a}_P/\mathfrak{a}_G)^*$. Now, consider $\hat{\mathcal{L}}$ the Hilbert space of collections of measurable functions F_P on Λ_P with values in $\mathcal{H}_P$ satisfying the above functional equations and are square integrable:

$$\sum_{\mathcal{P} \in \mathcal{A}} \sum_{P \in \mathcal{P}} \frac{1}{n(\mathfrak{a}_P)} \int_{\lambda \in \Lambda_P} \|F_P(\lambda)\|^2_{\mathcal{H}_P} \, d\lambda < \infty.$$

The spectral decomposition is usually formulated as follows (see [1] or [20] for example).

Theorem 12.2.10.3 *There is a dense subset of function $\varphi \in \mathcal{L}$ that can be expressed as*

$$\varphi(x) = \sum_{\mathcal{P} \in \mathcal{A}} \sum_{P \in \mathcal{P}} \frac{1}{n(\mathfrak{a}_P)} \int_{\lambda \in \Lambda_P} E(x; F_P(\lambda), \lambda) \, d\lambda.$$

The scalar product in $\mathcal{L}$ may be written

$$\langle \varphi, \varphi \rangle_{\mathcal{L}} = \sum_{\mathcal{P} \in \mathcal{A}} \sum_{P \in \mathcal{P}} \frac{1}{n(\mathfrak{a}_P)} \int_{\lambda \in \Lambda_P} \langle F_P(\lambda), F_P(\lambda) \rangle_{\mathcal{H}_P} \, d\lambda.$$

This extends to an isomorphism of Hilbert spaces $\hat{\mathcal{L}} \to \mathcal{L}$.

A comment is in order as regards the choice of Haar measures. We choose a Haar measure on $\mathfrak{A}_G \backslash G$ and endow Γ with the canonical measure for a discrete group. This defines the G-invariant measure on X_G. We take the invariant measure on N_P such that $\Gamma_P \cap N_P \backslash N_P$ has volume 1. This and the choice of a Haar measure da on $\mathfrak{A}_P/\mathfrak{A}_G$ produce an invariant measure on

$$X_P = \mathfrak{A}_P \Gamma_P N_P \backslash G$$

used to normalize the scalar product in $\mathcal{H}_P$. The real vector space $\Lambda_P = i(\mathfrak{a}_P/\mathfrak{a}_G)^*$ is identified with the Pontryagin dual of $\mathfrak{A}_P/\mathfrak{A}_G$ via the pairing

$$(\lambda, a) \mapsto e^{\langle \lambda, H_P(a) \rangle}$$

and $d\lambda$ is the canonical Pontryagin dual Haar measure of da.[4]

One sometimes uses the following variant of the above theorem.

---

[4] Observe that, instead of Haar measures on groups $\mathfrak{a}_P/\mathfrak{a}_G$ and $i(\mathfrak{a}_P/\mathfrak{a}_G)^*$ related by Pontryagin duality, surveys [16] and [1] use Lebesgue measures attached to dual basis in vector spaces $\mathfrak{a}_P/\mathfrak{a}_G$ and $(\mathfrak{a}_P/\mathfrak{a}_G)^*$ and hence powers of $2\pi i$ show up in their Fourier inversion formulas.

Theorem 12.2.10.4 *(i) For each parabolic subgroup P one may choose an orthonormal basis $\mathcal{B}(P)$ of $L^2_{\mathrm{disc}}(X_P)$ made of automorphic forms on P.*
(ii) There is a dense subset of functions function $\varphi \in \mathcal{L}$ that may be written (with some abuse of notation[5]):

$$\varphi(x) = \sum_{\mathcal{P} \in \mathcal{A}} \sum_{P \in \mathcal{P}} \frac{1}{n(\mathfrak{a}_P)} \sum_{\Phi \in \mathcal{B}(P)} \int_{\lambda \in \Lambda_P} \langle \varphi, E(\bullet; \Phi, \lambda) \rangle_{\mathcal{L}} E(x; \Phi, \lambda) \, d\lambda.$$

(iii) This extends to an isomorphism between Hilbert space $\mathcal{L}$ and the Hilbert space of measurable functions $\psi(\Phi, \lambda)$ such that

$$\sum_{\mathcal{P} \in \mathcal{A}} \sum_{P \in \mathcal{P}} \frac{1}{n(\mathfrak{a}_P)} \sum_{\Phi \in \mathcal{B}(P)} \int_{\lambda \in \Lambda_P} ||\psi(\Phi, \lambda)||^2 \, d\lambda < \infty.$$

(iv) In particular, given two smooth K-finite compactly supported functions φ_1 and φ_2 on X_G their scalar product can be written

$$\langle \varphi_1, \varphi_2 \rangle_{\mathcal{L}} = \sum_{\mathcal{P} \in \mathcal{A}} \sum_{P \in \mathcal{P}} \frac{1}{n(\mathfrak{a}_P)} \sum_{\Phi \in \mathcal{B}(P)} \int_{\lambda \in \Lambda_P} \hat{\varphi}_1(\Phi, \lambda) \overline{\hat{\varphi}_2(\Phi, \lambda)} \, d\lambda \qquad (\star)$$

or equivalently

$$\langle \varphi_1, \varphi_2 \rangle_{\mathcal{L}} = \sum_{P \in \mathcal{A}'} \frac{1}{w(P)} \sum_{\Phi \in \mathcal{B}(P)} \int_{\lambda \in \Lambda_P} \hat{\varphi}_1(\Phi, \lambda) \overline{\hat{\varphi}_2(\Phi, \lambda)} \, d\lambda \qquad (\star\star)$$

with

$$\hat{\varphi}_i(\Phi, \lambda) = \langle \varphi_i, E(\bullet; \Phi, \lambda) \rangle_{\mathcal{L}} = \int_{X_G} \varphi_i(x) \overline{E(x; \Phi, \lambda)} \, dx.$$

Assertion (i) is a consequence of Theorem 12.2.10.2. The equivalence of $(\star)$ and $(\star\star)$ follows from the following remarks: the value of the integral is constant when P varies in a given $\mathcal{P}$ since intertwining operators $M(s, \lambda)$ are unitary when λ is purely imaginary; it remains for us to observe that if we denote by $a(\mathcal{P})$ the cardinal of $\mathcal{P}$ one has: $n(\mathfrak{a}_P) = w(P)a(\mathcal{P})$.

The spectral decomposition allows us to express $\mathcal{L}$ as a Hilbert direct integral. Let π be a unitary representation of $P = \mathbf{P}(\mathbb{A})$ trivial on $N_P \mathfrak{A}_G$ and $\lambda \in \Lambda_P$. We denote by $\mathcal{I}_P^G(\pi, \lambda)$ the right regular representation of G in the space of function ϕ on G with values in the space of π satisfying

---

[5] Eisenstein series $E(\bullet; \Phi, \lambda)$ do not belong to $\mathcal{L}$ and hence the notation $\langle \varphi, E(\bullet; \Phi, \lambda) \rangle_{\mathcal{L}}$ makes sense a priori only when the implicit integral is convergent which is the case if φ is compactly supported for exemple. But then, the integral over λ may not be convergent. The isomorphism $\hat{\mathcal{L}} \to \mathcal{L}$ in Theorem 12.2.10.3 that extends the correspondence $\varphi \mapsto \{F_P\}$ defined by $\langle F_P(\lambda), \Phi \rangle = \langle \varphi, E(\bullet; \Phi, \lambda) \rangle_{\mathcal{L}}$ allows us to make sense of the statement. Another option is to integrate over (compact) balls in Λ_P and then take the limits in $\mathcal{L}$ when the radii of the balls tend to infinity (see theorem VI.2.1 of [20]). Except for some simple examples, giving precise conditions for the convergence of these integrals seems out of reach for the time being.

$$\Phi(px) = e^{\langle \lambda + \rho_P, H_P(p) \rangle} \pi(p) \Phi(x)$$

and that are square integrable on K. The "parabolically induced" representation $\mathcal{I}_P^G(\pi, \lambda)$ is unitary. Define $\mathcal{F}_P(\lambda)$ by:

$$\mathcal{F}_P(\lambda) = \mathcal{I}_P^G(L_{\text{disc}}^2(X_P), \lambda).$$

The unitarity of intertwining operators $M(s, \lambda)$ for $\lambda \in \Lambda_P$ shows that, for $s \in W(\mathfrak{a}_P, \mathfrak{a}_Q)$,

$$\mathcal{F}_P(\lambda) \simeq \mathcal{F}_Q(s\lambda).$$

Let $\mathcal{P}$ be the association class of P. Let $\mathcal{F}_\mathcal{P}$ be the Hilbert direct integral of representations $\mathcal{F}_P(\lambda)$:

$$\mathcal{F}_\mathcal{P} = \int_{\Lambda_P/W(\mathfrak{a}_P, \mathfrak{a}_P)}^{\oplus} \mathcal{F}_P(\lambda) \, d\lambda.$$

Up to isomorphism, it is independent of the choice of $P \in \mathcal{P}$. Theorems 12.2.10.3 and 12.2.10.4 imply that the right regular representation of G in $\mathcal{L} = L^2(X_G)$ can be written as the direct sum of direct integrals $\mathcal{F}_\mathcal{P}$.

Proposition 12.2.10.5

$$\mathcal{L} = \bigoplus_{\mathcal{P} \in \mathcal{A}} \mathcal{F}_\mathcal{P} = \bigoplus_{P \in \mathcal{A}'} \int_{\Lambda_P/W(\mathfrak{a}_P, \mathfrak{a}_P)}^{\oplus} \mathcal{F}_P(\lambda) \, d\lambda.$$

12.3 About the Proof

12.3.1 Pseudo-Eisenstein Series and their Scalar Product

Consider a parabolic subgroup P and the space

$$X_{P,G} = \mathfrak{A}_G \Gamma_P N_P \backslash G.$$

Denote by $\mathcal{D}_P$ the space of K-finite functions ϕ on $X_{P,G}$ such that $a \mapsto \phi(a \bullet k)$ is a compactly supported function on $\mathfrak{A}_P/\mathfrak{A}_G$ with values in a finite-dimensional space of cuspidal functions on M_P independent of $k \in K$. For $\phi \in \mathcal{D}_P$ the series

$$E_\phi(x) = \sum_{\gamma \in \Gamma_P \backslash \Gamma_G} \phi(\gamma x)$$

are absolutely convergent and, following [20], we call them pseudo-Eisenstein series.[6]

---

[6] These series are denoted $\hat{\phi}$ in [16], [1], and [12]. They appear as series θ_ϕ in [20] but there ϕ is the Fourier transform, on the torus, of our ϕ.

Let $\mathcal{A}$ be the set of associate classes of standard parabolic subgroups. For $\mathcal{P} \in \mathcal{A}$ we denote by $\mathcal{C}_\mathcal{P}$ the closure of the vector space generated by the E_ϕ for $\phi \in \mathcal{D}_P$ with $P \in \mathcal{P}$. Observe that $\mathcal{D}_G = \mathcal{C}_G$ is the space of cusp forms on G.

A first step toward the spectral decomposition is a direct sum decomposition indexed by association classes of standard parabolic subgroups. This is Lemma 2 in [16] and Proposition II.2.4 in [20].

Proposition 12.3.1.1 *One has a direct sum decomposition*

$$\mathcal{L} = \bigoplus_{\mathcal{P} \in \mathcal{A}} \mathcal{C}_\mathcal{P}.$$

Proof Lemma 12.2.4.1 and an inductive argument starting with the minimal parabolic subgroup show that a function on X_G that is orthogonal to all E_ϕ for all standard parabolic subgroups, including G, must vanish. This is Lemma 3.7, p. 55, and its Corollary, p. 58, of [14] (see also Theorem II.1.12 in [20]). It remains for us to show that $\mathcal{C}_\mathcal{P}$ and $\mathcal{C}_\mathcal{Q}$ are orthogonal whenever $\mathcal{P} \neq \mathcal{Q}$. Consider two standard parabolic subgroups P and Q; we have either
(i) for some element s of the Weyl group $\mathbf{M}_Q \subset s(\mathbf{M}_P)$
or
(ii) for any element s of the Weyl group $\mathbf{M}_Q \cap s(\mathbf{N}_P)$ is a nontrivial unipotent subgroup.
Consider P and Q that are not associated; then, up to exchanging the role of P and Q, we may assume that (ii) holds. Now consider functions ϕ and ψ cuspidal on P and Q respectively. The above formal computation yields

$$\langle E_\phi, E_\psi \rangle = \int_{X_{P,G}} \phi(x) \overline{E_{\psi,P}(x)} \, dx,$$

where $E_{\psi,P}$ is the constant term along P of E_ψ. Using Bruhat decomposition

$$\Gamma_P \backslash \Gamma / \Gamma_Q \simeq W^P \backslash W^G / W^Q,$$

where W^P denotes the Weyl group of M_P, we see that $E_{\psi,P}$ vanishes since ψ is cuspidal on Q.   □

Proposition 12.3.1.1 reduces the spectral decomposition of $\mathcal{L}$ to the spectral decomposition of spaces $\mathcal{C}_\mathcal{P}$. To proceed further, one needs to compute the scalar product of two pseudo-Eisenstein series when it does not vanish. Consider $\phi \in \mathcal{D}_P$; then the function Φ on $X_P \times (\mathfrak{a}_P / \mathfrak{a}_G)^* \otimes \mathbb{C}$ given by

$$\Phi(x, \lambda) = \int_{\mathfrak{A}_P / \mathfrak{A}_G} \phi(ax) e^{-\langle \lambda + \rho_P, H_P(ax) \rangle} \, da$$

is cuspidal on P in the first variable and is analytic of Paley–Wiener type in the second one. We identify the Pontryagin dual of $\mathfrak{A}_G \backslash \mathfrak{A}_P$ with $\Lambda_P = i(\mathfrak{a}_P / \mathfrak{a}_G)^*$. By Fourier inversion one recovers ϕ:

$$\phi(x) = \int_{\lambda \in \lambda_0 + \Lambda_P} \Phi(x, \lambda) e^{\langle \lambda + \rho_P, H_P(x) \rangle} \, d\lambda,$$

where λ_0 is arbitrary and $d\lambda$ is the Haar measure on $\Lambda_P = i(\mathfrak{a}_P / \mathfrak{a}_G)^*$ dual, for Pontryagin duality, to the Haar measure da, and hence the pseudo-Eisenstein series E_ϕ is equal to an integral of Eisenstein series:

$$E_\phi(x) = \int_{\lambda \in \lambda_0 + \Lambda_P} E(x; \Phi(\bullet, \lambda), \lambda) \, d\lambda$$

for any $\lambda_0 \in C_P + \rho_P$ (which one may also write $\lambda_0 > \rho_P$). We may now give the formula for the scalar product of two pseudo-Eisenstein series that is the key to the spectral decomposition.

Proposition 12.3.1.2 *Consider two associated parabolic subgroups P and Q and two functions $\phi_1 \in \mathcal{D}_P$ and $\phi_2 \in \mathcal{D}_Q$. When $\lambda_0 \in C_P + \rho_P$ one has, with the notation of Lemma 12.2.9.2:*

$$\langle E_{\phi_1}, E_{\phi_2} \rangle_{\mathcal{L}} = \sum_{s \in W(\mathfrak{a}_P, \mathfrak{a}_Q)} \int_{\lambda \in \lambda_0 + \Lambda_P} \langle M(s, \lambda) \Phi_1(\bullet, \lambda), \Phi_2(\bullet, -s\bar{\lambda}) \rangle_{\mathcal{L}_P} \, d\lambda.$$

Proof This is an immediate consequence of Lemmas 12.2.4.1 and 12.2.9.2.
□

It is useful to extend the space of cuspidal functions $\Phi(\bullet, \lambda)$ by asking that as a function of λ it is holomorphic and rapidly decreasing in vertical strips $\|\lambda\| < R$ for some R. Their Fourier transform builds a space denoted $\mathcal{D}_P(R)$. The above formula still holds provided that $\|\lambda_0\| < R$.

The spectral decomposition is obtained by shifting the integral to the purely imaginary space Λ_P, i.e. moving λ_0 to 0. To do this one needs to establish the meromorphic continuation of intertwining operators and to take into account the residues that show up. Analytic estimates are, moreover, necessary to allow such a contour shift. One establishes at the same time the analytic continuation and the functional equations of Eisenstein series.

12.3.2 Scalar Product of Truncated Eisenstein Series

In section 9 of [16], Langlands states, without detailed proof, a formula for the scalar product of two truncated Eisenstein series induced from cusp forms. We quote the result with the notation of [13]. Let P and Q be two standard parabolic subgroups and consider:

$$\lambda \in \mathfrak{a}_Q^* \otimes \mathbb{C} \qquad \text{and} \qquad \mu \in \mathfrak{a}_R^* \otimes \mathbb{C}$$

that coincide on $\mathfrak{a}_G$. One defines an operator valued function, meromorphic in both variables whenever intertwining operators are meromorphic,

$$\omega_{Q|P}^T(\lambda,\mu) = \sum_R \sum_{s \in W(\mathfrak{a}_P, \mathfrak{a}_R)}$$

$$\times \sum_{t \in W(\mathfrak{a}_Q, \mathfrak{a}_R)} e^{\langle s\lambda - t\mu, \, T \rangle} \varepsilon_R(s\lambda - t\mu) M(t, -\overline{\mu})^* M(s, \lambda),$$

where R runs over standard parabolic subgroups associated to P and Q. The function $\lambda \mapsto \varepsilon_R(\lambda)$ is the inverse of a product of monomials

$$\varepsilon_R(\lambda)^{-1} = V_R \prod_{\alpha \in \Delta_R} \langle \lambda, \check{\alpha} \rangle,$$

where the V_R is the volume of the parallelotope generated by Δ_R. The function $\omega_{Q|P}^T$ vanishes unless P and Q are associated. Langlands' formula is the following.

Theorem 12.3.2.1 *Assume that Φ and Ψ are cuspidal on P and Q. Then, provided λ and μ are in the convergence domain for Eisenstein series, i.e. the translate by ρ_P (resp. ρ_Q) of the Weyl chamber, we have*

$$\int_{X_G} \Lambda^T E(x, \Phi, \lambda) \overline{\Lambda^T E(x, \Psi, \mu)} dx = \langle \omega_{Q|P}^T(\lambda, -\overline{\mu}) \Phi, \Psi \rangle.$$

Proof We refer the reader to section 5.4 of [13] for a proof much shorter and more elementary than the one given in section 4 of [2].[7] $\qquad\square$

Theorem 12.3.2.2 *Again, Φ is assumed to be cuspidal. Assume we know that intertwining operators are holomorphic in some connected open set $\mathcal{O}$ containing the set of $\lambda \in \mathfrak{a}_P^* \otimes \mathbb{C}$, such that $\mathrm{Re}(\lambda) \in C_P + \rho_P$ (translated by ρ_P of the Weyl chamber) and satisfy the functional equations. Then Eisenstein series $E(x; \Phi, \lambda)$ have a holomorphic continuation in $\mathcal{O}$ and satisfy the functional equations*

$$E(x; \Phi, \lambda) = E(x; M(s, \lambda)\Phi, s\lambda).$$

---

[7] The proof in [2] proceeds directly, expanding the two truncated Eisenstein series: this yields a lot of rather complicated terms; fortunately cancellations occur but rely on subtle arguments using meromorphic continuation of partial expressions, growth estimates, contour shiftings, and computation of residues. In [13] one uses that Λ^T is an orthogonal projection and hence it is equivalent to show that

$$\int_{X_G} \Lambda^T E(x, \Phi, \lambda) \overline{E(x, \Psi, \mu)} dx = \langle \omega_{Q|P}^T(\lambda, -\overline{\mu}) \Phi, \Psi \rangle.$$

Fewer and simpler terms have to be dealt with, no cancellation is needed, and the computation is elementary.

Proof If we have at hand the functional equations for intertwining operators, one can show that the singularities arising from zeros of monomials in the denominator of the formula defining $\omega^T_{Q|P}(\lambda,\mu)$ are canceled by zeros of the numerator and that the only possible singularities arise from singularities of the intertwining operators $M(\bullet,\bullet)$ (cf. Propositions 1.10.4 and 5.3.3 of [13]). Now, if D is a holomorphic differential operator in the variable λ, the identity in Theorem 12.3.2.1 tells us that

$$\|D\Lambda^T E(\bullet,\Phi,\lambda)\|^2_{\mathcal{L}} = D\overline{D}\,\langle\omega^T_{P|P}(\lambda,-\overline{\lambda})\Phi,\Phi\rangle_{\mathcal{L}_P}.$$

This shows that if $\omega^T_{Q|P}(\lambda,-\overline{\lambda})$ is given by a convergent Taylor series in a neighborhood of some point λ_0, then the same is true for the function with values in the Hilbert space $\mathcal{L}$:

$$\lambda \mapsto \Lambda^T E(\bullet,\Phi,\lambda).$$

Since both sides in the formula in Theorem 12.3.2.1 are known to be equal in the convergence domain, they remain equal in $\mathcal{O}$. Now, if φ is a smooth compactly supported function on X_G it follows from Proposition 12.2.4.2 that, for T large enough,

$$\int_{X_G} \varphi(x)E(x;\Phi,\lambda)\,dx = \int_{X_G} \varphi(x)\Lambda^T E(x;\Phi,\lambda)\,dx.$$

This implies that Eisenstein series, when considered as distributions on X_G:

$$E(\bullet;\Phi,\lambda) : \varphi \mapsto \int_{X_G} \varphi(x)E(x;\Phi,\lambda)\,dx,$$

have a holomorphic continuation on $\mathcal{O}$. Moreover, since Φ is automorphic, one may find a compactly supported smooth function f on G such that $I_{P,\lambda}(f)\Phi = \Phi$. But, in the convergence domain, one has

$$E(x;I_{P,\lambda}(f)\Phi,\lambda) = E(x;\Phi,\lambda)\star f$$

and we get for $\lambda \in \mathcal{O}$ an equality of distributions:

$$E(\bullet;\Phi,\lambda) = E(\bullet;\Phi,\lambda)\star f.$$

This shows that the distributions obtained by analytic continuation of Eisenstein series are, in fact, smooth functions. Using the functional equation for intertwining operators and formula in Lemma 12.2.9.2, one checks that the constant terms of

$$E(x;\Phi,\lambda) \qquad \text{and} \qquad E(x;M(s,\lambda)\Phi,s\lambda)$$

along any proper parabolic subgroup Q are equal, and hence the difference

$$E(x;\Phi,\lambda) - E(x;M(s,\lambda)\Phi,s\lambda)$$

is cuspidal. But, at the same time, each term is orthogonal to cusp forms as follows from Lemma 12.2.9.1. The difference must vanish. This establishes the functional equation for Eisenstein series in the domain where they are analytic.   □

Theorem 12.3.2.2 reduces the proof of the analytic continuation and the functional equations of Eisenstein series built from cusp forms to the proof of the same properties for intertwining operators. That the analytic continuation of Eisenstein series yields automorphic forms needs a little more work.

12.3.3 Analytic Continuation: Cuspidal Case

Consider the case where $\mathcal{P}$ is an association class of maximal standard parabolic subgroups. There are two cases: either $\mathcal{P}$ has one element P and $W(\mathfrak{a}_P, \mathfrak{a}_P)$ has two elements, then we put $P_1 = P$; or $\mathcal{P}$ has two elements $P = P_1 \neq P_2 = Q$ and $W(\mathfrak{a}_P, \mathfrak{a}_Q)$ has one element. Let

$$\lambda_{P_i}(z) = z \frac{\alpha_{P_i}}{||\alpha_{P_i}||} \quad \text{with} \quad z \in \mathbb{C}$$

if we denote by α_{P_i} the unique positive root in Δ_{P_i}. Now define $M(z)$ by

$$M(z) = M(s, \lambda_P(z))$$

when $\mathcal{P}$ has one element and s is the nontrivial element in $W(\mathfrak{a}_P, \mathfrak{a}_P)$ or

$$M(z) = \begin{pmatrix} 0 & M(s^{-1}, \lambda_Q(z)) \\ M(s, \lambda_P(z)) & 0 \end{pmatrix}$$

if s belongs to $W(\mathfrak{a}_P, \mathfrak{a}_Q)$ with $Q \neq P$. Let $I = \{1\}$ or $I = \{1,2\}$ according to the cases. Define

$$\mathcal{L}_\mathcal{P} = \bigoplus_{i \in I} \mathcal{L}_{P_i}.$$

We have to study $M(z)$ and Eisenstein series

$$E(\bullet, \Phi, z) = \sum_{i \in I} E(\bullet, \Phi_i, z).$$

Proposition 12.3.3.1 (i) The functions $M(z)$ and $E(\bullet, \Phi, z)$ are meromorphic for $z \in \mathbb{C}$.
(ii) $M(z)M(-z) = 1$.
(iii) $E(\bullet, \Phi, z) = E(\bullet, M(z)\Phi, -z)$.

Proof We sketch an argument borrowed from sections 4 and 6 of [16] (see also Lemma 84 in [9]). Let $r = ||\rho_P||$. Given Φ_i and Ψ_i in $\mathcal{L}_{P_i}$ we define functions of z with values in $\mathcal{L}_\mathcal{P}$

$$\Phi(\bullet, z) = \oplus \Phi_i(\bullet, \lambda_{P_i}(z)) \quad \text{and} \quad \Psi(\bullet, z) = \oplus \Psi_i(\bullet, \lambda_{P_i}(z)).$$

The scalar product of pseudo-Eisenstein series E_ϕ and E_ψ attached to $\Phi(\bullet, z)$ and $\Psi(\bullet, z)$ is given by:

$$\langle E_\phi, E_\psi \rangle_\mathcal{L} = \int_{c+i\mathbb{R}} \langle \Phi(\bullet, z), \Psi(\bullet, -\bar{z}) \rangle_{\mathcal{L}_\mathcal{P}} + \langle M(z)\Phi(\bullet, z), \Psi(\bullet, \bar{z}) \rangle_{\mathcal{L}_\mathcal{P}} \, dz$$

provided $c > r$. Here dz is a suitably normalized Haar measure on $i\mathbb{R}$. Consider functions E_ϕ with $\phi_i \in \mathcal{D}_{P_i}(R)$. The above formula holds for

$$r < c < R \, .$$

Put $\Phi_i'(\bullet, \lambda) = \langle \lambda, \lambda \rangle \Phi_i(\bullet, \lambda)$ then, if we define $E_{\phi'}$ using the Φ_i', the densely defined linear operator $AE_\phi = E_{\phi'}$ is essentially self-adjoint (unbounded). Its spectrum is real and its resolvent

$$\mathbf{R}(\sigma, A) = (\sigma - A)^{-1}$$

is a holomorphic function of $\sigma \in \mathbb{C}$ off the interval $] - \infty, R^2]$. But since R can be arbitrarily close to r the resolvent is holomorphic off the interval $] - \infty, r^2]$ and

$$\langle \mathbf{R}(\sigma, A)E_\phi, E_\psi \rangle_\mathcal{L}$$

equals

$$\int_{c+i\mathbb{R}} \frac{1}{\sigma - z^2} \left(\langle \Phi(\bullet, z), \Psi(\bullet, -\bar{z}) \rangle_{\mathcal{L}_\mathcal{P}} + \langle M(z)\Phi(\bullet, z), \Psi(\bullet, \bar{z}) \rangle_{\mathcal{L}_\mathcal{P}} \right) dz$$

provided $\mathrm{Re}(\sigma) > c^2 > r^2$. For $t \in \mathbb{C}$, with $\mathrm{Re}(t) \neq \pm c$ while $r < c < R$, let

$$r(t, c) = \int_{c+i\mathbb{R}} \frac{1}{t^2 - z^2} \left(\langle \Phi(\bullet, z), \Psi(\bullet, -\bar{z}) \rangle_{\mathcal{L}_\mathcal{P}} + \langle M(z)\Phi(\bullet, z), \Psi(\bullet, \bar{z}) \rangle_{\mathcal{L}_\mathcal{P}} \right) dz.$$

This is a holomorphic function of t for $|\mathrm{Re}(t)| \neq c$ such that

$$r(t, c) = \langle \mathbf{R}(t^2, A)E_\phi, E_\psi \rangle_\mathcal{L} \quad \text{if} \quad \mathrm{Re}(t) > c > r.$$

Suppose now that

$$R > c_1 > \mathrm{Re}(t) > c > r$$

then shifting the above integral and evaluating the residue at $z = t$ we get

$$\langle \mathbf{R}(t^2, A)E_\phi, E_\psi \rangle_\mathcal{L} = s(t) + r(t, c_1),$$

where

$$s(t) = \frac{1}{2t} \left(\langle \Phi(\bullet, t), \Psi(\bullet, -\bar{t}) \rangle_{\mathcal{L}_\mathcal{P}} + \langle M(t)\Phi(\bullet, t), \Psi(\bullet, \bar{t}) \rangle_{\mathcal{L}_\mathcal{P}} \right).$$

Hence $s(t)$ has a holomorphic continuation on the subset $t^2 \notin] - \infty, r^2]$. Now suppose that

$$\Phi(z) = e^{z^2} \Phi_0 \quad \text{and} \quad \Psi(z) = e^{z^2} \Psi_0$$

with Φ_0 and Ψ_0 in $\mathcal{L}_\mathcal{P}$, we get

$$s(t) = \frac{e^{2t^2}}{2t}(\langle \Phi_0, \Psi_0 \rangle_{\mathcal{L}_\mathcal{P}} + \langle M(t)\Phi_0, \Psi_0 \rangle_{\mathcal{L}_\mathcal{P}})$$

and hence $M(t)$ has a holomorphic continuation whenever $t^2 \notin]-\infty, r^2]$ i.e. $\mathrm{Re}(t) > 0$ and $t \notin]0, r]$. The positivity of the inner product of Theorem 12.3.2.1 of a truncated Eisenstein series with itself tells us that the hermitian operator

$$\frac{1}{2\mathrm{Re}(t)}(I - M(t)M(t)^*) - \frac{1}{2i\,\mathrm{Im}(t)}(M(t) - M(t)^*)$$

is positive. The continuation for $\mathrm{Re}(t) = 0$ but $\mathrm{Im}(t) \neq 0$ and the functional equation

$$M(t)M(-t) = I$$

on this line are an easy consequence of this positivity provided one knows the operator remains bounded when $t = iy + \epsilon$ with $y \in \mathbb{R} - 0$ and $\epsilon > 0$ tends to 0. This fact will not be established here, being too technical: it uses estimates from section 5 of [14] (the case $\mathbf{G} = GL(2)$ is treated in appendix IV of [14], pages 326–331; see also lemma 98 in [9]). It also remains for us to show that only a finite number of singular points may show up in the segment $]0, \|\rho_P\|]$ (see, for example, sections 6, 7, and 8 of chapter IV of [9]). By symmetry one gets the meromorphic continuation to the full complex line. Meromorphic continuation and the functional equation for Eisenstein series now follow from Theorem 12.3.2.2.   □

We may now consider the case of arbitrary rank. Recall that, by definition, $N_w = wN_Pw^{-1} \cap N_Q\backslash N_Q$. Now if $s = s_1s_2$ is a factorization of elements in the Weyl group with length $\ell(s) = \ell(s_1) + \ell(s_2)$ and if w_i represent the s_i, there is an isomorphism

$$N_{w_1} \times N_{w_2} \twoheadrightarrow N_w,$$

which implies that

$$M(s, \lambda) = M(s_1, s_2\lambda)M(s_2, \lambda)$$

in the domain of convergence (cf. [22]). Decomposing s into a product of minimal length of simple reflection s_i we decompose $M(s, \lambda)$ into a product of intertwining operators $M(s_i, \lambda_i)$ and the meromorphic continuation of intertwining operators is reduced to the rank-one case. For more details we refer to section 7 of [16] and also [12], where the case of quasisplit groups is fully treated. Again, the meromorphic continuation of Eisenstein series induced from cusp forms and their functional equations now follows from this and Theorem 12.3.2.2 (cf. section 9 of [16]).

12.3.4 Contour Deformation Forgetting Residues

Start with the expression in Proposition 12.3.1.2 for the square norm of a
pseudo-Eisenstein series

$$\langle E_\phi, E_\phi \rangle_\mathcal{L} = \sum_{s \in W(\mathfrak{a}_P, \mathfrak{a}_P)} \int_{\lambda \in \lambda_0 + \Lambda_P} \langle M(s,\lambda)\Phi(\bullet,\lambda), \Phi(\bullet, -s\bar\lambda) \rangle_{\mathcal{L}_P} \, d\lambda.$$

Assume that the integrand is sufficiently decreasing at infinity so that we may
move the integration contour. By moving λ_0 to 0 following a path inside the
Weyl chamber we may cross singular hyperplanes. Assume for a while that the
residues coming from these crossings vanish. In such a case we get, at least
formally,

$$\langle E_\phi, E_\phi \rangle_\mathcal{L} = \sum_{s \in W(\mathfrak{a}_P, \mathfrak{a}_P)} \int_{\lambda \in \Lambda_P} \langle M(s,\lambda)\Phi(\bullet,\lambda), \Phi(\bullet, -s\bar\lambda) \rangle_{\mathcal{L}_P} \, d\lambda,$$

which is again

$$= \sum_{s \in W(\mathfrak{a}_P, \mathfrak{a}_P)} \int_{\lambda \in \Lambda_P} \langle M(s,\lambda)\Phi(\bullet,\lambda), \Phi(\bullet, s\lambda) \rangle_{\mathcal{L}_P} \, d\lambda.$$

This can be rewritten

$$= \frac{1}{w(P)} \sum_{s \in W(\mathfrak{a}_P, \mathfrak{a}_P)} \sum_{t \in W(\mathfrak{a}_P, \mathfrak{a}_P)} \int_{\lambda \in \Lambda_P} \langle M(s,t\lambda)\Phi(\bullet,t\lambda), \Phi(\bullet, st\lambda) \rangle_{\mathcal{L}_P} \, d\lambda,$$

but since

$$M(st,\lambda) = M(s,t\lambda)M(t,\lambda)$$

this is equal to

$$= \frac{1}{w(P)} \sum_{u \in W(\mathfrak{a}_P, \mathfrak{a}_P)} \sum_{t \in W(\mathfrak{a}_P, \mathfrak{a}_P)}$$

$$\times \int_{\lambda \in \Lambda_P} \langle M(t,\lambda)^{-1}\Phi(\bullet,t\lambda), M(u,\lambda)^{-1}\Phi(\bullet,u\lambda) \rangle_{\mathcal{L}_P} \, d\lambda.$$

Now if we let

$$F_P(\lambda) = \sum_{s \in W(\mathfrak{a}_P, \mathfrak{a}_P)} M(s,\lambda)^{-1}\Phi(\bullet, s\lambda)$$

we get the following.

Proposition 12.3.4.1 *Assume that the residues coming from the singularities
vanish and that functions are sufficiently decreasing at infinity so that we may
move the integration contour. Then*

$$\langle E_\phi, E_\phi \rangle = \frac{1}{w(P)} \int_{\lambda \in \Lambda_P} \langle F_P(\lambda), F_P(\lambda) \rangle \, d\lambda.$$

This is the expected contribution coming from P to the spectral decomposition, as given in Theorem 12.2.10.3, when residues do not contribute.

12.3.5 The Trivial Representation as a Residue

The trivial representation will show up when deforming the contour giving the contribution of the Eisenstein series induced from the constant function Φ on the minimal parabolic subgroup P_0 through the multiple residue at $\lambda = \rho$.

Assume that $\mathbf{G}$ is a split $\mathbb{Q}$-group and that $K = K_\infty \times \prod_p K_p$, where K_p is hyperspecial for all prime p. Consider series E_ϕ where ϕ is right-K-invariant on X_{P_0}. The expression in Proposition 12.3.1.2 for the scalar product of such a pseudo-Eisenstein series is

$$\langle E_\phi, E_\phi \rangle_{\mathcal{L}} = \sum_{s \in W^G} \int_{\lambda \in \lambda_0 + \Lambda_{P_0}} \langle M(s,\lambda)\Phi(\bullet, \lambda), \Phi(\bullet, -s\bar{\lambda}) \rangle_{\mathcal{L}_{P_0}} \, d\lambda.$$

The intertwining operators $M(s,\lambda)$ act on a constant function Φ by scalars

$$m(s,\lambda) = \prod_{\alpha > 0, \, s\alpha < 0} \frac{L(\langle \lambda, \check{\alpha} \rangle)}{L(1 + \langle \lambda, \check{\alpha} \rangle)},$$

where $L(\sigma)$ is the complete Zeta function:

$$L(\sigma) = \pi^{-\sigma/2} \Gamma(\sigma/2) \zeta(\sigma).$$

Recall that $L(\sigma)$ has a simple pole with residue 1 at $\sigma = 1$ and functional equation

$$L(\sigma) = L(1 - \sigma).$$

Observe that the half sum of positive roots ρ is also the sum of fundamental weights (i.e. the basis dual to the basis of simple coroots):

$$\rho = \frac{1}{2} \sum_{\alpha > 0} \alpha = \sum_{\alpha \in \Delta_{P_0}} \varpi_\alpha.$$

Hence ρ is the intersection point of the affine hyperplanes $\langle \lambda, \check{\alpha} \rangle = 1$, where $\check{\alpha}$ runs over simple coroots. This implies that the multiple residue at $\lambda = \rho$ is given by the term indexed by the longest element s in the Weyl group, i.e. the element that sends any positive roots to a negative root. Let us denote by

$$\prod_{\alpha > 0}{}'$$

the product over nonsimple positive root and define V by

$$\frac{1}{V} = \frac{\prod'_{\alpha > 0} L(\langle \rho, \check{\alpha} \rangle)}{\prod_{\alpha > 0} L(1 + \langle \rho, \check{\alpha} \rangle)}.$$

The multiple residue $\mathcal{T}$ at $\lambda = \rho$ is given by

$$\mathcal{T} = \frac{1}{V}\langle\Phi(\bullet,\rho),\Phi(\bullet,\rho)\rangle_{\mathcal{L}_{P_0}}.$$

Now if we denote by e_G the unit element in G we have

$$\langle E_\phi, 1\rangle_{\mathcal{L}} = \Phi(e_G,\rho)$$

if the measure on G is given by

$$e^{-\langle 2\rho, H_{P_0}(x)\rangle}\, dn\, dp\, dk$$

for $x = npk$, where $dp\, dk$ is normalized so that

$$\mathrm{vol}\ (X_{P_0}) = \mathrm{vol}\ (K) = 1,$$

while dn gives measure 1 to $\Gamma_{P_0} \cap N_{P_0}\backslash N_{P_0}$. This implies that for such a measure on G one has

$$\langle\Phi(\bullet,\rho),\Phi(\bullet,\rho)\rangle_{\mathcal{L}_{P_0}} = |\Phi(e_G,\rho)|^2 = |\langle E_\phi,1\rangle_{\mathcal{L}}|^2$$

and hence

$$\mathcal{T} = \frac{\langle E_\phi,1\rangle_{\mathcal{L}}\langle 1, E_\phi\rangle_{\mathcal{L}}}{V}.$$

Using this and a resolvent argument, Langlands shows in [15] that

$$\mathrm{vol}\ (X_G) = \langle 1,1\rangle_{\mathcal{L}} = V.$$

In other words, if Ψ_G is a constant function of norm 1 in $\mathcal{L}$

$$\mathcal{T} = \frac{1}{V}\langle\Phi(\bullet,\rho),\Phi(\bullet,\rho)\rangle_{\mathcal{L}_{P_0}} = \langle E_\phi,\Psi_G\rangle_{\mathcal{L}}\langle\Psi_G,E_\phi\rangle_{\mathcal{L}},$$

and hence $\mathcal{T}$ is the contribution to the spectral decomposition of the trivial representation.

This equation is the starting point for proving Weil's conjecture, which says that the Tamagawa number $\tau(G) = 1$ when G simply connected. The conjecture was proved by Langlands for split groups [15], and extended by K. F. Lai to arbitrary quasisplit groups [12]. As Langlands had expected, the proof of the general case, due to Kottwitz [10], reduces to Lai's result modulo the stabilization of the Trace Formula in a special case.

12.3.6 The General Case

It remains for us to deal with the most difficult part of the proof: to take into account more-general residues that show up when moving the contour. This is taken care of in chapter 7 of [14] and in chapters V and VI in [20]. A nice introduction is given by J. Arthur in [1] and since we could not do any better we refer the reader to this survey. Let us simply say that besides serious analytic difficulties the main obstacles come from the following facts: one does not know a priori the location of singularities of the intertwining operators. In the simplest cases intertwining operators are quotients of Riemann Zeta functions and, if their poles are known, their zeros are mysterious and may produce singularities. In the general case this is worse since one has to control intertwining operators that involve the most general L-functions attached to automorphic forms by Langlands, and very little is known about their singularities. Moreover, as will be seen in the $GL(3)$ example (in the proof of Proposition 12.4.4.3), many parasitic residues do cancel. A direct argument, like the one we shall use in our examples, would not be tractable in the general case and so a subtle induction process is necessary.

One has thus established two orthogonal decompositions: an easy one in Proposition 12.3.1.1

$$\mathcal{L} = \bigoplus_{\mathcal{P} \in \mathcal{A}} \mathcal{C}_{\mathcal{P}}$$

and a much deeper one in Proposition 12.2.10.5. The various projectors commutes with each other. Putting

$$\mathcal{G}_{\mathcal{P}}^{\mathcal{Q}} = \mathcal{C}_{\mathcal{P}} \cap \mathcal{F}_{\mathcal{Q}}$$

we get a finer decomposition.

Proposition 12.3.6.1

$$\mathcal{L} = \bigoplus_{\mathcal{P} \in \mathcal{A}} \bigoplus_{\{\mathcal{Q} \in \mathcal{A} | \mathcal{P} < \mathcal{Q}\}} \mathcal{G}_{\mathcal{P}}^{\mathcal{Q}}.$$

Here $\mathcal{P} < \mathcal{Q}$ means that, given $P \in \mathcal{P}$, there is a $Q \in \mathcal{Q}$ such that $P \subset Q$.

A representation occuring in $\mathcal{G}_{\mathcal{P}}^{\mathcal{Q}}$ is induced from a representation in the discrete spectrum of $Q \in \mathcal{Q}$, which in turn appears via a residue from an induced representation of a cuspidal representation of $P \in \mathcal{P}$ (i.e. a representation of the cuspidal spectrum on X_P).

12.4 Examples of Spectral Decomposition

When $\mathbf{G} = SL(2)$ the spectral decomposition is due to Selberg [23] (see [11] for a detailed treatment in the classical language). A treatment in the adèlic setting is given in [8]. For the quasisplit form of $U(3)$ attached to a quadratic extension E/F, the noncuspidal discrete spectrum is described in section 13.9 of [21].

The spectral decomposition for $GL(n)$ when $n = 2$ or 3, restricted to the K-invariant vectors, is treated below by following section 10 of [16]. We refer to appendix III of [14] for a similar study for groups G_2 and $SL(4)$ where new intrinsic difficulties show up.

12.4.1 Some Notation

Let $\mathcal{L}^K$ be the space of K-invariant functions in $\mathcal{L}$. This is a representation space for the spherical Hecke algebra, i.e. the convolution algebra of compactly supported left and right K-invariant functions. This is a commutative algebra and hence irreducible representations have dimension 1. Proposition 12.3.1.1 yields a first decomposition

$$\mathcal{L}^K = \sum_{\mathcal{P} \in \mathcal{A}} \mathcal{C}_{\mathcal{P}}^K,$$

where $\mathcal{C}_{\mathcal{P}}^K$ is the space of K-invariant functions in $\mathcal{C}_{\mathcal{P}}$. This is the closure of the space generated by pseudo-Eisenstein series constructed from K-invariant cuspidal functions

$$x \mapsto \Phi(x,\lambda)$$

on X_P for $P \in \mathcal{P}$. When $\mathcal{P} = \{P_0\}$ one has

$$X_{P_0} = \mathfrak{A}_{P_0} \Gamma_{P_0} N_{P_0} \backslash G = P_0 \backslash G$$

and $G = P_0 K$. In such a case functions $x \mapsto \Phi(x,\lambda)$ are simply constant on G and the scalar product in $\mathcal{L}_{P_0}$ is given by

$$\langle \Phi(\bullet,\lambda), \Phi(\bullet,\bar{\lambda}) \rangle_{\mathcal{L}_{P_0}} = \Phi(e_G,\lambda)\overline{\Phi(e_G,\bar{\lambda})}.$$

The intertwining operators $M(s,\lambda)$ act on such functions by scalars

$$m(s,\lambda) = \prod_{\alpha > 0, s\alpha < 0} \frac{L(\langle \lambda, \check{\alpha} \rangle)}{L(1 + \langle \lambda, \check{\alpha} \rangle)}.$$

12.4.2 The Case $\mathbf{G} = \mathbf{GL(2)}$

We now consider $\mathbf{G} = GL(2)$. The minimal parabolic subgroup P_0 can be taken to be the subgroup of upper triangular matrices

$$P_0 = \begin{pmatrix} \star & \star \\ 0 & \star \end{pmatrix}.$$

In G_∞ the maximal compact subgroup $K_\infty = O(2,\mathbb{R})$ while in G_p we take $K_p = GL(2,\mathbb{Z}_p)$ where $\mathbb{Z}_p$ is the ring of p-adic integers. When s is the nontrivial element in the Weyl group

$$m(s,\lambda) = \frac{L(\langle \lambda, \check{\alpha} \rangle)}{L(1 + \langle \lambda, \check{\alpha} \rangle)}.$$

We start from

$$\langle E_\phi, E_\phi \rangle_\mathcal{L} = \int_{\lambda \in \lambda_0 + \Lambda_{P_0}} \Phi(e_G, \lambda)\overline{\Phi(e_G, -\bar{\lambda})}$$

$$+ m(s, \lambda)\Phi(e_G, \lambda)\overline{\Phi(e_G, -s\bar{\lambda})}\, d\lambda.$$

Moving the contour from λ_0 to 0 we have to take into account the residue of $m(s, \lambda)$ at $\lambda = \rho = \alpha/2$ and we get

$$\langle E_\phi, E_\phi \rangle_\mathcal{L} = \int_{\lambda \in \Lambda_{P_0}} \left(\Phi_0(\lambda)\overline{\Phi_0(\lambda)} + m(s, \lambda)\Phi_0(\lambda)\overline{\Phi_0(s\lambda)} \right)\, d\lambda$$

$$+ \frac{1}{L(2)} \Phi_0(\rho)\overline{\Phi_0(\rho)},$$

where

$$\Phi_0(\lambda) = \Phi(e_G, \lambda).$$

Here Λ_{P_0} is of dimension 1. This yields the following.

Proposition 12.4.2.1 *For* $\mathbf{G} = GL(2)$ *the spectral decomposition of the space generated by K-invariant pseudo-Eisenstein series on P_0 is given by*

$$\langle E_\phi, E_\phi \rangle_\mathcal{L} = \frac{1}{2} \int_{\lambda \in \Lambda_{P_0}} \langle E_\phi, E(\bullet; \Psi_0, \lambda) \rangle_\mathcal{L} \overline{\langle E_\phi, E(\bullet; \Psi_0, \lambda) \rangle_\mathcal{L}}\, d\lambda$$

$$+ \langle E_\phi, \Psi_G \rangle_\mathcal{L} \langle \Psi_G, E_\phi \rangle_\mathcal{L}.$$

Here Ψ_0 is a constant function on G normalized so that $\langle \Psi_0, \Psi_0 \rangle_{\mathcal{L}_{P_0}} = 1$ and Ψ_G is a constant function on G normalized, so that $\langle \Psi_G, \Psi_G \rangle_\mathcal{L} = 1$. The orthogonal complement of $\mathcal{C}_{P_0}^K$ in $\mathcal{L}^K$ is $\mathcal{C}_G^K = \mathcal{L}_{\mathrm{cusp}}^K$.

12.4.3 The Case $\mathbf{G} = GL(3)$: Preparatory Material

When $\mathbf{G} = GL(3)$ our minimal parabolic subgroup P_0 is the subgroup of upper triangular matrices, and there are two associated standard maximal parabolic subgroups P_1 and P_2

$$P_0 = \begin{pmatrix} \star & \star & \star \\ 0 & \star & \star \\ 0 & 0 & \star \end{pmatrix} \qquad P_1 = \begin{pmatrix} \star & \star & \star \\ \star & \star & \star \\ 0 & 0 & \star \end{pmatrix} \qquad P_2 = \begin{pmatrix} \star & \star & \star \\ 0 & \star & \star \\ 0 & \star & \star \end{pmatrix}.$$

The maximal compact subgroup K_∞ in G_∞ is $O(3, \mathbb{R})$, while $K_p = GL(3, \mathbb{Z}_p)$. We denote by α_1 and α_2 the two simple roots, and by ϖ_1 and ϖ_2 the corresponding fundamental weights. Recall that ρ the half sum of positive roots is such that

$$\rho = \frac{1}{2}(\alpha_1 + \alpha_2 + \alpha_3) = \alpha_1 + \alpha_2 = \alpha_3 = \varpi_1 + \varpi_2$$

and hence

$$\langle \rho, \check{\alpha}_1 \rangle = \langle \rho, \check{\alpha}_2 \rangle = 1 \quad \text{while} \quad \langle \rho, \check{\rho} \rangle = 2.$$

Besides the three symmetries s_i defined by α_i the two other nontrivial elements in the Weyl group are rotations

$$r_1 = s_1 s_2 \quad \text{and} \quad r_2 = s_2 s_1.$$

Let $\delta_i = \frac{1}{2}\alpha_i$ and denote by σ_{ij} the unique element in the Weyl group which sends δ_i to $-\delta_j$. Consider vectors so that the angle between δ_i and e_i is $\pi/2$:

$$e_1 = -\varpi_2 \ , \quad e_2 = \varpi_1 \quad \text{and} \quad e_3 = z(\varpi_1 - \varpi_2).$$

We shall need the following.

Lemma 12.4.3.1 *When $\lambda_i = \delta_i + z e_i$ with $\bar{z} = -z \in \mathbb{C}$ we have*

$$\overline{-\sigma_{ij}(\lambda_i)} = \lambda_j.$$

Proof One has simply to observe that $\sigma_{ij}(e_i) = e_j$. This is obvious when $i = j$ and also when σ_{ij} is a rotation. It remains for us to check it when $\sigma_{ij} = s_3$ and $i < j$, which occurs only when $i = 1$ and $j = 2$ but $s_3(\varpi_1) = -\varpi_2$. □

We have

$$m(s_1, \lambda) = \frac{L(\langle \lambda, \check{\alpha}_1 \rangle)}{L(1 + \langle \lambda, \check{\alpha}_1 \rangle)}, \quad m(s_2, \lambda) = \frac{L(\langle \lambda, \check{\alpha}_2 \rangle)}{L(1 + \langle \lambda, \check{\alpha}_2 \rangle)}$$

$$m(r_1, \lambda) = \frac{L(\langle \lambda, \check{\alpha}_2 \rangle)}{L(1 + \langle \lambda, \check{\alpha}_2 \rangle)} \frac{L(\langle \lambda, \check{\rho} \rangle)}{L(1 + \langle \lambda, \check{\rho} \rangle)},$$

$$m(r_2, \lambda) = \frac{L(\langle \lambda, \check{\alpha}_1 \rangle)}{L(1 + \langle \lambda, \check{\alpha}_1 \rangle)} \frac{L(\langle \lambda, \check{\rho} \rangle)}{L(1 + \langle \lambda, \check{\rho} \rangle)}$$

and since $s_3 = s_1 s_2 s_1$

$$m(s_3, \lambda) = \frac{L(\langle \lambda, \check{\alpha}_1 \rangle)}{L(1 + \langle \lambda, \check{\alpha}_1 \rangle)} \frac{L(\langle \lambda, \check{\alpha}_2 \rangle)}{L(1 + \langle \lambda, \check{\alpha}_2 \rangle)} \frac{L(\langle \lambda, \check{\rho} \rangle)}{L(1 + \langle \lambda, \check{\rho} \rangle)}.$$

The singularities of functions $m(s, \lambda)$ in the domain

$$\mathcal{R}_s = \{\lambda \mid \mathrm{Re}(\langle \lambda, \check{\alpha}_i \rangle) \geq 0 \ \text{ for } \ \alpha_i > 0 \ \text{ and } \ s(\alpha_i) < 0\}$$

are along the affine subspaces

$$\Lambda_i = \{\lambda \mid \langle \lambda, \check{\alpha}_i \rangle = 1\}.$$

In fact, for $\lambda \in \mathcal{R}_s$, factors in the numerators are holomorphic except for poles whenever $\langle \lambda, \check{\alpha}_i \rangle = 1$ or 0 for some i, while denominators are holomorphic and nonzero whenever $\langle \lambda, \check{\alpha}_i \rangle \neq 0$; moreover, $m(s, \lambda)$ is holomorphic when $(\langle \lambda, \check{\alpha}_i \rangle) = 0$ for some i, since singularities of numerator and denominator cancel.

The subspaces Λ_i are the sets of $\lambda_i = \delta_i + ze_i$ with $z \in \mathbb{C}$. The residue of $m(\sigma_{ij}, \lambda)$ at $\lambda = \lambda_i = \delta_i + ze_i$ can be written

$$\frac{1}{L(2)} n_{ij}(z) = \frac{1}{L(2)} n(\sigma_{ij}, \lambda_i).$$

The matrix $\mathcal{N}(z)$ with entries $n_{ij}(z) = n(\sigma_{ij}, \lambda_i)$ is given by

$$\mathcal{N}(z) = \begin{pmatrix} 1 & n(s_3, \lambda_1) & n(r_2, \lambda_1) \\ n(s_3, \lambda_2) & 1 & n(r_1, \lambda_2) \\ n(r_1, \lambda_3) & n(r_2, \lambda_3) & n(s_3, \lambda_3) \end{pmatrix}.$$

Now, taking into account that

$$L(1 + \langle \lambda_1, \check{\alpha}_2 \rangle) = L(\langle \lambda_1, \check{\rho} \rangle) \quad \text{and} \quad L(1 + \langle \lambda_2, \check{\alpha}_1 \rangle) = L(\langle \lambda_2, \check{\rho} \rangle)$$

we see that

$$n(s_3, \lambda_1) = \frac{L(\langle \lambda_1, \check{\alpha}_2 \rangle)}{L(1 + \langle \lambda_1, \check{\alpha}_2 \rangle)} \frac{L(\langle \lambda_1, \check{\rho} \rangle)}{L(1 + \langle \lambda_1, \check{\rho} \rangle)} = \frac{L(\langle \lambda_1, \check{\alpha}_2 \rangle)}{L(1 + \langle \lambda_1, \check{\rho} \rangle)}$$

and a similar cancellation occurs for $n(s_3, \lambda_2)$. This shows that

$$\mathcal{N}(z) = \begin{pmatrix} 1 & \frac{L(\langle \lambda_1, \check{\alpha}_2 \rangle)}{L(1+\langle \lambda_1, \check{\rho} \rangle)} & \frac{L(\langle \lambda_1, \check{\rho} \rangle)}{L(1+\langle \lambda_1, \check{\rho} \rangle)} \\ \frac{L(\langle \lambda_2, \check{\alpha}_1 \rangle)}{L(1+\langle \lambda_2, \check{\rho} \rangle)} & 1 & \frac{L(\langle \lambda_2, \check{\rho} \rangle)}{L(1+\langle \lambda_2, \check{\rho} \rangle)} \\ \frac{L(\langle \lambda_3, \check{\alpha}_1 \rangle)}{L(1+\langle \lambda_3, \check{\alpha}_1 \rangle)} & \frac{L(\langle \lambda_3, \check{\alpha}_2 \rangle)}{L(1+\langle \lambda_3, \check{\alpha}_2 \rangle)} & \frac{L(\langle \lambda_3, \check{\alpha}_1 \rangle)}{L(1+\langle \lambda_3, \check{\alpha}_1 \rangle)} \frac{L(\langle \lambda_3, \check{\alpha}_2 \rangle)}{L(1+\langle \lambda_3, \check{\alpha}_2 \rangle)} \end{pmatrix}.$$

Since $\langle \delta_1, \check{\alpha}_2 \rangle = \langle \delta_2, \check{\alpha}_1 \rangle = -\frac{1}{2}$ and $\langle \delta_i, \check{\rho} \rangle = \frac{1}{2}$ we get

$$\mathcal{N}(z) = \begin{pmatrix} 1 & \frac{L(-z-\frac{1}{2})}{L(-z+\frac{3}{2})} & \frac{L(-z+\frac{1}{2}))}{L(-z+\frac{3}{2}))} \\ \frac{L(z-\frac{1}{2})}{L(z+\frac{3}{2})} & 1 & \frac{L(z+\frac{1}{2}))}{L(z+\frac{3}{2}))} \\ \frac{L(z+\frac{1}{2}))}{L(z+\frac{3}{2}))} & \frac{L(-z+\frac{1}{2}))}{L(-z+\frac{3}{2}))} & \frac{L(z+\frac{1}{2}))}{L(z+\frac{3}{2}))} \frac{L(-z+\frac{1}{2}))}{L(-z+\frac{3}{2}))} \end{pmatrix}.$$

Lemma 12.4.3.2 *The matrix $\mathcal{N}(z)$ is of rank one and $n_{ij}(z) = n_{ji}(-z)$.*

Proof That $\mathcal{N}(z)$ is of rank one can be checked using its explicit expression and the functional equation for $L(\sigma)$. $\qquad\square$

Lemma 12.4.3.3 *One has*

$$n_{ij}(z) = n_{ik}(z)\overline{n_{jk}(z)} \quad \text{if } \bar{z} = -z \text{ for } k = 1 \text{ or } 2.$$

Proof Since $\mathcal{N}(z)$ is of rank one there exist functions $c_{ij}(z)$ such that, for any i, j or k,

$$c_{ik}(z)n_{kj}(z) = n_{ij}(z) \quad \text{and} \quad c_{ik}(z)c_{kj}(z) = c_{ij}(z).$$

In particular

$$c_{ik}(z)c_{ki}(z) = 1.$$

But since $n_{kk} = 1$ for $k = 1$ or 2 we have $c_{ki}(z)n_{ik}(z) = 1$ and hence

$$c_{ik}(z) = n_{ik}(z) \quad \text{for } k = 1 \text{ or } 2.$$

This shows that

$$n_{ij}(z) = n_{ik}(z)n_{kj}(z)$$

for $k = 1$ or 2 and the lemma follows since

$$n_{jk}(z) = \overline{n_{kj}(z)} \quad \text{if } \bar{z} = -z. \qquad \square$$

12.4.4 Contour Deformation for $\mathcal{C}_{P_0}^K$

We start from

$$\langle E_\phi, E_\phi \rangle_{\mathcal{L}} = \sum_{s \in W(\mathfrak{a}_{P_0}, \mathfrak{a}_{P_0})} \int_{\lambda \in \lambda_0 + \Lambda_{P_0}} m(s,\lambda)\Phi(e_G,\lambda)\overline{\Phi(e_G, -s\bar{\lambda})}\, d\lambda.$$

When we move λ_0 to 0, ignoring for a while the singularities on affine lines $\langle \lambda, \check{\alpha}_i \rangle = 1$ of the intertwining operators, we get an integral over a space of dimension 2:

$$A = \sum_{s \in W(\mathfrak{a}_{P_0}, \mathfrak{a}_{P_0})} \int_{\lambda \in \Lambda_{P_0}} m(s,\lambda)\Phi_0(\lambda)\overline{\Phi_0(s\lambda)}\, d\lambda,$$

with $\Phi_0(\lambda) := \Phi(e_G,\lambda)$. Here Λ_{P_0} is of dimension 2.

We get, as a particular case of Proposition 12.3.4.1, the expected contribution to $\langle E_\phi, E_\phi \rangle_{\mathcal{L}}$ coming from P_0 to the spectral decomposition, when the contribution of residues is omitted. It will be denoted by A.

Proposition 12.4.4.1

$$A = \frac{1}{6} \int_{\lambda \in \Lambda_{P_0}} \langle E_\phi, E(\bullet; \Psi_0, \lambda) \rangle_{\mathcal{L}} \overline{\langle E_\phi, E(\bullet; \Psi_0, \lambda) \rangle_{\mathcal{L}}}\, d\lambda,$$

where Ψ_0 is a constant function on G normalized so that $\langle \Psi_0, \Psi_0 \rangle_{\mathcal{L}_{P_0}} = 1$.

Now, consider the residue on affine lines $\langle \lambda, \check{\alpha}_i \rangle = 1$ and move the contour to Λ_i where

$$\Lambda_i = \{\lambda_i \mid \langle \lambda_i, \check{\alpha}_i \rangle = 1, \operatorname{Re}(\lambda_i) = \delta_i\}$$

are affine subspaces of real dimension 1: $\lambda_i = \delta_i + z e_i$ with $\bar{z} = -z \in \mathbb{C}$. Ignoring the possible *"double residues"* that will be treated later on (see Proposition 12.4.4.3) and thanks to Lemma 12.4.3.1 we see that

$$\overline{-\sigma_{ij}(\lambda_i)} = \lambda_j$$

and we get a contribution of the form

$$B = \frac{1}{L(2)} \sum_{i=1}^{3} \sum_{j=1}^{3} \int_{z \in i\mathbb{R}} n_{ij}(z) \Phi_i(z) \overline{\Phi_j(z)} \, dz$$

if

$$\Phi_i(z) = \Phi(e_G, \lambda_i) = \Phi(e_G, \delta_i + z e_i).$$

Now, let us denote by Ψ_0 the function identically equal to 1. It yields an element of norm 1 in $\mathcal{L}_{P_0}$ since, by assumption, $\mathrm{vol}\,(K) = 1$. Let $E_i(x; \Psi_0, z e_i)$ be the residue of $E(x; \Psi_0, \lambda)$ on the affine complex line $\lambda = \delta_i + z e_i$; its constant term along P_0 is given by

$$\int_{\Gamma \cap N_0 \backslash N_0} E_i(nx; \Psi_0, z e_i) \, dn = \frac{1}{L(2)} \sum_{j=1}^{3} e^{\langle \sigma_{ij}(\delta_i + z e_i) + \rho, H_0(x) \rangle} n_{ij}(z) \Psi_0(x),$$

which, thanks to Lemma 12.4.3.1, is equal to

$$\frac{1}{L(2)} \sum_{j=1}^{3} e^{\langle -\delta_j + z e_j + \rho, H_0(x) \rangle} n_{ij}(z) \Psi_0(x).$$

Now

$$\langle E_\phi, E_i(\bullet; \Psi_0, z e_i) \rangle = \frac{1}{L(2)} \sum_{j=1}^{3} \int_{X_{P_0}} \phi(x) e^{\langle -\delta_i - z e_j + \rho, H_0(x) \rangle} \overline{n_{ij}(z) \Psi_0(x)} \, dx.$$

Since $\Psi_0(x) \equiv 1$ and $\overline{n_{ij}(z)} = n_{ji}(z)$ if $\overline{z} = -z$ we have

$$\langle E_\phi, E_i(\bullet; \Psi_0, z e_i) \rangle = \frac{1}{L(2)} \sum_{j=1}^{3} n_{ji}(z) \Phi_j(z).$$

Recall that

$$B = \frac{1}{L(2)} \sum_{i=1}^{3} \sum_{j=1}^{3} \int_{z \in i\mathbb{R}} n_{ij}(z) \Phi_i(z) \overline{\Phi_j(z)} \, dz.$$

Thanks to Lemma 12.4.3.3 this can be rewritten as

$$B = \frac{1}{L(2)} \sum_{i=1}^{3} \sum_{j=1}^{3} \int_{z \in i\mathbb{R}} n_{i1}(z) \overline{n_{j1}(z)} \Phi_i(z) \overline{\Phi_j(z)} \, dz,$$

and altogether we get

$$B = L(2) \int_{i\mathbb{R}} \langle E_\phi, E_1(\bullet; \Psi_0, z e_i) \rangle \overline{\langle E_\phi, E_1(\bullet; \Psi_0, z e_i) \rangle} \, dz \ .$$

Observe that we could have chosen E_2 instead of E_1. Let Ψ_1 to be a constant function of norm 1 in $\mathcal{L}_{P_1}$. We may take

$$\Psi_1(x) = \frac{\Psi_0(x)}{\sqrt{L(2)}} \equiv \frac{1}{\sqrt{L(2)}}$$

and hence the contribution B of simple residues to the spectral decomposition for the space of K-invariant vectors is given by the following.

Proposition 12.4.4.2

$$B = \int_{\lambda_1 \in \Lambda_{P_1}} \langle E_\phi, E_1(\bullet; \Psi_1, \lambda_1) \rangle_{\mathcal{L}} \overline{\langle E_\phi, E_1(\bullet; \Psi_1, \lambda_1) \rangle_{\mathcal{L}}} \, d\lambda_1,$$

where Λ_{P_1} is the line $z e_1$ with z imaginary.

Using the associate parabolic subgroup P_2 instead of P_1 one would get the same expression for B but with E_2 in place of E_1.

We have now to list elements s in the Weyl group and values of

$$\lambda = a \varpi_1 + b \varpi_2,$$

such that $m(s, \lambda)$ has a pole in each variable a and b, and compute the double residues. There are five possibilities. The function $m(r_1, \lambda)$ (resp. $m(r_2, \lambda)$) has a pole in each variable when $\lambda = \varpi_2$ (resp. $\lambda = \varpi_1$), and $m(s_3, \lambda)$ has also a pole in each variable when $\lambda = \varpi_2$ or $\lambda = \varpi_1$. The double residues are

$$\frac{1}{L(2)L(2)} \quad \text{for} \quad m(r_1, \lambda) \quad \text{or} \quad m(r_2, \lambda).$$

Since $r_1(\varpi_2) = -\varpi_1$ and $r_2(\varpi_1) = -\varpi_2$ we get as contributions from r_1 and r_2

$$\frac{1}{L(2)L(2)} \Phi(e_G, \varpi_2) \overline{\Phi(e_G, \varpi_1)} \quad \text{and} \quad \frac{1}{L(2)L(2)} \Phi(e_G, \varpi_1) \overline{\Phi(e_G, \varpi_2)}.$$

Now observe that

$$\lim_{\lambda \to \varpi_i} \frac{L(\langle \lambda, \check{\alpha}_i \rangle)}{L(1 + \langle \lambda, \check{\alpha}_i \rangle)} = -1$$

for $i = 1$ or 2. This shows that the double residues for $m(s_3, \lambda)$ at $\lambda = \varpi_i$ are equal to

$$\frac{-1}{L(2)L(2)}.$$

Since $s_3(\varpi_2) = -\varpi_1$ and $s_3(\varpi_1) = -\varpi_2$ the double residues of

$$m(s_3, \lambda) \Phi(e_G, \lambda) \overline{\Phi(e_G, -s\lambda)}$$

are

$$\frac{-1}{L(2)L(2)}\Phi(e_G,\varpi_2)\overline{\Phi(e_G,\varpi_1)} \quad \text{and} \quad \frac{-1}{L(2)L(2)}\Phi(e_G,\varpi_1)\overline{\Phi(e_G,\varpi_2)}.$$

These two contributions of s_3 cancel those coming from r_1 and r_2, and hence the double residues at $\lambda = \varpi_i$ do not contribute to the spectral decomposition. The fifth double residue occurs for $m(s_3, \lambda)$ at $\lambda = \rho$ and is given by

$$\frac{1}{L(2)L(3)}$$

so that, altogether, the contribution of the double residues is

$$C = \frac{1}{L(2)L(3)}\Phi(e_G,\rho)\overline{\Phi(e_G,\rho)}.$$

Using that

$$\langle 1,1\rangle_{\mathcal{L}} = \text{vol } (X_G) = L(2)L(3)$$

when the volume is computed using the Haar measure described above, we get the following.

Proposition 12.4.4.3

$$C = \langle E_\phi, \Psi_G\rangle_{\mathcal{L}}\langle \Psi_G, E_\phi\rangle_{\mathcal{L}}$$

if Ψ_G is a constant function of norm 1 in $\mathcal{L}$.

Summing up the contributions A, B, and C given in Proposition 12.4.4.1, 12.4.4.2, and 12.4.4.3, we have the following.

Proposition 12.4.4.4 *For $\mathbf{G} = GL(3)$ the spectral decomposition of the space $\mathcal{C}_{P_0}^K$ is given by*

$$\langle E_\phi, E_\phi\rangle_{\mathcal{L}} = \frac{1}{6}\int_{\lambda\in\Lambda_{P_0}}\langle E_\phi, E(\bullet; \Psi_0, \lambda)\rangle_{\mathcal{L}}\overline{\langle E_\phi, E(\bullet; \Psi_0, \lambda)\rangle_{\mathcal{L}}}\ d\lambda$$

$$+ \int_{\lambda_1\in\Lambda_{P_1}}\langle E_\phi, E_1(\bullet; \Psi_1, \lambda_1)\rangle_{\mathcal{L}}\overline{\langle E_\phi, E_1(\bullet; \Psi_1, \lambda_1)\rangle_{\mathcal{L}}}\ d\lambda_1$$

$$+ \langle E_\phi, \Psi_G\rangle_{\mathcal{L}}\langle \Psi_G, E_\phi\rangle_{\mathcal{L}}.$$

12.4.5 Other Contributions to $\mathcal{L}^K$

Recall that

$$\mathcal{L}^K = \mathcal{C}_{P_0}^K \oplus \mathcal{C}_{\mathcal{P}}^K \oplus \mathcal{C}_G^K,$$

where $\mathcal{P}$ is the association class of maximal standard parabolic subgroups. Since

$$\mathcal{C}_G^K = \mathcal{L}_{\text{cusp}}^K$$

it remains for us to analyze the space $\mathcal{C}_{\mathcal{P}}^K$ generated by the pseudo-Eisenstein series induced by K-invariant cusp form on the maximal parabolic subgroups $P_i \in \mathcal{P}$. Given $\phi_i \in \mathcal{D}_{P_i}$, the scalar product is given by

$$\langle E_{\phi_i}, E_{\phi_j} \rangle_{\mathcal{L}} = \sum_{s \in W(\mathfrak{a}_{P_i}, \mathfrak{a}_{P_j})} \int_{\lambda_i \in \lambda_0 + \Lambda_{P_i}} \langle M(s, \lambda_i) \Phi_1(\bullet, \lambda_i), \Phi_j(\bullet, -s\bar{\lambda}_i) \rangle_{\mathcal{L}_{P_i}} d\lambda_i.$$

We observe that $W(\mathfrak{a}_{P_i}, \mathfrak{a}_{P_j})$ is reduced to a single element. When $i = j = 1$ the deformation of the contour meets no singular point and yields a contribution of the form

$$\langle E_{\phi_1}, E_{\phi_1} \rangle_{\mathcal{L}} = \sum_{\Psi \in \mathcal{B}_{\text{cusp}}(P_1)} \int_{\lambda_1 \in \Lambda_{P_1}} \langle E_{\phi_1}, E(\bullet; \Psi, \lambda_1) \rangle_{\mathcal{L}} \overline{\langle E_{\phi_1}, E(\bullet; \Psi, \lambda_1) \rangle_{\mathcal{L}}} \, d\lambda_1,$$

where $\mathcal{B}_{\text{cusp}}^K(P_1)$ is an orthonormal basis of the K-invariant cuspidal spectrum for P_1 and $E(\bullet; \Psi, \lambda_1)$ is the Eisenstein series constructed from Ψ. No new contribution comes from the other scalar products. To show this, one has to analyze the intertwining $M(s, \lambda_1)$ operator attached to the unique element s in $W(\mathfrak{a}_{P_1}, \mathfrak{a}_{P_2})$. It turns out that this operator is built out of L-functions for cuspidal representations of $GL(2)$ that are holomorphic on the whole complex line and yield no further singularities when moving the contour.

References

[1] ARTHUR, JAMES *Eisenstein series and the trace formula.* Automorphic forms, representations and L-functions (Proc. Sympos. Pure Math., Oregon State University, Corvallis, Oregon, 1977), Part 1, pp. 253–274, Proc. Sympos. Pure Math., XXXIII, Amer. Math. Soc., Providence, R.I., 1979.

[2] ARTHUR, JAMES *A trace formula for reductive groups. II. Applications of a truncation operator.* Compositio Math. 40 (1980), no. 1, 87–121.

[3] BOREL, ARMAND *Reduction theory for arithmetic groups.* Algebraic Groups and Discontinuous Subgroups (Proc. Sympos. Pure Math., Boulder, Colorado, 1965) pp. 20–25 Amer. Math. Soc., Providence, R.I., 1966.

[4] BOREL, ARMAND *Introduction to automorphic forms.* Algebraic Groups and Discontinuous Subgroups (Proc. Sympos. Pure Math., Boulder, Colorado, 1965) pp. 199–210 Amer. Math. Soc., Providence, R.I., 1966.

[5] BOREL, ARMAND; JACQUET, HERVÉ. *Automorphic forms and automorphic representations. With a supplement "On the notion of an automorphic representation" by R. P. Langlands.* Proc. Sympos. Pure Math., XXXIII, Automorphic forms,

representations and L-functions (Proc. Sympos. Pure Math., Oregon State University, Corvallis, Oregon, 1977), Part 1, pp. 189–207, Amer. Math. Soc., Providence, R.I., 1979.

[6] BERNSTEIN, JOSEPH; LAPID, EREZ *On the meromorphic continuation of Eisenstein series* arXiv:1911.02342, 2019.

[7] FRANKE, JENS Harmonic analysis in weighted L2-spaces. *Ann. Sci. École Norm. Sup.* (4) 31 (1998), no. 2, 181–279.

[8] GODEMENT, ROGER *Analyse spectrale des fonctions modulaires.* [Spectral analysis of modular functions] Séminaire Bourbaki, Vol. 9, Exp. No. 278, 15–40, Soc. Math. France, Paris, 1995.

[9] HARISH-CHANDRA *Automorphic Forms on Semisimple Lie Groups.* Lecture Notes in Mathematics, Vol. 62. Springer-Verlag, Berlin-New York, 1968.

[10] KOTTWITZ, ROBERT E. Tamagawa numbers. *Ann. of Math.* (2) 127 (1988), no. 3, 629–646.

[11] KUBOTA, TOMIO *Elementary theory of Eisenstein series.* Kodansha Ltd., Tokyo; Halsted Press [John Wiley & Sons], New York-London-Sydney, 1973.

[12] LAI, K. F. Tamagawa number of reductive algebraic groups. *Compositio Math.* 41 (1980), no. 2, 153–188.

[13] LABESSE, JEAN-PIERRE; WALDSPURGER, JEAN-LOUP *La formule des traces tordue d'après le Friday Morning Seminar. With a foreword by Robert Langlands.* CRM Monograph Series, 31. American Mathematical Society, Providence, R.I., 2013.

[14] LANGLANDS, ROBERT P. *On the functional equations satisfied by Eisenstein series.* Lecture Notes in Mathematics, Vol. 544. Springer-Verlag, Berlin–New York, 1976.

[15] LANGLANDS, ROBERT P. *The volume of the fundamental domain for some arithmetical subgroups of Chevalley groups.* Algebraic Groups and Discontinuous Subgroups (Proc. Sympos. Pure Math., Boulder, Colorado, 1965) pp. 143–148 Amer. Math. Soc., Providence, R.I., 1966.

[16] LANGLANDS, ROBERT P. *Eisenstein series.* Algebraic Groups and Discontinuous Subgroups (Proc. Sympos. Pure Math., Boulder, Colorado, 1965) pp. 235–252 Amer. Math. Soc., Providence, R.I., 1966.

[17] LANGLANDS, ROBERT P. *Euler products.* A James K. Whittemore Lecture in Mathematics given at Yale University, 1967. Yale Mathematical Monographs, 1. Yale University Press, New Haven, Connecticut.

[18] LANGLANDS, ROBERT P. *Problems in the theory of automorphic forms.* Lectures in modern analysis and applications, III, pp. 18–61. Lecture Notes in Math., Vol. 170, Springer, Berlin, 1970.

[19] MŒGLIN, COLETTE; WALDSPURGER, JEAN-LOUP Le spectre résiduel de GL(n). *Ann. Sci. École Norm. Sup.* (4) 22 (1989), no. 4, 605–674.

[20] MŒGLIN, COLETTE; WALDSPURGER, JEAN-LOUP *Spectral decomposition and Eisenstein series.* Cambridge Tracts in Mathematics, 113. Cambridge University Press, Cambridge, 1995, xxviii+338 pp.

[21] ROGAWSKI, JONATHAN D. *Automorphic representations of unitary groups in three variables.* Annals of Mathematics Studies, 123. Princeton University Press, Princeton, NJ, 1990.

[22] SCHIFFMANN, GÉRARD Intégrales d'entrelacement et fonctions de Whittaker. *Bull. Soc. Math. France* 99 (1971), 3–72.

[23] SELBERG, ATLE *Discontinuous groups and harmonic analysis.* Proc. Internat. Congr. Mathematicians (Stockholm, 1962) pp. 177–189, 1963.

[24] TITS, JACQUES *Reductive groups over local fields.* Automorphic forms, representations and L-functions (Proc. Sympos. Pure Math., Oregon State University, Corvallis, Oregon, 1977), Part 1, pp. 29–69, Proc. Sympos. Pure Math., XXXIII, Amer. Math. Soc., Providence, R.I., 1979.

13

Automorphic Representations and L-Functions for $GL(n)$

Dorian Goldfeld and Hervé Jacquet

13.1 Introduction

Two of the main achievements of Hecke are the investigation of the L-function attached to a Grössencharacter and the L-function attached to a modular form. The modern view is that these are instances of the general notion of the L-function $L(s,\pi)$ attached to an automorphic representation π of the group $GL(n)$ over a number field F. The simplest method to obtain the analytic properties of this function is to imitate the construction of Tate in his thesis [33]. But we would like to stress that Hecke's method based on the Fourier expansion of modular forms gives the same result. Moreover Hecke's method generalizes to $GL(n)$.

Our goal in this chapter is to briefly review this method as explained in [16], [23], [24]. We refer to the book of Moeglin and Waldspurger [28] as a convenient reference for the general theory of automorphic forms. There is a huge literature on the subject (see the references in the above works). Here, in addition to the early work of Godement ([13],[14]) and Tamagawa [32], we quote the work of Maloletkin [27] who, like Godement, saw that in the Poisson formula, one can ignore the singular matrices when dealing with cuspforms.

Langlands was aware of the possibility of defining an L-function this way, even before the full theory was available, and alludes to it in his famous letter to Weil.

We cannot, in this elementary paper, get into the Langlands program. But we can at least state one conjecture, which is alluded to in the letter to Weil. Let E/F be an extension of number fields of degree n and let χ be an idèle class character for E. Then the L-function $L(s,\chi)$ attached to χ is equal to the L-function $L(s,\pi)$ attached to an automorphic representation π for the group $GL(n,F)$. This representation π needs not be cuspidal but the L-function $L(s,\pi)$ may be written as a product of L-functions $L(s,\pi_i)$ where the π_i are cuspidal automorphic representations for various groups $GL(n_i,F)$.

Dorian Goldfeld is partially supported by Simons Collaboration Grant 567168

Figure 13.1 Hervé Jacquet and Robert Langlands.
(Courtesy of the Simons Foundation).

In particular, we can take χ to be the trivial character. Then $L(s,\chi)$ is the Dedekind zeta function of E/F.

13.2 Local Non-Archimedean Theory

13.2.1 Smooth Representations

In this section F is a non-Archimedean local field. We denote by ψ a nontrivial additive character of F. We let $\mathcal{O}_F$ be the ring of integers of F, and by q_F or simply q, we mean the cardinality of the residual field.

We let G be the group $GL(n)$ regarded as an algebraic group. For $g \in G(F)$ we define its norm

$$\|g\| = \sup_{\substack{1 \le i \le n \\ 1 \le j \le n}} \left(\sup \left(|g_{ij}|, |(g^{-1})_{ij}| \right) \right).$$

We first describe the smooth representations of $G(F)$ (or more generally of $G(F)$ here G is a product of GL groups) ([1], [2], [3], [4], [21]). A representation π of $G(F)$ on a complex vector V is said to be smooth if the stabilizer of every vector $v \neq 0$ in V is an open subgroup of $G(F)$. A smooth representation π of $G(F)$ on V is admissible if, conversely, the space $V^{K'}$ of vectors fixed by a compact open subgroup K' of $G(F)$ is finite dimensional. If (π, V) is a smooth representation we denote by V^* the algebraic dual and by $\tilde{V}$ the subspace of those vectors (linear forms) fixed by some compact open subgroup. We denote by $\tilde{\pi}$ the representation of $G(F)$ on $\tilde{V}$. We say that $(\tilde{\pi}, \tilde{V})$ is the representation contragredient to (π, V). A smooth representation is said to be irreducible if it is algebraically irreducible. Any irreducible smooth representation is admissible. More precisely, given an open compact subgroup K' there is a constant c such that, for any irreducible representation (π, V), the dimension of $V^{K'}$ is bounded by c [1]. If π is an irreducible representation of $GL(n, F)$ then there is a character ω_π of $F^\times$ such that

$$\pi(zI_n) = \omega_\pi(z)$$

for all $z \in F^\times$. We call ω_π the central character of π. A function of the form

$$f(g) = \langle \pi(g)v, \tilde{v} \rangle, \ v \in V, \ \tilde{v} \in \tilde{V},$$

is a matrix coefficient of π. Then the function $\check{f}$ defined by

$$\check{f}(g) = f(g^{-1})$$

is a matrix coefficient of $\tilde{\pi}$.

If π is a unitary (topologically) irreducible representation of $G(F)$ on a Hilbert space H with scalar product $(\bullet, \bullet)$ then the space V of smooth vectors (i.e. fixed by some compact open subgroup of $G(F)$) is invariant under $G(F)$ and the representation (also denoted π) of $G(F)$ on V is algebraically irreducible and admissible. The representation $\tilde{\pi}$ is then the imaginary conjugate of π. In particular, for v_1, v_2 in V the function

$$f(g) = (\pi(g)v_1, v_2)$$

is a matrix coefficient of the admissible representation π. We say that an irreducible admissible representation is unitarizable if it is the space of smooth vectors in a topologically irreducible unitary representation of $G(F)$ on a Hilbert space.

We first review the definition of an induced representation. We let $P = P^{(m_1, m_2, \ldots, m_n)}$ be the upper parabolic subgroup of type $(m_1, m_2, \ldots, m_r)$ with $\sum_{1 \leq i \leq r} m_i = n$. This is the group of matrices of the form

$$p = \begin{pmatrix} g_1 & u_{12} & u_{13} & \cdots & u_{1j} & \cdots & u_{1r} \\ 0 & g_2 & u_{23} & \cdots & u_{2j} & \cdots & u_{2r} \\ 0 & 0 & g_3 & \cdots & u_{3j} & \cdots & u_{3r} \\ \cdots & \cdots & \cdots & \cdots & \cdots & \cdots & \cdots \\ 0 & 0 & 0 & \cdots & \cdots & g_{r-1} & u_{(r-1)r} \\ 0 & 0 & 0 & \cdots & \cdots & 0 & g_r \end{pmatrix},$$

where $g_i \in GL(m_i)$ and $u_{i,j}$ is a matrix with m_i rows and m_j columns. The unipotent radical $U = U^{(m_1, m, \ldots, m_n)}$ is the group of matrices with $g_i = 1$ for all i. We let $M = M_{(m_1, m_2, \ldots, m_n)}$ be the subgroup of matrices for which $u_{ij} = 0$ for all (i, j). So we have the Levi decomposition

$$P = MU,$$

and

$$G(F) = P(F)K = U(F)M(F)K,$$

where K is the standard maximal compact subgroup

$$K := GL(n, \mathcal{O}_F).$$

Let π_i, $1 \le i \le r$, be an irreducible (or simply admissible) representation of $GL(r_i, F)$ on a complex vector space V_i. We set

$$V = \bigotimes_{1 \le i \le r} V_i$$

and denote by $\sigma = \bigotimes \pi_i$ the tensor product representation of $M(F)$ on V.

We denote by δ_P the topological module of the locally compact group $P(F)$. Recall δ_P is trivial on $U(F)$ and is given on $M(F)$ by the formula

$$d(mum^{-1}) = \delta_P(m)du,$$

where du denotes a Haar measure on $U(F)$. In general, we denote by $\rho(g)$ the right translation of a function ϕ by g:

$$\rho(g)\phi(h) = \phi(hg).$$

The space of the corresponding induced representation

$$\text{Ind}(G, P; \pi_1, \pi_2, \ldots, \pi_r)$$

is the space of functions

$$\phi : G(F) \to V$$

such that

$$\phi(mug) = \delta_P^{1/2} \sigma(m)\phi(g)$$

for all $g \in G(F), m \in M(F), u \in U(F)$ and there is a compact open subgroup $K' \subset K$ such that, for all $k \in K'$,

$$\rho(k')\phi = \phi.$$

The representation π of $G(F)$ on the induced representation is by right-shifts. We can consider the representation

$$\text{Ind}(G, P; \tilde{\pi}_1, \tilde{\pi}_2, \ldots, \tilde{\pi}_r)$$

with $\tilde{V} = \bigotimes \tilde{V}_i, \tilde{\sigma} = \bigotimes \tilde{\pi}_i$. We have on $V \times \tilde{V}$ the invariant scalar product

$$\langle \otimes v_i, \otimes \tilde{v}_i \rangle = \prod_i \langle v_i, \tilde{v}_i \rangle$$

so that $(\widetilde{V}, \widetilde{\sigma})$ is contragredient to (V, σ). It follows that for $\phi \in \mathrm{Ind}(G, P; \pi_1, \pi_2, \ldots, \pi_r)$ and $\widetilde{\phi} \in \mathrm{Ind}(G, P; \widetilde{\pi}_1, \widetilde{\pi}_2, \ldots, \widetilde{\pi}_r)$, we have

$$\langle \phi(pg), \widetilde{\phi}(pg) \rangle = \delta_P(p) \langle \phi(g), \widetilde{\phi(g)} \rangle.$$

Hence if we set

$$\langle \phi, \widetilde{\phi} \rangle := \int_K \langle \phi(k), \widetilde{\phi}(k) \rangle dk$$

we obtain an invariant nondegenerate scalar product and $\widetilde{\pi}$ is indeed contragredient to π.

An irreducible representation of $GL(n, F)$ is said to be supercuspidal if it is not a component of an induced representation. A character of $GL(1, F)$ is by definition a supercuspidal representation. For $n > 1$ a matrix coefficient f of a supercuspidal representation transforms under the central character ω_π of π, that is,

$$f(zg) = \omega_\pi(z)f(g),$$

for all $z \in Z(F)$ and all g. The function f is compactly supported modulo the center. Moreover, if U is the unipotent radical of a proper parabolic subgroup of G, then

$$\int_{U(F)} f(g_1 u g_2) du = 0.$$

Any irreducible representation π is a subrepresentation of an induced representation

$$\mathrm{Ind}(G, P^{n_1, n_2, \ldots, n_r}; \sigma_1, \sigma_2, \ldots, \sigma_r),$$

where each σ_i is a supercuspidal representation of $GL(n_i, F)$.

13.2.2 The Main Theorem

Let π be an irreducible smooth representation of $GL(n, F)$. Let Φ be a Schwartz–Bruhat function on $M(n \times n, F)$ and f a matrix coefficient of π. We consider the integral

$$Z(\Phi, f, s) := \int_{GL(n, F)} \Phi(g) f(g) \, |\det g|^{s + \frac{n-1}{2}} dg.$$

We define the Fourier transform $\widehat{\Phi}$ of a Schwartz–Bruhat function Φ on $M(n \times n, F)$ by

$$\widehat{\Phi}(X) = \int_{M(n \times n, F)} \Phi(y) \psi(-\mathrm{tr} XY) dY.$$

The Haar measure dY is self-dual, that is, for all Φ,

$$\int \widehat{\Phi}(X)dX = \Phi(0).$$

Recall the notation $\check{f}(g) := f(g^{-1})$.

Theorem 13.2.1 *Let the notations be as above.*
(i) *The integral defining $Z(\Phi, f, s)$ converges absolutely for $\mathrm{Re}(s)$ sufficiently large ($\mathrm{Re}(s) > 0$ if π is tempered and $\mathrm{Re}(s) > \frac{n-1}{2}$ if π is unitary).*
(ii) *$Z(\Phi, f, s)$ is a rational function of q^{-s}, q^s. More precisely the space spanned by these integrals is a fractional ideal of $\mathbb{C}[q^{-s}, q^s]$ with a unique generator of the form*

$$L(s, \pi) = \frac{1}{P(q^{-s})}, \qquad \left(P \in \mathbb{C}[q^{-s}], P(0) = 1\right).$$

(iii) *There is a functional equation*

$$Z(1 - s, \widehat{\Phi}, \check{f}) = \gamma\,(s, \pi, \psi)\, Z(f, \Phi, s),$$

where $\gamma\,(s, \pi, \psi)$ is rational. Furthermore,

$$\gamma\,(s, \pi, \psi) = \frac{\epsilon(s, \pi, \psi) L(1 - s, \widetilde{\pi})}{L(s, \pi)},$$

where $\epsilon(s, \pi, \psi)$ has the form $c q^{-ms}$.

The factors $\epsilon(s, \pi, \psi)$ and $\gamma(s, \pi, \psi)$ depend on ψ. It is easily seen that if $\psi_a(x) := \psi(x)$, where $a \in F^\times$, then

$$\epsilon(s, \pi, \psi_a) = \omega_\pi(a)|s|^{n(s-\frac{1}{2})}\epsilon(s, \pi, \psi),$$

where ω_π is the central character of π.

From its definition it is clear that the factor $\epsilon(s, \pi, \psi)$ is a monomial in q^{-s}. If we apply the functional equation twice we find

$$\gamma(1 - s, \widetilde{\pi}, \overline{\psi}) \cdot \gamma(s, \pi, \psi) = 1,$$

or equivalently

$$\epsilon(1 - s, \widetilde{\pi}, \overline{\psi}) \cdot \epsilon(s, \pi, \psi) = 1.$$

In particular for $s = \frac{1}{2}$ we find

$$\epsilon\left(\frac{1}{2}, \widetilde{\pi}, \overline{\psi}\right) \epsilon\left(\frac{1}{2}, \pi, \psi\right) = 1.$$

If π is unitary we have

$$\epsilon\left(\frac{1}{2}, \widetilde{\pi}, \overline{\psi}\right) = \overline{\epsilon\left(\frac{1}{2}, \pi, \psi\right)},$$

and so

$$\left| \epsilon\left(\frac{1}{2}, \pi, \psi\right) \right| = 1.$$

13.2.3 Convergence

We first prove **(i)** for tempered representations. By definition an irreducible representation π is tempered if its central character is unitary and if any matrix coefficient of π is bounded by a constant multiple of the function Ξ defined as follows. Let $B = AN$ be the group of upper triangular matrices and δ_B its module function of the group B. Extend δ_B to be invariant under K on the right. Then

$$\Xi(g) = \int_K \delta_B^{1/2}(kg)dk.$$

Thus we only need to prove that an integral

$$\int_{G(F)} \Phi(g)\, \Xi(g)\, |\det g|^{s+\frac{n-1}{2}} dg$$

is absolutely convergent for $\mathrm{Re}(s) > 0$. Now Φ is bounded in absolute value by a function of the form

$$X \mapsto c\, \Phi_0(\varpi^m X),$$

where Φ_0 is the characteristic function of $M(n \times n, \mathcal{O}_F)$. So it suffices to prove that the integral

$$\int_{G(F)} \Phi_0(g)\, \Xi(g)\, |\det g|^{s+\frac{n-1}{2}} dg$$

is finite for $s > 0$. Since Φ_0 is K-invariant on the left, this integral, finite or infinite, is equal to

$$\int_{G(F)} \Phi_0(g)\, \delta_B^{1/2}(g)\, |\det g|^{s+\frac{n-1}{2}} dg.$$

This integral is computed below in Section 13.2.4 devoted to unramified representations and is equal to

$$(1 - q^{-s})^{-n},$$

which is finite for $s > 0$.

If π is unitary then its matrix coefficients are uniformly bounded. As in the previous case we are reduced to show that the integral

$$\int_{G(F)} \Phi_0(g)\, |\det g|^{s+\frac{n-1}{2}} dg$$

is finite for $s > \frac{n-1}{2}$. This integral can be computed as in the subsection devoted to unramified representations and is equal to

$$\prod_{k=1}^{k-n} \frac{1}{1 - q^{-(s-\frac{n}{2}+\frac{k}{2})}}.$$

Our assertion follows.

For a general representation π one can first prove that a matrix coefficient is majorized by a power of the norm and prove that an integral of the form

$$\int \Phi_0(g) \, \|g\|^m \, |\det g|^s dg$$

is finite for $s \gg 0$, or one can use the reduction step below.

For an arbitrary π, by taking suitable functions Φ with compact support contained in $G(F)$ we see we can choose Φ and f so that, for all s,

$$Z(\Phi, f, s) = 1.$$

Also we have, for $\mathrm{Re}(s) \gg 0$ and $h \in G(F)$,

$$\int \Phi(gh) f(gh) \, |\det g|^{s+\frac{n-1}{2}} dg = |\det h|^{-s-\frac{n-1}{2}} Z(\Phi, f, s).$$

This shows that if we prove the integrals are rational functions of q^{-s} then the complete assertion **(ii)** follows.

Consider now the case where π is a supercuspidal representation of $GL(n, F)$. If $n = 1$ this means that π is a one-dimensional character and the result follows from Tate's thesis. If $n > 1$ then the matrix coefficients of π are compactly supported modulo the center, and this can be used to prove the convergence of the integral for $\mathrm{Re}(s) \gg 0$ ($\mathrm{Re}(s) > 0$ if the central character is unitary) and also that the integrals are polynomials in q^{-s}, q^s, in other words that $L(s, \pi) = 1$. To prove the functional equation one can imitate Tate's argument.

One then uses a reduction step.

Lemma 13.2.2 (Reduction step) *Let $P = MU$ be a parabolic subgroup of type $(n_1, n_2, \ldots, n_r)$. For each i let π_i be an irreducible admissible representation of $GL(n_i, F)$. Let π be the induced representation*

$$\pi = \mathrm{Ind}(G, P; \pi_1, \pi_2, \ldots, \pi_r).$$

Suppose the assertions of the theorem are true for each π_i.
(i) Then they are true for any irreducible component σ of π.
(ii) Furthermore $\gamma(s, \sigma, \psi) = \prod_{1 \leq i \leq r} \gamma(s, \pi_i, \psi)$.

(iii) $L(s,\sigma) = R_\sigma(q^{-s}) \prod_{1 \le i \le r} L(s,\pi_i)$, *where R_σ is a polynomial, and*

$$L(s,\tilde{\sigma}) = \tilde{R}_\sigma(q^{-s}) \prod_{1 \le i \le r} L(s,\tilde{\pi}_i),$$

where $\tilde{R}_\sigma$ is the polynomial determined by

$$\tilde{R}_\sigma(q^{-s}) = R_\sigma(q^{-1+s}).$$

(iv) *If the induced representation is irreducible, so that σ is the induced representation, then $R_\sigma = 1$.*

Since every irreducible admissible representation of $GL(n,F)$ is induced by supercuspidal representations, Theorem 13.2.1 follows.

Lemma 13.2.2 gives the factor $\gamma(s,\pi,\psi)$ for any irreducible representation π. If π is tempered then $L(s,\pi)$ and $L(s,\tilde{\pi})$ are given by convergent integrals for $\mathrm{Re}(s) > 0$ and thus are holomorphic for $\mathrm{Re}(s) > 0$. It follows that the fraction

$$\frac{L(1-s,\tilde{\pi})}{L(s,\pi)}$$

is an irreducible fraction of the ring $\mathbb{C}[q^{-s},q^s]$. This observation determines completely the factors $L(s,\pi)$ and $L(s,\tilde{\pi})$.

For a complete computation of the L-factors see [24].

13.2.4 Unramified Representations

Because of its importance, we discuss the case of representations that have a vector fixed under $K := GL(n,\mathcal{O}_F)$. We first observe that a supercuspidal representation (π,V) of $GL(n,F)$ (for $n > 1$) cannot have a nonzero vector fixed under K. Indeed, assume that π has such a vector v. Then the contragredient representation $(\tilde{\pi},\tilde{V})$ has also a vector $\tilde{v} \ne 0$ fixed under K. The matrix coefficient

$$f(g) = \langle \pi(g)v,\tilde{v}\rangle$$

is bi-invariant under K, transforms under a character of $Z(F) = F^\times$, and is compactly supported modulo $Z(F)$. Moreover, because of the cuspidality, for all g

$$\int_{N(F)} f(ug)du = 0,$$

where we recall n is the group o upper triangular matrices with unit diagonal. By the Satake lemma [29] this implies $f = 0$, a contradiction. This result extends to a Levi subgroup M (which is a product of linear groups).

A supercuspidal representation σ of $M(F)$ can have a nonzero vector fixed under $K \cap M(F)$ only if M is a product of groups $GL(1)$, that is $M = A$, the group of diagonal matrices, and σ is a product of characters of $GL(1)$.

Now consider a general unramified representation π. It is a subrepresentation of an induced representation

$$\text{Ind}(G, MU; \sigma),$$

where σ is a supercuspidal representation of $M(F)$. The representation σ must have a vector fixed under $K \cap M(F)$. Thus it must be that $M = A$ (group of diagonal matrices) and σ is a product of one-dimensional unramified characters of $F^\times$. Hence π is an irreducible component of

$$\text{Ind}(G, AN; \chi_1, \chi_2, \ldots, \chi_n),$$

where each χ_i is an unramified character of $F^\times$. This representation may not be irreducible but it has a finite composition series. Since

$$G(F) = N(F)A(F)K$$

this representation has a unique irreducible component having a nonzero vector fixed under K. We denote it by $\pi(\chi_1, \chi_2, \ldots, \chi_n)$. We stress that it appears only once in the irreducible quotients of a composition series. If we permute the χ_i the character of the induced representation does not change, so the irreducible components do not change and, in particular, the representation $\pi(\chi_1, \chi_2, \ldots, \chi_n)$ does not change. An unramified character like χ_i is determined by its value $z_i = \chi_i(\varpi)$, where ϖ is a uniformizer. So we see that $\pi(\chi_1, \chi_2, \ldots, \chi_n)$ is determined by the conjugacy class in $GL(n, \mathbb{C})$ of the matrix

$$A = \text{diag}(z_1, z_2, \ldots, z_n).$$

This is the **Langlands conjugacy class** of the representation π.

Lemma 13.2.3 *Let π be an irreducible representation with a fixed vector under K. Then*

$$L(s, \pi) = \det\left(1_n - Aq^{-s}\right)^{-1},$$

where A is its Langlands conjugacy class of π. If, moreover, the conductor of ψ is $\mathcal{O}_F$ then $\epsilon(s, \pi, \psi) = 1$.

Proof We have $\pi = \pi(\chi_1, \chi_2, \ldots, \chi_n)$ for unramified characters χ_i. Let ϕ be the element of

$$\text{Ind}(G, AN; \chi_1, \chi_2, \ldots, \chi_n),$$

taking the value 1 on K. Define similarly $\widetilde{\phi}$ for the representation

$$\text{Ind}(G, AN; \chi_1^{-1}, \chi_2^{-1}, \ldots, \chi_n^{-1}).$$

Then the function

$$f(g) := \int_K \phi(kg)\widetilde{\phi}(k)dk = \int_K \phi(kg)dk$$

is a matrix coefficient of $\pi = \pi(\chi_1, \chi_2, \ldots, \chi_n)$. Let Φ be the characteristic function of $M(n \times n, \mathcal{O}_F)$. Then

$$Z(\Phi, f, s) = \int_G \Phi(g) f(g) |\det g|^{s + \frac{n-1}{2}} dg$$

$$= \int_{G(F) \times K} \Phi(g) \phi(kg) |\det g|^{s + \frac{n-1}{2}} dg dk.$$

Since Φ is K-invariant, this reduces to

$$\int_{G(F)} \Phi(g) \phi(g) |\det g|^{s + \frac{n-1}{2}} dg.$$

Using the Iwasawa decomposition $G(F) = A(F)N(F)K$ this reduces at once to

$$\prod_{1 \le i \le n} \int_{F^\times} \Phi_0(a_i)\chi_i(a_i)|a_i|^s \, d^\times a_i,$$

where Φ_0 is the characteristic function of $\mathcal{O}_F$ in F. This is equal to

$$\prod_{1 \le i \le n} L(s, \chi_i) = \det(1_n - Aq^s)^{-1},$$

which is the first assertion.

For the second assertion we remark that under the assumption on ψ we have $\widehat{\Phi} = \Phi$. Hence

$$Z(\widehat{\Phi}, \check{f}, 1 - s) = \int \Phi(g)\check{f}(g) |\det g|^{1 - s + \frac{n-1}{2}} dg.$$

Replacing f by $\check{f}$ amounts to exchange ϕ and $\widetilde{\phi}$. So this integral is equal to

$$\prod_{1 \le i \le n} L(1 - s, \chi^{-1}).$$

Hence $\epsilon(s, \pi, \psi) = 1$.   □

13.3 Local Theory for $GL(n, \mathbb{R})$ and $GL(n, \mathbb{C})$

In this section G denotes a product of groups $GL(n, \mathbb{R})$ and $GL(n, \mathbb{C})$ regarded as a real Lie group. We define the norm of an element of G. If $g \in GL(n, \mathbb{R})$ we set

$$||g|| = \sqrt{\sum_{i,j} \left(g_{ij}^2 + (g^{-1})_{ij}^2 \right)}.$$

We could also use the supremum of the absolute values of the entries of g and g^{-1}.

If $g \in GL(n, \mathbb{C})$ we set

$$||g|| = \sum_{i,j} \left(g_{ij}\overline{g_{i,j}} + (g^{-1})_{ij}\overline{(g^{-1})_{ij}} \right).$$

We could also use the supremum of the $g_{ij}\overline{g_{i,j}}$ and $(g^{-1})_{ij}\overline{(g^{-1})_{ij}}$.

The norm of an element of G is then the product of the norms of its components.

We denote by $\mathfrak{g}$ the Lie algebra of G and by $U(\mathfrak{g})$ the enveloping algebra of $\mathfrak{g}$. We also denote by $Z(\mathfrak{g})$ the center of $U(\mathfrak{g})$. The standard maximal compact subgroup of $GL(n, \mathbb{R})$ is the orthogonal group $\mathbf{O}(n)$ and the standard maximal subgroup of $GL(n, \mathbb{C})$ is the unitary group $\mathbf{U}(n)$. The standard maximal compact subgroup of G is the product of the standard maximal subgroups of the factors. It is denoted K. We note that because G is contained in a product of groups $GL(n, \mathbb{C})$ the center $Z(\mathfrak{g})$ is equal to $Z_G(\mathfrak{g})$, the set of elements of $U(\mathfrak{g})$ fixed by the operators $\mathrm{Ad}g$, $g \in G$.

We assume the reader is familiar with the notion of the $(\mathfrak{g}, K)$ module. A $(\mathfrak{g}, K)$ module is said to be admissible if any irreducible representation of K appears with finite multiplicity. We denote by $\mathcal{H}$ the category of admissible, finitely generated $(\mathfrak{g}, K)$ modules.

Lemma 13.3.1 (Harish-Chandra) *Consider a $(\mathfrak{g}, K)$-module V that is finitely generated. If V is annihilated by an ideal of finite codimension in $Z(\mathfrak{g})$ then V is admissible.*

Lemma 13.3.2 *Let σ be an irreducible representation of K. Then the multiplicity of σ in an irreducible admissible $(\mathfrak{g}, K)$ module is bounded by the dimension of σ.*

Let (π_0, V_0) be an admissible finitely generated $(\mathfrak{g}, K)$ module. Then there exists a locally convex complete topological vector space V and a continuous representation π of G on V such that V_0 can be identified with the space of K-finite vectors in V and π_0 is the corresponding representation of $(\mathfrak{g}, K)$. There are many choices for the topological vector space V. If (π, V_0) is an admissible algebraically irreducible $(\mathfrak{g}, K)$ module then, for any choice of V, the representation of G on V is topologically irreducible and the center of G operates by a scalar. So if $G = GL(n, \mathbb{R})$ or $Gl(n, \mathbb{C})$ we can define the central character ω_π. It depends only on V_0 and not on the choice of V.

Let $\mathcal{H}(G, K)$ be the convolution algebra of bi-K-finite smooth functions of compact support on G. Then the operators $\pi(f)$, $f \in \mathcal{H}(G, K)$ leave V_0

invariant. We have thus a representation of $\mathcal{H}(G, K)$ on V_0. This representation does not depend on V but only on V_0. The algebra $\mathcal{H}(G, K)$ does not have a unity but it has an approximation of unity. In particular, given vectors $v_0, v_1, \ldots v_n$ in V_0 there is $f \in \mathcal{H}(G, K)$ such that $\pi_0(f)v_i = v_i$ for all i. The following lemma (Harish-Chandra [20]) follows from the above considerations.

Lemma 13.3.3 *Let G be as above. Given $\mathbb{C}^\infty$ functions*

$$f_1, f_2, \ldots, f_r : G \to \mathbb{C},$$

which are K-finite and annihilated by an ideal of finite codimension in $Z(\mathfrak{g})$, then there exists $h \in \mathbb{C}_c^\infty (G)$ such that

$$f_i = \rho(h) f_i, \qquad (for \ i = 1, 2, \ldots, r).$$

We recall a lemma of Dixmier and Malliavin [9] that similarly can be used to show that some functions can be written as convolutions. Again let G be a Lie group, say a product of $GL(n, \mathbb{R})$ and $GL(n, \mathbb{C})$. Let π be a unitary representation of G on a Hilbert space H. Then let V be the subspace of C^∞ vectors in H. The space V is equipped with the topology defined by the seminorms $v \mapsto ||\pi(X)v||$ where X is in the enveloping algebra of the Lie algebra of G. It is complete for this topology. The group G operates on V.

Lemma 13.3.4 *Any vector $v \in V$ can be written as a finite sum*

$$v = \sum_{1 \le j \le r} \pi(f_j) v_j,$$

where the vectors v_j are in V and the functions f_j are C^∞ functions of compact support on G.

Again let (π, V_0) be a finitely generated admissible $(\mathfrak{g}, K)$ module. There is a completion V of v_0 with the following properties (Casselman-Wallach, see [5], [34]). The space V is a Frechet space and let the representation π of G be C^∞. This means that for each vector v the map $v \mapsto \pi(g)v$ is C^∞. Finally, we demand for any continuous seminorm λ on V there is another continuous seminorm ν_λ and $m > 0$ such that

$$\lambda(\pi(g)v) \le ||g||^m \nu_\lambda(v)$$

for all v and g. The representation (π, V) is uniquely determined by these conditions. We call it the canonical completion of (π_0, V_0).

Moreover, let $\widetilde{V}_0$ be the contragredient module: this is the vector space of K-finite linear forms on V_0. Let $\widetilde{V}$ be the canonical completion of $\widetilde{V}_0$. Then the natural bilinear form on $V_0 \times \widetilde{V}_0$ extends to a continuous, invariant bilinear form on $V \times \widetilde{V}$. Usually, this bilinear form is denoted $\langle v, \widetilde{v} \rangle$. The functions

$$g \mapsto \langle \pi(g)v, \tilde{v} \rangle$$

are the matrix coefficients of π.

Finally, let (π, H) be a unitary (topologically) irreducible representation of G on a Hilbert space H with Hermitian scalar product (v_1, v_2) and norm $||v|| = (v, v)^{\frac{1}{2}}$. Let H_K be the space of K-finite vectors. Every vector v in H_K is C^∞ so that $\mathfrak{g}$ operates on H_K and H_k is a $(\mathfrak{g}, K)$ module admissible and irreducible. Let V be the space of C^∞ vectors in H. The space V equipped with the topology defined by the seminorms

$$v \to ||\pi(X)v||, \quad \left(X \in U(\mathfrak{g})\right),$$

is the canonical completion of the $(\mathfrak{g}, K)$ module H_K. The space $\tilde{V}$ is simply the space imaginary conjugate of V; that is, the same space, with the same addition and the same topology and scalar multiplication defined

$$\lambda._{\tilde{V}} v = \overline{\lambda}._V v.$$

Thus a matrix coefficient of V has the form

$$g \mapsto (\pi(g)v_1, v_2), \quad \left(v_1, v_2 \in H_K\right).$$

If the ground field is $\mathbb{R}$ we let ψ be a nontrivial additive character. We write ψ in the form $\psi(x) = \exp(2i\pi ax)$, $a \in \mathbb{R}^\times$. We denote by $\mathcal{S}_0(M(n \times n, \mathbb{R}))$ the subspace of $\mathcal{S}(M(n \times n, \mathbb{R}))$ spanned by the functions of the form

$$\Phi(X) = \exp\left(-\pi \mathrm{tr}({}^t XX)\right) P(X),$$

where P is a polynomial.

If the ground field is $\mathbb{C}$, we let ψ be a nontrivial additive character. We write ψ in the form $\psi(z) = \exp\left(2i\pi(az + \overline{az})\right)$, $a \in \mathbb{C}^\times$. We denote by $\mathcal{S}_0(M(n \times n, \mathbb{C}))$ the subspace of $\mathcal{S}(M(n \times n, \mathbb{C}))$ spanned by the functions of the form

$$\Phi(X) = \exp\left(-2\pi \mathrm{tr}({}^t \overline{X}X)\right) P(X, \overline{X}),$$

where P is a polynomial. Often, we write $\mathcal{S}$ and $\mathcal{S}_0$ for these spaces.

We define the Fourier transform

$$\widehat{\Phi}(X) = \int \Phi(Y)\psi(-\mathrm{Tr}XY)\, dY$$

of a function Φ. The Haar measure dX is self-dual, that is, for all Φ,

$$\int \widehat{\Phi}(X)\, dX = \Phi(0).$$

The space $\mathcal{S}$ (resp. $\mathcal{S}_0$) is invariant under the Fourier transform (resp. if $a = \pm 1$).

From now on we do not distinguish between a $(\mathfrak{g}, K)$ module and its canonical completion.

Theorem 13.3.5 *Let π be an irreducible $(\mathfrak{g}, K)$ and (π, V) its canonical completion. Let f be a smooth matrix coefficient of π and let $\Phi \in S(M(n \times n, \mathbb{R}))$.*
(i) *The integral*

$$Z(\Phi, f, s) := \int_{GL(n,\mathbb{R})} \Phi(g) f(g) |\det g|^{s+\frac{n-1}{2}} dg$$

converges absolutely for $\mathrm{Re}(s) \gg 0$ ($\mathrm{Re}(s) > 0$ if π is tempered and $\mathrm{Re}(s) > \frac{n-1}{2}$ if π is unitary).
(ii) *If $P(s)$ is any polynomial, then*

$$P(s)Z(\Phi, f, s) = \sum_{1 \le i \le r} Z(\Phi_i, f_i, s)$$

for suitable Φ_i and f_i. If Φ is in S_0 and f is bi-K-finite then one can take $\Phi_i \in S_0$ and f_i bi-K-finite.
(iii) *The integrals $Z(\Phi, f, s)$ extend to meromorphic function of s. There is a meromorphic function $L(s, \pi)$, that never vanishes, with the following properties. The integrals $Z(\Phi, f, s)$ are entire multiples of $L(s, \pi)$. If Φ is in S_0 and f is K-finite then $Z(\Phi, f, s)$ is a polynomial multiple of $L(s, \pi)$. Conversely, if P is any polynomial then one can find $\Phi_i \in S_0$ and K-finite coefficients f_i such that*

$$P(s)L(s, \pi) = \sum_i Z(\Phi_i, f_i, s).$$

(iv) *As a meromorphic function of s, the integral $Z(\Phi, f, s)$ satisfies the functional equation:*

$$Z(\widehat{\Phi}, \check{f}, 1 - s) = \gamma\,(s, \sigma, \psi)\, Z(\Phi, f, s),$$

where $\gamma\,(s, \sigma, \psi)$ is a suitable meromorphic function.
   The factor γ has the form

$$\gamma\,(s, \sigma, \psi) = \frac{\epsilon(\pi, s, \psi)L(1 - s, \widetilde{s})}{L(s, \pi)},$$

where $\epsilon(\pi, s, \psi)$ is an exponential function of s.

These conditions determine the factor $L(s, \pi)$ up to a scalar factor. It will turn out to be a product of Γ factors. In a vertical strip it has only finitely many poles.

We move to assertion **(ii)**. We have a representation of $G(F) \times G(F)$ on S, the action of (g_1, g_2) being given by

$$\lambda(g_1)\rho(g_2)\Phi[X] := \Phi(g_1^{-1} X g_2).$$

This action is C^∞ so we have a corresponding action of $\mathfrak{g} \times \mathfrak{g}$. For instance, let $X \in \mathfrak{g}$. Then

$$\rho(X)\Phi(X) = \frac{d}{dt}\Phi[Xe^{tX}]\bigg|_{t=0}.$$

The space $\mathcal{S}_0$ is invariant under the action of $K \times K$ and its elements are $K \times K$ finite. Furthermore the space $\mathcal{S}_0$ is invariant under $\mathfrak{g} \times \mathfrak{g}$. Finally, if Φ is in $\mathcal{S}_0$ and f is a matrix coefficient, then

$$Z(\Phi, f, s) = Z(\Phi, f_0, s),$$

where f_0 is a bi-K-finite coefficient.

Now let $X \in \mathfrak{g}$. Then

$Z(\Phi, \rho(X)f, s)$

$$= \frac{d}{dt}\int \Phi(g) f(ge^{tX}) |\det g|^{s+\frac{n-1}{2}} dg\bigg|_{t=0}$$

$$= \frac{d}{dt}\int \Phi(ge^{-tX}) f(g) \left|\det ge^{-tX}\right|^{s+\frac{n-1}{2}} dg\bigg|_{t=0}$$

$$= -\int \rho(X)\,\Phi(g) f(g) |\det g|^{s+\frac{n-1}{2}} dg - \left(s + \frac{n-1}{2}\right) Z(\Phi, f, s).$$

Assertion (ii) follows.

For $n = 1$ our assertions are essentially contained in Tate's thesis. In the context of $(\mathfrak{g}, K)$ modules or their canonical completion, we have a notion of induced representations that we take for granted. Then we have again a reduction step.

Lemma 13.3.6 *Let π_i, $1 \leq i \leq r$ be irreducible representations of $GL(n_i, F)$. Suppose the assertions of the theorem are true for each representation π_i. Let σ be an irreducible component of the induced representation*

$$\text{Ind}(G, P; \pi_1, \pi_2, \ldots, \pi_r).$$

(i) *The assertions of the theorem are true for the representation σ.*
(ii) *We have*

$$\gamma(s, \sigma, \psi) = \prod_{1 \leq i \leq r} \gamma(s, \pi_i, \psi),$$

$$L(s, \sigma) = P(s) \prod_{1 \leq i \leq r} L(s, \pi_i),$$

$$L(s, \tilde{\sigma}) = \tilde{P}(s) \prod_{1 \leq i \leq r} L(s, \tilde{\pi}_i).$$

where P and $\widetilde{P}$ are polynomials and

$$\widetilde{P}(s) = P(1 - s).$$

(iii) *If the induced representation is irreducible (and equal to σ) then $P = 1$.*

We use the reduction step in the following way. Let π be an irreducible admissible $(\mathfrak{g}, K)$ module (or its canonical completion). Then there are n characters $\pi_i : F^{\times} \to \mathbb{C}$ with the following property. Consider the induced representation

$$\mathrm{Ind}(G, P; \pi_1, \pi_2, \ldots, \pi_n),$$

where P is the group of upper triangular matrices. The space of this induced representation is the space of C^{∞} functions

$$f : GL(n, F) \to \mathbb{C}$$

such that

$$f(nag) = \delta_P(a)^{1/2} \mu(a) f(g),$$

where

$$\mu(a) = \mu_1(a_{1,1})\mu_2(a_{2,2}) \cdots \mu_n(a_{n,n}).$$

The canonical completion of π is a subquotient of this induced representation, and the $(\mathfrak{g}, K)$ module π is then a subquotient of the $(\mathfrak{g}, K)$ module of K-finite functions in the induced representation.

One then proves that the integrals $Z(\Phi, f, s)$ for f a matrix coefficient of π extend to meromorphic functions that are entire multiples of

$$\prod_{i=1}^{n} L(s, \pi_i)$$

bounded at infinity in vertical strips. This space of meromorphic functions has a natural topology defined as follows. Consider a strip $A \leq \mathrm{Re}(s) \leq B$ and a polynomial $P(s)$, which cancels the poles of $\prod L(s, \mu_i)$ in the strip. We define then a seminorm

$$\sup_{A \leq \mathrm{Re}(s) \leq B} |P(s)f(s)|.$$

The topology is then the one defined by these seminorms. The map

$$\Phi \mapsto Z(\Phi, f, s)$$

is then continuous for this topology. If we write $f(g) = \langle \pi(g)v, \widetilde{v} \rangle$, the bilinear form

$$(v, \widetilde{v}) \mapsto Z(\Phi, f, s)$$

is also continuous. We have also the functional equation:

$$Z(\widehat{\Phi}, \check{f}, 1 - s) = \prod_{i=1}^{n} \gamma(s, \mu_i, \psi) Z(\Phi, f, s).$$

If we take f to be K-finite and $\Phi \in S_0$ then the integrals are polynomial multiples of $\prod_{i=1}^{n} L(s, \mu_i)$. The vector space spanned by these polynomials is an ideal with a generator P_π, well defined up to a scalar multiple. We set $L(s, \pi) = P_\pi(s) \prod_{i=1}^{n} L(s, \mu_i)$. So $L(s, \pi)$ is defined up to multiplication by a constant. We define similarly $L(s, \widetilde{\pi})$. We have also a functional equation. A density argument implies that the integrals $Z(\Phi, f, s)$ for Φ arbitrary are again holomorphic multiples of $L(s, \pi)$. Since given f and s_0 one can choose Φ of compact support on $GL(n, F)$, such that $Z(\Phi, f, s_0) \neq 0$, one concludes that $L(s_0, \pi) \neq 0$. In other words the zeros of P_π must cancel poles of $\prod_{i=1}^{n} L(s, \mu_i)$. We have a similar polynomial $P_{\widetilde{\pi}}$ and the factor $L(s, \widetilde{\pi})$. Moreover, in the functional equation

$$\frac{Z(\widehat{\Phi}, \check{f}, 1 - s)}{\prod L(1 - s, \mu_i^{-1})} = \prod \epsilon(s, \mu_i, \psi) \frac{Z(\Phi, f, s)}{\prod L(s, \mu_i)}$$

the product of the epsilon factors is, in fact, a constant. If f is a K-finite matrix coefficient of π and Φ in S_0 then the right-hand side is a polynomial multiple of $P_\pi(s)$ and the left-hand side is a polynomial multiple of $P_{\widetilde{\pi}}(1 - s)$. We conclude that $P_{\widetilde{\pi}}(1 - s) = c P_\pi(s)$ for a suitable constant c. Finally, we can write the functional equation in the form

$$Z(\widehat{\Phi}, \check{f}, 1 - s) = \gamma(s, \pi, \psi) Z(\Phi, s, f),$$

where

$$\gamma(s, \pi, \psi) = \frac{\epsilon(s, \pi, \psi) L(1 - s, \widetilde{\pi})}{L(s, \pi)},$$

and $\epsilon(s, \pi, \psi)$ is an exponential function of s (in fact a constant with our choice of ψ).

In principle, the above considerations determine the factor $\gamma(s, \pi, \psi)$. It remains to compute the factor $L(s, \pi)$. Thus we need to consider the case of a representation π of $GL(n, F)$, square integrable (modulo the center) other than a character of $GL(1)$ of module 1. Such a representation exists only if $F = \mathbb{R}$ and $n = 2$. There exist two characters μ_1, μ_2 of $\mathbb{R}^\times$ such that π is a subrepresentation of the induced representation

$$\text{Ind}(G, P; \mu_1, \mu_2).$$

Since the representation π is tempered, the integrals $Z(\Phi, f, s)$ and $Z(\Phi, \check{k}, s)$ (where f is a matrix coefficient of π) converge for $\text{Re}(s) > 0$. Thus the products

$$P_\pi(s)L(s,\mu_1)L(s,\mu_2), \qquad P_{\widetilde{\pi}}(s)L(s,\mu_1^{-1})L(s,\mu_2^{-1}),$$

are holomorphic for $\mathrm{Re}(s) > 0$. This added condition determines the polynomials and the factors $L(s,\pi)$, $L(s,\widetilde{\pi})$. See [24] for a computation of the L factor in all cases.

13.4 Tensor Product of Representations

Let F be number field and let G be the group $GL(n)$ regarded as an algebraic group over F. Let $\mathbb{A}$ be the ring of adèles of F. Let π be a (topologically) irreducible unitary representation of $G(\mathbb{A})$ on a Hilbert space H.

If v is a finite place we let $\mathcal{O}_v$ be the ring of integers of F_v and we set

$$K_v := GL(n, \mathcal{O}_v).$$

If v is a real place we set $K_v := \mathbf{O}(n)$. If v is a complex place we set $K_v := \mathbf{U}(n)$. We set

$$K_\infty := \prod_{v \in \infty} K_v,\ K_f := \prod_{v \notin \infty} K_v,\ K := K_\infty \cdot K_f.$$

We can restrict this representation to the maximal compact subgroup K. This representation decomposes into a discrete sum of unitary irreducible representations of K. In particular, the space of K-finite vectors is dense in H. Consider an irreducible representation σ of K. Then there is a finite set of places S containing all the Archimedean places, for each $v \in S$ an irreducible representation σ_v of G_v such that σ is the tensor product of the σ_v, $v \in S$ and the trivial representation of $K^S := \prod_{v \notin S} K_v$. It follows that the union of the closed subvector spaces H^{K^S} is dense in H. Consider one of them H^{K^S}, say. Then the product group $G_S := \prod_{v \in S} G_v$ leaves that space invariant and so does the Hecke algebra $\mathcal{H}^S$. Fix a vector $v_0 \neq 0 \in H^{K^D}$. If v is any other vector in the same space then v can be approached by vectors of the form

$$\sum_i c_i \pi_i(g_i)\pi(h_i)v_0$$

with c_i in $\mathbb{C}$, $g_i \in G_S$, $h_i \in G^S$. Since V and v_0 are in H^{K^S} we see that v can be approached by vectors the form

$$\sum_i c_i \pi_i(g_i) \int_{K^S} \pi(k)dk\pi(h_i) \int_{K^S} \pi(k)dkv_0$$

and $\int_{K^S} \pi(k)dk\pi(h_i) \int_{K^S} \pi(k)dk = \pi(\phi)$ for some ϕ in $\mathcal{H}^S$. So H^{K^S} must be irreducible under the action of G_S and $\mathcal{H}^S$.

But $\mathcal{H}^S$ is commutative and the operators $\pi(\phi)$, $\phi \in \mathcal{H}^S$ commute to the operators $\pi(g)$, $g \in G_S$. So the operators $\pi(\phi)$, $\phi \in \mathcal{H}^S$ must be scalars. It follows that the representation of G_S on H^{K^S} is topologically irreducible. Concretely, because the local groups G_v are of type I the representation must be the tensor product $\bigotimes_{v \in S} \pi_v$, where the π_v are irreducible unitary representations. If $T \supset S$ then we get unitary irreducible representations π_t', $t \in T$. For $s \in S$ we have $\pi_s \simeq \pi_s'$. For $t \in T - S$ the representation is π_t; it contains a unit vector e_t invariant under K_t and $\bigotimes_{t \in T} H_t' \simeq \bigotimes_{s \in S} H_s \bigotimes_{t \in T-S} e_t$. Finally, we have obtained for every place v a unitary irreducible representation (π_v, H_v). For almost all v, the space H_v contains a unit vector fixed under K_v (unique up to a scalar factor of module 1). If S is sufficiently large and $T \supset S$

$$\bigotimes_{t \in T} H_t \supset \bigotimes_{v \in S} H_v \bigotimes_{v \in T-S} e_t \simeq \bigotimes_{v \in S} H_v.$$

We can define the algebraic inductive limit of the spaces $\bigotimes_{v \in S} H_v$, and H is the completed space of the algebraic limit. In a more concrete way, choose for almost all places V a unit vector invariant under K_v. We have the pure tensor vectors

$$\bigotimes_{\text{all } v} u_v$$

with $u_v = e_v$ for almost all v. The linear span of the pure tensors is dense in H. The scalar product of two pure tensors is given by

$$\left(\bigotimes_{\text{all } v} u_v, \bigotimes_{\text{all } v} u_v' \right) = \prod_{\text{all } v} (u_v, u_v').$$

Concretely, the matrix coefficients $(\pi(g)u, u')$ for u and u' pure tensors are given by the infinite product:

$$(\pi(g)u, u') = \prod_{\text{all } v} (\pi_v(g_v)u_v, u_v').$$

For a given g almost all factors are equal to 1. This description applies to the space V of K-finite vectors (see [15] for a discussion in the case of $GL(2)$). The space V is invariant and irreducible under the action of $(\mathfrak{g}, K_\infty)$ and $G(\mathbb{A}_f)$. It is also admissible in the sense that any irreducible representation of K appears with finite multiplicity.

More generally, consider a $(\mathfrak{g}_\infty, K_\infty) \times GL(n, \mathbb{A}_f)$ module (π, V). This means that V is a $(\mathfrak{g}_\infty, K_\infty)$ module and a $GL(n, \mathbb{A}_f)$ module and the actions commute. We assume that each vector in V is fixed under some compact open subgroup of $GL(n, \mathbb{A}_f)$. We say that V is admissible if each irreducible representation of $K_\infty \times K_f$ appears with finite multiplicity. We say that

(π, V) is irreducible if there are no nontrivial invariant subspaces. Then π is isomorphic to a restricted infinite product

$$\bigotimes_v (\pi_v, V_v).$$

For v infinite, (π_v, V_v) is an irreducible $(\mathfrak{g}_v, K_v)$. For v finite, (π_v, V_v) is an irreducible (admissible) representation of $GL(n, F_v)$. For almost all finite v the vector space contains a nonzero vector e_v fixed by $K_v := GL(n, \mathcal{O}_v)$. We have a similar description of the contragredient representation $(\widetilde{\pi}, \widetilde{V})$ as the infinite tensor product $\bigotimes \widetilde{\pi}_v, \bigotimes \widetilde{V}_v)$. One can choose $\widetilde{e}_v$ to be such that $\langle e_v, \widetilde{e}_v \rangle = 1$. See [18] and [22] for a detailed discussion in the case of $GL(2)$, and [10] for the general case.

13.5 Reduction Theory for $GL(n)$

Let F be a number field and $\mathbb{A}_F$ or simply $\mathbb{A}$ its ring of adèles. We denote by $\mathbb{A}_{>0}^{\times}$ the group of idèles whose finite components are 1 and whose infinite components are all equal to the same positive number. We can identify this group with $\mathbb{R}_{>0}$. Now let

$$x = (t, t, \ldots, t, 1, 1, \ldots) \in \mathbb{A}_{>0}^{\times}, \qquad (t \in \mathbb{R}_{>0}).$$

We must remember that if v is a real place then $|x_v|_v = t$ and if v is a complex place then $|x_v|_v = t^2$. In particular,

$$|x| = t^{r+2c},$$

where r is the number of real places, and c is the number of complex places. Define $\mathbb{A}^1$ to be the group of idèles of norm one, and let $|x|$ be the usual absolute value on $\mathbb{A}$. The we have the decomposition

$$\mathbb{A}^{\times} = \mathbb{A}_{>0}^{\times} \cdot \mathbb{A}^1.$$

Now $F^{\times} \subset \mathbb{A}^1$ and $\mathbb{A}^1/F^{\times}$ is compact (reduction theory for $GL(1)$).

Let $G = GL(n)$ be regarded as an algebraic group over F. We let G^1 be the set of $g \in G(\mathbb{A})$ such that $|\det g| = 1$. We have, of course, $G(F) \subset G^1$. We let Z be the center of G. We define $Z_{>0} \subset Z(\mathbb{A})$ as the subgroup of elements whose entries are in $\mathbb{A}_{>0}^{\times}$. We have (direct product)

$$G(\mathbb{A}) = Z_{>0} \cdot G^1.$$

Let A be the group of diagonal matrices, regarded as an algebraic group. Let A^1 be the subgroup of elements in $A(\mathbb{A})$ whose entries have absolute value 1, and let $A_{>0}$ be the subgroup of elements whose entries are in $\mathbb{A}_{>0}^{\times}$. We have

$$A(\mathbb{A}) = A_{>0} \cdot A^1.$$

Finally we let A_1 be the subgroup of elements of $A_{>0}$ whose determinant is 1. Then we have

$$A_{>0} = Z_{>0} \cdot A_1$$

and

$$A(\mathbb{A}) = Z_{>0} \cdot A_1 \cdot A^1$$

as well as

$$A(\mathbb{A}) \cap G^1 = A_1 \cdot A^1.$$

We let N be the group of upper triangular matrices with unit diagonal.
We have the Iwasawa decomposition

$$G(\mathbb{A}) = N(\mathbb{A}) \cdot A(\mathbb{A}) \cdot K$$

and

$$G^1 = N(\mathbb{A}) \cdot A_1 \cdot A^1 \cdot K.$$

Recall that $N(F)\backslash N(\mathbb{A})$ and $A(F)\backslash A^1$ are compact. We also recall the simple roots:

$$\alpha_i : A \to GL(1)$$

defined by

$$\alpha_i(a) := a_{i,i}/a_{i+1,i+1}.$$

If $t > 0$ we denote by $A(t)$ the subset of elements a of A_1 satisfying

$$|\alpha_i(a)| \geq t, \qquad 1 \leq i \leq n-1.$$

Let Ω_N be a compact subset of $N(\mathbb{A})$, let Ω_A be a compact subset of A^1, and let $t > 0$. We denote by $\mathfrak{S}_{t,\Omega_N,\Omega_A}$ the set of $g \in G^1$ of the form

$$g = \omega_N \cdot a \cdot \omega_A \cdot k$$

with $a \in A(t)$, $\omega_N \in \Omega_N$, $\omega_A \in \Omega_A$, $k \in K$. Such a set is called a Siegel set. It is elementary that a Siegel set has finite volume for the Haar measure of G^1. Moreover, if g is as above then $a^{-1} \cdot \omega_N \cdot a$ remain in a compact set of $N(\mathbb{A})$ so that

$$g = a \cdot \omega_G,$$

where ω_G remain in a compact set Ω_G of G^1.
The basic result of reduction theory is as follows (see [12]).

Theorem 13.5.1 *For any Siegel set* $\mathfrak{S}$ *the set*

$$X_{\mathfrak{S}} := \{\gamma \in G(F) \mid \gamma\mathfrak{S} \cap \mathfrak{S} \neq \emptyset\}$$

is finite.
There is a Siegel set $\mathfrak{S}$ *such that*

$$G^1 = G(F)\,\mathfrak{S}.$$

As a consequence we see the volume of $G(F)\backslash G^1$ is finite. More precisely we have the following result.

Theorem 13.5.2 *There is a Siegel set* $\mathfrak{S}$ *such that*

$$\mathrm{Vol}(G(F)\backslash G^1) \leq \mathrm{Vol}(\mathfrak{S}).$$

For any Siegel set $\mathfrak{S}$, *there is a constant c such that*

$$\mathrm{Vol}(\mathfrak{S}) \leq c\mathrm{Vol}(G(F)\backslash G^1).$$

Proof Since we may always replace a Siegel set by a larger one, it suffices to consider a Siegel set $\mathfrak{S}$ such that $G^1 = G(F)\mathfrak{S}$. Let $\mathfrak{S}'$ be a measurable section of $G(F)\backslash G^1$ contained in $\mathfrak{S}$. We have then (disjoint union)

$$G^1 = \bigsqcup_{\gamma \in G(F)} \gamma\mathfrak{S}'$$

and (finite disjoint union)

$$\mathfrak{S} \subset \bigsqcup_{\gamma \in X_{\mathfrak{S}}} \gamma\mathfrak{S}'.$$

Let c be the cardinality of X_σ. Then

$$\mathrm{Vol}(G(F)\backslash G^1) = \mathrm{Vol}(\mathfrak{S}') \leq \mathrm{Vol}(\mathfrak{S}) \leq c\mathrm{Vol}(\mathfrak{S}') = c\mathrm{Vol}(G(F)\backslash G^1).$$

$\square$

We have also the weak approximation theorem.

Theorem 13.5.3 *Let K' be an open compact subgroup of K_f. Then there are finitely many elements c_i, $1 \leq i \leq r$, of $G(\mathbb{A}_f)$ such that we have a disjoint union*

$$G(\mathbb{A}) = \bigcup_{1 \leq i \leq r} G(F)G_\infty c_i K'.$$

Finally, for $g \in G(\mathbb{A})$, we let $\|g\| = \prod_v \|g_v\|_v$ denote the norm of the element g.

Lemma 13.5.4 *Given a Siegel set $\mathfrak{S}$ then for every $g \in \mathfrak{S}$ we have*

$$\|g\| \asymp \inf_{\gamma \in G(F)} \|\gamma g\|.$$

Proof Of course we have, for all $g \in G(\mathbb{A})$,

$$\inf_{\gamma \in G(F)} \|\gamma g\| \le \|g\|.$$

We prove an inequality in the reverse direction for g in a Siegel set. If g is in a Siegel set then it has the form

$$g = a\omega,$$

where ω is in a compact set and $a \in A(t)$. Thus it suffices to prove that there is a constant c such that

$$\|a\| \le c \cdot \|\gamma a\|$$

for all $\gamma \in G(F)$ and $a \in A_{>0}$. At this point we may use the supremum norm at each infinite place and, for any place v, we adopt the convention that for an adèle $a \in \mathbb{A}$

$$|a|_v := |a_v|_v.$$

Similarly, for $g \in G(\mathbb{A})$ and any place v, we set

$$\|g\|_v := \|g_v\|_v.$$

We have now for any index j

$$a_{j,j} = \left(t_j, t_j, \ldots, t_j, 1, 1, \ldots, 1 \ldots \right) \quad (t_j > 0)$$

and

$$\|a\| = \left(\sup_j \sup(t_j, t_j^{-1}) \right)^r \cdot \left(\sup_j \sup(t_j^2, t_j^{-2}) \right)^c$$

Let j be an index. There is an index i such that $\gamma_{i,j} \ne 0$ (*i*th row and *j*th column) For a real place v

$$t_j |\gamma_{i,j}|_v = |\gamma_{i,j} a_{j,j}| \le \|\gamma a\|_v.$$

For v a complex place, we have

$$t_j^2 |\gamma_{i,j}|_v = |\gamma_{i,j} a_{j,j}|_v \le \|\gamma a\|_v.$$

For v finite, we have

$$|\gamma_{i,j}|_v = |\gamma_{i,j} a_{j,j}|_v \le \|\gamma a\|_v.$$

Taking the product of these inequalities we get

$$t_j^{r+2c} \le \|\gamma a\|.$$

Similarly,

$$t_j^{-r-2c} \le \|a^{-1}\gamma^{-1}\| = \|\gamma a\|.$$

So we get

$$||a|| \leq ||\gamma a||.$$

We are done. □

Thus to check that a function ϕ on G^1 invariant on the left under $G(F)$ is of moderate growth, that is bounded by a constant multiple of the power of the norm, it suffices to check it is of moderate growth on a Siegel set.

For $a \in A(\mathbb{A})$ define

$$\beta(a) := \prod_{1 \leq i \leq n-1} |\alpha_i(a)|.$$

Lemma 13.5.5 *Given $t > 0$, there exist $c_1, c_2, m_1, m_2 > 0$ such that*

$$c_1 \beta(a)^{m_1} \leq ||a|| \leq c_2 \beta(a)^{m_2}$$

for all $a \in A(t)$.

Proof This follows from the fact that

$$a_{1,1}^n = \prod_{i=1}^{n-1} \alpha_i(a)^{n_i}$$

with $n_i, (i = 1, \ldots, n-1)$ positive integers and for $j \geq 2$

$$a_{j,j} = a_{1,1} \prod_{i=1}^{n-1} \alpha_i(a)^{-u_i},$$

where $u_i \geq 0$ are integers.

□

Now consider a Siegel set $\mathfrak{S}_{t, \Omega_N, \Omega_A}$. Since Ω_N, Ω_A, K are compact sets, it follows that for $\omega_N \in \Omega_N$, $\omega_A \in \Omega_A$, $a \in A(t)$, and $k \in K$,

$$||\omega_N a \omega_A k|| \asymp ||a||.$$

It immediately follows from Lemma 13.5.5 that on the Siegel set $\mathfrak{S}_{t, \Omega_N, \Omega_A}$ we have

$$\beta(a)^{m_1} \ll ||\omega_N a \omega_A k|| \ll \beta(a)^{m_2}.$$

Finally, we see that a function ϕ on G^1 invariant on the left under $G(F)$ is of moderate growth if and only if, for every $t > 0$, there is a constant m such that, for every compact set Ω, there is a constant c with

$$|\phi(a\omega)| \leq c\beta(a)^m$$

for $a \in A(t)$ and $\omega \in \Omega$.

Another application is the following lemma.

Lemma 13.5.6 *Let C be a compact subset of $G(\mathbb{A})$. Then there is $c > 0$ and $m > 0$ such that, for all x in $G(\mathbb{A})$, the cardinality of the set*

$$G(F) \cap xCx^{-1}$$

is bounded by $c||x||^m$.

Proof Let Ω be a compact subset of $G(\mathbb{A}_f)$ and $t > 0$. Let $G_{t,\Omega}$ be the set of $g \in G_\infty \Omega$ such that $||g|| \leq t$. It easy to see that the volume of $G_{t,\Omega}$ for a Haar measure of $G(\mathbb{A})$ is bounded by ct^m for suitable $c > 0$ and $m > 0$.

Now, as a function of x the cardinality of the set $G(F) \cap xCx^{-1}$ is invariant under $G(F)$ on the left. Thus, to prove our contention we have assumed that x is in a Siegel set and, a fortiori, in the set $G_\infty C'$, where C' is a compact set of $G(\mathbb{A}_f)$. Replacing the set C by the set $C'CC'^{-1}$ we see we may assume that x is in G_∞. For $\gamma \in G(F) \cap xCx^{-1}$ we have

$$||\gamma|| \leq c||x|| \cdot ||x^{-1}|| = c||x||^2.$$

Now let V be a compact neighborhood of 1 in $G(\mathbb{A})$ such that

$$G(F) \cap (V \cdot V^{-1}) = \{1\}.$$

For $v \in V$ and $\gamma \in G(F) \cap xCx^{-1}$ we have

$$||\gamma v|| \leq c_1||\gamma|| \leq c_2||x||^2.$$

On the other hand,

$$\gamma v \in xCx^{-1}V.$$

Since C and V are contained in the product of a compact set of G_∞ and a compact set of $G(\mathbb{A}_f)$, the set $xCx^{-1}V$ is contained in a product $G_\infty \Omega$, where Ω is a compact set of $G(\mathbb{A}_f)$. We see that if V is as above and γ is in the intersection then

$$\gamma V \subset G_{c_2||x||^2, \Omega}.$$

The disjoint union

$$\cup_{\gamma \in G(F) \cap xCx^{-1}} \gamma V$$

is contained in

$$G_{c_2||x||^2, \Omega}.$$

Thus

$$\text{Vol}(\cup \gamma V) \leq \text{Vol}(G_{c_2||x||^2, \Omega}) \leq c||x||^{2m}.$$

But the volume on the left is $\text{Vol}(V)$ times the cardinality we are trying to bound. $\qquad\square$

Lemma 13.5.7 *Let* $x, y \in G(\mathbb{A})$ *and let* C *be a compact set of* $G(\mathbb{A})$. *Then the cardinality of the set*

$$G(F) \cap (xCy)$$

is bounded by $c||x||^m$ *for suitable* $c > 0$ *and* $m > 0$.

Proof Fix an element δ in the set in question. Then for any other element γ we have

$$\delta^{-1}\gamma \in xCC^{-1}x^{-1}.$$

Since CC^{-1} is a compact set, it suffices to apply the previous lemma. $\qquad\square$

On the group $G(\mathbb{A})$ (and in general for any group) we denote by $\rho(x)$ the right translation by x:

$$\rho(x)\phi(h) = \phi(hx).$$

Moreover, if f is a function on $G(\mathbb{A})$, we set

$$\rho(f)\phi(h) = \int_{G(\mathbb{A})} \phi(hg)dh,$$

where dx is a Haar measure on $G(\mathbb{A})$.

Lemma 13.5.8 *Suppose* f *is a continuous function of compact support on* $G(\mathbb{A})$. *Then there are constants* $c > 0$ *and* $m > 0$ *such that, for any* $\phi \in L^2(Z_{>0}G(F)\backslash G(\mathbb{A}))$ *and every* $x \in G(\mathbb{A})$,

$$|\rho(f)\phi(x)| \leq c||x||^m ||\phi||_2.$$

Proof Set

$$f_1(g) := \int_{Z_{>0}} f(zg)dz.$$

Then f_1 is a continuous function of compact support on $Z_{>0}\backslash G(\mathbb{A})$. We have

$$\rho(f)\phi(g) = \int_{Z_{>0}\backslash G(\mathbb{A})} \phi(hg)f_1(h)dh$$

$$= \int_{Z_{>0}\backslash G(\mathbb{A})} \phi(h)f_1(g^{-1}h)dh$$

$$= \int_{Z_{>0}G(F)\backslash G(\mathbb{A})} \phi(h) \sum_{\gamma \in G(F)} f_1(g^{-1}\gamma h)dh.$$

But the inner integral is bounded in absolute value by $\sup|f_1|$ times the cardinality of the set

$$\left\{ \gamma \in G(F) \,\middle|\, g^{-1}\gamma h \in \Omega \right\}$$

where Ω is the support of f_1. Since this is also the set

$$G(F) \cap g\Omega h^{-1}$$

we can apply the previous lemma. We find

$$|\rho(f)\phi(g)| \leq c||x||^m \int |\phi(h)|dh \leq c||x||^m ||\phi||_2 \mathrm{Vol}(Z_{>0}G(F)\backslash G(\mathbb{A})).$$

$\square$

13.6 Definition of Automorphic Forms

In the adèlic setting of $G(\mathbb{A})$, an automorphic form is a function

$$\phi : G(\mathbb{A}) \to \mathbb{C}$$

invariant under $G(F)$ on the left and K-finite on the right. Further, we demand that ϕ be C^∞ and $Z(\mathfrak{g})$-finite. Finally, we demand that ϕ be of moderate growth, that is,

$$|\phi(g)| \leq c||g||^M$$

for some $c > 0$ and some $M > 0$.

Since

$$0 \leq ||g_1 g_2|| \leq ||g_1|| \cdot ||g_2||$$

for all $g_1, g_2 \in G(\mathbb{A})$, the right translates of ϕ or the convolution of ϕ on the right with a smooth function of compact support are of moderate growth with the same exponent m.

An automorphic form ϕ is thus annihilated by an ideal $\mathfrak{i}$ of $Z(\mathfrak{g})$ of finite codimension and the space V_0 of its right translates by K is finite dimesional. The K-type of ϕ is the set θ of irreducible representations of K which appears in V_0. The pair $(\mathfrak{i}, \theta)$ is the type of ϕ.

Lemma 13.6.1 *Suppose ϕ is an automorphic form. Then there is a smooth function of compact support f on $G(\mathbb{A})$ such that*

$$\rho(f)\phi = \phi.$$

Proof Indeed, the theorem of weak approximation asserts that we have a finite disjoint union

$$G(\mathbb{A}) = \bigcup_{1 \leq i \leq r} G(F) \cdot G_\infty \cdot g_i \cdot K'$$

with $g_i \in G(\mathbb{A}_f)$. We apply Lemma 13.3.3 to the functions

$$g_\infty \mapsto \phi(g_\infty c_i).$$

Thus there is a C^∞ function of compact support f_∞ on G_∞ such that

$$\int_{G^\infty} \phi(g_\infty h c_i) f_\infty(h) dh = \phi(g_\infty c_i)$$

for all i. Now define a function f on $G(\mathbb{A})$ by

$$f(g_\infty g_f) = \begin{cases} \frac{1}{\text{Vol}(K')} f_\infty(g_\infty) & \text{if } g_f \in K', \\ 0 & \text{if } g_f \notin K'. \end{cases}$$

We claim that

$$\int_{G((\mathbb{A})} \phi(gh) f(h) dh = \phi(g)$$

for all $g \in G(\mathbb{A})$. Since the functions of g on the left-hand side and the right-hand side are invariant under K' on the right it suffices to check this relation for $g = g_\infty c_i$ for some i. But then it reduces to

$$\int_{G_\infty} \phi(g_\infty h_\infty c_i) f_\infty(h_\infty) dh\infty = \phi(g_\infty c_i),$$

which is true by the choice of f_∞. □

By Lemma 13.6.1 there is a smooth function of compact support f on $G(\mathbb{A})$ such that $\phi = \rho(f)\phi$. Then for every $X \in U(\mathfrak{g})$ we have

$$\rho(X)\phi = \rho(\rho(X)f)\phi$$

and so $\rho(X)\phi$ is still of moderate growth with the same exponent M.

For $n = 1$ the condition of moderate growth is superfluous. An automorphic form on $GL(1, \mathbb{A}) = \mathbb{A}^\times$ is a finite sum

$$\phi(x) = \sum_j \chi_j(x) P_j(\log |x|),$$

where each χ_j is an idèle class character and each P_j is a polynomial.

For $n > 1$ an automorphic form ϕ is $Z(\mathbb{A})$ finite. In particular, it is $Z_{>0}$ finite. We write $|z| = |z_{i,i}|$ (recall all the $z_{i,i}$ are equal and in $\mathbb{A}^\times_{>0}$.). Then, for $z \in Z_{>0}$,

$$\phi(zg) = \sum_{1 \le j \le r} |z|^{s_j} \left(\sum_{1 \le i \le M_{i,j}} (\log |z|)^{m_{i,j}} \phi_{i,j}(g) \right),$$

where the functions $\phi_{i,j}$ are automorphic forms. We will be mostly concerned in the case where, for $z \in Z_{>0}$,

$$\phi(zg) = |z|^s \phi(g).$$

In fact, by multiplying by a power of $|\det g|$ we may reduce ourselves to the case where

$$\phi(zg) = \phi(g)$$

for all $z \in Z_{>0}$ and this will be the case of interest.

We can also consider square integrable automorphic forms. Those are elements of $L^2(Z_{>0}G(F)\backslash G(\mathbb{A}))$, K_∞ finite on the right and annihilated by an ideal of finite codimension of $Z(\mathfrak{g})$. A priori, this last condition must be taken in the distribution sense. But the two conditions together imply that such a function is real analytic, so the differential equation can be taken in the ordinary sense. The conclusion of Lemma 13.6.1 applies. Thus there is a smooth function of compact support f such that

$$\rho(f)\phi = \phi.$$

For any $X \in U(\mathfrak{g})$ we have

$$\rho(X)f = \rho\left((\rho(X)f)\right)\phi.$$

This implies that for all $X \in U(\mathfrak{g})$ the function $\rho(X)\phi$ is still square integrable. Moreover the function ϕ is of moderate growth. Indeed by Lemma 13.5.8 we have

$$|\phi(x)| = |\rho(f)\phi(x)| \le c||x||^m ||f||_2.$$

Thus ϕ is a slowly increasing automorphic form.

We could also consider more generally functions transforming on the left under a unitary character of $Z_{>0}$.

We have also the following result.

Lemma 13.6.2 *Let V be the space of C^∞ vectors in $L^2(Z_{>0}G(F)\backslash G(\mathbb{A}))$. Every $v \in V$ can be written as a finite sum*

$$v = \sum_{1 \le i \le r} \rho(f_i)v_i,$$

where $v_i \in V$ and the f_i are smooth functions of compact support on $G(\mathbb{A})$.

Proof One argues as in Lemma 13.6.1 using Lemma 13.3.4 instead of Lemma 13.3.3. $\square$

We comment briefly on the relation with Harish-Chandra's notion of automorphic forms [20]. Let ϕ be an adèlic automorphic form. In particular, it is invariant under a compact open subgroup K' of $G(\mathbb{A}_f)$. Consider the intersection subgroup

$$\Gamma := K' \cap G(F),$$

where $G(F)$ is embedded diagonally into $G(\mathbb{A}_f)$. If we embed $G(F)$ into $G\infty$ we can regard Γ as a discrete subgroup of G_∞. It is an arithmetic subgroup. Let ϕ_0 be the restriction of ϕ to G_∞. Then we have

$$\phi_0(\gamma g) = \phi_0(g)$$

for all $\gamma \in \Gamma$, and then ϕ_0 is an automorphic form in the sense of Harish-Chandra for the group Γ. Suppose we translate ϕ on the right by an element $C \in G(\mathbb{A}_f)$. Then the function

$$\phi^c(g) := \phi(gc)$$

is invariant under the open compact subgroup $c^{-1}K'c$, and its restriction ϕ_0^c to G_∞ is invariant under another discrete subgroup Γ^c of G_∞. By the weak approximation theorem we have (disjoint union)

$$G(\mathbb{A}) = \bigcup G(F)G_\infty c_i K'$$

with $c_i \in G(\mathbb{A}_f)$. Thus we see that the adèlic form ϕ is completely determined by the Harish-Chadra automorphic forms $\phi_0^{c_i}$ for different arithmetic groups Γ^{c_i}. In favorable circumstances ϕ is determined by one single Harish-Chandra automorphic form.

13.7 Two Lemmas of Functional Analysis

We recall two lemmas of functional analysis.

Lemma 13.7.1 *Let X be a locally compact space, and let μ be a Borel measure on X such that $\mu(X) < +\infty$. Suppose V is a closed subspace of $L^2(X,\mu)$ such that any element $f \in V$ is a uniformly bounded continuous function. Then V is finite dimensional.*

This lemma is due to Godement. A proof due to Hörmander can be found in [7], Lemma 8.3, or [20], pp. 17–18.

Lemma 13.7.2 *Let X be a locally compact space, countable at infinity, with a countable dense subset. Let μ be a Borel measure on X such that $\mu(X) < +\infty$. Suppose T is a continuous operator*

$$T : L^2(X,\mu) \to C_b(X),$$

where C_b is the space of bounded continuous functions with sup norm. Then T viewed as an operator $T : L^2(X,\mu) \to L^2(X,\mu)$ is a Hilbert–Schmidt operator and, in particular, a compact operator.

A complete proof can be found in [25], XII Section 3, Theorem 6.

13.8 Cusp Forms and Square Integrable Forms

A continuous function ϕ on $G(F)\backslash G(\mathbb{A})$ is said to be cuspidal if

$$\int_{U(F)\backslash U(\mathbb{A})} \phi(ug)du = 0$$

each time U is the unipotent radical of a proper parabolic subgroup of G and $g \in G(\mathbb{A})$. Here du is a Haar masure on $U(\mathbb{A})$.

If ϕ is invariant under $Z_{>0}$ and square integrable on $Z_{>0}G(F)\backslash G(\mathbb{A})$ we say it it is cuspidal if, for every smooth function of compact support f, the continuous function

$$g \mapsto \int \phi(gh)f(h)dh$$

is cuspidal. Thus the space $L^2_{\text{cusp}}(Z_{>0}G(F)\backslash G(\mathbb{A}))$ of cuspidal elements of $L^2(Z_{>0}G(F)\backslash G(\mathbb{A}))$ is a closed subspace.

We are going to see that a cusp form that is invariant under $Z_{>0}$ is, in fact, square integrable.

Lemma 13.8.1 *Suppose ϕ is a cusp form invariant under $Z_{>0}$. Then ϕ is bounded and in particular square integrable.*

We review the elegant proof of Godement in [11] (which is somewhat incomplete).

Proof We need only to prove that ϕ is bounded on any Siegel set. In fact we prove that it is rapidly decreasing in an appropriate sense. Since a Siegel set is contained in a set of the form $A(t)\Omega$, where Ω is a compact set, it will suffice to prove that, for any $m \geq 1$, there is a constant c such that

$$|\phi(a\omega)| \leq c\beta(a)^{-m},$$

for $a \in A(t)$ and $\omega \in \Omega$. We recall the definition

$$\beta(a) := \prod_{1 \leq i \leq n-1} |\alpha_i(a)|.$$

Indeed, we can write

$$\phi(g) = \int_{G^1} \phi(gh)f(h)dh = \int_{G^1} \phi(h)f(g^{-1}h)dh,$$

where f is a smooth function of compact support on G^1. Using the Iwasawa decomposition for G^1, we get

$$\phi(a\omega) = \int \phi(ubk)f(\omega^{-1}a^{-1}ubk)du\delta(b)^{-1}dbdk$$

with $u \in N(\mathbb{A}), b \in A(\mathbb{A}) \cap G^1, k \in K$. Since f has compact support, we see that, in the above integral, the element

$$a^{-1}ubk = (a^{-1}ua)(a^{-1}b)k$$

remains in a fixed compact set. This implies that $a^{-1}b$ remains in a compact set Ω_A of $A(\mathbb{A})$.

Since ϕ is invariant on the left under $N(F)$ we can write this as

$$\int \left(\int_{N(F) \backslash N(\mathbb{A})} \sum_{\gamma \in N(F)} f(\omega^{-1}a^{-1}\gamma ubk)\phi(ubk)du \right) \delta(b)^{-1}dbdk.$$

We now use the cuspidality of ϕ.

If α and β are sums of positive roots we write $\alpha \succeq \beta$ if $\alpha - \beta$ is a sum (possibly empty) of positive roots. Let $\Delta = \{\alpha_i, 1 \le i \le n\}$ be the set of simple positive roots. For every subset $\theta \subseteq \Delta$ we denote by V^θ the subgroup of N defined by the following condition. For each positive root α, the one-dimensional subgroup N_α is contained in V^θ if and only if there is a simple root $\alpha_i \in \theta$ such that $\alpha \succeq \alpha_i$. Thus V^θ is the unipotent radical of a parabolic subgroup $P^\theta = M^\theta V^\theta$. We set $M^\theta \cap N = N^\theta$. We have a semidirect product where V^θ is normal:

$$N = N^\theta V^\theta.$$

Thus if $\theta = \emptyset$ then $P^\emptyset = G$, $N^\emptyset = N$, $V^\emptyset = \{e\}$. For $\theta = \Delta$ then P^Δ is the minimal parabolic subgroup, i.e. the group of triangular matrices and $V^\Delta = N$, $N^\Delta = \{e\}$. For instance, for $n = 3$, we have $\Delta = \{\alpha_1, \alpha_2\}$ and

$$V^{\{\alpha_1\}} = \left\{ \begin{pmatrix} 1 & * & * \\ 0 & 1 & 0 \\ 0 & 0 & 1 \end{pmatrix} \right\}, N^{\alpha_1} = \left\{ \begin{pmatrix} 1 & 0 & 0 \\ 0 & 1 & * \\ 0 & 0 & 1 \end{pmatrix} \right\},$$

$$V^{\{\alpha_2\}} = \left\{ \begin{pmatrix} 1 & 0 & * \\ 0 & 1 & * \\ 0 & 0 & 1 \end{pmatrix} \right\}, N^{\{\alpha_2\}} = \left\{ \begin{pmatrix} 1 & * & 0 \\ 0 & 1 & 0 \\ 0 & 0 & 1 \end{pmatrix} \right\}.$$

Now we consider the following alternating sum

$$\sum_{\theta \subseteq \Delta} (-1)^{|\theta|} \int_{V^\theta(\mathbb{A})} \sum_{\gamma \in N^\theta(F)} f(\omega^{-1}a^{-1}\gamma vubk)dv,$$

as a function of $u \in N(\mathbb{A})$. Here $|\theta|$ denotes the cardinality of θ. The term corresponding to $\theta = 0$ is just our original expression, namely

$$\sum_{\gamma \in N(F)} f(\omega^{-1}a^{-1}\gamma ubk).$$

In this sum, for a given $\theta \neq 0$ the corresponding term

$$\int_{V^\theta(\mathbb{A})} \sum_{\gamma \in N^\theta(F)} f(\omega^{-1}a^{-1}\gamma vubk)dv,$$

as a function of u, is invariant on the left under $V^\theta(\mathbb{A})$ and $N^\theta(F)$, so is also invariant under $N(F)$. If now we integrate against ϕ on $N(F)\backslash N(\mathbb{A})$, we get

$$\int_{N(F)\backslash N(\mathbb{A})} \left(\int_{V^\theta(\mathbb{A})} \sum_{\gamma \in N^\theta(F)} f(\omega^{-1}a^{-1}\gamma vubk)dv \right) \phi(ubk)du.$$

But the integral over $N(F)\backslash N(\mathbb{A})$ can be decomposed as an integral over $V^\theta(F)\backslash V^\theta(\mathbb{A})$ followed by an integral over $N^\theta(F)\backslash N^\theta(\mathbb{A})$ (because V^θ is a normal subgroup). Since ϕ is cuspidal, this integral is 0.

Thus our expression for $\phi(a\omega)$ can be replaced by

$$\int_{N(F)\backslash N(\mathbb{A}) \times A^1 \times K} \text{Alt} \cdot \delta^{-1}(b) \cdot \phi(ubk) \cdot dudbdk,$$

where

$$\text{Alt} := \sum_\theta (-1)^\theta \int_{V^\theta(F)\backslash V^\theta(\mathbb{A})} dv \sum_{\gamma \in N^\theta(F)} f(\omega^{-1}a^{-1}\gamma vubk).$$

We now want to use a Poisson summation formula on the Lie algebra of N. For a general group we would have to use the exponential function, but on $GL(n)$ we can dispense with it. Indeed, the Lie algebra of N denoted $\mathfrak{n}$ can identified with the space of upper triangular matrices with 0 diagonal. The dual vector space ${}^t\mathfrak{n}$ is the space of lower triangular matrices with 0 diagonal. The duality is given by

$$(x, y) \mapsto \text{tr}xy.$$

If we use the standard basis (X_α) of $\mathfrak{n}$ the dual basis is $(X_{-\alpha})$. For $x = \sum x_\alpha X_\alpha$, $y = \sum y_{-\alpha} X_{-\alpha}$ we have $\text{tr}xy = \sum x_\alpha y_{-\alpha}$. Similarly, the Lie algebra $\text{Lie}(N^\theta) = \mathfrak{n}^\theta$ and $\text{Lie}(V^\theta) = \mathfrak{v}^\theta$ are vector subspaces of $\mathfrak{n}$, and we have a direct sum decomposition

$$\mathfrak{n} = \mathfrak{v}^\theta \oplus \mathfrak{n}^\theta$$

and an orthogonal decomposition of the dual vector space

$${}^t\mathfrak{n} = {}^t\mathfrak{v}^\theta \oplus {}^t\mathfrak{n}^\theta.$$

Our alternating sum can also be written as

$$\text{Alt} = \sum_\theta (-1)^{|\theta|} \int_{\mathfrak{v}^\theta(\mathbb{A})} \sum_{\xi \in \mathfrak{n}^\theta(F)} \int f(\omega^{-1}a^{-1}(1+\xi)(1+X)ubk)dX.$$

But

$$(1 + \xi)(1 + X) = 1 + \xi + (1 + \xi)X$$

and we can change X into $(1 + \xi)^{-1}X$. So we get

$$\mathrm{Alt} = \sum_\theta (-1)^{|\theta|} \int_{\mathrm{Lie}V^\theta(\mathbb{A})} \sum_{\xi \in \mathrm{Lie}N^\theta(F)} \int f(\omega^{-1}a^{-1}(1 + \xi + X)ubk)dX.$$

Now let us now introduce a Fourier transform. It is a function on ${}^t\mathrm{n}(\mathbb{A})$:

$$Y \mapsto \int_{\mathrm{n}(\mathbb{A})} f(\omega^{-1}a^{-1}(1 + X)ubk)\psi(\mathrm{tr}XY)dX.$$

Using a Poisson summation formula we get

$$\mathrm{Alt} = \sum_\theta (-1)^{|\theta|} \sum_{\lambda \in {}^t\mathrm{n}^\theta(F)} \int_{\mathrm{n}(\mathbb{A})} f(\omega^{-1}a^{-1}(1 + X)ubk)\psi(\mathrm{tr}X\lambda)dX.$$

After taking into account the cancellation we find this reduces to

$$\mathrm{Alt} = \sum_{\lambda \in {}^t\mathrm{n}(F)}^{\bullet} \int_{\mathrm{n}(\mathbb{A})} f(\omega^{-1}a^{-1}(1 + X)ubk)\psi(\mathrm{tr}X\lambda)dX,$$

where the $\bullet$ means we sum only for those λ that do not belong to some ${}^t\mathrm{n}^\theta$ with $\theta \neq 0$. If we write

$$\lambda = \sum X_{-\alpha}\lambda_{-\alpha}$$

we sum only for those λ such that

$$\sum_{\lambda_{-\alpha} \neq 0} \alpha \succeq \sum_{1 \leq i \leq n-1} \alpha_i.$$

For instance, for $n = 3$, the sum is over the elements

$$\lambda = X_{-\alpha_1}\lambda_{-\alpha_1} + X_{-\alpha_3}\lambda_{-\alpha_2} + X_{-\alpha_1-\alpha_2}\lambda_{-\alpha_1-\alpha_2}$$

such that

$$\lambda_{-\alpha_1} \neq 0 \text{ and } \lambda_{-\alpha_2} \neq 0$$

or

$$\lambda_{-\alpha_1-\alpha_2} \neq 0.$$

Now let us majorize

$$\delta(b)^{-1}\mathrm{Alt} = \delta(b)^{-1} \sum_{\lambda \in {}^t\mathrm{n}(F)}^{\bullet} \int_{\mathrm{n}(\mathbb{A})} f(\omega^{-1}a^{-1}(1 + X)ubk)\psi(\mathrm{tr}X\lambda)dX.$$

It can be written as

$$\delta(b)^{-1} \sum_{\lambda \in {}^t\mathrm{n}(F)}^{\bullet} \int_{\mathrm{n}(\mathbb{A})} f(\omega^{-1}(1 + a^{-1}Xa)a^{-1}uaa^{-1}bk)\psi(\mathrm{tr}X\lambda)dX.$$

After changing variables, and recalling that $\mathrm{tr}\, Xa^{-1}\lambda = \mathrm{tr}X(a^{-1}\lambda a)$, we find

$$\delta(b)^{-1}\mathrm{Alt}$$

$$= \delta(ab^{-1}) \sum_{\lambda \in {}^t\mathrm{n}(F)}^{\bullet} \int_{\mathrm{n}(\mathbb{A})} f(\omega^{-1}(1 + X)a^{-1}uaa^{-1}bk)\psi(\mathrm{tr}Xa^{-1}\lambda a)dX.$$

Since ω, $a^{-1}ua$, $a^{-1}b$ and K remain in compact sets, the functions

$$X \mapsto \delta(ab^{-1})f(\omega^{-1}(1 + X)a^{-1}uaa^{-1}bk)\delta(ab^{-1})$$

remain in a compact set of the space of Schwartz–Brunat functions. So do their Fourier transforms. We now appeal to the following lemma.

Lemma 13.8.2 *Suppose B is a compact set of the space of Schwartz–Bruhat functions on ${}^t\mathrm{n}(\mathbb{A})$ and let $m > 1$. Then there exist $c > 0$ such that, for all $\Phi \in B$ and $a \in A(t)$,*

$$\left| \sum_{\lambda \in {}^t\mathrm{n}(F))}^{\bullet} \Phi(a^{-1}\lambda a) \right| \le c\beta(a)^{-m}.$$

Taking the lemma for granted at the moment, we have

$$\delta(b)^{-1}\mathrm{Alt} =\prec \beta(a)^{-m}$$

and

$$\phi(a\omega) = \int_{N(F)\backslash N(\mathbb{A}) \times A^1 \times K} \mathrm{Alt} \cdot \delta^{-1}(b) \cdot \phi(ubk) \cdot dudbdk.$$

Now u, ba^{-1} and K are in compact sets we have

$$|\phi(ubk)| \prec ||b||^{m_0} \prec ||a||^{m_0}$$

for some $m_0 > 0$. Thus we find

$$|\phi(a\omega)| \prec \beta^{-m}(a)||a||^{m_0}$$

for some m_0 and any $m > 0$. Since a is in $A(t)$ by taking m large enough we obtain

$$|\phi(a\omega)| \prec \beta^{-m_1}(a)$$

for any m_1.

It remains to prove the lemma. We may assume that

$$\Phi = \Phi_\infty \prod_{v \text{ finite}} \Phi_v,$$

where each Φ_v is the characteristic function of

$$\omega_v^{-r_v} M(n \times n, \mathcal{O}_v)$$

and Φ_∞ remains in a compact set. Then the sum takes the form

$$\sum_{\lambda \in \Lambda}^{\bullet} \Phi_\infty(a^{-1}\lambda a),$$

where Λ is a lattice in $M(n \times n, F)$ and Φ_∞ a Schwartz function that remains in a compact set. For any Archimedean place v, there is $c_v > 0$ such that, for $\lambda \in \Lambda$ and $\lambda_{-\alpha} \neq 0$, we have

$$|\lambda_{-\alpha}|_v \geq c_v.$$

There is also a constant d_v such that, for $a \in A(t)$ and all positive root α

$$|\alpha(a)|_v \geq d_v.$$

Now we have

$$|\Phi_\infty(x)| \leq C \prod_{v \text{ real}} \frac{1}{\left(1 + x_{-\alpha,v}^2\right)^{2m}} \prod_{v \text{ complex}} \frac{1}{\left(1 + (x_{-\alpha,v}\overline{x_{-\alpha,v}})^2\right)^{2m}}.$$

Now take

$$x = a^{-1}\lambda a = \sum_\alpha \alpha(a)\lambda_{-\alpha} X_{-\alpha}.$$

which appears in our $\bullet$ sum. For v real, we have

$$(1 + \alpha(a)_v^2\lambda_{-\alpha,v}^2)^{2m} \geq |\alpha(a)|_v^m c_v^m \cdot (1 + d_v^2\lambda_{-\alpha,v}^2)^m.$$

For v complex, we have

$$(1 + \alpha(a)_v^4(\lambda_{-\alpha v}\overline{\lambda_{-\alpha,v}})^2)^{2m} \geq |\alpha(a)|_v^m c_v^m \cdot (1 + d_v^2(\lambda_{-\alpha v}\overline{\lambda_{-\alpha,v}})^2)^m.$$

So for λ in our sum we get

$$|\Phi_\infty(a^{-1}\lambda a)| \prec$$

$$\prod_\alpha |\alpha(a)|^{-2m} \prod_{v \text{ real}} \frac{1}{(1 + d_v^2\lambda_{-\alpha,v}^2)^m} \prod_{v \text{ complex}} \frac{1}{(1 + d_v^2(\lambda_{-\alpha v}\overline{\lambda_{-\alpha,v}})^2)^m}.$$

The first product is over those α for which $\lambda_{-\alpha} \neq 0$. By assumption, summing over those α we have

$$\sum_\alpha \alpha \succeq \sum_{1 \leq i \leq n-1} \alpha_i.$$

and thus $\prod |\alpha(a)| \succeq \beta(a)$. Finally, we find

$$\left| \sum^{\bullet} \Phi_\infty(a^{-1}\lambda a) \right|$$

$$\prec \beta(a)^{-m} \sum_{\lambda \in \Lambda} \prod_{v \text{ real}} \frac{1}{(1 + d_v^2 \lambda_{-\alpha,v}^2)^m} \prod_{v \text{ complex}} \frac{1}{(1 + d_v^2 (\lambda_{-\alpha v} \overline{\lambda_{-\alpha,v}})^2)^m},$$

where in the new sum we have no restriction on λ. For m large enough this sum is finite and we are done. $\qquad\square$

Thus the space of automorphic cusp forms of a given type and invariant under $Z_{>0}$ is a closed subspace of $L^2(G(F)\backslash G^1)$ whose members are continuous bounded functions. Hence it is finite dimensional by Lemma 13.7.1.

This result can be easily extended to any space of cuspidal automorphic forms of a given type.

Suppose f is an automoprhic form on $G(\mathbb{A})$ of a given type i, θ. Let $P = MU$ be a proper parabolic subgroup of G. We claim that, for any $k \in K$, the function

$$m \mapsto f_U(mk) := \int_{U(F)\backslash U(\mathbb{A})} f(umk) \, du$$

on $M(\mathbb{A})$ is an automorphic form. (We have to extend the discussion to the case of a product of linear groups.) Indeed it is invariant under $M(F)$. It is of moderate growth since

$$|f_U(mk)| \le c \int_{U(F)\backslash U(\mathbb{A})} \|umk\|^r du \prec \|m\|^r \int_{U(F)\backslash U(\mathbb{A})} du.$$

It is $K \cap M(\mathbb{A})$ finite of a type determined by θ. Finally we have

$$Z(\mathfrak{g}) \subset Z(\mathfrak{m}) + \mathfrak{u}U(\mathfrak{g}).$$

For $X \in Z(\mathfrak{g})$ call $r(X)$ its projection on $Z(\mathfrak{m})$. Then r is an homomorphism and each function $m \mapsto f_U(mk)$ is annihilated by $r(\mathfrak{i})$. In addition, $Z(\mathfrak{m})$ is a $Z(\mathfrak{g})$ module of finite type. Thus $r(\mathfrak{i})$ is an ideal of finite codimension in $Z(\mathfrak{m})$. A simple inductive argument shows that the dimension of the space of automorphic forms in a given type is finite (see [20]).

Finally, let us consider the space $L^2_{\text{cusp}}(Z_{>0}G(F)\backslash G(\mathbb{A}))$.

Theorem 13.8.3 *Suppose f is a smooth function of compact support on $G(\mathbb{A})$. For any $\phi \in L^2_{\text{cusp}}(Z_{>0}G(F)\backslash G(\mathbb{A}))$ the function $\rho(f)\phi$ is a bounded continuous function, in fact a rapidly decreasing function on $G(F)\backslash G^1$.*

Proof We use the notations of the proof of Lemma 13.8.1. It suffices to estimate $\rho(f)\phi(a\omega)$, where $a \in A(t)$ and ω is in a compact set. We have

$$\rho(f)\phi(g) = \int_{G^1} f^1(h)\phi(gh)dh,$$

where $f^1(g) = \int_{Z_{>0}} f(zg)dz$. We find

$$\rho(f)\phi(a\omega) = \int \text{Alt} \cdot \delta(b)^{-1}\phi(ubk)dudbdk,$$

where

$$|\text{Alt}| \prec \delta(a)\beta^{-m}(a)$$

and $a^{-1}b$ is in a compact set. Thus ubk is in a Siegel set that is itself contained in a finite union of translates by elements of $G(F)$ of a section $\mathfrak{S}'$ of $G(F)\backslash G^1$. Thus the above expression is majorized by

$$\delta(a)\beta(a)^{-m} \int |\phi(ubk)|\delta(b)^{-1}dudbdk$$

$$\prec \delta(a)\beta(a)^{-m} \int_{G(F)\backslash G^1} |\phi|(g)dg$$

$$\leq \delta(a)\beta(a)^{-m}\text{Vol}(G(F)\backslash G^1)^{1/2}\|\phi\|_2.$$

Taking m large enough, $\delta(a)\beta(a)^{-m}$ is bounded independently of a and our assertion follows. ☐

Lemma 13.7.2 implies that the operator $\rho(f)$ on $L^2_{\text{cusp}}(Z_{>0}G(F)\backslash G(\mathbb{A}))$ is a compact operator. It follows that this space decomposes as a discrete sum of unitary irreducible representations, each occurring with finite multiplicity. In fact for $GL(n)$ the multiplicity is (at most) 1 but we will not need this fact.

We also have the following result.

Lemma 13.8.4 *Let V be the space of smooth vectors in $L^2(Z_{>0}G(F)\backslash G(\mathbb{A}))$. Every ϕ in V is bounded (in fact rapidly decreasing on a Siegel set as in the proof of Lemma 13.8.1).*

Proof The proof is similar to the proof of the previous theorem using Lemma 13.6.2. ☐

13.9 Global Theory of L-Functions for Cusp Forms

Theorem 13.9.1 *Let π be a unitary irreducible representation of $G(\mathbb{A})$. Suppose π occurs in $L^2_{\text{cusp}}(Z_{>0}G(F)\backslash G(\mathbb{A})$. Define*

$$L(s,\pi) = \prod_v L(s,\pi_v), \qquad \epsilon(s,\pi) = \prod_v \epsilon(s,\pi_v,\psi_v).$$

Then the Eulerian product $L(s,\pi)$ converges absolutely for $\mathrm{Re}(s) \gg 0$, and can be analytically continued as an entire function of s bounded at infinity in vertical strips. As such, it satisfies the functional equation

$$L(1-s,\tilde{\pi}) = \epsilon(s,\pi)L(s,\pi).$$

Proof We first observe that the contragredient $\tilde{\pi}$ is the imaginary conjugate of π and occurs in the space of cusp forms. Moreover, because the central character ω of π is automorphic, the factor $\epsilon(s,\pi)$ does not depend on the choice of ψ, which justifies the notation.

For the proof we consider a matrix coefficient f of π given by the formula

$$f(g) = \int_{G(F)\backslash G^1} \phi(hg)\tilde{\phi}(h)dh,$$

where ϕ and $\tilde{\phi}$ are K-finite vectors (or even smooth vectors) in the space of π and $\tilde{\pi}$ respectively. Then we consider the global Zeta integral

$$Z(\Phi, f, s) := \int_{G(\mathbb{A})} \Phi(g)f(g) |\det g|^{s+\frac{n-1}{2}}dg,$$

where Φ is a Schwartz–Bruhat function on $M(n \times n, \mathbb{A})$. We assume that Φ is a product

$$\Phi(g) = \prod_v \Phi_v(g_v),$$

where Φ_v is the characteristic function of $M(n \times n, \mathcal{O}_v)$ for almost all v. We will see that this integral converges for $\mathrm{Re}(s) \gg 0$.

Replacing f by its definition we find

$$\int_{G(\mathbb{A})} \Phi(g) \left(\int_{G(F)\backslash G^1} \phi(hg)\tilde{\phi}(h)dh \right) |\det g|^{s+\frac{n-1}{2}}dg.$$

Exchanging the order of integration and changing g to $h^{-1}g$ we find

$$\int_{G(F)\backslash G^1} \tilde{\phi}(h) \left(\int_{G(\mathbb{A})} \Phi(h^{-1}g)\,\phi(g) |\det g|^{s+\frac{n-1}{2}}dg \right) dh.$$

We further decompose the integral over g and we find

$$\int_{G(F)\backslash G^1 \times G(F)\backslash G^1} \tilde{\phi}(h_1)\phi(h_2)dh_1 dh_2 \int_{Z_{>0}} \sum_{\gamma \in G(F)} \Phi\left(h_1^{-1}z\gamma h_2\right)|z|^{ns+\frac{n(n-1)}{2}}d^\times z.$$

Here z has the form

$$z = xI_n, x = (y, y, \ldots, y, 1, 1, \ldots), y > 0, |z| = y^{r+2c}.$$

We need to majorize the sum over γ for h_1 and h_2 in a Siegel set. Then

$$h_1 = a\omega_1, \qquad h_2 = b\omega_2,$$

where ω_1 and ω_2 remains in compact sets while a and b are in $A(t)$.

We need a lemma.

Lemma 13.9.2 *Let $1 \leq r \leq n$, be an integer. With the previous notations, we have the following majorizations.*

(i) *There is a constant c such that*

$$\left| \sum_{\mathrm{rank}(\gamma)=r} \Phi(h_1^{-1} z \gamma h_2) \right| \leq c \|a\| \cdot \|b\| \cdot |z|^{-n^2}$$

for $|z| \leq 1$.

(ii) *For every $M \gg 0$ there is a constant c_M such that*

$$\left| \sum_{\mathrm{rank}(\gamma)=r} \Phi(h_1^{-1} z \gamma h_2) \right| \leq c_M \|a\|^{n^2(M+1)} \cdot \|b\|^{n^2(M+1)} \cdot |z|^{-M}$$

for $|z| \geq 1$.

Proof The functions

$$X \mapsto \Phi(\omega_1^{-1} X \omega_2)$$

remain in a compact set of the space of Schwartz–Bruhat functions. Thus are dominated in absolute value by a fixed Schwartz–Bruhat function $\Phi_0 \geq 0$. Thus it suffices to estimate the sums

$$\sum_{\mathrm{rank}(\gamma)=r} \Phi(a^{-1} z \gamma b)$$

with $\Phi \geq 0$. Each one of these sums is bounded by

$$\sum_{\gamma \in M(n \times n, F) \neq 0} \Phi(a^{-1} z \gamma b).$$

In turn we may assume Φ is majorized by a sum of decomposable functions. So we may as well assume that

$$\Phi(x) = \prod_{(i,j) \in [1,n] \times [1,n]} \phi_{ij}(x_{ij})$$

with $\phi_{ij} \geq 0$. The sum is then equal to a sum over all nonempty subsets S of the product $[1,n] \times [1,n]$:

$$\sum_S \prod_{(i,j) \in S} \left(\sum_{\xi \in F^\times} \phi_{i,j}(a_i^{-1} \xi z b_j) \right) \prod_{(i,j) \notin S} \phi_{ij}(0). \tag{13.1}$$

In general if $\phi \geq 0 \in \mathcal{S}(\mathbb{A})$, and $y \in \mathbb{A}_{>0}$ then, for any $M > 0$, we have, for a suitable $c > 0$,

$$\sum_{\xi \in F^\times} \phi(y\xi) \leq c \frac{|y|^{-1}}{1 + |y|^M}.$$

Thus the term corresponding to a subset S in (13.1) is bounded by a constant times

$$|z|^{-|S|} \prod_{(i,j) \in S} |a_i||b_j|^{-1}$$

and by a constant times

$$|z|^{-|S|-M|S|} \prod_{(i,j) \in S} |a_i|^{1+M} |b_j|^{-1-M}.$$

Now

$$|a_i| \prec ||a||, \ |a_i|^{-1} \prec ||a||, \ |b_j| \prec ||b||, \ |b_j|^{-1} \prec ||b||.$$

So the sums of the terms corresponding to S are dominated by a constant times

$$|z|^{-|S|} \cdot ||a|| \cdot ||b||$$

for $|z| \leq 1$ or since $|S| \leq n^2$

$$|z|^{-n^2} \cdot ||a|| \cdot ||b||.$$

For $|z| \geq 1$ the sums of the terms corresponding to S are dominated by a constant times

$$|z|^{-|S|(M+1)}||a||^{|S|(1+M)}||b||^{|S|(1+M)} \leq |z|^{-M}||a||^{n^2(1+M)}||b||^{n^2(1+M)}.$$

Our assertion follows. $\qquad\qquad\qquad\qquad\qquad\qquad\qquad\qquad\qquad\square$

Thus for $|z| \leq 1$ we have

$$\left| \widetilde{\phi}(h_1)\phi(h_2) \sum_{\gamma \in G(F)} \Phi(h_1^{-1} z\gamma h_2) \right| \prec \beta(a)^{-M} \beta(b)^{-M} \cdot ||a|| \cdot ||b|| \cdot |z|^{-n^2},$$

where M is arbitrary large. On the other hand, $||a|| \cdot ||b|| \prec \beta(a)^{m_1} \beta(b)^{m_1}$ for some m_1. We conclude that, for $|z| \leq 1$,

$$\left| \widetilde{\phi}(h_1)\phi(h_2) \sum_{\gamma \in G(F)} \Phi(h_1^{-1} z\gamma h_2) \right| \prec |z|^{-n^2}.$$

So the integral over $|z| \leq 1$ converges for $\mathrm{Re}(s) \gg 0$.

On the other hand, for $|z| \geq 1$ we have

$$\left| \widetilde{\phi}(h_1)\phi(h_2) \sum_{\gamma \in G(F)} \Phi(h_1^{-1}z\gamma h_2) \right| \prec \beta(a)^{-M}\beta(b)^{-M}||a||^{M_2} \cdot ||b||^{M_2} \cdot |z|^{-M_1},$$

where M and M_1 are arbitrarily large but independent, whereas M_2 depends on M_1. In turn this is dominated by

$$\beta(a)^{-M}\beta(b)^{-M}\beta(a)^{M_2 m_1}\beta(b)^{M_2 m_1} \cdot |z|^{-M_1}.$$

We conclude that, for $|z| \leq 1$,

$$\left| \widetilde{\phi}(h_1)\phi(h_2) \sum_{\gamma \in G(F)} \Phi(h_1^{-1}z\gamma h_2) \right| \prec |z|^{-M_1}.$$

So the integral for $|z| \geq 1$ converges for all s.

Now we apply Poisson summation formula. We have

$$\sum_{\gamma \in G(F)} \Phi(h_1^{-1}\gamma h_2) = \sum_{\gamma \in G(F)} \widehat{\Phi}(h_2^{-1}\gamma z^{-1}h_1)|z|^{-n^2}$$

$$+ \sum_{1 \leq r \leq n-1} \sum_{\text{rank}(\gamma)=r} \widehat{\Phi}(h_2^{-1}\gamma z^{-1}h_1)|z|^{-n^2}$$

$$- \sum_{1 \leq r \leq n-1} \sum_{\text{rank}(\gamma)=r} \Phi(h_1^{-1}\gamma z h_1)$$

$$+ \widehat{\Phi}(0)|z|^{-n^2} - \Phi(0).$$

We integrate this expression against

$$\widetilde{\phi}(h_2)\phi(h_1)|z|^{s+\frac{n(n-1)}{2}}$$

for $|z| \leq 1$. Using the same argument as before we see that the integral of the first term (over $\gamma \in G(F)$) converges for all s. Similarly, the integral over matrices of rank r for $\widehat{\Phi}$ converges for all s. The integral over matrices of rank r for Φ converges for $\text{Re}(s) \gg 0$. Finally, the integral of the term for $\Phi(0)$ and $\widehat{\Phi}(0)$ converges for $\text{Re}(s) \gg 0$.

Because

$$\int \widetilde{\phi}_2(h_2)\phi(h_1)dh_2dh_1 = 0,$$

the terms containing $\Phi(0)$ and $\widehat{\Phi}(0)$ give a zero integral. We claim that the integral

$$\int_{G(F)\backslash G^1 \times G(F)\backslash G^1} \widetilde{\phi}(h_1)\phi(h_2)dh_1dh_2 \sum_{\text{rank}(\gamma)=r} \Phi(h_1^{-1}z\gamma h_2)$$

is 0. Indeed, the matrices of rank r can be written as

$$\gamma = \gamma_1^{-1} \begin{pmatrix} 1_r & 0 \\ 0 & 0_{n-r \times n-r} \end{pmatrix} \gamma_2$$

with $\gamma_1, \gamma_2 \in G(F)$. Call M the group of pairs (γ_1, γ_2) such that

$$\gamma_1^{-1} \begin{pmatrix} 1_r & 0 \\ 0 & 0_{n-r \times n-r} \end{pmatrix} \gamma_2 = \begin{pmatrix} 1_r & 0 \\ 0 & 0_{n-r \times n-r} \end{pmatrix}.$$

Then M is the set of pairs (h_2, h_1) of the form

$$h_2 = u \begin{pmatrix} a & 0 \\ 0 & a_2 \end{pmatrix}, h_1 = v \begin{pmatrix} a & 0 \\ 0 & a_1 \end{pmatrix},$$

where $a \in GL(r)$, $a_1, a_2 \in GL(n-r)$, while u is in the group

$$U = \left\{ \begin{pmatrix} 1_r & * \\ 0 & 1_{n-r} \end{pmatrix} \right\},$$

and v in the group

$$V = \left\{ \begin{pmatrix} 1_r & 0 \\ * & 1_{n-r} \end{pmatrix} \right\}.$$

The integral

$$\int_{G(F)\backslash G^1 \times G(F)\backslash G^1} \widetilde{\phi}(h_1)\phi(h_2) dh_1 dh_2 \sum_{\mathrm{rank}(\gamma)=r} \Phi(h_1^{-1} z \gamma h_2)$$

becomes the integral

$$\int_{M(F)\backslash G^1 \times G^1} \widetilde{\phi}(h_1)\phi(h_2) \Phi\left(h_1^{-1} z \begin{pmatrix} 1_r & 0 \\ 0 & 0_{n-r} \end{pmatrix} h_2\right) dh_1 dh_2.$$

This integral factors through an integral over $M(F)\backslash M(\mathbb{A})$ against the left-invariant measure on $M(\mathbb{A})$. Because the group $U \times V$ is a normal subgroup of M, in turn, this integral factors through an integral over $(U(F)\backslash U(\mathbb{A})) \times V(F)(\backslash V(\mathbb{A}))$. That is, to compute our integral we first compute the integral

$$\int \int_{(U(F)\backslash U(\mathbb{A})) \times V(F)(\backslash V(\mathbb{A}))} \widetilde{\phi}(vh_1)\phi(uh_2) \Phi\left(h_1^{-1} z \begin{pmatrix} 1_r & 0 \\ 0 & 0_{n-r} \end{pmatrix} h_2\right) du dv$$

and then further integrate over (h_1, h_2) against certain measures. Because ϕ and $\widetilde{\phi}$ are cuspidal the integrals over u and v are 0, which proves our claim. Similarly the terms containing $\widehat{\Phi}$ and matrices of rank r give a 0 integral.

Finally we see

$$\int_{G(\mathbb{A})} \Phi(g) f(g) \, |\det g|^{s+\frac{n-1}{2}} dg$$

$$= \int_{|z|\geq 1} \int_{G(F)\backslash G^1 \times G(F)\backslash G^1} \widetilde{\phi}(h_1)\phi(h_2) dh_1 dh_2$$

$$\sum_{\gamma \in G(F)} \Phi\left(h_1^{-1} z\gamma h_2\right) |z|^{ns+\frac{n(n-1)}{2}} d^\times z$$

$$+ \int_{|z|\geq 1} \int_{G(F)\backslash G^1 \times G(F)\backslash G^1} \widetilde{\phi}(h_1)\phi(h_2) dh_1 dh_2$$

$$\sum_{\gamma \in G(F)} \widehat{\Phi}\left(h_2^{-1} z\gamma h_1\right) |z|^{n(1-s)s+\frac{n(n-1)}{2}} d^\times z.$$

We have changed z into z^{-1} on the second integral.

In this expression both integrals converge for all s. This shows that they represent entire functions of s. The proof also shows that these functions of s are bounded at infinity in vertical strips. Moreover, we clearly have the functional equation

$$\int_{G(\mathbb{A})} \Phi(g) f(g) \, |\det g|^{s+\frac{n-1}{2}} dg = \int_{G(\mathbb{A})} \widehat{\Phi}(g) \check{f}(g) \, |\det g|^{1-s+\frac{n-1}{2}} dg.$$

Now we are ready to use the local theory of L-functions. First we write the Haar measure on $G(\mathbb{A})$ as a tensor product of local Haar measure, it being understood that for almost all (or even for all) finite places v the measure of $GL(n, \mathcal{O}_v)$ is 1. Note that here we do not need to normalize the Haar measure, because the **same** Haar measure appears on both sides of our functional equation. We can take ϕ and $\widetilde{\phi}$ to be pure tensors. Then

$$f(g) = \prod_v f_v(g_v),$$

where, for all v, the function f_v is a matrix coefficient of π_v and, for almost all finite v, the function f_v is the spherical coefficient of π_v, and in particular takes the value 1 at e. Formally we have

$$Z(\Phi, f, s) = \prod_v Z(\Phi_v, f, s).$$

If we take Res $> \frac{n-1}{2}$ each local integral converges. Moreover, we have seen that the integral on the left converges for $\mathrm{Re}(s) \gg 0$. This implies that the infinite product on the right converges absolutely for $\mathrm{Re}(s) \gg 0$ (by replacing f by the constant function one can see the infinite product converges for $\mathrm{Re}(s) > 1 + \frac{n-1}{2}$). Almost all factors in the product are equal to $L(s, \pi_v)$.

Now using the local theory we can choose the functions Φ_i and matrix coefficients f_i so that

$$L(s,\pi) = \sum_{i=1}^{r} Z(\Phi_i, f_i, s).$$

This shows that $L(s,\pi)$ has an analytic continuation as an entire function of s. Then

$$\epsilon(s,\pi)L(1-s,\tilde{\pi}) = \sum_{i=1}^{r} Z(\widehat{\Phi}_i, \check{f}_i, 1-s),$$

and our assertion follows.   □

13.10 General Automorphic Forms

We can also define the notion of an irreducible automorphic representation [6]. Such a representation is really a representation of $(\mathfrak{g}, K_\infty)$ and a representation $G(\mathbb{A}_F)$ commuting to one another on a complex vector space V. The space V has no nontrivial invariant subspace. Furthermore the representation is admissible in the sense that an irreducible representation of K appears with finite multiplicity. We say that such a representation is autormorphic if there exist two invariant subspaces $V_0 \subset V_1$ of the space of automorphic forms such that the representation π is the representation on the quotient V_1/V_0.

One can show (Langlands, [6]) that any such π is an irreducible component of an induced representation

$$I(G, P; \pi_1, \pi_2, \ldots, \pi_r),$$

where each representation π_i is automorphic and cuspidal. This means in fact that, for any place v, the representation π_v is a component of the induced representation

$$I(G_v, P_v; \pi_{1,v}, \pi_{2,v}, \ldots, \pi_{r,v}).$$

Furthermore, for almost all finite v, the induced representation has a unique irreducible component with a vector fixed by K_v, and π_v is this irreducible component. This implies that

$$L(s,\pi) := \prod_v L(s,\pi_v)$$

is equal to

$$P(s) \prod_{1 \le i \le r} L(s,\pi_i),$$

where

$$P(s) = \prod_v P_v(s),$$

and $P_v(s)$ is a polynomial in s if v is infinite, a polynomial in q^{-s} if v is finite, and $P_v = 1$ for almost all v. Similarly for the contragredient representations. On the other hand

$$\gamma(s, \pi, \psi_v) = \prod_{1 \le i \le r} \gamma(s, \pi_i, \psi_v)$$

for all v. We conclude that $L(s, \pi)$ is meromorphic with the functional equation

$$L(1 - s, \tilde{\pi}) = \epsilon(s, \pi) L(s, \pi).$$

Finally, the theory of Eisenstein series (Langlands, [6]) shows that any irreducible component of such an induced representation is automorphic.

13.11 $GL(2)$ Examples

The earliest examples of automorphic forms were holomorphic modular forms for $G = GL(2)$. Let $K = O(2, \mathbb{R})$ be the maximal compact subgroup of $G(\mathbb{R})$. By the Iwasawa decomposition, the upper half-plane

$$\mathfrak{h}^2 := \{x + iy \mid x \in \mathbb{R}, \ y > 0\}$$

can be identified with

$$\mathfrak{h}^2 \cong G(\mathbb{R})/(K \cdot \mathbb{R}^\times) \cong \left\{ \begin{pmatrix} y & x \\ 0 & 1 \end{pmatrix} \ \middle| \ x \in \mathbb{R}, \ y > 0 \right\}.$$

Indeed, under the action of $GL(2, \mathbb{R})$ on the upper half-plane given by

$$gz := \frac{az + b}{cz + d}, \qquad \left(\text{for } g = \begin{pmatrix} a & b \\ c & d \end{pmatrix} \in GL(2, \mathbb{R}), \ z \in \mathfrak{h}^2 \right),$$

we see that $\begin{pmatrix} y & x \\ 0 & 1 \end{pmatrix} i = x + iy$ establishes that $\mathfrak{h}^2 \cong G(\mathbb{R})/(K \cdot \mathbb{R}^\times)$.

One of the most famous examples of a classical holomorphic modular form is the Ramanujan cusp form of weight 12 given by:

$$\Delta(z) := e^{2\pi iz} \prod_{n=1}^{\infty} \left(1 - e^{2\pi inz}\right)^{24}$$

$$= e^{2\pi iz} - 24e^{4\pi iz} + 252e^{6\pi iz} - 1472e^{8\pi iz} + \cdots$$

for $z = x + iy \in \mathfrak{h}^2$. The Ramanujan cusp form statisfies the modular relations

$$\Delta\left(\frac{az+b}{cz+d}\right) = (cz+d)^{12}\Delta(z)$$

for all $\left(\begin{smallmatrix} a & b \\ c & d \end{smallmatrix}\right) \in SL(2,\mathbb{Z})$.

We would like to define a modular form purely in group theoretic terms. For modular forms for the group $SL(2,\mathbb{Z})$ one might make the following definition. Define an automorphic form for $SL(2,\mathbb{Z})$ as a function

$$\phi : G \to \mathbb{C},$$

which is invariant under $SL(2,\mathbb{Z})$ on the left, K-invariant on the right, and is invariant under the center $\mathbb{R}^\times$ of $G(\mathbb{R})$. Further, we demand that $\phi\left(\left(\begin{smallmatrix} y & x \\ 0 & 1 \end{smallmatrix}\right)\right)$ is $\mathbb{C}^\infty$ and has moderate growth, that is

$$\left| \phi\left(\begin{pmatrix} y & x \\ 0 & 1 \end{pmatrix}\right) \right| \le c \cdot y^M$$

for some $c, M > 0$, and $\left(\begin{smallmatrix} y & x \\ 0 & 1 \end{smallmatrix}\right)$ in a Siegel set, i.e. $0 \le x < 1$, $y > \frac{\sqrt{3}}{2}$. We term this the *"group theoretic upper half-plane model."*

Note that Δ does not satisfy the above definition because Δ is not invariant on the left under $SL(2,\mathbb{Z})$. To get around this difficulty we need to make the following modification.

We introduce the cocycle $j : GL(2,\mathbb{R}) \times \mathbb{C} \to \mathbb{C}$, which is defined by

$$j(\gamma,\tau) := c\tau + d \qquad \left(\text{for } \gamma = \left(\begin{smallmatrix} a & b \\ c & d \end{smallmatrix}\right) \in GL(2,\mathbb{R}), \ \tau \in \mathfrak{h}^2\right).$$

One easily checks that j satisfies the cocycle relation

$$j(\gamma\gamma',\tau) = j(\gamma,\gamma'\tau) \cdot j(\gamma',\tau).$$

Define

$$\phi(gkd) := \Delta(g\,i) \cdot j(g,i)^{-12}$$

for all $g = \left(\begin{smallmatrix} y & x \\ 0 & 1 \end{smallmatrix}\right) \in \mathfrak{h}^2$, all $k \in K = O(2,\mathbb{R})$, and all $d = \left(\begin{smallmatrix} r & 0 \\ 0 & r \end{smallmatrix}\right)$ with $r \in \mathbb{R}^\times$.

Clearly

$$\phi(g) = \Delta(x + iy) = \Delta(z)$$

for $g = \left(\begin{smallmatrix} y & x \\ 0 & 1 \end{smallmatrix}\right) \in \mathfrak{h}^2$. Note that we are forcing ϕ to be K-invariant and also invariant under the center of $GL(2,\mathbb{R})$ to conform with the upper half-plane model.

It follows from the Iwasawa decomposition that, for $\gamma = \begin{pmatrix} a & b \\ c & d \end{pmatrix} \in SL(2,\mathbb{Z})$ and $g = \begin{pmatrix} y & x \\ 0 & 1 \end{pmatrix} \in \mathfrak{h}^2$, with $z = x + iy$, we have

$$
\begin{aligned}
\phi(\gamma g) &= \Delta(\gamma z) \cdot j(\gamma g, i)^{-12} \\
&= (cz+d)^{12} \cdot \Delta(z) \cdot j(\gamma, g\, i)^{-12} \cdot j(g,i)^{-12} \\
&= (cz+d)^{12} \Delta(z) j(\gamma, z)^{-12} j(g,i)^{-12} \\
&= \Delta(z) j(g,i)^{-12} \\
&= \phi(g).
\end{aligned}
$$

This shows that $\phi(g)$ is invariant under $SL(2,\mathbb{Z})$ on the left. Furthermore, by the Fourier expansion it is clear that ϕ is $\mathbb{C}^\infty$ and that, for $y > \frac{\sqrt{3}}{2}$ and $0 < x < 1$, we have

$$
|\phi(g)| \ll e^{-2\pi y} \cdot y^{-12},
$$

so ϕ has moderate growth. Therefore, ϕ is a modular form for $SL(2,\mathbb{Z})$ for the group theoretic upper half-plane model.

More generally, if

$$
f(z) = \sum_{n=0}^{\infty} a_n e^{2\pi i n z}
$$

is a classical holomorphic modular form of weight $\ell \equiv 0 \pmod 2$ (with an integer $\ell \geq 0$) for $SL(2,\mathbb{Z})$, then the function $\phi : \mathfrak{h}^2 \to \mathbb{C}$ defined by

$$
\phi(g) := f(g\, i) \cdot j(g,i)^{-\ell}
$$

will satisfy the general definition of an automorphic form for $SL(2,\mathbb{Z})$ in the group theoretic upper half-plane model.

We have thus shown that one may replace the classical definition of a holomorphic modular form $f(z)$ (with $z = x + iy$ in the upper half-plane) by defining a new function $\phi(g)$, where g is a matrix of the form $\begin{pmatrix} y & x \\ 0 & 1 \end{pmatrix}$. Unfortunately, this definition is too restrictive and loses information. We, therefore, drop the assumption that ϕ be K-invariant and replace it with another function $\widetilde{\phi}$, which will turn out to be both K-finite and invariant under the center $\mathbb{R}^\times$ of $GL(2,\mathbb{R})^+$, where the superscript $+$ indicates that the matrices are of positive determinant. In this case we define

$$
\widetilde{\phi}(g) := \operatorname{Im}(gi)^{\frac{\ell}{2}} \cdot f(gi) \cdot \left(\frac{j(g,i)}{|j(g,i)|} \right)^{-\ell}
$$

for all $g \in GL(2,\mathbb{R})^+$. Here, again, we have

$$
\widetilde{\phi}(\gamma g) = \widetilde{\phi}(g)
$$

for all $\gamma \in GL(2, \mathbb{Z})$. This is because $\text{Im}\left(\frac{az+b}{cz+d}\right) = \frac{(ad-bc)}{|cz+d|^2} \cdot y$ for all $\left(\begin{smallmatrix} a & b \\ c & d \end{smallmatrix}\right) \in SL(2, \mathbb{Z})$ and $z = x + iy$ in the upper half-plane. Note that inserting the ratio $\frac{j(g,i)}{|j(g,i)|}$ ensures that $\widetilde{\phi}$ is invariant under the center of $g \in GL(2, \mathbb{R})^+$.

Now every $g \in GL(2, \mathbb{R})^+$ has a unique Iwasawa decomposition

$$g = \begin{pmatrix} y & x \\ 0 & 1 \end{pmatrix} \begin{pmatrix} r & 0 \\ 0 & r \end{pmatrix} \begin{pmatrix} \cos\theta & \sin\theta \\ -\sin\theta & \cos\theta \end{pmatrix},$$

with $x, y, r, \theta \in \mathbb{R}$, $y, r > 0$, and $0 \leq \theta < 2\pi$. It follows that

$$\widetilde{\phi}(g) = \left(\cos\theta + i\sin\theta\right)^\ell y^{\frac{\ell}{2}} f(x+iy) = e^{i\ell\theta} y^{\frac{\ell}{2}} f(x+iy).$$

Consider the character $\rho_\ell : SO(2, \mathbb{R}) \to \mathbb{C}^\times$ defined by

$$\rho_\ell\left(\begin{pmatrix} \cos\theta & \sin\theta \\ -\sin\theta & \cos\theta \end{pmatrix}\right) := (\cos\theta + i\sin\theta)^\ell.$$

We then see that

$$\widetilde{\phi}(gzk) = \rho_\ell(k)\widetilde{\phi}(g)$$

for all $g \in GL(2, \mathbb{R})^+$, all $z \in Z$ (here Z is the center of $GL(2, \mathbb{R})^+$), and all $k \in K$. This establishes that $\widetilde{\phi}$ is Z-invariant and K-finite.

If we assume that f, or equivalently that $\widetilde{\phi}$, is an eigenfunction of the Hecke operators, then associated to $\widetilde{\phi}$ one has the Hecke L-function [31]

$$L\left(s, \widetilde{\phi}\right) := \sum_{n=1}^{\infty} a_n n^{-s} = \prod_p \left(1 - a_p p^{-s} + p^{k-1-2s}\right)^{-1},$$

the product ranging over all rational primes, where for every prime p, the complex number a_p is the eigenvalue of the pth Hecke operator. The above series and product converge absolutely for $\text{Re}(s) > (k+1)/2$ by the work of Deligne [8], who proved the Ramanujan conjecture that

$$|a_p| \leq 2p^{\frac{k-1}{2}}.$$

It is well known that $L\left(s, \widetilde{\phi}\right)$ has meromorphic continuation to all $s \in \mathbb{C}$ with, at most, a simple pole at $s = 1$ (only if $a_0 \neq 0$) and satisfies the functional equation

$$(2\pi)^{-s}\Gamma(s)L\left(s, \widetilde{\phi}\right) = \pm(2\pi)^{-(k-s)}\Gamma(k-s)L(k-s, \phi).$$

In addition to holomorphic modular forms there are infinitely many non-holomorphic forms, first found by Maass [26]. The simplest examples are of weight zero. A Maass form of weight zero is an automorphic form $f : \mathfrak{h}^2 \to \mathbb{C}$ that is left invariant under $GL(2, \mathbb{Z})$ and is also an eigenfunction of the

Laplacian with Laplace eigenvalue $v(1 - v)$ ($v \in \mathbb{C}$). For $z = x + iy \in \mathfrak{h}^2$, the Maass form has Fourier expansion of the form

$$f(z) = \sum_{n \neq 0} a_n \sqrt{2\pi y} K_{v-\frac{1}{2}}(2\pi |n| y) e^{2\pi i n x},$$

where, for $v \in \mathbb{C}$ and $y > 0$,

$$K_v(y) = \frac{1}{2} \int_0^\infty e^{-\frac{1}{2} \cdot y(u + u^{-1})} u^v \, du$$

is the modified Bessel function of the second kind.

As before we may lift the Maass form f to a function $\widetilde{\phi} : GL(2, \mathbb{R})^+ \to \mathbb{C}$ defined by

$$\widetilde{\phi}(g) := f(g\,i) \qquad \left(g \in GL(2, \mathbb{R})^+\right).$$

If the Maass form $\widetilde{\phi}$ is also an eigenfunction of the Hecke operators then the L-function associated to $\widetilde{\phi}$ is given by

$$L\left(s, \widetilde{\phi}\right) = \sum_{n=1}^\infty a_n n^{-s} = \prod_p \left(1 - a_p p^{-s} + p^{-2s}\right)^{-1} \qquad (\mathrm{Re}(s) > 3/2),$$

where, for each prime p, the coefficient $a_p \in \mathbb{C}$ is the eigenvalue of the pth Hecke operator. Furthermore, $L\left(s, \widetilde{\phi}\right)$ is an entire function and satisfies the functional equation

$$\Lambda\left(s, \widetilde{\phi}\right) := \pi^{-s} \Gamma\left(\frac{s - \frac{1}{2} + v}{2}\right) \Gamma\left(\frac{s + \frac{1}{2} - v}{2}\right) L\left(s, \widetilde{\phi}\right) = \Lambda\left(1 - s, \widetilde{\phi}\right).$$

We have now shown some simple examples $\widetilde{\phi}$ of automorphic forms for the real group $GL(2, \mathbb{R})^+$. It is, then, possible to define ([18], Section 4.12) an adèlic automorphic form $\phi_{\mathrm{adèlic}}((g_\infty, g_2, g_3, \ldots))$ on $GL(2, \mathbb{A}_\mathbb{Q})$ that is identical to $\widetilde{\phi}(g_\infty)$ when the finite adèle $(g_2, g_3, \ldots, g_p, \ldots)$ is just $(I_2, I_2, I_2, \ldots)$ and I_2 is the 2×2 identity matrix.

More generally, one may consider a classical modular form f, which has integer weight $\ell \geq 0$, level $N \geq 1$, character χ (mod N), and is an eigenfunction of the Hecke operators as well as the Laplacian. Then f is a smooth function of moderate growth on the upper half-plane $\{z = x + iy \mid x \in \mathbb{R}, y > 0\}$ that satisfies

$$f(\gamma z) = \chi(d)(cz + d)^\ell f(z), \qquad \left(\forall \gamma = \begin{pmatrix} a & b \\ c & d \end{pmatrix} \in \Gamma_0(N)\right).$$

Again, each of these classical modular forms can be lifted to an adèlic form (see [18], Section 4.12).

One may ask whether the space of adèlic automorphic forms for $GL(2, \mathbb{A}_\mathbb{Q})$ contains new objects in addition to the lifts of the classical automorphic forms. We now show that it is also possible to go in the other direction and establish that, in every irreducible automorphic cuspidal representation, there is a vector that is an idèlic lift of a classical modular form of weight ℓ, level N, and character χ (mod N), as described above.

Fix an integer $N \geq 1$. The Iwahori subgroup $K_0(N) \subset GL(2,\mathbb{A}_\mathbb{Q})$ is defined as $K_0(N) = \prod_p K_0(N)_p$ (with the product ranging over all primes p), where

$$K_0(N)_p = \left\{ \begin{pmatrix} a & b \\ N \cdot c & d \end{pmatrix} \in GL(2,\mathbb{Z}_p) \,\middle|\, c \in \mathbb{Z}_p \right\}.$$

We have the strong approximation theorem ([18], section 4.11)

$$GL(2,\mathbb{A}_\mathbb{Q}) = GL(2,\mathbb{Q})\, GL(2,\mathbb{R})^+ \, K_0(N),$$

where $GL(2,\mathbb{Q}) \cap \left(GL(2,\mathbb{R})^+ K_0(N)\right) = \Gamma_0(N)$.

Recall that an adèlic automorphic form ϕ for $GL(2,\mathbb{A}_\mathbb{Q})$ with central character ω is left invariant under $GL(2,\mathbb{Q})$, right K-finite, $\mathbb{Z}(\mathfrak{g})$-finite, and has moderate growth. If we assume, in addition, that ϕ is a suitable vector in an irreducible automorphic cuspidal representation then ϕ will be invariant under an Iwahori subgroup, and we will have

$$\phi(gk) = \psi(k) \cdot \phi(g)$$

for all $g \in GL(2,\mathbb{A}_\mathbb{Q})$ and all $k \in K_0(N)$ for some Iwahori subgroup $K_0(N)$, and where $\psi : K_0(N) \to \mathbb{C}^\times$ is a character of the Iwahori subgroup. Furthermore, at the Archimedean place we must have

$$\phi\left(g \cdot \left(\begin{pmatrix} \cos\theta & \sin\theta \\ -\sin\theta & \cos\theta \end{pmatrix}, I_2, I_2, I_2, \ldots\right)\right) = e^{\pi i \ell}\phi(g)$$

for all $g \in GL(2,\mathbb{A}_\mathbb{Q})$, where I_2 is the identity matrix.

We may then define the classical modular form $f : \mathfrak{h} \to \mathbb{C}$ by

$$f(x+iy) := \phi\left(\left(\begin{pmatrix} y^{\frac{1}{2}} & xy^{-\frac{1}{2}} \\ 0 & y^{-\frac{1}{2}} \end{pmatrix}, I_2, I_2, I_2, \ldots\right)\right),$$

which satisfies

$$f(\gamma z) = \chi(d)(cz+d)^\ell f(z)$$

for all $z \in \mathfrak{h}$, all $\gamma \in \Gamma_0(N)$, and where χ is a Dirichlet character (mod N) determined by the character ψ of the Iwahori subgroup $K_0(N)$. For the precise determination of the Dirichlet character χ, see ([18], Section 5.5.6).

If ϕ is a Hecke cusp form on the upper half-plane, then we first lift ϕ to a function $\tilde{\phi}$ on the real group $GL(2,\mathbb{R})^+$, and then lift this function to an adèlic automorphic form $\phi_{\text{adèlic}}$ as above. We may then associate to ϕ an irreducible unitary infinite-dimensional automorphic representation π_ϕ of

$GL(2, \mathbb{A}_\mathbb{Q})$. This can be done as follows. We consider the following actions (denoted $\mathcal{A}$) on the adèlic automorphic form $\phi_{\text{adèlic}}$.

- *The action of the finite adèles $GL(2, \mathbb{A}_f)$ of $\mathbb{A}_\mathbb{Q}$ by right translation.*
- *The action of the universal enveloping algebra $\mathfrak{U}$ by differential operators $D \in \mathfrak{U}$ (at the real place g_∞).*

Now, define the vector space

$$V_\phi := \left\{ \sum_{\ell=1}^{N} c_\ell \cdot D_\ell \cdot \phi_{\text{adèlic}} \left(g \cdot h_\ell \right) \,\middle|\, N \geq 0,\ c_l \in \mathbb{C},\ h_\ell \in GL(2, \mathbb{A}_f),\ D_\ell \in \mathfrak{U} \right\}.$$

Then V_ϕ is clearly invariant under the actions $\mathcal{A}$. The space V_ϕ with the actions $\mathcal{A}$ define the automorphic representation π_ϕ. Further, it can be shown that the Godement–Jacquet L-function $L\left(s, \pi_\phi\right) = L(s, \phi)$.

13.12 $GL(n)$ Examples

We shall now present some examples of $G = GL(n)$ automorphic forms over $\mathbb{A}_\mathbb{Q}$ for $n > 2$. It is enough to present examples for the real group $GL(n, \mathbb{R})$, since these may be lifted to adèlic automorphic forms. We may define the generalized upper half-plane

$$\mathfrak{h}^n \cong G(\mathbb{R})/(K \cdot \mathbb{R}^\times),$$

where $K = O(n, \mathbb{R})$ is the maximal compact subgroup. By the Iwasawa decomposition, every $g \in \mathfrak{h}^n$ is an element of the form $g = xy$, where

$$x = \begin{pmatrix} 1 & x_{1,2} & x_{1,3} & \cdots & & x_{1,n} \\ & 1 & x_{2,3} & \cdots & & x_{2,n} \\ & & \ddots & & & \vdots \\ & & & 1 & & x_{n-1,n} \\ & & & & & 1 \end{pmatrix}, \quad y = \begin{pmatrix} y_1 y_2 \cdots y_{n-1} & & & \\ & \ddots & & \\ & & y_1 y_2 & \\ & & & y_1 \\ & & & & 1 \end{pmatrix},$$

with $x_{i,j} \in \mathbb{R}$ for $1 \leq i < j \leq n$ and $y_i > 0$ for $1 \leq i \leq n - 1$.

If we consider the discrete subgroup $G(\mathbb{Z})$, then an automorphic form is a function

$$\phi : G \to \mathbb{C},$$

which is invariant under $G(\mathbb{Z})$ on the left, K-invariant on the right, and is invariant under the center $\mathbb{R}^\times$ of $G(\mathbb{R})$. Further, we demand that ϕ is C^∞ and has moderate growth, that is

$$|\phi(xy)| \le c \prod_{i=1}^{n-1} y_i^M$$

for some $c, M > 0$, and xy in a Siegel set, i.e. $0 \le x_{\ell,j} < 1$, $y_i > \frac{\sqrt{3}}{2}$, for $1 \le \ell < j \le n$ and $1 \le i < n$.

The space $\mathfrak{h}^n$ does not have a complex structure for $n > 2$, so there will be no holomorphic automorphic forms. There will, however, be Maass forms, which we now describe. A Maass form is defined to be a complex-valued function $\phi : \mathfrak{h}^n \to \mathbb{C}$, which is an automorphic form (as defined above) and, in addition, is an eigenfunction of the center of the universal enveloping algebra of $\mathfrak{g}$ (denoted by $\mathcal{D}^n$), which is just the ring of $GL(n,\mathbb{R})$ invariant differential operators on $\mathfrak{h}^n$.

Since $\mathcal{D}^n$ is commutative we may construct a basis of simultaneous eigenfunctions of all $\delta \in \mathcal{D}^n$. The eigenvalues of such eigenfunctions can be expressed in terms of Langlands parameters

$$\alpha = \{\alpha_1, \alpha_2, \dots, \alpha_n\} \in \mathbb{C}^n$$

with $\sum_{i=1}^n \alpha_i = 0$. We shall now explicitly describe the representation of eigenvalues of $\mathcal{D}^n$ in terms of Langlands parameters.

Let $\alpha = \{\alpha_1, \alpha_2, \dots, \alpha_n\} \in \mathbb{C}^n$ denote a set of Langlands parameters. We define a character $I_\alpha : U_n(\mathbb{R})\backslash\mathfrak{h}^n \to \mathbb{C}$ by

$$I_\alpha(g) := \prod_{i=1}^{n-1}\prod_{j=1}^{n-1} y_i^{b_{i,j} \frac{\alpha_j - \alpha_{j+1}}{n}}, \qquad b_{i,j} = \begin{cases} ij & \text{if } i + j \le n, \\ (n-i)(n-j) & \text{if } i + j \ge n. \end{cases}$$

Here, the powers of the y_i are chosen to simplify later formulae.

Then I_α is an eigenfunction of all $\delta \in \mathcal{D}^n$, so we may write

$$\delta I_\alpha = \lambda_\delta \cdot I_\alpha,$$

where λ_δ denotes the Harish-Chandra character. The Laplace eigenvalue λ_Δ can be represented in the form (see [30])

$$\lambda_\Delta = \frac{n^3 - n}{24} - \frac{\alpha_1^2 + \alpha_2^2 + \cdots \alpha_n^2}{2}.$$

Consider Maass forms $\phi : \mathfrak{h}^n \to \mathbb{C}$. Since $\mathcal{D}^n$ is a commutative ring, we may take a basis of Maass forms consisting of Laplace eigenfunctions, which are also common eigenfunctions of all $\delta \in \mathcal{D}^n$. Then ϕ will be an eigenfunction of the Laplacian Δ for $\mathfrak{h}^n$, i.e.

$$\Delta\phi = \lambda_\Delta\phi \qquad \text{(for some } \lambda_\Delta \in \mathbb{C}).$$

Each such Maass form ϕ will have an associated Langlands parameter $\alpha \in \mathbb{C}^n$ with associated Laplace eigenvalue $\lambda_\Delta = \frac{n^3-n}{24} - \frac{\alpha_1^2+\alpha_2^2+\cdots\alpha_n^2}{2}$.

Given Langlands parameters $\alpha \in \mathbb{C}^n$ (with Harish-Chandra character λ_δ as described above) and a character ψ of the unipotent subgroup $U_n(\mathbb{R}) \subset G(R)$, then there exists a unique (up to a constant multiple) Whittaker function

$$W_\alpha : \mathfrak{h}^n \to \mathbb{C},$$

which satisfies the following properties:

- $\delta W_\alpha = \lambda_\delta \cdot W_\alpha, \quad (\forall \delta \in \mathcal{D}^n)$,
- $W_\alpha(ug) = \psi(u) \cdot W_\alpha(g), \quad (\forall u \in N(\mathbb{R}), g \in GL(n,\mathbb{R}))$,
- W_α is invariant under all permutations of $\alpha = \{\alpha_1, \ldots, \alpha_n\}$,
- W_α has holomorphic continuation to all $\alpha \in \mathbb{C}^n$,
- $W_\alpha(y)$ has rapid decay in $y_i \to \infty$ where $y = \text{diag}(y_1, y_2, \cdots y_n)$.

Let $M = (m_1, \ldots, m_{n-1}) \in \mathbb{Z}^{n-1}$, $\Gamma_{n-1} = \text{SL}(n-1,\mathbb{Z})$, and $U_{n-1} = U_{n-1}(\mathbb{Z})$. It was proved by Shalika and Piatetski–Shapiro (see [17], (9.1.2)) that every Maass form with Langlands parameter α has a Fourier–Whittaker expansion of type

$$\phi(g) = \sum_{\gamma \in U_{n-1} \backslash \Gamma_{n-1}} \sum_{M \neq 0} \frac{A(M)}{\prod_{k=1}^{n-1} |m_k|^{\frac{k(n-k)}{2}}} \, W_\alpha \left(M^* \begin{pmatrix} \gamma & 0 \\ 0 & 1 \end{pmatrix} g \right),$$

where $g \in \mathfrak{h}^n$ and

$$M^* = \begin{pmatrix} m_1 \cdots m_{n-2}|m_{n-1}| & & & \\ & \ddots & & \\ & & m_1 & \\ & & & 1 \end{pmatrix}.$$

Here $A(m_1, \ldots, m_{n-1}) \in \mathbb{C}$ is called the Mth Fourier coefficient of ϕ.

We may associate to ϕ the Godement–Jacquet L-function

$$L(s,\phi) = \sum_{m=1}^\infty \frac{A(m,1,\ldots,1)}{m^s}.$$

If the Maass form ϕ is also an eigenfunction of the Hecke operators then it has the following Euler product representation (see [17])

$$\prod_p \left(1 - \frac{A(p,1,\ldots,1)}{p^s} + \frac{A(1,p,1,\ldots,1)}{p^{2s}} - \frac{A(1,1,p,\ldots,1)}{p^{3s}} \right.$$
$$\left. + \cdots + (-1)^{n-1} \frac{A(1,\ldots,1,p)}{p^{(n-1)s}} + \frac{(-1)^n}{p^{ns}} \right)^{-1}.$$

Now $L(s,\phi)$ is a degree-n L-function, which means the completed L-function has n local factors at every place and satisfies the following functional equation (see [17], Theorem 12.3.6):

$$\Lambda(s,\phi) := \pi^{-\frac{ns}{2}} \prod_{i=1}^{n} \Gamma\left(\frac{s-\alpha_i}{2}\right) L(s,\phi) = \Lambda(1-s,\widetilde{\phi}),$$

where $\widetilde{\phi}$ denotes the dual form, which has an Mth Fourier coefficient (for $M = (m_1, m_2, \ldots, m_{n-1})$) given by $A(m_{n-1}, m_{n-2}, \ldots, m_1)$.

More generally, we may also consider automorphic forms of arbitrary weight, level, and character for the real group $GL(n,\mathbb{R})^+$ that acts on $\mathfrak{h}^n$ by left matrix multiplication. This action determines a function

$$\kappa : GL(n,\mathbb{R})^+ \times \mathfrak{h}^n \longrightarrow SO(n,\mathbb{R})$$

as follows.

By the Iwasawa decomposition, every $g \in GL(n,\mathbb{R})^+$ has a unique decompostion

$$g = \widetilde{g} \cdot d \cdot k$$

with $\widetilde{g} \in \mathfrak{h}^n$, $d = r \cdot I_n (r > 0)$, and $k \in K = SO(n,\mathbb{R})$. Then, for any $\gamma \in GL(n,\mathbb{R})^+$ and $g \in GL(n,\mathbb{R})$, we define $\kappa(\gamma,g)$ by

$$\gamma g = \widetilde{\gamma g} \cdot d \cdot \kappa(\gamma,g),$$

where $d = rI_n$ for some real number $r > 0$. Then $\kappa(\gamma,g)$ satisfies the cocycle identity

$$\kappa(\gamma'\gamma, g) = \kappa\left(\gamma, \widetilde{\gamma'g}\right) \cdot \kappa(\gamma',g).$$

One would like to generalize the notion of *"weight"* to the higher-rank situation of $GL(n,\mathbb{R})^+$ with $n > 2$. In this case, the *"weight"* may be realized as a finite irreducible representation ρ of $SO(n,\mathbb{R})$ that generalizes the $GL(2)$-weight, which corresponds to an irreducible representation of $SO(2,\mathbb{R})$. Of course, since $SO(2,\mathbb{R})$ is abelian, then it can have only one-dimensional representations, i.e. characters.

Let $\rho : SO(n,\mathbb{R}) \to GL(r,\mathbb{C})$ be an irreducible representation. We define a function $J_\rho : GL(n,\mathbb{R})^+ \times \mathfrak{h}^n \to GL(r,\mathbb{C})$ as follows. Let $\gamma \in GL(n,\mathbb{R})^+$ and $g \in \mathfrak{h}^n$. Then we define

$$J_\rho(\gamma,g) := \rho\left(\kappa(\gamma,g)^{-1}\right).$$

We now prove that J_ρ is a one-cocycle satisfying

$$J_\rho(\gamma\gamma', g) = J_\rho(\gamma',g) J_\rho\left(\gamma, \widetilde{\gamma'g}\right)$$

for all $\gamma, \gamma' \in GL(n,\mathbb{R})^+$ and all $g \in \mathfrak{h}^n$.

Proof We have

$$
\begin{aligned}
J_\rho(\gamma\gamma', g) &= \rho\left(\kappa(\gamma\gamma', g)^{-1}\right) \\
&= \rho\left(\kappa(\gamma', g)^{-1} \cdot \kappa\left(\gamma, \widetilde{\gamma'g}\right)^{-1}\right) \\
&= \rho\left(\kappa(\gamma', g)^{-1}\right) \cdot \rho\left(\kappa\left(\gamma, \widetilde{\gamma'g}\right)^{-1}\right) \\
&= J_\rho(\gamma', g)\, J_\rho(\gamma, \widetilde{\gamma'g}).
\end{aligned}
$$
$\qquad\square$

Since the *"weight"* is a representation into $GL(r, \mathbb{C})$ it is necessary to consider vector-valued automorphic forms of the type

$$
\Phi(g) := \begin{pmatrix} \phi_1(g) \\ \vdots \\ \phi_r(g) \end{pmatrix} \qquad \left(g \in \mathfrak{h}^n\right),
$$

where each $\phi_i : \mathfrak{h}^n \to \mathbb{C}$, $(1 \le i \le r)$ is smooth. We say Φ has weight ρ for a discrete subgroup $\Gamma \subset GL(n, \mathbb{R})^+$ if

$$
\Phi(\gamma g) = J_\rho(\gamma, g) \cdot \Phi(g)
$$

for all $\gamma \in \Gamma$ and all $g \in \mathfrak{h}^n$.

Next, we consider vector-valued automorphic functions for the real group $GL(n, \mathbb{R})^+$ with level N and character. For an integer $N \ge 2$, we define the congruence subgroup $\Gamma_0(N) \subset SL(n, \mathbb{Z})$ to be the multiplicative group of all matrices of the form:

$$
\begin{pmatrix} A & B \\ C & d \end{pmatrix} \text{ with } \begin{cases} A \text{ is an } (n-1) \times (n-1) \text{ matrix with entries in } \mathbb{Z}, \\ B \text{ is a column vector with entries in } \mathbb{Z}, \\ C \text{ is a row vector with entries in } N \cdot \mathbb{Z}, \\ d \in \mathbb{Z}. \end{cases}
$$

In addition, we define $\Gamma_0(1) := SL(n, \mathbb{Z})$.

We call N the " *level.*" For a given level N we may consider introduce a *"character,"* which we take to be a Dirichlet character χ (mod N). We say a vector-valued automorphic function of the type Φ above has weight ρ, level N, and character χ if

$$
\Phi(\gamma g) = \chi(d)\, J_\rho(\gamma, g)\, \Phi(g)
$$

for all

$$
\gamma = \begin{pmatrix} A & B \\ C & d \end{pmatrix} \in \Gamma_0(N).
$$

Next, consider a vector-valued automorphic function Φ on the real group $GL(n, \mathbb{R})^+$ of weight ρ, level N, and character χ for some r-dimensional representation of $SO(n, \mathbb{R})$, which is Z-finite, $\mathbb{Z}(\mathfrak{g})$-finite, and has moderate growth. We will show that Φ can be lifted to an adèlic automorphic form on $GL(n, \mathbb{A}_\mathbb{Q})$. One immediate problem that arises is the fact that a vector-valued automorphic function takes values in $\mathbb{C}^r$ while an adèlic automorphic form always takes values in $\mathbb{C}$.

Fix an integer $N \geq 1$. The Iwahori subgroup $K_0(N) \subset GL(n, \mathbb{A}_\mathbb{Q})$ is defined as $K_0(N) = \prod_p K_0(N)_p$ (with the product ranging over all primes p) where

$$K_0(N)_p = \left\{ \begin{pmatrix} A & B \\ N \cdot C & d \end{pmatrix} \in GL(2, \mathbb{Z}_p) \right\},$$

where A is an $(n-1) \times (n-1)$ matrix with entries in $\mathbb{Z}_p$, where B is a column vector with entries in $\mathbb{Z}_p$, while C is a row vector with entries in Z_p, and $d \in \mathbb{Z}_p$.

We have the strong approximation theorem ([19], Proposition 13.3.3)

$$GL(n, \mathbb{A}_\mathbb{Q}) = GL(n, \mathbb{Q}) \, GL(n, \mathbb{R})^+ \, K_0(N),$$

where $GL(n, \mathbb{Q}) \cap \big(GL(n, \mathbb{R})^+ K_0(N)\big) = \Gamma_0(N)$.

Strong approximation can be used to define the adèlic lift

$$\Phi_{\text{adèlic}} : GL(n, \mathbb{A}_\mathbb{Q}) \to \mathbb{C}^r$$

given by

$$\Phi_{\text{adèlic}}(\gamma g_\infty k) := \psi(k) \, J_\rho(g_\infty, I_n) \, \Phi(g_\infty) \qquad (\text{for all } g_\infty \in GL(n, \mathbb{R})^+),$$

where $k \in K_0(N)$ and $\gamma = (\alpha, \alpha, \alpha, \ldots) \in GL(n, \mathbb{A}_\mathbb{Q})$, where we have $\alpha \in GL(n, \mathbb{Q})$. Here ψ will be a character of the Iwahori subgroup $K_0(N)$. One may show that (see [19], Lemma 13.4.8) that

$$\Phi_{\text{adèlic}}(g) = \begin{pmatrix} \phi_1^*(g) \\ \vdots \\ \phi_r^*(g) \end{pmatrix} \qquad \big(\text{for all } g \in GL(n, \mathbb{A}_\mathbb{Q})\big),$$

where each ϕ_i^* $(i = 1, 2, \ldots, r)$ is an adèlic automorphic form.

References

[1] I. N. Bernšteĭn. All reductive p-adic groups are of type I. *Funkcional. Anal. i Priložen.*, 8(2):3–6, 1974.

[2] I. N. Bernšteĭn and A. V. Zelevinskiĭ. Induced representations of the group $GL(n)$ over a p-adic field. *Funkcional. Anal. i Priložen.*, 10(3):74–75, 1976.

[3] I. N. Bernšteĭn and A. V. Zelevinskiĭ. Representations of the group $GL(n, F)$, where F is a local non-Archimedean field. *Uspehi Mat. Nauk*, 31(3(189)):5–70, 1976.

[4] I. N. Bernstein and A. V. Zelevinsky. Induced representations of reductive p-adic groups. I. *Ann. Sci. École Norm. Sup. (4)*, 10(4):441–472, 1977.

[5] J. Bernstein and B. Krötz. Smooth Fréchet globalizations of Harish-Chandra modules. *Israel J. Math.*, 199(1):45–111, 2014.

[6] A. Borel and H. Jacquet. Automorphic Forms and automorphic representations. In *Automorphic forms, representations and L-functions (Proc. Sympos. Pure Math., Oregon State University, Corvallis, Oregon, 1977), Part 1*, Proc. Sympos. Pure Math., XXXIII, pages 189–207. Amer. Math. Soc., Providence, R.I., 1979. With a supplement "On the notion of an automorphic representation" by R. P. Langlands.

[7] A. Borel. *Automorphic forms on* $\mathrm{SL}_2(\boldsymbol{R})$, volume 130 of *Cambridge Tracts in Mathematics*. Cambridge University Press, Cambridge, 1997.

[8] P. Deligne. La conjecture de Weil, I. *Publications Mathématiques de l'IHES*, 43: 273–307, 1974.

[9] J. Dixmier and P. Malliavin. Factorisations de fonctions et de vecteurs indéfiniment différentiables. *Bull. Sci. Math. (2)*, 102(4):307–330, 1978.

[10] D. Flath. Decomposition of representations into tensor products. In *Automorphic forms, representations and L-functions (Proc. Sympos. Pure Math., Oregon State University, Corvallis, Oregon, 1977), Part 1*, Proc. Sympos. Pure Math., XXXIII, pages 179–183. Amer. Math. Soc., Providence, R.I., 1979.

[11] R. Godement. The spectral decomposition of cusp-forms. In *Algebraic Groups and Discontinuous Subgroups (Proc. Sympos. Pure Math., Boulder, Colo., 1965)*, pages 225–234. Amer. Math. Soc., Providence, R.I., 1966.

[12] R. Godement. Domaines fondamentaux des groupes arithmétiques [MR0191899 (33 #126)]. In *Séminaire Bourbaki, Vol. 8*, pages Exp. No. 257, 201–225. Soc. Math. France, Paris, 1995.

[13] R. Godement. Les fonctions ζ des algèbres simples. I. In *Séminaire Bourbaki, Vol. 5*, pages Exp. No. 171, 27–49. Soc. Math. France, Paris, 1995.

[14] R. Godement. Les fonctions ζ des algèbres simples. II. In *Séminaire Bourbaki, Vol. 5*, pages Exp. No. 176, 109–128. Soc. Math. France, Paris, 1995.

[15] R. Godement. *Notes on Jacquet–Langlands' theory*, volume 8 of *CTM. Classical Topics in Mathematics*. Higher Education Press, Beijing, 2018. With commentaries by R. Langlands and H. Jacquet.

[16] R. Godement and H. Jacquet. *Zeta functions of simple algebras*. Lecture Notes in Mathematics, Vol. 260. Springer-Verlag, Berlin–New York, 1972.

[17] D. Goldfeld. *Automorphic forms and L-functions for the group* $\mathrm{GL}(n, \mathrm{R})$, volume 99 of *Cambridge Studies in Advanced Mathematics*. Cambridge University Press, Cambridge, 2015. With an appendix by K. A. Broughan, Paperback edition of the 2006 original.

[18] D. Goldfeld and J. Hundley. *Automorphic representations and L-functions for the general linear group. Volume I*, volume 129 of *Cambridge Studies in Advanced Mathematics*. Cambridge University Press, Cambridge, 2011. With exercises and a preface by X. Faber.

[19] D. Goldfeld and J. Hundley. *Automorphic representations and L-functions for the general linear group. Volume II*, volume 130 of *Cambridge Studies in Advanced Mathematics*. Cambridge University Press, Cambridge, 2011. With exercises and a preface by X. Faber.

[20] Harish-Chandra. *Automorphic forms on semisimple Lie groups*. Notes by J. G. M. Mars. Lecture Notes in Mathematics, No. 62. Springer-Verlag, Berlin–New York, 1968.

[21] Harish-Chandra. *Harmonic analysis on reductive p-adic groups*. Lecture Notes in Mathematics, Vol. 162. Springer-Verlag, Berlin–New York, 1970. Notes by G. van Dijk.

[22] H. Jacquet and R. P. Langlands. *Automorphic forms on* GL(2). Lecture Notes in Mathematics, Vol. 114. Springer-Verlag, Berlin–New York, 1970.

[23] H. Jacquet. Zeta functions of simple algebras (local theory). In *Harmonic analysis on homogeneous spaces (Proc. Sympos. Pure Math., Vol. XXVI, Williams College, Williamstown, Massachusetts, 1972)*, pages 381–385. Amer. Math. Soc., Providence, R.I., 1973.

[24] H. Jacquet. Principal *L*-functions of the linear group. In *Automorphic forms, representations and L-functions (Proc. Sympos. Pure Math., Oregon State University, Corvallis, Oregon, 1977), Part 2*, Proc. Sympos. Pure Math., XXXIII, pages 63–86. Amer. Math. Soc., Providence, R.I., 1979.

[25] S. Lang. $SL_2(\mathbf{R})$, volume 105 of *Graduate Texts in Mathematics*. Springer-Verlag, New York, 1985. Reprint of the 1975 edition.

[26] H. Maass. Über eine neue Art von nichtanalytischen automorphen Funktionen und di Bestimmung Dirichletscher Reihen durch Funktionalgleichungen. Ann. Math. 122, 141–183.

[27] G. N. Maloletkin. Zeta-functions of parabolic forms. *Mat. Sb. (N.S.)*, 86(128):622–643, 1971.

[28] C. Moeglin and J. L. Waldspurger. *Spectral Decomposition and Eisenstein Series. Cambridge Tracts in Mathematics 113*, Cambridge University Press, Cambridge, 1995.

[29] I. Satake. Theory of spherical functions on reductive algebraic groups over p-adic fields. *Inst. Hautes Études Sci. Publ. Math.*, (18):5–69, 1963.

[30] S. D. Miller. The highest lowest zero and other applications of positivity. *Duke Math. J.*, 112(1):83–116, 2002.

[31] G. Shimura. *Introduction to the arithmetic theory of automorphic functions*. Princeton University Press 1971.

[32] T. Tamagawa. On the ζ-functions of a division algebra. *Ann. of Math. (2)*, 77:387–405, 1963.

[33] J. T. Tate. Fourier analysis in number fields, and Hecke's zeta-functions. In *Algebraic Number Theory (Proc. Instructional Conf., Brighton, 1965)*, pages 305–347. Thompson, Washington, D.C., 1967.

[34] N. R. Wallach. *Real reductive groups. II*, volume 132 of *Pure and Applied Mathematics*. Academic Press, Inc., Boston, MA, 1992.

14

Automorphic L-Functions

Freydoon Shahidi

14.1 Introduction

Zeta functions and L-functions have played a central role in the development of number theory and arithmetic algebraic geometry. Their simplest examples are the Riemann zeta function $\zeta(s)$ and its extensions, Dirichlet L-functions, $L(s, \chi)$ with χ a Dirichlet character. They are meromorphic functions on the complex plane with a simple pole at $s = 1$ and only when $\chi = 1$ or for $\xi(s)$. They also satisfy a functional equation of the same type, although a bit more complicated, as $\xi(s)$, i.e., of the type

$$L(1 - s) = L(s),$$

sending s to $1 - s$, where

$$L(s) = \pi^{-s/2}\,\Gamma(s/2)\xi(s),$$

with the standard Γ-function.

A more sophisticated class of zeta functions comes from classical modular forms on upper half-plane. They are obtained as the Mellin transform of the form along the half-imaginary axis, giving the full L-function, i.e., with Γ-functions included. If the form is cuspidal, i.e., vanishes at infinity cusp and thus its Fourier expansion at infinity has no constant term, then the zeta function defines an entire function on $\mathbb{C}$. Moreover, the finite part of the zeta function can be written as a Dirichlet series whose coefficients are the Fourier coefficients of this expansion.

There is again a functional equation under the change $s \mapsto 1 - s$. This is evident from the behavior of the form under the action of the corresponding congruence subgroup within the Mellin transform.

For the zeta function to have an Euler product, one needs to make the further assumption that the form is an eigenfunction for all the Hecke operators, as well as the hyperbolic Laplacian. In this case, the local factors in the Euler

The author is partially supported by NSF grant DMS 1801273.

product at different primes $p < \infty$ are inverses of quadratic polynomials in p^{-s}. These facts are rather classical and are due to Hecke and Maass (weight zero forms, functions on upper half-plane, invariant under fractional linear transformations by a congruence subgroup of $SL_2(\mathbb{Z})$). Among several books on these facts, which treat the subject as a whole, one can mention Shimura's classical book [31]. We refer to [14] for a treatment of this theory within Langlands program.

Similar classical theories existed for modular forms on Siegel half-spaces, the theory of Siegel modular forms, before Langlands embarked on his program. As discussed in the writing of Goldfeld and Jacquet for $GL(n)$, reference [12] in this volume, all of these cusp forms can be considered as subrepresentations of $L^2(G(\mathbb{Q})\backslash G(\mathbb{A}))$ under the action of $G(\mathbb{A})$ on this Hilbert space by right translations, where $G = Sp_{2n}$, the symplectic group of rank n. Here $\mathbb{A}$ is the ring of adèles of $\mathbb{Q}$. The case of classical modular forms on upper half-plane is the special case when $n = 1$. We refer to [11] for an excellent exposition in the case of $G = GL_2$, as well as [6] for the general case. Also see [12].

The corresponding L-functions are defined as a product of local factors upon decomposing the representations as a restricted tensor product of local representations whenever the representation is irreducible, which is equivalent to the generating form on the corresponding upper half-plane being an eigenfunction for Hecke operators.

This was the idea by means of which Langlands defined these local factors (L-functions) in full generality for unramified representations of $G(\mathbb{Q}_p)$, or more generally $G(F)$, where F is any nonarchimedean field and G is any unramified connected reductive group over F, i.e., a quasisplit group to split over an unramified extension of F. All the split groups are unramified. This includes all the classical cases discussed above.

An unramified group is automatically defined over the ring of integers O_F of F and an unramified representation is one which has a vector fixed by $G(O_F)$.

Our aim in this chapter is to explain what Langlands L-functions are and how he was led to their definition, which requires the introduction of certain concepts introduced by him for the first time in [20] and that have been fundamental in formulating his program and conjectures.

One of the main ingredients in his definition is that of an L-group, which we define with some care in Section 14.3. These are complex reductive groups defined uniquely through a structural duality.

For historical reasons it is important to know how Langlands was led to the definition of L-groups. They were noticed as part of the package defining his L-functions for the first time in his Yale Lectures manuscript "Euler Products" [20] upon calculating the constant terms of Eisenstein series, within which they were disguised. In fact, what he originally noticed was a ratio of Euler products that included all the classical L-functions as special cases. His attempts to understand what these Euler products meant led him to the definition of L-groups in the split cases. The definition in general was later given in his important paper [21], where he stated the body of his main conjectures, including his "Functoriality Principle" [1] and those on analytic properties of his L-functions. Here we treat his definitions through the customary language of algebraic groups and Chevalley–Grothendieck root datum, which were introduced later into the construction by Borel [5]. To make the construction more transparent for nonexperts, we immediately implement the definition in full detail in the case of GL_n and recover $GL_n(\mathbb{C})$ as its L-group.

As pointed out earlier, Langlands' L-functions are defined locally for unramified representations of any unramified group, concepts which will be defined shortly. To build them into the definition of L-functions, Langlands introduced a semisimple conjugacy class $c(\pi)$, attached to a representation π, in the L-group. This he accomplished by identifying the image of the Satake isomorphism [27] with certain restriction algebra of the representation ring of the L-group. As we explain later, this is the first time one sees the L-group appearing in his definition of L-functions or anything else in his program. Moreover, Satake isomorphism by itself would not be enough to define the L-functions. One needs to introduce what we like to call the "Langlands isomorphism" to bridge the gap. For the reasons given above, Langlands calls this semisimple conjugacy class "the Frobenius–Hecke conjugacy class" attached to the unramified representation as opposed to the customary name of "Satake parameter" used by others. This is discussed in Section 14.4 in detail.

An (unramified) Langlands L-function attached to π is the inverse $L(s,\pi,\rho)$ of the characteristic polynomial for $\rho(c(\pi))$ in the variable q^{-s}, where ρ is a finite-dimensional representation of the L-group $^L G$. We discuss this in Section 14.5 and give a number of examples for $G = GL_2$.

Then, in Section 14.6, we discuss the genesis of the definition of L-functions: His constant term calculations for the Eisentein series in his "Euler Products" manuscript [20], a matter that was pointed out by him at different occasions through his personal postings, letters, emails, and communications with the present author, as well as his extended interview in [26]. We refer to the first two paragraphs in Section 14.6 here for the explanation of how [20] has been crucial in the formulation of his famous letter to Weil [22] in 1967.

To this end, we briefly introduce Eisenstein series, relying on Labesse's contribution [19] to this volume for any detail. We then state the main result on constant terms and their connection with L-functions and conclude with a number of examples, including a discussion of his proof of meromorphy for the third symmetric power L-functions for GL_2, which he labeled as "particularly striking" about 50 years ago, when our knowledge of L-functions were quite rudimentary.

We conclude in Section 14.7 by remarking that his calculations in [20] were also the genesis of what is now called the "Langlands–Shahidi method," which has been responsible for establishing a number of exotic and out of reach cases of functoriality principle with many applications in number theory and representation theory. The paper ends with the author's personal reflections on Professor Langlands. Finally, we recommend Casselman's writing "The L-group" [8] as another exposition of material treated in this article.

14.2 What is an Automorphic L-Function?

When F is a nonarchimedean field and G is a connected reductive group defined over O_F, the ring of integers of F, we may consider irreducible admissible representations of $G(F)$ that are unramified, i.e., have a vector fixed under $G(O_F)$. This is the case exactly when G is quasisplit over F (to be defined later) to split over an unramified extension of F, called an *unramified* group. The subgroup $G(O_F)$ is among the maximal compact subgroups of $G(F)$ and, more specifically, a hyperspecial one [33, 34]. The classical cases we discussed in Section 14.1 are all attached to groups which are as such. They are in fact split and thus unramified at every finite prime.

The significance of unramified representations is that if G is a group over a global field k and π is an irreducible admissible representation of $G(\mathbb{A}_k)$, where G is a connected reductive group over k, then $\pi = \otimes_v \pi_v$ as v runs over all the places of k and for almost all v, G is unramified over k_v, and π_v is an unramified representation of $G(k_v)$, (cf. [9, 33]). Here $\mathbb{A}_k$ is the ring of adèles of k.

What Langlands did was to define a far more general class of L-functions by defining their local factors, using the unramified data available at almost all places v of k. The definition is motivated by his calculation of the constant terms of Eisenstein series that appeared in his Euler Products monograph [20], a consequence of the computations he did in the Fall of 1966. (See paragraph (3.15) in [26].) In this very same manuscript [20], Langlands originated the idea of an L-group, which he discussed at the time, both in [20] and in his famous paper "Problems in the theory of automorphic forms" [21], and called the "associate group" later in [24].[1] The associate group or the L-group plays

---

[1] The manuscript [24] classifies all the irreducible admissible representations of a real reductive group by identifying their spaces of K-finite vectors as unique quotients of certain parabolically induced representations, called the "Langlands quotient," with equivariant Lie algebra actions. Using this he establishes in the same paper the local Langlands correspondence for real groups, parameterizing these representations by "admissible" homomorphisms of the archimedean Weil group into the L-group. While the classification was a breakthrough in representation theory of real groups in the 1970s and was labeled as "Langlands classification," the manuscript remained unpublished until 1989.

a central role in his formulation of both L-functions and his "functoriality principle" (cf. [1]), stated first in [21]. This is what we discuss next.

14.3 L-Groups

Langlands' notion of associate group can be put into the setting of root datum of Chevalley–Grothendieck as it was done by Borel [5], who called it the L-group. Let us briefly review the construction.

Let G be a connected reductive algebraic group over a field k, either local or global, i.e., a closed subgroup of some $GL_n(\bar{k})$ for which the intersection of all its maximal connected solvable subgroups, its *radical*, contains no unipotent matrices. Here $\bar{k}$ is an algebraic closure of k. To be defined over k is equivalent to having a set of polynomials defining the algebraic variety G with coefficients in k. Note that G is always defined over $\bar{k}$. Let $B = TU$ be a *Borel subgroup* of G as a group over $\bar{k}$, a maximal connected solvable subgroup with unipotent radical U and the reductive quotient $T = B/U$, which is in fact a maximal torus. The group G is called quasisplit if it has a Borel subgroup defined over k.

A first example of such objects is the general linear group GL_n. For a Borel subgroup we can take the subgroup of upper triangulars. Its unipotent radical consists of those that have 1 on their diagonals. We can take T in this case to be the subgroup of diagonals since the quotient B/U can be identified with the subgroup T of diagonals. This is true in general and for any connected reductive group G with a Borel subgroup B, every maximal torus $T \subset B$ allows a splitting of $B \longrightarrow B/U$ and thus $B = TU$. All maximal tori in B are conjugate under elements of U.

The torus T acts on U by conjugation. Its eigenfunctions on $\mathrm{Lie}(U)$, the Lie algebra of U, provide us with a set of positive roots Φ^+ and

$$\mathrm{Lie}(U) = \bigoplus_{\alpha \in \Phi^+} \mathfrak{u}_\alpha$$

with each $\mathfrak{u}_\alpha$ a one-dimensional root space (eigenspace), on which $t \in T$ acts by the scalar $\alpha(t) \in \bar{k}^*$. Then $\Phi = \Phi^+ \cup (-\Phi^+)$ is called the set of all the *roots* of T.

One also has a notion of a *coroot* $\alpha^\vee$, a homomorphism from $\bar{k}^*$ into $T = T(\bar{k})$, for each root α, satisfying

$$\alpha(\alpha^\vee(t)) = t^{\langle \alpha, \alpha^\vee \rangle} = t^2.$$

We denote by $\Phi^\vee$ the set of all coroots.

Roots and coroots are subsets of the groups of characters $X = X(T)$ and cocharacters $X^\vee = X^\vee(T)$ of T, respectively. Characters are just homomorphism from $T(\bar{k})$ into $\bar{k}^*$ whose values are rational functions in the coordinates of the tori. Recall that $T(\bar{k}) \cong (\bar{k}^*)^n$ for some positive integer n. Cocharacters are then rational homomorphism from $\overline{k}^*$ into $T(\overline{k})$, i.e., the coordinates of the image are rational functions. The set

$$\psi(G) = (X, \Phi, X^\vee, \Phi^\vee)$$
$$= \psi(G, T)$$

is called the *root datum* for G, since every other maximal torus is conjugate to T, rendering an isomorphism on the datum.

On the other hand, one can define a root datum, a quadruple $\psi = (X, \Phi, X^\vee, \Phi^\vee)$ in which X and $X^\vee$ are free abelian groups of finite rank in duality by a pairing

$$\langle\,,\,\rangle : X \times X^\vee \longrightarrow \mathbb{Z},$$

with finite subsets $\Phi \subset X$ and $\Phi^\vee \subset X^\vee$ and a bijection $\alpha \mapsto \alpha^\vee$ of Φ onto $\Phi^\vee$ satisfying $\langle \alpha, \alpha^\vee \rangle = 2$. Moreover, one can define endomorphism s_α and $s_{\alpha^\vee}$ of X and $X^\vee$ by

$$s_\alpha(x) = x - \langle x, \alpha^\vee \rangle \alpha, \quad x \in X$$

and

$$s_{\alpha^\vee}(x^\vee) = x^\vee - \langle \alpha, x^\vee \rangle \alpha^\vee, \quad x^\vee \in X^\vee.$$

One must then have $s_\alpha(\Phi) = \Phi$ and $s_{\alpha^\vee}(\Phi^\vee) = \Phi^\vee$. We refer to Section 7.4.1 of [32] for details.

With the definition of root datum in hand we now recall the Chevalley–Grothendieck classification theorem (Theorem 10.1.1 in [32]): *Given a root datum ψ, there exists a connected reductive group G over $\overline{k}$, unique up to isomorphism, with a maximal torus T such that $\psi(G, T) = \psi$.*

Let $\psi = (X, \Phi, X^\vee, \Phi^\vee)$ be a root datum. Then

$$\psi^\vee = (X^\vee, \Phi^\vee, X, \Phi)$$

is called the *dual* datum.

Given a connected reductive group G (over $\overline{k}$) with a maximal torus T, we will denote by $\hat{G}$, the complex reductive group with a maximal torus $\hat{T}$ for which $\psi(\hat{G}, \hat{T}) = \psi(G, T)^\vee$. The group $\hat{G}$ is called the *dual* group of G.

Now assume G is defined over k. Then there exists a split group G_s such that

$$\eta : G \simeq G_s,$$

with η an isomorphism over $\overline{k}$, but not necessarily over k. Let $\Gamma_k = Gal(k_s/k)$, where k_s is the separable closure of k. Then as σ runs over Γ_k, $\eta \cdot \eta^{-\sigma}$ defines an element in $H^1(\Gamma_k, \mathrm{Aut}_{\overline{k}}(G_s))$, the set of 1-cocycles with values in $\mathrm{Aut}_{\overline{k}}(G_s) = \mathrm{Aut}_{\overline{k}}(G)$. As explained in [5] this cocycle defines a homomorphism:

$$\mu_G : \Gamma_k \longrightarrow \mathrm{Aut}(G, B, T, \{x_\alpha\}_\alpha) \subset \mathrm{Aut}_{\overline{k}}(G),$$

where $\mathrm{Aut}(G, B, T, \{x_\alpha\}_\alpha)$ is the subgroup of all the $\overline{k}$-automorphisms of G, preserving B, T, and the set $\{x_\alpha\}_\alpha$. To explain what the set $\{x_\alpha\}_\alpha$ is, let us recall that Φ^+ was defined by fixing B. We then have a subset of *simple roots* $\Delta \subset \Phi^+$, where every element in Φ^+ is a nonnegative $\mathbb{Z}$-linear combination of elements in Δ, and Δ has minimal cardinality as such. The set $\{x_\alpha\}_\alpha, \alpha \in \Delta$, is then defined by a set of root vectors X_α in $\mathrm{Lie}(U)$ with each $X_\alpha \in \mathfrak{u}_\alpha, \alpha \in \Delta$, through $x_\alpha = \exp(X_\alpha)$.

Now that the set of simple roots Δ is introduced one can introduce the *based root datum*

$$\psi_0(G) = (X, \Delta, X^\vee, \Delta^\vee)$$
$$= \psi_0(G, B, T)$$

and its dual

$$\psi_0(G)^\vee = (X^\vee, \Delta^\vee, X, \Delta).$$

Our dual group $\hat{G}$ satisfies

$$\psi_0(\hat{G}) = \psi_0(G)^\vee.$$

Note that the elements of $\mathrm{Aut}(\psi_0(G))$ are graph automorphisms of the Dynkin diagram of G and thus the same as the outer automorphisms of G, giving quasisplit forms of G_s. We therefore have the exact sequence

$$1 \longrightarrow \mathrm{Int}(G) \longrightarrow \mathrm{Aut}(G) \longrightarrow \mathrm{Aut}(\psi_0(G)) \longrightarrow 1,$$

where $\mathrm{Int}(G)$ is the group of inner automorphisms, i.e., conjugations, and $\mathrm{Aut}(G) = \mathrm{Aut}_{\overline{k}}(G)$. Moreover,

$$\mathrm{Aut}(\psi_0(G)) \simeq \mathrm{Aut}(G, B, T, \{x_\alpha\}_\alpha),$$

and thus $\mathrm{Aut}(G, B, T, \{x_\alpha\}_\alpha)$ gives a splitting of the above exact sequence by realizing $\mathrm{Aut}(\psi_0(G))$ as a subgroup of $\mathrm{Aut}(G)$. For this reason, a choice of root vectors $\{x_\alpha\}_\alpha$ is usually called a *splitting*.

We can now choose a Borel subgroup $\hat{B}$ of $\hat{G}$ and a maximal torus $\hat{T}$ thereof, as well as a set of root vectors $\{x_{\alpha^\vee}\}_{\alpha^\vee}, \alpha^\vee \in \Delta^\vee$, to get

$$\mathrm{Aut}(\psi_0(\hat{G})) \simeq \mathrm{Aut}(\hat{G}, \hat{B}, \hat{T}, \{x_{\alpha^\vee}\}_{\alpha^\vee}).$$

Using μ_G and the definition of $\psi_0(G)^\vee$, we now have a homomorphism

$$\mu'_G : \Gamma_k \longrightarrow \mathrm{Aut}(\psi_0(G))^\vee \simeq \mathrm{Aut}(\psi_0(\hat{G})).$$

Composing this with the inclusion

$$\mathrm{Aut}(\hat{G}, \hat{B}, \hat{T}, \{x_{\alpha^\vee}\}_{\alpha^\vee}) \hookrightarrow \mathrm{Aut}(\hat{G}, \hat{B}, \hat{T}),$$

we now get a homomorphism

$$\mu_G^\vee : \Gamma_k \longrightarrow \mathrm{Aut}(\hat{G}, \hat{B}, \hat{T}).$$

The L-group $^L G$ is defined by

$$^L G = \hat{G} \rtimes \Gamma_k,$$

where the semidirect product, i.e., the action of Γ_k on $\hat{G}$, is defined by $\mu_{\hat{G}}^\vee$. Note that $\hat{G}$ is the connected component of identity of $^L G$ and therefore sometimes denoted by $^L G^0$ (cf. [5]).

Examples; the Case of GL_n We now construct the L-group of GL_n which is split over any field, i.e., has a maximally split torus over any field, namely the diagonal subgroup. Let us take B as the subgroup of upper triangulars and let T be the subgroup of diagonals inside B as we mentioned earlier.
 Write

$$T = T(\bar{k}) = \{t = \mathrm{diag}(t_1, \ldots, t_n), \ t_i \in \bar{k}^*\}.$$

A character χ of T is of the form

$$\chi(t) = \prod_{i=1}^{n} t_i^{m_i}$$

with $m_i \in \mathbb{Z}$. Consequently, the character group $X(T)$ of T is a $\mathbb{Z}$-module isomorphic onto $\mathbb{Z}^n$. If we let e_i, $1 \le i \le n$, be the ith coordinate characters, i.e.,

$$e_i(t) = t_i,$$

then

$$X(T) = \bigoplus_{i=1}^{n} \mathbb{Z} e_i \simeq \mathbb{Z}^n.$$

 The group of cocharacters will be generated by homomorphism

$$e_i^\vee : \bar{k}^* \longrightarrow T(\bar{k})$$

defined as

$$e_i^\vee(t) = \mathrm{diag}(1, \ldots, t, 1 \ldots),$$

where t is the (i, i) entry of $e_i^\vee(t)$, and thus

$$X^\vee = \bigoplus_{i=1}^{n} \mathbb{Z} e_i^\vee \simeq \mathbb{Z}^n.$$

 The roots are the subset

$$\Phi = \{e_i - e_j \mid 1 \le i \ne j \le n\}$$

of X. With the choice of B to be the upper triangulars,

$$\Phi^+ = \{e_i - e_j \mid 1 \le i < j \le n\}$$

and $\Phi^- = -\Phi^+$, giving $\Phi = \Phi^+ \sqcup \Phi^-$. Note that

$$(e_i - e_j)(t) = t_i t_j^{-1}.$$

The simple roots inside Φ^+ are just

$$\Delta = \{e_i - e_{i+1} \mid 1 \le i \le n - 1\}.$$

We can take the coroots to be the subset

$$\Phi^\vee = \{e_i^\vee - e_j^\vee \mid 1 \le i \ne j \le n\}$$

of $X^\vee$. Again simple coroots are just

$$\Delta^\vee = \{e_i^\vee - e_{i+1}^\vee \mid 1 \le i \le n - 1\}.$$

Note that the bijection $\alpha \mapsto \alpha^\vee$ of Φ onto $\Phi^\vee$ is just $e_i - e_j \mapsto e_i^\vee - e_j^\vee$. Moreover,

$$(e_i - e_j)((e_i^\vee - e_j^\vee)(t)) = t^2.$$

Note that the Cartesian pairing $\langle \, , \, \rangle$ defined on $\mathbb{Z}^n$, upon identifying X and $X^\vee$ with $\mathbb{Z}^n$, when applied to $e_i - e_j$ and $e_i^\vee - e_j^\vee$ gives

$$\langle e_i - e_j, \, e_i^\vee - e_j^\vee \rangle = 2,$$

since $\langle e_i, e_j^\vee \rangle = \delta_{ij}$. We can therefore take the standard Cartesian inner product on $\mathbb{Z}^n \times \mathbb{Z}^n$ for our pairing $\langle \, , \, \rangle$.

Now we have the root datum for G as

$$\psi(G, T) = (\mathbb{Z}\langle e_i \rangle, \Phi, \mathbb{Z}\langle e_i^\vee \rangle, \Phi^\vee),$$

where $\mathbb{Z}\langle e_i \rangle$ means the $\mathbb{Z}$-module generated by $\{e_i\}$ and with the same meaning for $\mathbb{Z}\langle e_i^\vee \rangle$. Thus the dual datum is

$$\psi(G, T)^\vee = (\mathbb{Z}\langle e_i^\vee \rangle, \Phi^\vee, \mathbb{Z}\langle e_i \rangle, \Phi).$$

If we identify the basic cocharacters $\{e_i^\vee\}$ of T with basic characters $\{\hat{e}_i\}$ of $\hat{T} \simeq (\mathbb{C}^*)^n$ and similarly the basic characters $\{e_i\}$ of T with basic cocharacters $\{(\hat{e}_i)^\vee\}$ of $\hat{T}$, then $\Phi^\vee \longleftrightarrow \hat{\Phi}$, through $e_i^\vee - e_j^\vee \longleftrightarrow \hat{e}_i - \hat{e}_j$, and $\Phi \longleftrightarrow (\hat{\Phi})^\vee$, through $e_i - e_j \longleftrightarrow (\hat{e}_i)^\vee - (\hat{e}_j)^\vee$. Moreover $\mathbb{Z}\langle e_i^\vee \rangle \longleftrightarrow \mathbb{Z}\langle \hat{e}_i \rangle$, while $\mathbb{Z}\langle e_i \rangle \longleftrightarrow \mathbb{Z}\langle (\hat{e}_i)^\vee \rangle$. The dual group $\hat{G}$ now has the root datum of GL_n and consequently $\hat{G} = GL_n(\mathbb{C})$.

Since GL_n is split over any field the maps μ_G and $\mu_G^\vee$ are identity for all $\sigma \in \Gamma_k$, giving a direct product

$$^L GL_n = GL_n(\mathbb{C}) \times \Gamma_k$$

for the L-group. It is customary to drop Γ_k for split groups and just consider $\hat{G}$ for the L-group.

The Case of Semisimple Groups A reductive group G is *semisimple* if it has a finite center. Then $G(\overline{k})$ equals the $\overline{k}$-points of its derived group. (The same is not true for the k-points and in general the k-points of the derived group are larger than the derived group of the k-points. To wit, consider $G = PGL_2$ as a group over $\mathbb{R}$. Note that $PGL_2(\mathbb{C}) = PSL_2(\mathbb{C})$. Then $\mathrm{diag}(1, -1)$ lies in $PGL_2(\mathbb{R})$ modulo $Z(G)$ but not in $PSL_2(\mathbb{R})$. On the other hand, $\mathrm{diag}(\sqrt{-1}, -\sqrt{-1}) \equiv \mathrm{diag}(1, -1) (\mathrm{mod}\, Z(G))$ lies in $PSL_2(\mathbb{C})$. We conclude by pointing out that $PSL_2(\mathbb{R})$ is the connected component of identity in $PGL_2(\mathbb{R})$.)

The pairing $\langle\, ,\, \rangle$ between X and $X^\vee$ can be obtained from a nondegenerate bilinear form $(\, ,\,)$ on $X \otimes_{\mathbb{Z}} \mathbb{R}$ which we will furthur assume is an isometry for the action of the *Weyl group* $W(G, T) = N_G(T)/T$ of T in G. We can define the pairing $\langle\, ,\, \rangle$ as

$$\langle x, y^\vee \rangle = \left(x, \frac{2y}{(y, y)} \right) \qquad (x, y \in X)$$

$$= 2\frac{(x, y)}{(y, y)},$$

if we identify $X^\vee \otimes \mathbb{R}$ with $X \otimes \mathbb{R}$ through

$$x \longmapsto x^\vee = 2x/(x, x) \in X \otimes \mathbb{R}.$$

Clearly $\langle \alpha, \alpha^\vee \rangle = 2$ for $\forall \alpha \in \Phi$.

We can define the adjoint pairing on $X^\vee \otimes \mathbb{R}$ as follows. If $x^\vee, y^\vee \in X^\vee \otimes \mathbb{R}$, we define a bilinear nondegenerate form on $X^\vee \otimes \mathbb{R}$ through the identification we just discussed, i.e.,

$$(x^\vee, y^\vee) = \left(\frac{2x}{(x, x)}, \frac{2y}{(y, y)} \right)$$

and thus

$$(x^\vee, y^\vee) = \frac{4(x, y)}{(x, x)(y, y)}.$$

This implies

$$2\frac{(y^\vee, x^\vee)}{(x^\vee, x^\vee)} = 2\frac{(x, y)}{(y, y)}. \qquad (14.3.1)$$

If we define

$$\langle x, y \rangle := 2\frac{(x, y)}{(y, y)} \qquad (x, y \in X \otimes \mathbb{R}),$$

which in terms of our pairing $\langle\, ,\, \rangle$ on $X \times X^\vee$ equals $\langle x, y^\vee \rangle$, $x, y \in X$, and similarly

$$\langle y^\vee, x^\vee \rangle := 2\frac{(y^\vee, x^\vee)}{(x^\vee, x^\vee)} \qquad (x^\vee, y^\vee \in X^\vee \otimes \mathbb{R}),$$

then

$$\langle x, y \rangle = \langle y^\vee, x^\vee \rangle$$

by (14.3.1). In particular $\langle \alpha, \beta \rangle = \langle \beta^\vee, \alpha^\vee \rangle$, $\forall \alpha, \beta \in \Phi$. If $\alpha_1, \ldots, \alpha_r$ are the simple roots, then the matrix

$$C = (\langle \alpha_i, \alpha_j \rangle)_{1 \le i, \, j \le r}$$

is called the Cartan matrix of G, which for a semisimple group determines it up to isogeny, i.e., C gives the Lie algebra uniquely. Clearly

$$\hat{C} = (\langle \alpha_i^\vee, \alpha_j^\vee \rangle)$$
$$= (\langle \alpha_j, \alpha_i \rangle)$$
$$= {}^t C,$$

i.e., the Cartan matrix $\hat{C}$ of $\hat{G}$ is the transpose of that of G. In particular, this gives a duality between the group G and its dual group $\hat{G}$ if G is semisimple.

We recall that the only isogeny classes of simple groups that are dual to a different one than themselves are those of odd special orthogonal groups (B-type) and symplectic groups (C-type). In particular, $SO_{2n+1}^\vee = Sp_{2n}(\mathbb{C})$ and $Sp_{2n}^\vee = SO_{2n+1}(\mathbb{C})$. More precisely, one has to pick $\hat{G}$ in its isogeny class, those which have surjections with finite kernel among themselves. They all have the same Cartan matrix. Let us consider only the case of two extremes in the isogeny class: those of *adjoint* type or with trivial center, where their character group $X(T)$ is equal to the root lattice of T, i.e., the $\mathbb{Z}$-lattice generated by the roots of T; and those of *simply connected* type, i.e., those where $X(T)$ is equal to the weight lattice of T, the $\mathbb{Z}$-lattice of all $\lambda \in X \otimes \mathbb{R}$ such that $\langle \lambda, \alpha \rangle \in \mathbb{Z}$ for all $\alpha \in \Delta$, the weights of T. They have the largest center in the class. The two lattices are in duality and, in our duality process, adjoint groups go to simply connected ones and vice versa. This applies to all isogeny classes. For example, $SL_n^\vee = PGL_n(\mathbb{C})$ and $PGL_n^\vee = SL_n(\mathbb{C})$. We refer to [5] for details.

14.4 Satake and Langlands Isomorphisms; Frobenius–Hecke Conjugacy Classes

To explain how Langlands defined his L-functions we need to explain some of the tools that he used. These L-functions were defined for unramified groups, starting with Satake isomorphism to which he contributed his own isomorphism and interpretation.

14.4.1 Satake Isomorphism

Let k be a nonarchimedean local field with the ring integer O_k. Let G be an unramified reductive group defined over k. Then it is defined over O_k and is a quasisplit group to split over an unramified (cyclic) extension of k. Let

$K = G(O_k)$. It is a maximal compact subgroup of $G(k)$ which is of "hyperspecial" type, i.e., the stabilizer of a hyperspecial point on the Tits building of the group [34].

Let $\mathcal{H}(G(k), K)$ be the **spherical Hecke algebra** of G, i.e., the algebra of all the (locally constant), compactly supported functions on $G(k)$ satisfying $f(k_1 g k_2) = f(g), k_1, k_2 \in K$, under the convolution:

$$f_1 * f_2(h) = \int_{G(k)} f_1(g) f_2(gh^{-1}) dg, \tag{14.4.1}$$

$f_i \in \mathcal{H}(G(k), K)$, where dg is a Haar measure on $G(k)$, normalized by $dg(K) = 1$.

Let B be a Borel subgroup of G that is defined over k and exists since G is quasisplit. Write $B = TU$, where T is a maximal torus of B, $T \simeq B/U$. Let $A_0 \subset T$ be the maximal split torus of T, and let $W_0 = W(G, A_0)$ be the corresponding Weyl group. More precisely, $W_0 \simeq N_G(A_0)/Z_G(A_0)$, a finite group, where N_G and Z_G denote the corresponding normalizer and centralizer in G, respectively. As in Section 14.3, let $\hat{T}$ and $\hat{A}_0$ be the corresponding dual groups. Let $^\circ T$ be the largest compact subgroup of $T(k)$. Assume $^\circ T = T(k) \cap K$. Then

$$\Lambda := T(k)/^\circ T \simeq X_*(T)^{\Gamma_k}$$
$$= X_*(A_0) \tag{14.4.2}$$
$$= X^*(\hat{A}_0),$$

where $X_*(T)^{\Gamma_k}$ is the Γ_k-fixed elements of $X_*(T)$. A complex character χ of $T(k)$ is called **unramified** if it factors through $T(k)/^\circ T$ and is thus trivial on $^\circ T$. We therefore have the group $X_{un}(T(k))$ of unramified characters of $T(k)$ equal to

$$X_{un}(T(k)) = \mathrm{Hom}(X^*(\hat{A}_0), \mathbb{C}^*)$$
$$= X_*(\hat{A}_0) \otimes_{\mathbb{Z}} \mathbb{C}^* \tag{14.4.3}$$
$$= \hat{A}_0(\mathbb{C}) = \hat{A}_0,$$

where $\mathbb{C}^*$ is realized as a $\mathbb{Z}$-module by $n \cdot z := z^n$, $z \in \mathbb{C}^*$ ([7], page 134). The isomorphism (14.4.2) allows us to attach to every $\lambda \in \Lambda$, the characteristic function f_λ of its preimage in $T(k)$. If we now replace G with T and K with $^\circ T$ and define the convolution for $\mathcal{H}(T(k), ^\circ T)$ as in (14.4.1), then $f_{\lambda_1} * f_{\lambda_2}$ will be the characteristic function of the preimage of $\lambda_1 + \lambda_2$, $\lambda_i \in \Lambda$. The correspondence $\lambda \mapsto f_\lambda$ allows us to realize $\mathcal{H}(T(k), ^\circ T)$ as the group algebra of Λ, i.e., $\mathcal{H}(T(k), ^\circ T) = \mathbb{C}[\Lambda]$.

Let $\mathfrak{u} = \mathrm{Lie}(U)$. Denote by $Ad_\mathfrak{u}$, the **adjoint** representation (conjugation) of T on $\mathfrak{u}$ whose eigenfunctions were the positive roots discussed in Section 14.3. Set

$$\delta(t) = |\det Ad_\mathfrak{u}(t)|, \qquad (14.4.4)$$

and define the Satake transformation as the linear map

$$S : \mathcal{H}(G(k), K) \to \mathcal{H}(T(k), {}^\circ T) = \mathbb{C}[\Lambda]$$

by

$$Sf(t) = \delta(t)^{1/2} \int_{U(k)} f(tu)du$$
$$= \delta(t)^{-1/2} \int_{U(k)} f(ut)du. \qquad (14.4.5)$$

Satake proved the following in [27].

Theorem (Satake [27]) *The linear transformation S is an algebra isomorphism from $\mathcal{H}(G(k), K)$ onto $\mathbb{C}[\Lambda]^{W_0}$, the W_0-invariants of $\mathbb{C}[\Lambda]$, the group algebra of $\Lambda = X^*(\hat{A}_0)$.*

Note that $\mathbb{C}[\Lambda]$, the group algebra of $\Lambda = X^*(\hat{A}_0)$, is in fact the polynomial ring $\mathbb{C}[\hat{A}_0]$ of $\hat{A}_0$.

In [21], Langlands uses $f \mapsto \hat{f}$ to denote the Satake isomorphism, and $\hat{M}$ to denote $X^*(\hat{A}_0) = X_*(T)^{\Gamma_k}$. He then writes an arbitrary element in $\mathbb{C}[\hat{A}_0]^{W_0}$ as

$$\sum_{\lambda \in \hat{M}} \hat{f}(\lambda)\lambda. \qquad (14.4.6)$$

Every character of $\mathcal{H}(G(k), K)$ is of the form

$$\omega_\chi(f) = \int_{T(k)} \hat{f}(t)\chi(t)dt \qquad (14.4.7)$$

for a $\chi \in X_{un}(T(k))$. Using (14.4.3), choose $t \in \hat{A}_0$ corresponding to χ. Write $\chi = \chi_t, t \in \hat{A}_0$. Then the isomorphism in (14.4.3) implies

$$\chi_t\left(\sum_{\lambda \in \hat{M}} \hat{f}(\lambda)\lambda\right) = \sum_{\lambda \in \hat{M}} \hat{f}(\lambda)\lambda(t). \qquad (14.4.8)$$

Let $\hat{N}$ be the normalizer of $\hat{T}$ in $\hat{G}$ and $\hat{W} = \hat{N}/\hat{T}$. We may identify $\hat{W} \simeq W := W(G, T)$, by considering them as Weyl groups of root systems for G and $\hat{G}$, i.e., the groups generated by endomorphism s_α and $s_{\alpha^\vee}$ defined in Section 14.3. The group G being quasisplit implies that $W_0 = W(G, A_0)$

is a subgroup of W. Let $\hat{N}_0$ be the inverse image of W_0 in $\hat{N}$. Let σ be a Frobenius element, i.e., an element of Γ_k that sends $x \in O_k$ to x^q up to an element in $\mathcal{P}_k$. Here O_k and $\mathcal{P}_k$ are the ring of integers and its maximal ideal and q is the cardinality of the residue field $O_k/\mathcal{P}_k$. In the unramified case the set of Frobenius elements makes a full conjugacy class.

There are maps between conjugacy classes

$$\overline{\mu} : (\hat{T} \rtimes \sigma)/\operatorname{Int} \hat{N}_0 \simeq (\hat{G} \rtimes \sigma)_{ss}/\operatorname{Int} \hat{G} \tag{14.4.9}$$

and

$$\overline{\nu} : (\hat{T} \rtimes \sigma)/\operatorname{Int} \hat{N}_0 \simeq \hat{A}_0/W_0, \tag{14.4.10}$$

where $(\hat{G} \rtimes \sigma)_{ss}$ is the set of semisimple (diagonalizable) elements in $\hat{G} \rtimes \sigma$ and Int means conjugation. Both maps are defined by the natural surjection $\nu : \hat{T} \to \hat{A}_0$, dual to the embedding of $X^*(\hat{A}_0)$ into $X^*(\hat{T})$ (cf. [5, 29]). The fact that $\overline{\mu}$ is a surjection is rather subtle and Langlands needed to use results of F. Gantmacher [10], as he refers to in [21]. We thus get an isomorphism

$$\alpha = \overline{\mu} \cdot \overline{\nu}^{-1} : \hat{A}_0/W_0 \simeq (\hat{G} \rtimes \sigma)_{ss}/\operatorname{Int} \hat{G}. \tag{14.4.11}$$

Let $\operatorname{Rep}({}^L G) \subset \mathbb{C}({}^L G)$ be the subalgebra generated by characters of finite-dimensional representations of ${}^L G$. Denote by $\mathfrak{A}$ the algebra of restrictions of elements of $\operatorname{Rep}({}^L G)$ to $(\hat{G} \rtimes \sigma)_{ss}/\operatorname{Int} \hat{G}$. Then α induces an isomorphism

$$\beta : \mathfrak{A} \simeq \mathbb{C}[\hat{A}_0]^{W_0}. \tag{14.4.12}$$

In fact, if ρ is a finite-dimensional representation of ${}^L G$, let f_ρ be a function on $\hat{T}$ defined by

$$f_\rho(t) := \operatorname{trace} \rho(t \rtimes \sigma). \tag{14.4.13}$$

Then $f_\rho \cdot \overline{\mu}^{-1}$ is in $\mathfrak{A}$ and

$$\beta(f_\rho \cdot \overline{\mu}^{-1}) := f_\rho \cdot \overline{\nu}^{-1} \in \mathbb{C}[\hat{A}_0]^{W_0} \tag{14.4.14}$$

and thus β is induced by α.

If we write

$$f_\rho = \sum_{\lambda \in X^*(\hat{T})} c_\lambda \lambda \quad (c_\lambda \in \mathbb{C}), \tag{14.4.15}$$

then the fact that f_ρ is a class function implies

$$f_\rho(s^{-1}t\sigma(s)) = f_\rho(t) \tag{14.4.16}$$

for all $s, t \in \hat{T}$. Together with linear independence of characters of irreducible representations, this implies that if $c_\lambda \neq 0$, then $\lambda \in X^*(\hat{T})^\Gamma = X^*(\hat{A}_0)$ and thus f_ρ may be considered to be in $\mathbb{C}[\hat{A}_0]^{W_0}$. We refer to [5, 21] for details.

Langlands proved in [21] that β defined by (14.4.14) is an isomorphism. It is clear that with the Satake isomorphism between $\mathcal{H}$ and $\mathbb{C}[\hat{A}_0]^{W_0}$ alone, one would never see the objects Rep $(^L G)$ and $\mathfrak{A}$, and it is in the next step of relating $\mathbb{C}[\hat{A}_0]^{W_0}$ to $\mathfrak{A}$, that one can see the significance of the L-group $^L G = \hat{G} \rtimes \Gamma$. Moreover, if $A(\pi) \in \hat{A}_0/W_0$ represents the unramified representation π of $G(k)$ through $X_{un}(T(k)) \simeq \hat{A}_0$, then we have the conjugacy class $c(\pi)$ in $(\hat{G} \rtimes \sigma)_{ss}/\operatorname{Int}(\hat{G})$ defined by

$$c(\pi) := \alpha(A(\pi)). \qquad (14.4.17)$$

Recall that $\alpha := \overline{\mu} \cdot \overline{\nu}^{-1}$. This conjugacy class is what Langlands calls the **Frobenius–Hecke conjugacy class** attached to π. It is clear that this definition requires Langlands analysis of the map β, and thus beyond Satake's reach, which has led mistakenly to calling it the same conjugacy class as Satake parameter of π.

14.5 Automorphic L-Functions

With the definition of the Frobenius–Hecke conjugacy class $c(\pi)$ attached to an unramified representation π of $G(k)$ for an unramified group G defined over a local field k in hand, we can now define the corresponding unramified L-functions. We need only one more ingredient: a finite-dimensional complex (analytic) representation ρ of $^L G$. (Finite dimensionality and being analytic are equivalent.)

Given a complex variable $s \in \mathbb{C}$, Langlands defines his L-function attached to π and ρ as

$$L(s, \pi, \rho) = \det(I - \rho(c(\pi))q^{-s})^{-1}, \qquad (14.5.1)$$

where q is the cardinality of residue field O_k/P_k. This may be considered as the characteristic polynomial of the endomorphism $\rho(c(\pi))$. This definition is quite significant as it leads to the most general definition of an L-function over a global field, since local components of an irreducible automorphic representation are unramified at almost all finite places. Let $\pi = \otimes_v \pi_v$ be an automorphic representation of $G(\mathbb{A}_k)$, where G is a connected reductive group defined over a global field k. Then almost all π_v are unramified representations of $G(k_v)$ (cf. [9, 33]). Let S be a finite set of places of k such that for each $v \notin S$, G as a group over k_v and π_v are both unramified. Let ρ be a finite-dimensional representation of $^L G$. One then defines

$$L^S(s, \pi, \rho) := \prod_{v \notin S} L(s, \pi_v, \rho_v), \qquad (14.5.2)$$

where ρ_v is the representation of $^L G_v$ through the natural map $^L G_v \to {}^L G$, where $^L G_v$ is the L-group of G as a group over $k_v \supset k$. In [20, 21] Langlands proves that the infinite (Euler) product (14.5.2) converges absolutely as a function of s in some right half-plane in $\mathbb{C}$, giving a holomorphic function there. It is the analytic properties of these "automorphic L-functions" $L^S(s, \pi, \rho)$ for different ρ, a functional equation and meromorphic continuation to all of $\mathbb{C}$, which is a central part of Langlands' program [21, 29]. It vastly generalizes any classical theory of L-functions such as those of Hecke L-functions and those attached to modular forms [11, 30].

Examples; the Case of GL_2 Under the isomorphisms in (14.4.3) a character $\chi \in X_{unr}(T(k))$ giving the unramified representation π corresponds to the W_0-orbit $A(\pi)$ in $\hat{A}_0 / W_0$. Let γ be a root of A_0 giving the coroot $\gamma^\vee$. Then

$$\chi(\gamma^\vee(\varpi)) = \gamma^\vee(A(\pi)), \tag{14.5.3}$$

where $\gamma^\vee$ on the right is considered as a root of $\hat{A}_0$. We should remark that in [20] Langlands uses another normalization for (14.5.3) in which $\gamma^\vee$ on the left is changed to $-\gamma^\vee = (\gamma^\vee)^{-1}$.

Now assume $G = GL_2$. Then $\hat{G} = GL_2(\mathbb{C})$. The equation (14.5.3) for $\chi(\mathrm{diag}(a_1, a_2)) := \chi_1(a_1)\chi_2(a_2)$, becomes

$$\chi\left(\begin{pmatrix} \varpi & 0 \\ 0 & \varpi^{-1} \end{pmatrix}\right) = \chi_1(\varpi)/\chi_2(\varpi)^{-1}$$
$$= \gamma^\vee\left(\begin{pmatrix} \chi_1(\varpi) & 0 \\ 0 & \chi_2(\varpi) \end{pmatrix}\right), \tag{14.5.4}$$

where $\gamma^\vee$ on the right is the simple root of $GL_2(\mathbb{C})$. Here $\varpi \in k$ is an element for which $|\varpi| = q^{-1}$. Consequently, we may take $A(\pi) = \mathrm{diag}(\chi_1(\varpi), \chi_2(\varpi))$.

Irreducible finite-dimensional representations of $SL_2(\mathbb{C})$ are given by their highest weights as multiples of the fundamental weight δ_1 of $SL_2(\mathbb{C})$, i.e., the one such that $\langle \delta_1, \alpha^\vee \rangle = 1$. They will then have $m\delta_1$, $m \in \mathbb{N}$, as their highest weights and are of dimension $m + 1$, and for each m the only one of dimension $m + 1$.

Such a unique structure is not true for $GL_2(\mathbb{C})$. But still there is a family of irreducible finite dimensional representations of $GL_2(\mathbb{C})$ called the symmetric powers, $\rho_m = \mathrm{Sym}^m = \mathrm{Sym}^m(\rho_1)$, of the standard representation ρ_1 of $GL_2(\mathbb{C})$, of dimension $m + 1$. The restriction of ρ_m to $SL_2(\mathbb{C})$ is the one of highest weight $m\delta_1$ and the analytic properties of the corresponding L-functions have important consequences in number theory (Ramanujan–Selberg and Sato–Tate conjectures).

There is an explicit way to define Sym^m. Let $P = P(x, y)$ be a homogeneous polynomial (form) of degree m in two variables x and y. Given $g \in GL_2(\mathbb{C})$, consider the polynomial $P_1(x, y) = P((x, y)g)$. The coefficients of the form P_1, can be obtained from those of $P(x, y)$ by multiplication by a matrix of size $m + 1$. We denote this matrix by $\mathrm{Sym}^m(g)$. The map

$g \mapsto \mathrm{Sym}^m(g)$ is a homomorphism and defines an $(m+1)$-dimensional irreducible representation of $GL_2(\mathbb{C})$ equivalent to $\rho_m = \mathrm{Sym}^m(\rho_1)$. Now let π be an unramified representation of $GL_2(k)$ given by the character (χ_1, χ_2) of $T(k) = A_0(k) = GL_1(k) \times GL_1(k)$. If we take the defining representation of $GL_2(\mathbb{C})$ in the definition of the Langlands L-function to be ρ_m, then

$$L(s, \pi, \rho_m) = \prod_{\substack{i+j=m \\ 1 \le i, j \le m}} (1 - \chi_1(\varpi)^i \chi_2(\varpi)^j q^{-s})^{-1}$$

since

$$\rho_m(A(\pi)) = \mathrm{diag}\,(a^m, a^{m-1} b, \ldots, b^m),$$

where $a = \chi_1(\varpi)$ and $b = \chi_2(\varpi)$.

When $\Pi = \otimes_v \Pi_v$ is a cuspidal representation of $GL_2(\mathbb{A}_k)$, with k a number field, the analytic properties of

$$L^S(s, \Pi, \rho_m) = \prod_{v \notin S} L(s, \Pi_v, \rho_m)$$

play a central role in the study of Ramanujan–Selberg conjectures for Maass forms on $GL_2(\mathbb{A}_k)$. In particular, as Langlands explains in [21], if the L-functions $L^S(s, \Pi, \rho_m)$ are holomorphic for $\mathrm{Re}\,(s) > 1$ and all m, then $|a_v| = |b_v| = 1$ for all $v \notin S$. In particular, each π_v, $v \notin S$, will be tempered since χ_1 and χ_2 will become unitary characters. Even analytic properties of $L^S(s, \Pi, \rho_m)$ for finite sets of $\{\rho_m\}$ can provide us with estimates on $|a_v|$ and $|b_v|$. We refer to [15, 16] for the state of art estimate

$$q_v^{-7/64} < |a_v| \quad \text{and} \quad |b_v| < q_v^{7/64}$$

for the Frobenius–Hecke eigenvalues.

14.6 Eisenstein Series and L-Functions

Langlands was led to the definition of L-functions (Section 14.5) through his calculations of the constant terms of the Eisenstein series. This has been pointed out by him at different occasions, including his interview [26], as well as his personal postings, letters, emails, and conversation with the author and others.

While the notion of Langlands' L-functions, the formulation of their properties, and their connection to functoriality was explained in his famous letter to Weil [22], the key to their discovery, the Eisenstein series, their constant terms, and their calculation were not mentioned. To wit, let me quote a paragraph from an email communication from Langlands to myself in February of 2014: "It is my own fault that I never explained clearly the relation between the calculations of the Yale lecture and the Weil letter, especially the definition of the 'L-group', of the local parameters, whatever one calls them, and of the L-functions. I confess that I thought the connection is obvious."

We start with a brief review of the Eisenstein series, a subject that Langlands developed over the period 1962–1964, culminating in a full theory for every connected reductive group over a number field. Eisenstein series are needed for understanding the continuous spectrum of these groups as well as their residual parts. What Langlands discovered was that their constant terms were deeply connected to Euler products, which became the source of inspiration for his definition of L-functions.

Langlands' mimeographed notes on Eisenstein series presenting his work remained unpublished until 1976 when it appeared as a Springer Lecture Note [23]. There have been a number of books written since Langlands developed his theory, among them one should mention [13, 25]. We refer to Labesse's article [10] in this volume for further details.

To explain, let k be a number field and G a connected reductive group over k. Let $P = MN$ be a parabolic subgroup of G, containing a fixed minimal one $P_0 = M_0 N_0$, with $M \supset M_0$ and $N_0 \supset N$. The Levi subgroup M is now uniquely chosen, subject to $M \supset M_0$ as soon as M_0 is fixed. When the group is quasisplit, the minimal parabolics are exactly the Borel subgroups, introduced earlier. We recall that a closed subgroup P of G is called *parabolic* if G/P is a projective variety, i.e., a closed subvariety of a projective space, a notion easier to comprehend than the usual definition that G/P be a "complete variety," since the two notions are the same, G/P being "quasiprojective," i.e., an open subvariety of a projective space (cf. paragraphs AG7.3, AG7.4, 6.8, 11.2 in [4]).

When $G = GL_n$, each partition $n_1 + n_2 + \cdots + n_r = n$ defines a parabolic subgroup P whose Levi subgroup is isomorphic to $GL_{n_1} \times \cdots \times GL_{n_r}$ with

$$P = \left\{ \begin{pmatrix} \boxed{g_1} & & & & * \\ & \boxed{g_2} & & & \\ & & \ddots & & \\ 0 & & & & \\ & & & & \boxed{g_r} \end{pmatrix} \middle| g_i \in GL_{n_i},\ 1 \le i \le r \right\}.$$

The minimal (Borel) subgroup in this case can be taken to be the subgroup of upper triangulars. Every other parabolic subgroup is a conjugate of some P by an element in $GL_n(\bar{k})$. For $G = SL_n$, one only needs to restrict the parabolic subgroups of GL_n to SL_n to get all the parabolic subgroups of SL_n.

Let T be a maximally split torus in M_0 with a maximal split torus A_0. Moreover, let A be the maximal split torus in the connected component of the center of M(split component of M). Then $A \subset A_0$. The roots of A in N, the eigenfunctions of conjugation action of A on $\mathrm{Lie}(N)$, are among those of A_0

on $\mathrm{Lie}(N) \subset \mathrm{Lie}(N_0)$. But there will be multiplicity for root spaces. If there exists a unique simple root α for the action of A on $\mathrm{Lie}(N)$, then P will be called a maximal parabolic. This is equivalent to rank $(A/A_G) = 1$, where A_G is the split component of G.

For GL_n, maximal parabolics correspond to partitions $n_1 + n_2 = n$, i.e., Levi subgroups with two blocks, and for SL_n, restrictions of maximal ones of GL_n.

The Levi subgroup M also normalizes N and thus acts by conjugation, which we can denote by $Ad_{\mathfrak{n}}(m)$, $m \in M(\mathbb{A}_k)$, $\mathfrak{n} = \mathrm{Lie}(N)$. We set

$$\delta_P(m) = |\det Ad_{\mathfrak{n}}(m)|,$$

where $|\ | = \prod_v |\ |_v$ is the idèle-norm of $\mathbb{A}_k^*$, idèles of k (units of $\mathbb{A}_k$). We extend δ_P to all of $G(\mathbb{A}_k)$ by setting

$$\delta_P(x) = \delta_P(m),$$

where $x = mnk$, $m \in M(\mathbb{A}_k)$, $n \in N(\mathbb{A}_k)$, $k \in K$. Here $K = \prod_v K_v$ with each K_v a maximal compact subgroup of $G(k_v)$, "adapted to T/k_v", with $K_v = G(O_v)$ for almost all v.

Now let $\pi = \otimes_v \pi_v$ be a cuspidal representation of $M(\mathbb{A}_k)$. Let φ be a cusp form in the space $\mathcal{H}(\pi)$ realized in $L^2_{\mathrm{cusp}}(M(k)\backslash M(\mathbb{A}_k), \omega_\pi)$, where ω_π is the restriction of the central character of π to $A(\mathbb{A}_k)$, i.e., those functions in the cuspidal spectrum L^2_{cusp} of $M(\mathbb{A}_k)$ which transform under $A(\mathbb{A}_k)$ according to ω_π.

One can then extend φ to all of $G(\mathbb{A}_k)$ and define a function ϕ,

$$\phi : N(\mathbb{A}_k)M(k)\backslash G(\mathbb{A}_k) \longrightarrow \mathbb{C}$$

for which

$$\phi_x : m \mapsto \phi(mx) \qquad (m \in M(k) \backslash M(\mathbb{A}_k))$$

belongs to $L^2_{\mathrm{cusp}}(M(k) \backslash M(\mathbb{A}_k), \omega_\pi)$, and more precisely into $\mathcal{H}(\pi)$. This is done using the Iwasawa decomposition $G(\mathbb{A}_k) = M(\mathbb{A}_k)N(\mathbb{A}_k)K$. It is fairly obvious how one can do this if ϕ is right K-invariant, i.e., spherical at every place v. It is just the extension trivial on $N(\mathbb{A}_k)K$. This is the case Langlands treats in [20]. In general, more work is needed using K-matrix coefficients as is explained in Section 6.3 of [29].

To explain Langlands' computation of the constant terms in [20] we will now assume P is maximal. Given a complex number $s \in \mathbb{C}$, we define the Eisenstein series on $G(\mathbb{A}_k)$ attached to ϕ as

$$E(x, \phi, s) = \sum_{\delta \in P(k)\backslash G(k)} \phi(\delta x) \delta_P^{s+\frac{1}{2}}(\delta x). \qquad (14.6.1)$$

This series converges absolutely when $\mathrm{Re}(s) >> 0$. This is nontrivial and requires a rather involved proof even for low-rank cases.

Two remarks are in order. In [20], Langlands treated the case $k = \mathbb{Q}$. But extending it to any number field is no more difficult. One may consider G as a group over $\mathbb{Q}$ by restriction of scalars from k to $\mathbb{Q}$, a standard construction in algebraic geometry, and in fact, Arthur's treatment of his trace formula and his writings on Eisenstein series are always done for groups over $\mathbb{Q}$. One nice consequence of this approach is that the number of archimedean places reduce to one, i.e., $k_\infty = \mathbb{R}$.

The second remark is that Langlands treatment of the Eisenstein series [23] is quite general and for all parabolic subgroups of G and not only the maximal ones in [20]. But the study of the rank-one (maximal) cases plays an important role in both the general case and all the arithmetic consequence discussed in [21, 29].

Now, if $P' \supset P_0$ is another maximal parabolic subgroup of G, one can consider the **constant term of** $E(x,\phi,s)$ **along** P' defined as

$$E_{P'}(x,\phi,s) = \int_{N'(k)\backslash N'(\mathbb{A}_k)} E(nx,\phi,s)dn. \qquad (14.6.2)$$

Note that $N'(k)\backslash N'(\mathbb{A}_k)$ is compact, basically being a product of compact sets $k\backslash\mathbb{A}_k$ and thus the integration is always convergent so long as $E(\cdot,\phi,s)$ has no poles at s. More precisely, Langlands proved that $E(\cdot,\phi,s)$ extends to a meromorphic function of s on all of $\mathbb{C}$ with a finite number of simple poles for $\mathrm{Re}(s) > 0$ and none on the imaginary axis $\mathrm{Re}(s) = 0$. For simplicity of exposition and to represent the work in [20], we will now assume G is split over k.

Write $P' = M'N'$, $M' \supset M_0$ and $N' \subset N_0$. Note that by our assumption $P_0 = B = TU$, $N_0 = U$, $M_0 = T = A_0$. Let $W = W(G,T)$ be the Weyl group of $T = A_0$ in G. Assume there exists an element $\widetilde{w} \in W(G,T)$ but not in $W(M,T) = W_M$, such that $M' = w^{-1}Mw$, where w is a representative of $\widetilde{w}$ in $G(k)$. Note that this fact is independent of the choice of representative w for $\widetilde{w} \in W = N_G(T)/T$ since $M \supset T$.

The constant term is zero unless $P = P'$, or P and $P' \neq P$ are "associated." This means there exists a $\widetilde{w} \in W \setminus W_M$ such that $(P')^- = w^{-1}Pw$, where $(P')^-$ is the opposite of P', i.e., $(P')^- = M(N')^-$, with $(N')^-$ the unipotent group generated by the negatives of the roots in N. This could happen even if $P = P'$, with a $\widetilde{w} \notin W_M$ in which case we will call P "self-associate." The parabolic subgroup defined by partition $n + n = 2n$ for GL_{2n} whose Levi subgroup is isomorphic to $GL_n \times GL_n$ is one such, while any parabolic with Levi $GL_n \times GL_m$ inside GL_{n+m} with $m \neq n$ is not.

At any rate, using cuspidality of π, it can be shown, and Langlands showed the following in [20].

Theorem 14.6.1 *Assume P and P' are associated when $P' \neq P$. Let*

$$\phi_s(x) = \phi(x)\delta_P^S(x)$$

and choose $w \in G(k)$, but not in $M(k)$, representing $\widetilde{w}$ as above. Then

$$E_{P'}(x, \phi, s) = \phi_s(x) \cdot \delta_{P, P'} + \int\limits_{N'(\mathbb{A})} \phi_s(wnx)dn. \qquad (14.6.3)$$

Here $\delta_{P, P'}$ is non-zero and equals 1 only if $P' = P$, i.e., when P is self-associate. If P is not self-associate, then

$$E_P(x, \phi, s) = \phi_s(x). \qquad (14.6.4)$$

Langlands then shows in [20] that the "global intertwining operator"

$$M_w(\phi_s)(x) = \int\limits_{N'(\mathbb{A})} \phi_s(wnx)dn \qquad (14.6.5)$$

is a product of ratios of *L*-functions as we now explain; and this is how he was led to the definition of his *L*-functions explained in Section 14.5 since originally he only noticed a ratio of some Euler products appearing in $M_w(\phi_s)(\varphi)$, and the definition was inspired by his attempts to understand these Euler products.

Let $^L\mathfrak{n}$ be the Lie algebra of the complex group ^LN, "*L*-group" of N (cf. [5]). The *L*-group ^LM of M is a Levi subgroup of a parabolic subgroup $^LP = {}^LM\,{}^LN$ of ^LG. The adjoint (conjugation) action r of ^LM on $^L\mathfrak{n}$ is a finite-dimensional complex representation

$$r : {}^LM \longrightarrow \text{Aut}(^L\mathfrak{n}),$$

which in general is reducible and there exist positive integers $0 < a_1 < a_2 < \cdots < a_m$, with $a_i | a_{i+1}$, such that

$$r = \bigoplus_{i=1}^{m} r_i \qquad (14.6.6)$$

The integers a_i may be considered as eigenvalues of a central element in $^LA \subset {}^LM$.

Now let S be a finite set of places of k such that for $v \notin S$, π_v is spherical and if $\phi = \otimes_v \phi_v$, then $\phi_v = \phi_v^0$, the spherical vector in $\mathcal{H}(\pi_v)$ for all $v \notin S$, giving the restricted tensor product decomposition $\pi = \otimes_v \pi_v$ of π (cf. [9]).

Following a suggestion of Tits, Langlands formulated his calculations of $M_w(\phi_s)(e)$ in [20] as

$$M_w(\phi_s)(e) = \prod_{v \in S} M_w(\phi_{s,v})(e_v) \cdot \prod_{v \notin S} \prod_{i=1}^{m} L(a_i s, \pi_v, \tilde{r}_i)/L(1 + a_i s, \pi_v, \tilde{r}_i),$$

(14.6.7)

where $\tilde{r}_i$ denotes the contragredient of r_i, $1 \leq i \leq m$. (In fact, he assumed everything to be unramified, i.e., $S = \phi$, and did not need to worry about places in S.)

Using meromorphy of $M_w(\phi_s)(\varphi)$ for all $s \in \mathbb{C}$, which Langlands already established in general in [23], and nonvanishing and meromorphy of local intertwining operators $M_w(\phi_{s,v})$, he concluded that

$$\prod_{i=1}^{m} L^S(a_i s, \pi, \tilde{r}_i)/L^S(1 + a_i s, \pi, \tilde{r}_i) \qquad (14.6.8)$$

is a meromorphic function of s on all of $\mathbb{C}$. (Again for him $S = \phi$, the empty set.) As he pointed out in [20], when $m = 1$, a simple shift argument on $a_1 s$ starting with a line of absolute convergence for $L^S(a_1 s, \pi, \tilde{r}_1)$ implies that $L^S(s, \pi, r_1)$ is meromorphic on all of $\mathbb{C}$. Otherwise, as he stated in [20], if $m = 2$ and we know the meromorphy of $L^S(s, \pi, r_2)$ on all of $\mathbb{C}$, then we will get that of $L^S(s, \pi, r_1)$.

As it was shown in general in [2, 28, 30], there exists an induction that implies each $L^S(s, \pi, r_i)$ is in fact meromorphic on all of $\mathbb{C}$ for all $i = 1, \ldots, m$. It should be remarked that in general m could go as high as $m = 6$ which will happen if G is an exceptional group of type E_8 with the derived group of M, a simply connected group, isomorphic to $SL_3 \times SL_2 \times SL_4$.

In [20], Langlands only considered the cases when the derived group of M is a simple group. His list includes all the standard L-functions of simple split groups, except for when the group was of type G_2, F_4, or E_8 as they could not be given as Levi subgroups of finite-dimensional reductive groups. Complete list of r_i for any quasisplit group G and any rank-one Levi subgroup M was later given in [30]; also see [29] for both lists together.

Example 14.6.1 The first example is that of Dirichlet/Hecke L-function attached to a Dirichlet character, or more generally a grössencharacter χ of k, a character of $k^* \backslash \mathbb{A}_k^*$. Then $\chi = \otimes_v \chi_v$, with each χ_v a character of k_v^*. If we take $G = GL_2$ and $M = GL_1 \times GL_1$ and choose π on $M(\mathbb{A}_k)$ to be $\chi^{-1} \otimes \mathbf{1}$, then $m = 1$ and $L(s, \pi, \tilde{r}_1) = L(s, \chi)$, the Hecke L-function attached to χ. Recall that

$$L^S(s, \chi) = \prod_{v \notin S} (1 - \chi_v(\varpi_v) q_v^{-s})^{-1}.$$

This as a special case, which includes the Dedekind zeta function (attached to k) and the Riemann zeta function when $\chi = 1$, or $k = \mathbb{Q}$ and $\chi = 1$, respectively.

Example 14.6.2   Now assume $\pi = \otimes_v \pi_v$ is a cuspidal automorphic representation of $GL_2(\mathbb{A}_k)$. Note that when $k = \mathbb{Q}$, these are in a one–one correspondence with normalized classical Hecke and Maass eigenforms ϕ on the upper half-plane [11, 30]. The standard L-functions attached to them are the Mellin transform of the corresponding form ϕ. These L-functions can also be found in his list in [20]. One can take $G = GL_3$ and $M = GL_2 \times GL_1$, and consider $\tilde{\pi} \otimes \mathbf{1}$ as a representation of $M(\mathbb{A}_k)$. Then again $m = 1$ and $L(s, \tilde{\pi} \otimes \mathbf{1}, \tilde{r}_1)$ is the standard L-function attached to π, i.e., $L(s, \tilde{\pi} \otimes \mathbf{1}, \tilde{r}_1) = L(s, \pi, \mathrm{std})$. When $k = \mathbb{Q}$ and π corresponds to a normalized eigenform ϕ, then $L(s, \pi, \mathrm{std}) = L(s, \phi)$, the Hecke L-function attached to ϕ.

Example 14.6.3   Next one can consider Siegel modular eigenforms ϕ of rank n as cuspidal representations π of $Sp_{2n}(\mathbb{A}_\mathbb{Q})$. We note that $M = Sp_{2n} \times GL_1$ can be considered as a Levi subgroup of $G = Sp_{2n+2}$ and the machinery of [20] applies (case (xx) of [20] or $C_{n,i}$ of [29]). Then again $m = 1$ and $L(s, \pi, r_1) = L(s, \pi, \mathrm{std}) = L(s, \phi)$, an L-function of degree $2n + 1$ which is attached to the standard $(2n+1)$–dimensional representation of $SO_{2n+1}(\mathbb{C}) = {}^L Sp_{2n}$. In the case $n = 2$, this is the so called "degree 5" L-function attached to the Siegel modular form ϕ. The more subtle "spinor" L-function of Sp_4 of degree 4 is also in the list, case (xxi) of [20], where $PSp_4 \simeq SO_5$ may be considered as a Levi subgroup of SO_7 and r_1 is the standard four-dimensional representation of $Sp_4(\mathbb{C}) = {}^L SO_5$.

A more interesting case of spinor L-function of Sp_6, an L-function of degree 8, can be obtained by using the Levi subgroup of an exceptional group of type F_4 whose derived group is isomorphic to Sp_6. (The group F_4 is both simply connected and adjoint.) This is the case (xxii) of [20] or the case $(F_{4,i})$ of [29]. This was addressed in detail in [3], a great reading which also compares the classical and modern treatment of Siegel modular forms in full detail.

Example 14.6.4   We will conclude our examples by looking at the case (xv) in [20] (the case $G_{2,ii}$ in [29]), which Langlands considered "particularly stricking". Here the group G is an exceptional group of type G_2 and $M \simeq GL_2$ is the Levi subgroup generated by the long simple root. We let π be a cuspidal representation of $GL_2(\mathbb{A}_k)$. In this case $m = 2$ and $r_1 = \mathrm{Sym}^3(\rho_1) \otimes (\Lambda^2 \rho_1)^{-1}$ while $r_2 = \Lambda^2 \rho_1$, where $(\Lambda^2 \rho_1)(m) = \det m$, $m \in GL_2(\mathbb{C})$, and $\mathrm{Sym}^3(\rho_1)$ is the third symmetric power representation of $GL_2(\mathbb{C})$ introduced in Section 14.5. Consequently, if ω_π is the central character of π, then

$$L(s, \pi \otimes \omega_\pi, r_1) = L(s, \pi, \mathrm{Sym}^3(\rho_1))$$

and

$$L(s, \pi \otimes \omega_\pi, r_2) = L(s, \omega_\pi^3),$$

where the latter is the Hecke L-function for ω_π^3. This was the first time that the meromorphy of the third symmetric power L-functions of GL_2 was proved and at the time a sign of how deep and consequential the work in [20, 21] were. This was later pursued in [17] where all the analytic properties of $L(s, \pi, \mathrm{Sym}^3(\rho_1))$, including the location of its poles, were established.

14.7 Concluding Remarks and Reflections

The Euler Products manuscript [20] is in fact the genesis of a method, later called the "Langlands–Shahidi method," within which a rather complete theory of L-functions was established for all $L(s, \pi, r_i)$ attached to every triple (G, M, r_i) in [28, 29, 30]. This has led to many important results in number theory and automorphic forms, including new and out of reach cases of functoriality [15, 18] and striking estimates towards the Ramanujan and Selberg conjectures for Maass forms [16]. We refer to [29] for an account of the method with an expanded bibliography on the related literature.

As a personal reflection. I would now like to recall how all this started and express my deep gratitude and appreciation for Langlands mentorship during the academic year 1975–1976 when I was a fresh postdoctoral member at the Institute for Advanced Study (IAS) directly after my PhD from Johns Hopkins under Joseph Shalika in 1975. Those who have spent time at IAS, especially as postdoctoral members, appreciate the immense pressure of being in this amazing center of knowledge and scholarship. It was Langlands who made me feel at ease, not only with his reassuring kindness, but by introducing me to what became a lifelong career, allowing me to develop a mathematical direction that is now usually addressed as the Langlands–Shahidi method. It was during that year that he directed my attention towards his Euler Products manuscript and provided me with a copy of his letter to Godement where a number of his ideas were explained. The fruit of that year and my first research paper "Functional equation satisfied by certain L-functions" was also kindly solicited by him for *Compositio Mathematica*, where he was an editor. Throughout the years we remained friends and after my two full year visits to IAS, my wife, Guity, and I became close friends of Charlotte and Bob Langlands, a friendship that we have cherished over the years. I look forward to many more years of friendship!

References

[1] J. Arthur, An introduction to Langlands functoriality, this volume.

[2] J. Arthur, Endoscopic L-functions and a combinatorial identity, Dedicated to H.S.M. Coxeter, *Canad. J. Math.* **51** (1999), 1135–1148.

[3] M. Asgari and R. Schmidt, Seigel modular forms and representations, *Manuscripta Math.* **104** (2001), 173–200.

[4] A. Borel, Automorphic L-functions, in *Automorphic Forms and Automorphic Representations, Proc. Sympos. Pure Math.* **33; II**, Amer. Math. Soc., Providence, RI, 1979, pp. 27–61.

[5] A. Borel, *Linear Algebraic Groups*, GTM126, Springer-Verlag, New York, 1997.

[6] A. Borel and H. Jacquet, Automorphic forms and automorphic representations, in *Automorphic Forms and Automorphic Representations, Proc. Sympos. Pure Math.* **33; I**, AMS, Providence, RI, 1979, pp. 189–202.

[7] P. Cartier, Representations of p-adic groups: A survey, in *Automorphic Forms and Automorphic Representations, Proc. Sympos. Pure Math.* **33; I**, Amer. Math, Soc., Providence, RI, 1979, pp. 111–155.

[8] W. Casselman, The L-group, Class field theory – its centenary and prospect (Tokyo, 1998), 217–258, *Adv. Stud. Pure Math.* **30**, Math. Soc. Japan, Tokyo, 2001 – https://www.math.ubc.ca/~ cass/research/pdf/miyake.pdf

[9] D. Flath, Decomposition of representations into tensor products, in *Automorphic Forms and Automorphic Representations, Proc. Sympos. Pure Math.* **33; I**, AMS, Providence, RI, 1979, pp. 179–183.

[10] F. Gantmacher, Canonical representations of automorphisms of a complex semi-simple Lie group, *Math. Sb.* **47** (1939).

[11] S. Gelbart, *Automorphic Forms on Adèle Groups*, Annals of Math. Studies, Vol. 83, Princeton University Press, 1975.

[12] D. Goldfeld and H. Jacquet, Automorphic representations and L-functions for the group $GL(n)$, this volume.

[13] Harish-Chandra, *Automorphic Forms on Semisimple Lie Groups*, SLN **62**, Berlin–Heidelberg–New York, 1968.

[14] H. Jacquet and R. P. Langlands, *Automorphic Forms on $GL(2)$*, SLN **114**, Berlin–Heidelberg–New York, 1970.

[15] H. H. Kim, Functoriality for the exterior square of GL_4 and symmetric fourth of GL_2, *Journal of AMS* **16** (2002), 139–183.

[16] H. H. Kim and P. Sarnak, Refined estimates towards the Ramanujan and Selberg conjectures, Appendix 2 to [15].

[17] H. H. Kim and F. Shahidi, Symmetric cube L-functions for GL_2 are entire, *Ann. of Math.* **150** (1999), 645–662.

[18] H. H. Kim and F. Shahidi, Functorial products for $GL_2 \times GL_3$ and the symmetric cube for GL_2, *Annals of Math.* **155** (2002), 837–893.

[19] J.-P. Labesse, The Langlands spectral decomposition, this volume.

[20] R. P. Langlands, *Euler Products*, Yale University Press, 1971.

[21] R. P. Langlands, *Problems in the theory of automorphic forms*, in Lecture Notes in Math., vol. 170, Springer-Verlag, Berlin–Heidelberg–New York, 1970, pp. 18–86.

[22] R. P. Langlands, Letter to André Weil, January 1967.

[23] R. P. Langlands, *On the functional equations satisfied by Eisenstein series*, Lecture Notes in Math., vol. 544, Springer-Verlag, 1976.

[24] R. P. Langlands, On the classification of irreducible representations of real algebraic groups, *Representation Theory and Harmonic Analysis on Semisimple Lie Groups* (P. J. Sally, Jr. and D. A. Vogan, eds.), vol. 31, Math. Surveys and Monographs, AMS, 1989, pp. 101–170.

[25] C. Moeglin and J.-L. Waldspurger, *Spectral decomposition and Eisenstein series*, Cambridge Tracts in Math. **113**, Cambridge University Press, 1995.

[26] J. Mueller, On the genesis of Robert P. Langlands' conjectures and his letter to André Weil, *Bulletin of AMS* **55** (2018), 493–528; also in this volume.

[27] I. Satake, Theory of spherical functions on reductive algebraic groups over p–adic fields, *IHES Publ. Math.* **18** (1963), 1–69.

[28] F. Shahidi, A proof of Langlands Conjecture on Plancherel measures; Complementary series for p-adic groups, *Annals of Math.* **132** (1990), 273–330.

[29] F. Shahidi, *Eisenstein Series and Automorphic L–functions*, Colloquium Publications, vol. 58, AMS, 2010.

[30] F. Shahidi, On the Ramanujan conjecture and finiteness of poles for certain L-functions, *Annals of Math.* **127** (1988), 547–584.

[31] G. Shimura, *Introduction to the Arithmetic Theory of Automorphic Functions*, Princeton University Press, 1994.

[32] T. A. Springer, *Linear Algebraic Groups, PM9*, Birkhäuser, Boston, 1981.

[33] T. A. Springer, Reductive groups, in *Automorphic Forms and Automorphic Representations, Proc. Sympos. Pure Math.* **33; I**, AMS, Providence, RI, 1979, pp. 3–27.

[34] J. Tits, Reductive groups over local fields, in *Automorphic Forms and Automorphic Representations, Proc. Sympos. Pure Math.* **33; I**, AMS, Providence, RI, 1979, pp. 29–69.

15

Langlands Reciprocity: *L*-Functions, Automorphic Forms, and Diophantine Equations

Matthew Emerton

Abstract. This chapter gives a description of the theory of reciprocity laws in algebraic number theory and its relationship to the theory of *L*-functions. It begins with a historical overview of the work of Euler, Dirichlet, Riemann, Dedekind, Hecke, and Tate. It then proceeds through the development of class field theory, before turning to the theory of Artin *L*-functions, and then of general motivic *L*-functions. It culminates in a description of Langlands's very general reciprocity conjecture, relating arbitrary motivic *L*-functions to the automorphic *L*-functions that he defined. The reciprocity conjecture is closely bound up with another fundamental conjecture of Langlands, namely his functoriality conjecture, and we give some indication of the relationship between the two. We conclude with a brief discussion of some recent results related to reciprocity for curves of genus one and two.

The goal of this chapter is to discuss the theory of ζ- and *L*-functions from the viewpoint of number theory, and to explain some of the relationships (both established and conjectured) with Langlands's theory of automorphic *L*-functions and functoriality. The key conjectural relationship, also due to Langlands, is that of *reciprocity*. The word *reciprocity* is a storied one in number theory, beginning with the celebrated theorem of quadratic reciprocity, discovered by Euler and Legendre, and famously proved by Gauss (several times over, in fact, and many more times since by others). The meaning of *reciprocity*, both in its precise technical usage and in its more general, intuitive sense, has evolved in a complicated way since it was coined (by Legendre) in the context of quadratic reciprocity in the eighteenth century. We hope to give the reader some sense of its modern meaning, but also to give the reader an indication of how this modern meaning evolved from the original meaning of Legendre.[1] To this end, we have taken a quasihistorical approach to our subject

The author was supported in part by the NSF grants DMS-1601871 and DMS-1902307.

[1] Although it only alludes to the general framework laid out by Langlands at its very end, the article [Wym72] of Wyman, titled "What is a Reciprocity Law?", remains an excellent introduction to the subtle notion of a reciprocity law in number theory. We strongly recommend it.

matter in the earlier parts of the chapter, although our historical review is by no means complete.

One of Langlands's earliest statements regarding automorphic L-functions and reciprocity appeared in his Yale lectures (delivered in 1967, published as [Lan71]), in which he wrote

> Before beginning the substantial part of these lectures let me make, without committing myself, a further observation. The Euler products mentioned above are defined by means of the Hecke operators. Thus they are defined in an entirely different manner than those of Artin or Hasse–Weil. An assertion that an Euler product of the latter type is equal to one of those associated to an automorphic form is tantamount to a reciprocity law (for one equation in one variable in the case of the Artin L-series and for several equations in several variables in the case of the Hasse–Weil L-series).

The title of Langlands's lecture series was "Euler products." For this reason too it seems appropriate that our discussion should begin with the work of Euler, and trace the thread of ideas leading from that work to the work of Langlands.

15.1 Introduction

We elaborate slightly on our goals in writing this chapter, before giving a brief overview of its contents.

15.1.1 The Langlands Program, and Unification in Number Theory

As will be clear to any reader, the view of history adopted in this note is not a neutral one. Rather, in addition to explaining some of the mathematics related to the Langlands program, I hope that my presentation of the historical development will convince the reader of the essential unity of number theory as a discipline over time, despite the changes in language, in technique, and in apparent focus. In fact, more than just unity, I hope to illustrate to the reader that the trend in number theory is towards one of *unification*. Apparently disparate phenomena become related to one another over time, and the theories used to explain them become more overarching. The Langlands program currently stands as one these overarching theories. Its apparent abstraction, and air of inaccessibility to the novice, are manifestations of the grand scope of the ideas and problems (old and new) which it encompasses. I hope that this note will help some nonexpert readers overcome some of the difficulties of appreciation that necessarily attend such a broad and general theory.

15.1.2 Problems in Number Theory

To counterbalance the abstract sentiments of the preceding paragraph, it is good to recall that number theory has always been, and continues to be, driven by attempts to solve concrete problems. One of the beauties of the subject is that these problems seem almost inevitably to lead to theoretical and structural considerations of the deepest nature, whether or not the necessity of such considerations was apparent in the original question.

As a historical example, one might mention the law of quadratic reciprocity itself, which as we already noted lies at the root of all the mathematics we will be discussing. Actually, the point is illustrated even if one looks just at the early history of the quadratic reciprocity law: in trying to prove this law, Legendre[2] was led to state (nowadays we might say "conjecture") the theorem (proved later by Dirichlet) on the infinitude of primes in arithmetic progressions. Conversely, as we will recall in more detail below, Dirichlet's proof of that result itself relies on quadratic reciprocity (which had been proved in the meantime by Gauss, along different lines to the one pursued by Legendre) – and so these two seemingly distinct threads of research are intertwined (one example of unity in number theory). As more recent examples, one can think of the modularity conjecture for elliptic curves, or the Sato–Tate conjecture, originally formulated on their own terms, but now seen as instances of the general conjectural framework of Langlands reciprocity.

We hope that the occasional examples we have included, as well as the various proof sketches that we have given, will enable the reader to see some of the concrete mathematics underlying the generalities to be discussed.

15.1.3 Overview

The first half of our chapter discusses (what are now called) *abelian* L-functions, beginning at the beginning, with the work of Euler, Riemann, and Dirichlet, and moving on to discuss the contributions of Dedekind and Hecke. This initial discussion culminates in a brief presentation of Tate's thesis. One important thread woven through this history is the notion of *Euler product*. Originally a discovery of Euler, beautiful but seemingly contingent, by the time one comes to Tate's adèlic (re)formulation, it is built into the underlying structure of the entire theory.

Reciprocity was discovered in this abelian setting; from our perspective (i.e. in the context of studying L-functions) it arises first when one attempts

---

[2] In fact Euler stated this result as a theorem, at least for progressions of the form $1 + mN$ [Eulb]; this particular case is indeed known to admit a fairly direct proof. In his introductory discussion of the problem [Dir37], Dirichlet treats the statement as a generally observed phenomenon, and doesn't attribute it to any particular mathematician. He does discuss Legendre's attempts to prove it, though (and does not mention Euler in this context).

to compare Dirichlet's formula for the value $L(1, \chi)$, when χ is a quadratic character, with Dedekind's general class number formula. The point is that the former L-function is "automorphic," while (in the context of this particular comparison) the latter should be thought of as "motivic." A reciprocity law (in this particular case, quadratic reciprocity) is required for the comparison to be effected.

The consideration of abelian reciprocity laws culminates in the general statements of class field theory. We briefly recall these general statements, emphasizing the formulations in terms of L-functions, and illustrating them with the special cases of quadratic and cyclotomic extensions of the rational numbers. In anticipation of our dicussion of compatible families of ℓ-adic Galois representations, we also explain how compatible families of characters can be constructed from algebraic Hecke characters.

We then turn to describing motivic L-functions in general. Historically, these L-functions were first discovered somewhat implicitly, in conjunction with the investigations of abelian reciprocity: they arise (implicitly) in the factorization of $\zeta_K(s)$ into a product of Hecke L-functions for F (as in (15.32) below), when K/F is an abelian extension. It was Artin who singled out the factorization of $\zeta_K(s)$ into a product of L-functions for arbitrary extension K/F (see (15.43) below), where the L-functions now depend on (typically greater than one-dimensional) representations of (typically nonabelian) Galois groups, rather than on characters (i.e. one-dimensional representations) of ray class or idèle class groups.

We discuss Artin's L-functions, and then go on to discuss ζ-functions of finite-type $\mathbb{Z}$-algebras (or, more generally, of finite-type schemes over $\mathbb{Z}$). We give a brief overview of the theory in the case of varieties over finite fields, and then turn to the case of varieties over number fields, where we encounter the Hasse–Weil ζ-functions and their factorization into L-functions associated now to a *compatible family* of ℓ-adic Galois representations; these are the motivic L-functions – a vast generalization of Artin's L-functions.

Along the way, we also briefly describe the theory of local root numbers (or ϵ factors) in this nonabelian context. Although the problem of developing such a theory was considered by several mathematicians before him, it was Langlands who ultimately solved it [Lanc].

We conclude our discussion of motivic L-functions by introducing the conjectural Langlands group. This is a hypothetical object, whose existence Langlands proposed in [Lan79].[3] It mediates the relationship between automorphic forms and motives – that is, it mediates reciprocity. Although it is usually introduced first in the automorphic context, we introduce it in the motivic context, by analogy with the construction of compatible families of Galois characters from algebraic Hecke characters. To provide motivation (and

---

[3] Actually, we follow Kottwitz [Kot84] and Arthur [Art02] in modifying Langlands' construction.

hopefully not too much confusion), our discussion of the Langlands group is intertwined with an explanation of the Sato–Tate conjecture for elliptic curves; in particular, we explain its interpretation as a Tschebotareff density-type theorem for the image of the Langlands group.

We then turn to our discussion of automorphic forms and Langlands reciprocity itself. We first briefly recall the basic notions related to automorphic forms, and also Hecke's work associating L-functions to modular forms. We then describe Langlands's construction of automorphic L-functions in general – and observe that the very definition is in terms of an Euler product! This already suggests (and already suggested to Langlands, as our quote above indicates) that these L-functions could be related to the L-functions attached to motives. The reciprocity conjecture, which we state next, formalizes this suggestion. The reciprocity conjecture is closely related to another conjecture of Langlands, his functoriality conjecture, and we briefly recall this relationship as well. Indeed, if understood in a sufficiently broad fashion, the latter conjecture subsumes the former.

Finally, we discuss some of the progress on the reciprocity conjecture, though with no pretense at completeness.

15.1.4 A Remark on Terminology and Notation

There is no precise rule for distinguishing between the usage of the expressions "ζ-function" and "L-function." Roughly, one speaks of the ζ-function of a variety, and the L-function of a motive (so ζ-functions can be broken up into products of L-functions, analogously to the way that varieties can be decomposed into motives). In the automorphic context, one speaks primarily of L-functions (although Tate uses the notation ζ in his thesis). The reader unfamiliar with the nuances of contemporary usage shouldn't worry overly much about trying to discern any distinction between the two choices of terminology and notation. Of course, the attribution of proper names to ζ- and L-functions is an even more fraught enterprise. In the discussion that follows, I have tried to observe both tradition and contemporary practice as best I know how.

15.2 The Early History of ζ- and L-Functions – Euler, Riemann, and Dirichlet

The problems that Langlands is concerned with, in his theory of automorphic L-functions and his reciprocity conjecture, have their origins in the very beginnings of modern number theory, and I hope that it will be useful to the reader to review some of the history of this "early modern" number theory, if

only to convince them of the essential continuity between the contemporary
Langlands program and the number theory of previous eras.

15.2.1 Euler

The story of ζ- and L-functions begins with the study of the series

$$\sum_{n=1}^{\infty} \frac{1}{n^k} = 1 + \frac{1}{2^k} + \frac{1}{3^k} + \cdots. \tag{15.1}$$

A comparison with the improper integral $\int_1^{\infty} \frac{1}{x^k}$ shows that (15.1) converges
if $k > 1$, while it diverges if $k = 1$; indeed,

$$\sum_{n=1}^{N} \frac{1}{n} \sim \log N. \tag{15.2}$$

The investigation of this series for various values of k seems to date back to
the very beginning of the study of mathematical analysis in the seventeenth
century. For example, the Basel Problem, posed in 1650, asked for the value of
the sum

$$1 + 1/4 + 1/9 + \cdots$$

(i.e. the value of (15.1) when $k = 2$). Around 80 years later, this problem was
solved by Euler [Eulc], who showed that the value was equal to $\frac{\pi^2}{6}$. His method
of proof was to factor the function $\sin z$ as

$$\sin z = z \prod_{n=1}^{\infty} \left(1 - \frac{z}{\pi n}\right)\left(1 + \frac{z}{\pi n}\right) = z \prod_{n=1}^{\infty} \left(1 - \frac{z^2}{\pi^2 n^2}\right)$$

(via a consideration of its zeros, and via the analogy with polynomials);
expanding out the product and equating the resulting power series with the
usual Taylor series for $\sin z$ yields the formula

$$1 + 1/4 + 1/9 + \cdots = \frac{\pi^2}{6} \tag{15.3}$$

(equate the coefficients of z^3 on either side of the equation) and, with some
effort, analogous formulas for all higher even values of k. (As is well known,
one obtains rational multiplies of π^k.)

In fact, in the same paper, Euler considered other analogous factorizations
derived in the same manner, beginning with the factorization

$$1 - \sin z = \prod_{n=0}^{\infty} \left(1 - \frac{z}{(4n+1)\pi/2}\right)^2 \prod_{n=1}^{\infty} \left(1 - \frac{z}{(4n-1)\pi/2}\right)^2,$$

which yields, for example, the formula

$$1 - \frac{1}{3} + \frac{1}{5} - \frac{1}{7} + \cdots = \frac{\pi}{4} \tag{15.4}$$

(by identifying the coefficients of z on either side of the equation), due originally to Leibniz, as well as another (Euler's first) proof of (15.3) (by identifying the coefficients of z^2).

An even more consequential factorization, more directly related to the series (15.1), was discovered by Euler soon afterwards. Namely, Euler [Eulf] observed that the fundamental theorem of arithmetic (which states that each natural number admits a unique factorization into a product of prime powers) leads to the identity

$$\sum_{n=1}^{\infty} \frac{1}{n^k} = \prod_{p \text{ prime}} \left(1 + \frac{1}{p^k} + \frac{1}{p^{2k}} + \frac{1}{p^{3k}} + \cdots\right) = \prod_{p \text{ prime}} \left(1 - \frac{1}{p^k}\right)^{-1}. \tag{15.5}$$

As an immediate consequence, by considering the case $k = 1$, Euler deduced that there must be infinitely many primes, since a minimal necessary condition for the product over p to diverge (as it must when we set $k = 1$, since the sum over n does) is that it involve an infinite number of factors. More quantitatively, one may take logarithms of both sides of this formula when $k = 1$; taking into account (15.2), one deduces that

$$\sum_{p \text{ prime} \leq N} \frac{1}{p} \sim \log\log N. \tag{15.6}$$

Euler obtains an analogous factorization of Leibniz's formula (15.4), to the effect that

$$\prod_{\substack{p \text{ prime} \equiv 1 \bmod 4}} \left(1 - \frac{1}{p}\right)^{-1} \prod_{\substack{p \text{ prime} \equiv -1 \bmod 4}} \left(1 + \frac{1}{p}\right)^{-1} = \frac{\pi}{4}. \tag{15.7}$$

Taking logarithms, one finds that

$$\sum_{\substack{p \text{ prime} \equiv 1 \bmod 4 \\ p \leq N}} \frac{1}{p} - \sum_{\substack{p \text{ prime} \equiv -1 \bmod 4 \\ p \leq N}} \frac{1}{p} \tag{15.8}$$

converges as $N \to \infty$. Comparing this with (15.6), one finds that each of the sums appearing in (15.8), considered individually, diverges as $N \to \infty$, and thus concludes that there are infinitely many primes congruent to each of $\pm 1 \bmod 4$.[4]

---

[4] Much of Euler's results on what are now called *Euler products* are also discussed in Chapter 15 of his text [Euld]. The study of the particular difference appearing in (15.8) is the subject of his paper [Eulb]. This method of argument was vastly generalized by Dirichlet, as we recall below.

In his paper [Eula], Euler attempted to evaluate (15.1) at odd values of k. He was unsuccessful at this, but instead obtained generalizations of the formula (15.4), to the effect that

$$1 - \frac{1}{3^k} + \frac{1}{5^k} - \cdots = \text{a rational multiple of } \pi^k \qquad (15.9)$$

when k is an odd natural number. In the subsequent paper [Eule], Euler studied the series (15.1), as well as the series appearing in (15.9), at *negative* integral values of k. Of course, these latter series diverge, but Euler assigned them values via a regularization technique ("Abelian summation"): he considered the series $\sum_{n=1}^{\infty} (-1)^{n-1} \frac{t^n}{n^k}$ and $\sum_{n=0}^{\infty} (-1)^n \frac{t^{2n+1}}{(2n+1)^k}$ as functions of t (functions that can be explicitly determined using formulas for geometric series and their derivatives, if $k \leq 0$), and then let $t \to 1$.[5] Euler then found explicit formulas relating the resulting values of (15.1), as well as the series appearing in (15.9), at k and at $1 - k$, and he conjectured that these formulas hold for arbitrary values of k. (And he rigorously verified his conjecture at $k = 1/2$ for the series (15.1), and verified it numerically at additional values of k.)

15.2.2 Riemann

Riemann [Rie] took up the investigation of the series (15.1) where Euler left off. He introduced the notation

$$\zeta(s) = \sum_{n=1}^{\infty} \frac{1}{n^s} = 1 + \frac{1}{2^s} + \frac{1}{3^s} + \cdots, \qquad (15.10)$$

and considered $\zeta(s)$ as a function of a complex variable, initially for $\Re s > 1$ (where the series (15.10) converges absolutely), but then for all s, after effecting a meromorphic continuation. Indeed, he showed that $\zeta(s)$ is holomorphic on $\mathbb{C} \setminus \{1\}$, and has a simple pole at $s = 1$, with residue 1.[6] He also established the famous functional equation for $\zeta(s)$, namely that

$$\pi^{-\frac{s}{2}} \Gamma\left(\frac{s}{2}\right) \zeta(s) = \pi^{-\frac{1-s}{2}} \Gamma\left(\frac{1-s}{2}\right) \zeta(1-s). \qquad (15.11)$$

---

[5] Note that $1 - 1/2^k + 1/3^k - 1/4^k + \cdots = \left(1 - 2^{1-k}\right)\left(1 + 1/2^k + 1/3^k + 1/4^k + \cdots\right)$; thus one may regard this series as being obtained from the series (15.1) via a modification of the factor at $p = 2$ in the product formula (15.5). For a modern explication of Euler's method, and an indication of its relationship to more contemporary ideas in number theory, the reader can consult e.g. the first part of the paper [Kat75] of Katz.

[6] If one considers the function $\eta(s) = (1 - 2^{1-s})\zeta(s) = 1 - 1/2^s + 1/3^s - 1/4^s + \cdots$, then $\eta(s)$ is holomorphic on the entire complex plane – the zero at $s = 1$ of the factor $1 - 2^{1-s}$ cancels out the simple pole of $\zeta(s)$. The determination of the residue at $s = 1$ of $\zeta(s)$ is then seen to be equivalent to another formula of classical analysis, namely that $1 - 1/2 + 1/3 - 1/4 + \cdots = \log 2$.

This recovered the results of Euler for integral values of s already mentioned, and amounts to Euler's conjectural generalization of those results (for arbitrary complex values of s!).

Riemann gave two proofs of the analytic continuation and functional equation of $\zeta(s)$. His second proof established a connection between $\zeta(s)$ and the Jacobi theta function, beginning an intimate relationship between the theories of L-functions and of automorphic forms which continues to this day.

Jacobi's theta function is defined by the formula

$$\vartheta(\tau) = \sum_{n=-\infty}^{\infty} e^{\pi i n^2 \tau} = 1 + 2\sum_{n=1}^{\infty} e^{\pi i n^2 \tau} = 1 + 2q + 2q^4 + 2q^9 + \cdots,$$

where $q = e^{\pi i \tau}$ and $\Im \tau > 0$ (equivalently, $|q| < 1$). An application of the Poisson summation formula to the Gaussian e^{-x^2} (which is essentially its own Fourier transform) yields the functional equation

$$\vartheta\left(\frac{-1}{\tau}\right) = \sqrt{\frac{\tau}{i}}\,\vartheta(\tau). \tag{15.12}$$

Combining this with the integral formula

$$\pi^{-\frac{s}{2}}\Gamma\left(\frac{s}{2}\right)\zeta(s) = \int_0^{i\infty} \frac{\vartheta(\tau) - 1}{2}\left(\frac{\tau}{i}\right)^{s/2}\frac{d\tau}{\tau} \tag{15.13}$$

yields both the analytic continuation and functional equation.[7]

We can rewrite the formula (15.5) as

$$\zeta(s) = \prod_{p \text{ prime}} (1 - p^{-s})^{-1} \quad (\Re s > 1); \tag{15.14}$$

this is now referred to as the *Euler product* formula for $\zeta(s)$. Riemann famously employed (15.14) to study the distribution of the prime numbers (building on Euler's application of it to reprove the infinitude of the set of primes). His ideas were further developed by Hadamard and de la Vallée Poussin to prove the Prime Number Theorem (that the number of primes $\leq x$ is asymptotic to $x/\log x$; note that this can be regarded as a strengthening of Euler's asymptotic (15.6), since it immediately implies that result). A fundamental conjecture of Riemann is that the zeros of $\zeta(s)$ (other than those at the negative even integers, which were already discovered by Euler) all lie on the line $\Re s = \frac{1}{2}$ – this is the famous *Riemann Hypothesis*. The crux of both Hadamard and de la Vallée Poussin's proofs of the Prime Number Theorem is a weaker

---

[7] The functional equation (15.12) lets us rewrite the integral appearing in (15.13) in the form
$$\frac{1}{s(s-1)} + \int_i^{i\infty} \frac{\vartheta(\tau)-1}{2}\left(\left(\frac{\tau}{i}\right)^{\frac{s}{2}} + \left(\frac{\tau}{i}\right)^{\frac{1-s}{2}}\right)\frac{d\tau}{\tau},$$ from which the meromorphic continuation and functional equation both immediately follow.

result in the direction of the Riemann Hypothesis, namely that $\zeta(s)$ is zero-free on the line $\Re s = 1$.

We remark that the two properties of $\zeta(s)$ that we have discussed – (i) its analytic continuation and functional equation (which are mediated via its relationship to the automorphic form $\vartheta(\tau)$), and (ii) its Euler product – are *completely unrelated* in Riemann's treatment of them. This apparently technical point is in fact of a more than merely technical nature; it is crucial, and we will return to it below.

15.2.3 Dirichlet

Dirichlet introduced what are now called *Dirichlet L-functions* in his celebrated investigation of primes in arithmetic progressions [Dir37].

Following contemporary notation,[8] if $\chi : (\mathbb{Z}/N\mathbb{Z})^\times \to \mathbb{C}^\times$ is a character (i.e. a homomorphism between the indicated multiplicative groups), then we write

$$L(s,\chi) = \sum_{\substack{n \geq 1 \\ n \text{ coprime to } N}} \frac{\chi(n)}{n^s} = \prod_{\substack{p \text{ prime} \\ p \nmid N}} \left(1 - \frac{\chi(p)}{p^s}\right)^{-1}, \tag{15.15}$$

where the validity of the Euler product follows from the multiplicative property of χ. Dirichlet cites chapter 15 of Euler's text [Euld] as an inspiration for his consideration of such Euler products.

If χ is *not* primitive, i.e. if χ can be written as a composite $(\mathbb{Z}/N\mathbb{Z})^\times \to (\mathbb{Z}/M\mathbb{Z})^\times \xrightarrow{\psi} \mathbb{C}^\times$ for some proper divisor M of N and some character ψ (more colloquially, if the *period* of χ is a proper divisor of N), then $L(s,\chi)$ and $L(s,\psi)$ coincide, up to a finite number of factors in their Euler product related to primes that divide N but not M. Thus we may, and typically do, restrict attention to *primitive* L-functions, i.e. those $L(s,\chi)$ for which χ is primitive.

Note that if $N = 1$, so that χ is trivial, then $L(s,\chi) = \zeta(s)$, which has a simple pole at $s = 1$. On the other hand, it is easily seen that $L(s,\chi)$ extends holomorphically over the entire complex plane if χ is nontrivial.[9] Dirichlet then proves that $L(1,\chi) \neq 0$ if χ is non-trivial. Passing to logarithms and taking into account the Euler product, he obtains the result that

$$\sum_{p \text{ prime}} \frac{\chi(p)}{p} \tag{15.16}$$

---

[8] This notation differs from that of Dirichlet, who wrote simply L, or L with a subscript if he wished to indicate a particular character.

[9] We mention, though, that Dirichlet considers only s as a real variable.

converges if χ is a nontrivial character. Combining this with (15.2) (the corresponding statement for the trivial χ), an application of finite Fourier theory shows that $\sum_{p\equiv a \bmod N} \frac{1}{p}$ diverges if $(a, N) = 1$, proving in particular that there are infinitely many primes in the arithmetic progression $a + mN$.

Dirichlet proves that $L(1, \chi) \neq 0$ (for nontrivial χ) by first considering the sum of the expressions (15.16), namely

$$\sum_{\substack{\chi \text{ a char.} \\ \bmod N}} \sum_{p \text{ prime}} \frac{\chi(p)}{p} = \sum_{p \text{ prime}} \frac{\sum_\chi \chi(p)}{p}. \tag{15.17}$$

The orthogonality relations for characters imply that this series has nonnegative terms, and hence if it diverges, it must diverge to $+\infty$. On the other hand, each expression (15.16) is either finite (if $L(1, \chi)$ is finite and nonzero), diverges to $+\infty$ (which happens precisely when χ is trivial), or diverges to $-\infty$ (if $L(1, \chi) = 0$). Making an analysis of the rates of convergence, one finds that at most one $L(1, \chi)$ can possibly equal zero (since the sum of at least two divergences to $-\infty$ would overwhelm the divergence to $+\infty$ that comes from the contribution of the trivial character, contradicting the positivity already noted). Since $L(1, \overline{\chi}) = \overline{L(1, \chi)}$, this shows that $L(1, \chi) \neq 0$ if χ is a genuinely complex character (i.e. if $\chi \neq \overline{\chi}$).

If χ is a real- (i.e. ± 1-) valued character (one also refers to such characters as *quadratic*), then Dirichlet's proof that $L(1, \chi) \neq 0$ is more involved, and is number-theoretic in nature – if χ is a primitive quadratic character mod N, then he shows [Dir] that $L(1, \chi)$ is a certain positive multiple of the class number of the field $\mathbb{Q}(\sqrt{\pm N})$ (here the sign is chosen equal to $\chi(-1)$).[10] Since this class number is nonzero, so is $L(1, \chi)$.

As a special case, when $N = 4$ and χ is the unique nontrivial character, he recovers Leibniz's formula (15.4); from Dirichlet's point-of-view, this formula can be interpreted as showing that $\mathbb{Q}(i)$ has class number 1.

15.3 Dedekind ζ-Functions and Hecke L-Functions

Over the course of the nineteenth century, algebraic number theory emerged as a topic of central importance, and many preceding preoccupations of number theorists (such as the theory of Diophantine equations, and the theory of quadratic forms) became, to a large extent, absorbed into the general apparatus of algebraic number theory. This is in particular true of the theory of L-functions.

---

[10] Dirichlet actually employs the older language of quadratic forms, due to Legendre and Gauss. The transition to the language of algebraic number theory comes about in the work of Kummer and Dedekind; see e.g. this MathOverflow answer by Lemmermeyer [Lem]. See also (15.24) below for the explicit form of Dirichlet's result.

Two of the principal contributors to these developments, both the development of algebraic number theory in general, and the positioning of the theory of L-functions at the center of it, were Kummer and Dedekind. Lack of space prohibits us from discussing Kummer's seminal contributions in any detail (though we refer the reader to the review [Maz77] for a wonderful overview), and we instead limit ourselves to recalling Dedekind's definition of the ζ-function of a number field, and the statement of his class number formula. This formula in a sense generalizes Dirichlet's formula for the special value $L(1,\chi)$ when χ is a quadratic character, but there is a fundamental point that has to be understood to make *precise* the sense in which this is true; indeed, as we already remarked in the introduction, comparing these two formulas is a first example of *reciprocity*, and we discuss it more fully in Section 15.4.1 below.

In this section, after briefly recalling Dedekind's contributions, we move onto the key (for our purposes) further developments of the analytic theory of L-functions in the first half of the twentieth century – first those of Hecke (whose results generalized those of both Riemann and Dirichlet from the field $\mathbb{Q}$ to a general number field), and then those of Tate, as famously developed in his thesis.

15.3.1 Dedekind

If K is a number field (i.e. a finite extension of $\mathbb{Q}$), with ring of algebraic integers $\mathcal{O}_K$, then we may imitate the definition of the Riemann zeta function, and define a ζ-function attached to K, via the formula

$$\zeta_K(s) = \sum_{\mathfrak{a}\neq 0} \frac{1}{N(\mathfrak{a})^s};$$

here the sum is over all nonzero ideals $\mathfrak{a} \subseteq \mathcal{O}_K$, and $N(\mathfrak{a})$ denotes the norm of $\mathfrak{a}$, i.e. the order of the (finite) quotient ring $\mathcal{O}_K/\mathfrak{a}$. We call this the *Dedekind ζ-function* of K. If we take into account the unique factorization of ideals into products of prime ideals, along with the multiplicativity of the norm, we find that $\zeta_K(s)$ admits an Euler product

$$\zeta_K(s) = \prod_{\mathfrak{p}} \left(1 - \frac{1}{N(\mathfrak{p})^s}\right)^{-1}; \tag{15.18}$$

here the product runs over nonzero prime ideals (equivalently, maximal ideals) of $\mathcal{O}_K$.

We remark that the series and Euler product converge, and hence these formulas make sense, when $\Re s > 1$, and indeed Dedekind regarded $\zeta_K(s)$ as a function on the halfline $s > 1$.[11]

Following Dirichlet's arguments for evaluating $L(1, \chi)$ for quadratic characters χ, Dedekind proved the famous *class number formula*

$$\lim_{s \to 1} (s - 1)\zeta_K(s) = \frac{2^{r_1}(2\pi)^{r_2} R h}{w\sqrt{|D|}}. \tag{15.19}$$

where r_1 and r_2 denote the number of real and complex places of K respectively, h and R respectively denote the class number and regulator of K, w denotes the number of roots of unity contained in K, and $|D|$ is the absolute value of the discriminant of K.

Subsequently, Landau [Lan03] proved that $\zeta_K(s)$ has an analytic continuation to the left of $\Re s = 1$, holomorphic except for a simple pole at $s = 1$ (whose residue is then computed by (15.19)). He was then able to follow the method of Hadamard–de la Vallée Poussin to establish the *Prime Ideal Theorem* (that the number of prime ideals of norm $\leq X$ is asymptotic to $X/\log X$), which is the number field analogue of the Prime Number Theorem.[12]

It is natural to consider number field analogues of Dirichlet's theorem on primes in arithmetic progression as well. Already Dedekind suggested considering (what we would now call) L-functions

$$L(s, \chi) = \sum_{\mathfrak{a} \neq 0} \frac{\chi(\mathfrak{a})}{N(\mathfrak{a})^s},$$

where χ is a character of the class group of K [Ded, §184]. One might hope to use such L-functions to prove a result analogous to Dirichlet's theorem on primes in arithmetic progression, to the effect that each ideal class contains infinitely many prime ideals. This idea was taken up by Hecke.

---

[11] Dedekind developed the theory of $\zeta_K(s)$ – which he denotes $\Omega(s)$, in line with his notational choice of Ω for the number field that we are denoting K – in §184 of his Supplement XI to Dirichlet's lectures on number theory [Ded].

[12] We recall one particular consequence of this theorem. Suppose that K/F is a Galois extension of number fields. As is well-known, 100% (in the sense of density) of prime ideals of K are split completely over F. (A non-split prime $\mathfrak{q}$ of K lying over some prime $\mathfrak{p}$ of F has norm at least $N(\mathfrak{p})^2$, and so the contribution of these primes to $\lim_{s \to 1} \sum_{\mathfrak{q}} N(\mathfrak{q})^{-s}$ is actually convergent.) Since there are $[K : F]$ primes of K lying over a given prime of F that splits in K, we then deduce – by comparing the prime ideal theorems for F and for K – that the primes in F that split completely in K have density $1/[K : F]$ among all the primes in F. This proves the Tschebotareff density theorem (see Section 15.5.3) for the trivial conjugacy class in $\mathrm{Gal}(K/F)$.

15.3.2 Hecke

Hecke [Hec17, Hec20] proved the analytic continuation of and functional equation for Dedekind ζ-functions, and for a more general class of L-functions

$$L(s,\chi) = \sum_{\substack{\mathfrak{a} \\ (\mathfrak{a},\mathfrak{m})=1}} \frac{\chi(\mathfrak{a})^s}{N(\mathfrak{a})},$$

where χ is (what is now called) a *Hecke character*, or equivalently a *Grössencharacter*, or (again equivalently, after an appropriate reformulation) an *idèle class character*, of conductor $\mathfrak{m}$.

If $\mathfrak{m}$ is a nonzero ideal in $\mathcal{O}_K$, and if S denotes the finite set of prime ideals dividing $\mathfrak{m}$, then a Hecke character of conductor[13] $\mathfrak{m}$ is a homomorphism of multiplicative groups

$$\chi : \{\text{fractional ideals coprime to } S\} \to \mathbb{C}^\times$$

for which there is some continuous character of multiplicative groups[14]

$$\theta : (\mathbb{R} \otimes_{\mathbb{Q}} K)^\times \to \mathbb{C}^\times$$

such that for any $\alpha \in K^\times$ with $\alpha \equiv 1 \bmod \mathfrak{m}$, we have $\chi((\alpha)) = \theta(\alpha)$ (where we think of α as lying in $(\mathbb{R} \otimes_{\mathbb{Q}} K)^\times$ via the natural embedding of $K^\times$ into the former group).

If we take θ to be the trivial character (so that we simply insist that $\chi((\alpha)) = 1$ if $\alpha \equiv 1 \bmod \mathfrak{m}$), then we obtain the notion of a *ray class character* of conductor $\mathfrak{m}$. These are precisely the Hecke characters of finite order.

The multiplicative property of a Hecke character χ ensures that Hecke's L-functions again admit an Euler product

$$L(s,\chi) = \prod_{\mathfrak{p} \notin S} \left(1 - \frac{\chi(\mathfrak{p})}{N(\mathfrak{p})^s}\right)^{-1}.$$

Example 15.3.2.1 If $K = \mathbb{Q}$, then conductors are either of the form N or $N\infty$ (where $N > 0$ is a natural number, and ∞ denotes the embedding $\mathbb{Q} \hookrightarrow \mathbb{R}$). Ray class characters of conductor N are precisely the *even* primitive

---

[13] In fact, the notion of conductor should be broadened in the usual way: a conductor $\mathfrak{m}$ should consist of a nonzero ideal (the finite, or nonarchimedean part of the conductor) together with any subset of the set of real embeddings of K (the infinite, or archimedean, part of the conductor). If $\sigma : K \hookrightarrow \mathbb{R}$ is a real embedding, and if $\alpha \in K^\times$, then the condition $\alpha \equiv 1 \bmod \sigma$ is interpreted as meaning that $\sigma(\alpha) > 0$ (i.e. that $\sigma(\alpha)$ and 1 lie in the same connected component of $\mathbb{R}^\times$).

[14] Note that $\mathbb{R} \otimes_{\mathbb{Q}} K$ is just a more functorial way of writing $\mathbb{R}^{r_1} \times \mathbb{C}^{r_2}$, and so we are simply considering characters $(\mathbb{R}^\times)^{r_1} \times (\mathbb{C}^\times)^{r_2} \to \mathbb{C}^\times$. The canonical embedding of K into $\mathbb{R} \otimes_{\mathbb{Q}} K$ (and hence of $K^\times$ into $(\mathbb{R} \otimes_{\mathbb{Q}} K)^\times$) is just a functorial way of describing the embedding of K into $\mathbb{R}^{r_1} \times \mathbb{C}^{r_2} \cong \mathbb{R}^{r_1+2r_2}$ that is traditionally considered in algebraic number theory.

characters $\chi : (\mathbb{Z}/N\mathbb{Z})^{\times} \to \mathbb{C}^{\times}$, while those of conductor $N\infty$ are the *odd* primitive characters $\chi : (\mathbb{Z}/N\mathbb{Z})^{\times} \to \mathbb{C}^{\times}$. (We say a character is *even* or *odd* according to whether $\chi(-1) = 1$ or -1.) Thus, in this case, Hecke's ray class L-functions reduce to Dirichlet's L-functions.

The notion of Hecke character in this case (i.e. when $K = \mathbb{Q}$) is scarcely more general: any Hecke character has the form $(a) \mapsto \chi(a)|a|^{s_0}$ for some mod N character χ and some $s_0 \in \mathbb{C}$, in which case the corresponding L-function is simply $L(s - s_0, \chi)$ – i.e. a Dirichlet L-function with the variable shifted. In particular, the only *unitary* Hecke characters (i.e. those that land in the group of complex numbers of modulus 1) are products of the finite-order characters that already appear in Dirichlet's theory with a unitary character $a \mapsto |a|^{iy}$ of $\mathbb{R}^{\times}$.

Example 15.3.2.2 If $K = \mathbb{Q}(i)$, then the ring of integers is $\mathbb{Z}[i]$, which is well known to have trivial class group. The unique prime ideal above 2 is $(1 + i)$; all other prime ideals (which we might call *odd*) have a unique generator π satisfying the congruence $\pi \equiv 1 \bmod (1 + i)^3$. The character $\psi :$ $(\pi) \to \pi$ (where the generator is chosen to satisfy the preceding congruence) is an infinite-order Hecke character of conductor $(1 + i)^3$. (The corresponding character θ is simply the identity character $z \mapsto z$ on $\mathbb{C}^{\times}$.) The character ψ is not unitary, but it is easy enough to modify it so as to make it so, e.g. by defining $\chi\big((\pi)\big) = \frac{\pi}{\sqrt{N(\pi)}}$.

(More generally, for any number field K, we always have the one-parameter family of Hecke characters of trivial conductor $\mathfrak{a} \mapsto |N(\mathfrak{a})|^s$. If ψ is any Hecke character, then $|\psi|$ belongs to this one-parameter family, and $\chi = \psi|\psi|^{-1/2}$ is unitary.)

Hecke proved the analytic properties of his L-functions by combining Riemann's arguments using theta functions with Dirichlet's and Dedekind's arguments (appearing in their proofs of the class number formula) involving sums over ideal classes. As in Riemann's case, the analytic aspects of his theory apparently bear no relation to the Euler product structure of the L-series. Rather than writing more about Hecke's arguments, though, we now turn to Tate's reworking of them, since this marked a decisive shift in the analytic theory of L-functions.

15.3.3 Adéles and Idéles

In his thesis [Tat67], Tate gives a reformulation of the theory of Hecke L-functions in terms of adèles and idèles. We begin by briefly recalling the definitions of these latter objects.

As usual, we let $\widehat{\mathbb{Z}}$ denote the profinite completion of $\mathbb{Z}$, so

$$\widehat{\mathbb{Z}} = \varprojlim_N \mathbb{Z}/N\mathbb{Z} = \prod_p \mathbb{Z}_p,$$

the second equality holding by virtue of the Chinese Remainder Theorem. By its construction, $\widehat{\mathbb{Z}}$ is a profinite (and, in particular, a compact) commutative ring. We may extend scalars to $\mathbb{Q}$ to obtain a locally compact $\mathbb{Q}$-algebra[15] $\widehat{\mathbb{Z}}\otimes_{\mathbb{Z}}\mathbb{Q}$, which we also denote by $\mathbb{A}_{\mathbb{Q}}^{\infty}$, and refer to as the ring of *finite adèles* of $\mathbb{Q}$. We may also describe $\mathbb{A}_{\mathbb{Q}}^{\infty}$ as the *restricted product* $\prod_p{}'\mathbb{Q}_p$ of the various p-adic fields $\mathbb{Q}_p$: it consists of those tuples $(x_p) \in \prod_p \mathbb{Q}_p$ for which x_p is an integer for all but finitely many p. We then define the full ring of *adèles* of $\mathbb{Q}$ to be the product

$$\mathbb{A}_{\mathbb{Q}} = \mathbb{R} \times \mathbb{A}_{\mathbb{Q}}^{\infty};$$

it is again a locally compact $\mathbb{Q}$-algebra, and also admits a description as the restricted product $\prod_v{}'\mathbb{Q}_v$, where v now ranges over *all* places of $\mathbb{Q}$, including the infinite (archimedean) place.

If K is any number field, then we define

$$\mathbb{A}_K^{\infty} = \mathbb{A}_{\mathbb{Q}}^{\infty} \otimes_{\mathbb{Q}} K = \widehat{\mathbb{Z}} \otimes_{\mathbb{Z}} K = \prod_{\mathfrak{p}}{}' K_{\mathfrak{p}},$$

the final description here being the restricted product of the various non-archimedean completions $K_{\mathfrak{p}}$ of K (in the same sense as used above, i.e. consisting of those elements of the product for which all but finitely many of their components are integers). We define

$$\mathbb{A}_K = \mathbb{A}_{\mathbb{Q}} \otimes_{\mathbb{Q}} K = (\mathbb{R} \otimes_{\mathbb{Q}} K) \times \mathbb{A}_K^{\infty} = \mathbb{R}^{r_1} \times \mathbb{C}^{r_2} \times \mathbb{A}_K^{\infty} = \prod_v{}' K_v,$$

where now the restricted product is over *all* places of K. We refer to these two K-algebras as the ring of *finite adèles* over K and the ring of *adèles* over K, respectively. They are both locally compact K-algebras. (As a $\mathbb{Q}$-vector space, i.e. forgetting the K-algebra structure, $\mathbb{A}_K$ is just a product of $[K : \mathbb{Q}]$ copies of $\mathbb{A}_{\mathbb{Q}}$.)

If G is any affine algebraic group over K, then we may consider its group $G(\mathbb{A}_K)$ of $\mathbb{A}_K$-valued points; the hypothesized Zariski closed embedding of G into some d-dimensional affine space over K induces an embedding of $G(\mathbb{A}_K)$ as a closed subset of $\mathbb{A}_K^d$, giving $G(\mathbb{A}_K)$ the structure of a locally compact group. For now, we apply this to the group

$$\mathbb{G}_m = \mathrm{GL}_1 = \{(x,y) \in \mathbb{A}^2 \mid xy = 1\}.$$

As a group, $\mathrm{GL}_1(\mathbb{A}_K)$ is simply the group of units $\mathbb{A}_K^{\times}$; the algebraic group viewpoint is important for endowing it with the correct topology. We refer

---

[15] Writing $\mathbb{Q} = \bigcup_{N>0} \frac{1}{N}\mathbb{Z}$, we find that $\mathbb{A}_{\mathbb{Q}}^{\infty} = \bigcup_{N>0} \frac{1}{N}\widehat{\mathbb{Z}}$ is simply a union of copies of $\widehat{\mathbb{Z}}$, in a manner analogous to that in which $\mathbb{Q}_p = \mathbb{Z}_p \otimes_{\mathbb{Z}} \mathbb{Q} = \bigcup_{n\geq 0} \frac{1}{p^n}\mathbb{Z}_p$ is a union of copies of $\mathbb{Z}_p$.

to the abelian group $\mathbb{A}_K^\times$, endowed with the locally compact topology just discussed (so we identify it with the closed subset $\{(x, y) \in \mathbb{A}_K^2 \mid xy = 1\}$ of $\mathbb{A}_K^2$) as the group of idèles of K. We may also write $\mathbb{A}_K^\times$ as the restricted product $\prod_v' K_v^\times$ consisting of elements in the product that are integral units at all but finitely many finite places.

More important for our purposes is not the idèle group of K itself, but the idèle class group of K. Before defining this, we note that the natural (diagonal) embedding $\mathbb{Q} \hookrightarrow \mathbb{A}_\mathbb{Q}$ has discrete image.[16] Extending scalars we obtain the discrete embedding $K \hookrightarrow \mathbb{A}_K$, and consequently a discrete embedding $K^\times \hookrightarrow \mathbb{A}_K^\times$. The *idèle class group* of K is defined to be the quotient[17] $K^\times \backslash \mathbb{A}_K^\times$. We are quotienting a locally compact abelian group by a discrete subgroup; the result is again a locally compact abelian group.

In fact, the group $K^\times \backslash \mathbb{A}_K^\times$ is very close to being compact. Namely, the product formula for global fields shows that the surjective homomorphism

$$| \, | : \mathbb{A}_K^\times = \prod_v' K_v^\times \to \mathbb{R}_{>0}^\times$$

defined by $(x_v) \mapsto \prod_v |x_v|_v$ (where $| \, |_v$ is an appropriately normalized absolute value on $K_v^\times$) contains $K^\times$ in its kernel. If we let $\mathbb{A}_K^1$ denote this kernel, then $K^\times \backslash \mathbb{A}_K^1$ *is compact*,[18] and we have the short exact sequence

$$1 \to K^\times \backslash \mathbb{A}_K^1 \to K^\times \backslash \mathbb{A}_K^\times \to \mathbb{R}_{>0}^\times \to 1.$$

The first key reinterpretation of Hecke's theory in terms of these objects is that a Hecke character is *the same thing as* a continuous character

$$\chi : K^\times \backslash \mathbb{A}_K^\times \to \mathbb{C}^\times.$$

Indeed, composing χ with the inclusion $K_v^\times \to \mathbb{A}_K^\times \to K^\times \backslash \mathbb{A}_K^\times$ yields a continuous character $\chi_v : K_v^\times \to \mathbb{C}^\times$, and we may factor χ as

$$\chi = \prod_v \chi_v \tag{15.20}$$

The continuity of χ ensures that for all v outside a finite set S (which we now take to include the infinite places) the character χ_v is *unramified*, in that it

---

[16] This embedding is an analogue of the more classical discrete embedding $\mathbb{Z} \hookrightarrow \mathbb{R}$, and one of the technical jobs that the adèles perform is to mimic this latter embedding, while nevertheless allowing one to work with structures over $\mathbb{Q}$ – such as $\mathbb{Q}$-vector spaces, or groups like $\mathrm{GL}_n(\mathbb{Q})$, rather than having to contend with more general abelian groups, or groups like $\mathrm{GL}_n(\mathbb{Z})$, whose technical properties are typically not as pleasant.

[17] It is traditional at this point, if disorienting for the uninitiated, to write the group by which we are taking the quotient on the left.

[18] This is a reformulation of two key theorems of elementary algebraic number theory, namely the finiteness of the class group and Dirichlet's unit theorem.

factors through the quotient $\mathbb{Z} = K_v^\times / \mathcal{O}_{K_v}^\times$ – i.e. $\chi_v(x)$ depends only on the v-adic valuation of x. We now define a character[19]

$$\tilde{\chi} : \{\text{fractional ideals coprime to } S\} \to \mathbb{C}^\times$$

via

$$\mathfrak{a} = \prod_{\mathfrak{p} \notin S} \mathfrak{p}^{n_\mathfrak{p}} \mapsto \prod_{\mathfrak{p} \notin S} \chi_\mathfrak{p}(\pi_\mathfrak{p}^{-1})^{n_\mathfrak{p}},$$

where $\pi_\mathfrak{p}$ is some choice of uniformizer in $K_\mathfrak{p}$ (it doesn't matter which). If we let θ denote

$$\prod_{v \text{ infinite}} \chi_v : (\mathbb{R} \otimes_\mathbb{Q} K)^\times \to \mathbb{C}^\times,$$

then the continuity of χ will ensure that $\tilde{\chi}$ satisfies the requirements to be a Hecke character with respect to the character θ and *some* conductor $\mathfrak{m}$ involving just the places in S. Note in particular that although the definition of $\tilde{\chi}$ involves only the χ_v for $v \notin S$, we may recover χ, and in particular the remaining χ_v for $v \in S$, from $\tilde{\chi}$, using the fact that χ is continuous and is defined on the quotient $K^\times \setminus \mathbb{A}_K^\times$.

The finite-order idèle class characters correspond to Hecke's ray class characters. These necessarily factor through a finite quotient of $K^\times \setminus \mathbb{A}_K^\times$. In fact, the group of connected components $\pi_0(K^\times \setminus \mathbb{A}_K^\times)$ of $K^\times \setminus \mathbb{A}_K^\times$ is naturally a profinite group, and may be described as the inverse limit, over all conductors $\mathfrak{m}$, of the ray class groups of conductor $\mathfrak{m}$. The connected component of the identity in $K^\times \setminus \mathbb{A}_K^\times$ is the product of a copy of $\mathbb{R}_{>0}^\times$ with the connected component of the identity in $K^\times \setminus \mathbb{A}_K^1$, this latter group being a connected compact group. It is trivial in the simplest case of $K = \mathbb{Q}$, but is otherwise nontrivial.

Example 15.3.3.1   One easily computes that

$$\mathbb{Q}^\times \setminus \mathbb{A}_\mathbb{Q}^\times = \mathbb{R}_{>0}^\times \times \widehat{\mathbb{Z}}^\times.$$

This recovers the description of Hecke characters for $\mathbb{Q}$ given in Example 15.3.2.1 above.

Example 15.3.3.2   It is also easy to compute that

$$\mathbb{Q}(i)^\times \setminus \mathbb{A}_{\mathbb{Q}(i)}^\times = \mathbb{C}^\times \times \{x \in \widehat{\mathbb{Z}[i]}^\times \mid x \equiv 1 \bmod (1+i)^3\}.$$

---

[19]   The exponent -1 on $\pi_\mathfrak{p}$ is chosen to ensure a compatibly between various conventions and constructions; e.g. if we identify a character $\chi : (\mathbb{Z}/N\mathbb{Z})^\times \to \mathbb{C}^\times$ with an idèle class character via the discussion of Example 15.3.3.1, then the character $\tilde{\chi}$ coincides with the ray class character associated to χ via the discussion of Example 15.3.2.1.

(More precisely, there is an obvious map from the right-hand side to the left, which is an isomorphism.) In these terms, the Hecke character ψ of Example 15.3.2.2 may be described simply as the projection onto the $\mathbb{C}^\times$ factor.

Example 15.3.3.3   Recalling that $\mathbb{Q}(\sqrt{2})$ has class number 1, that $1 + \sqrt{2}$ is a fundamental unit of $\mathbb{Q}(\sqrt{2})$, and that $N(1 + \sqrt{2}) = -1$, one finds that

$$\mathbb{Q}(\sqrt{2})^\times \setminus \mathbb{A}^\times_{\mathbb{Q}(\sqrt{2})} = \langle (1 + \sqrt{2})^2 \rangle \setminus \left(\mathbb{R}^\times_{>0} \times \mathbb{R}^\times_{>0} \times \widehat{\mathbb{Z}[\sqrt{2}]}^\times \right).$$

In particular, the connected component of the identity is equal to

$$\varprojlim_n \left(\langle (1 + \sqrt{2})^{2n} \rangle \setminus \left(\mathbb{R}^\times_{>0} \times \mathbb{R}^\times_{>0} \right) \right),$$

which is the product of $\mathbb{R}^\times_{>0}$ with a *solenoid* (the inverse limit of an inverse system of circles with respect to nontrivial covering maps).

15.3.4 Tate

In his thesis [Tat67], Tate reproved Hecke's results on L-functions of Hecke characters, using the formalism of the idèles and idèle class characters. As in Riemann's and Hecke's arguments, he uses an integral transform to obtain the analytic continuation and functional equation. However, the integral is now taken over the group of idèles $\mathbb{A}^\times_K$ (from this point of view, the integral in Riemann's formula (15.13) should be viewed as taking place over the purely archimedean group $\mathbb{R}^\times_{>0}$), and the application of Poisson summation that underlies the proof of the functional equation is now performed with respect to the discrete group $K \subset \mathbb{A}_K$. The fact that the integral is taken over $\mathbb{A}^\times_K$ has a crucial consequence: the Euler product for the Hecke L-functions appears naturally, as a part of the same sequence of arguments that proves the analytic continuation and functional equation!

In a little more detail:[20] Tate considers a *unitary* idèle class character χ, and then considers integrals of the form

$$\zeta(f, \chi \mid |^s)) = \int_{\mathbb{A}^\times_K} f(x)\chi(x)|x|^s d^\times x,$$

[20] We sketch the contents of Tate's thesis here, since it played, and continues to play, such a key role in the development of the the theory of automorphic L-functions. For more details, we encourage the interested reader to study the thesis itself, as well as Weil's important Seminaire Bourbaki [Wei95], which gives a distribution-theoretic interpretation of Tate's arguments; suitable extensions of the ideas in Weil's lecture are basic to the contemporary analysis of integral formulas in the theory of automorphic forms. There are also many excellent commentaries on Tate's thesis available; among these, we particularly recommend the article [Kud03] of Kudla.

where f is a function on $\mathbb{A}_K$ that decays suitably at infinity (i.e. belongs to an appropriately defined Schwarz space of functions on $\mathbb{A}_K$), and $d^\times x$ is a suitably normalized Haar measure on the idèle group $\mathbb{A}_K^\times$, obtained as a product of Haar measures $d^\times x_v$ on each of the multiplicative groups $K_v^\times$.

If the function f factors as a product $f = \otimes_v f_v$ of (appropriately Schwarz) functions f_v on each K_v, then Tate's ζ-integral admits a factorization

$$\zeta(f, \chi \mid |^s) = \int_{\mathbb{A}_K^\times} f(x)\chi(x)|x|^s d^\times x$$

$$= \prod_v \int_{K_v^\times} f_v \chi_v(x_v)|x_v|_v^s d^\times x_v = \prod_v \zeta_v(f_v, \chi_v \mid |_v^s), \qquad (15.21)$$

where the local ζ-integrals are defined according to the indicated formula. Directly evaluating these local ζ-integrals, and then (via the factorization (15.21)) analyzing the global ζ-integrals, Tate finds that the local ζ-integrals converge for $\Re s > 0$ – this directly corresponds to the fact that the geometric series

$$\sum_{n=0}^\infty \frac{1}{p^{ns}} = \left(1 - \frac{1}{p^s}\right)^{-1}$$

defining the Euler factors in (15.5) converge when $\Re s > 0$ – and that the global ζ-integrals converge for $\Re s > 1$ – directly corresponding to the fact the Euler product (15.5) itself converges[21] precisely when $\Re s > 1$.

Tate establishes two crucial results. The first, local, result, concerns the ratio

$$\frac{\zeta_v(f_v, \chi_v \mid |_v^s)}{\zeta_v(\hat{f}_v, \overline{\chi}_v \mid |^{1-s})}, \qquad (15.22)$$

where $\hat{f}_v$ denotes the additive Fourier transform of f_v (with respect to some chosen additive character ψ_v); note that both integrals are defined when $0 < \Re s < 1$, and so this ratio makes sense (provided that the denominator does not vanish). What Tate shows is that the ratio (15.22) is in fact independent of the choice of the function f_v, and analytically continues to a meromorphic function of s. This is proved by an elementary manipulation of integrals.

The second, global, result, states that $\zeta(f, \chi \mid |^s)$ has a meromorphic continuation to the entire complex plane, and satisfies a functional equation

$$\zeta(f, \chi \mid |^s) = \zeta(\hat{f}, \overline{\chi} \mid |^{1-s});$$

[21] It is important that χ be unitary for these bounds on the region of convergence to be correct. As we already saw in Example 15.3.2.1, and as is clear from the formulas defining the ζ-integrals, if we multiply χ by some power $||^{s_0}$ so as to make it possibly nonunitary, then this amounts to shifting the variable s in the ζ-integral by s_0, and correspondingly translating the regions of convergence by $\Re s_0$.

here $\hat{f}$ denotes the additive Fourier transform of f, computed with respect to an additive character of $\mathbb{A}_K$ that is chosen to be trivial on K. The proof of this result follows Riemann's proof of (15.11) via the integral transform of a theta series; in particular, as already remarked, Poissson summation (but now with respect to the discrete subgroup K of $\mathbb{A}_K$) plays a key role.

By choosing $f = \otimes_v f_v$ to be (essentially) Fourier self-dual (so at an infinite place, f_v will be an appropriate Gaussian – and so we connect back to Riemann's proof via Jacobi's ϑ – while at a finite place, f_v will be the characteristic function of $\mathcal{O}_K$), we find that $\zeta(f, \chi \mid \mid^s)$ is essentially Hecke's L-function $L(s, \chi)$. In fact, we essentially obtain the *completed* L-function $\Lambda(s, \chi)$ – i.e. with the appropriate Γ factors, powers of π, etc., included – and (via (15.21)) we obtain it directly in its *Euler product form*! We say "essentially," because at the primes dividing the conductor of χ the local ζ-integrals are a little more finicky (they involve Gauss sum-like expressions); and we also have factors at the infinite places. When we put everything together, we obtain Hecke's functional equation[22]

$$\Lambda(1 - s, \overline{\chi}) = \epsilon(\chi)\Lambda(s, \chi),$$

where the quantity $\epsilon(\chi)$, which mediates the functional equation, is itself written as an "Euler product"

$$\epsilon(\chi) = \prod_{v \in S} \epsilon_v(\chi, \psi_v); \tag{15.23}$$

here S denotes the set of ramified primes for χ, together with the infinite places, and the local factors ϵ_v arise from an explicit evaluation of the ratios of (15.22) at the places $v \in S$.[23] (As the notation indicates, the local ϵ factors depend on the additive character ψ_v used to compute the local additive Fourier transforms – although in an easily understood way – whereas the global ϵ factor does not. Indeed, since the functional equation is what it is, regardless

---

[22] In the case of the Riemann, or more generally Dedekind, ζ-function, which in Hecke's optic is the case of the trivial character, one has the simplest possible functional equation: if $\xi_K(s)$ denotes the appropriately completed form of $\zeta_K(s)$, then $\xi_K(1 - s) = \xi_K(s)$. For quadratic (equivalently, real-valued unitary) characters, the functional equation again takes a simple form, namely

$$\Lambda(1 - s, \chi) = \pm\Lambda(s, \chi)$$

(a factor of ± 1 is the only possibility that is consistent with the involutive nature of the symmetry $s \mapsto 1 - s$). But for genuinely complex characters, the functional equation relates $\Lambda(s, \chi)$ and $\Lambda(1 - s, \overline{\chi})$; since χ and $\overline{\chi}$ are now distinct, the constant in the functional equation is less constrained; a priori it is merely a complex number of absolute value 1.

[23] It is more correct to include the variable s in the global and local ϵ factors as well; the factor $(|D|N(\mathfrak{m}))^{s/2}$, which appears in the classical completed L-function is then incorporated into the ϵ factors, and written as the product of its various local contributions. The ϵ factors in our formula are obtained by evaluating at the point of symmetry, i.e. at $s = 1/2$. When we evaluate the ϵ factors at a specific value of s in this way, it might be better to refer to them as local and global *root numbers*.

of how we normalize the Fourier transforms used to establish it, the global ϵ factor cannot depend on the choice of ψ; more concretely, one sees that this is so via an application of the global product formula.)

If we take χ to be the trivial character, Tate's ζ-integral essentially computes the Dedekind ζ-function $\zeta_K(s)$. In this optic, the class number formula (15.19) emerges quite directly from a computation of the volume of the compact group $K^\times \backslash \mathbb{A}_K^1$.

15.3.5 Applications to Density

Given the analytic continuation of his L-series, Hecke was able to generalize Dirichlet's arguments (and indeed the subsequent improvements on them by Landau) to prove the infinitude (and indeed, the appropriate density) of prime ideals in a given ray class (by working with $L(s,\chi)$ for χ running over all characters of the appropriate ray class group), and even more general density theorems (by using $L(s,\chi)$ for infinite order χ), such as results on the density of Gaussian primes lying in a given angular sector of the complex plane. (One uses Fourier theory on $\mathbb{C}^\times$ to approximate the indicator function of the sector in terms of characters θ on $\mathbb{C}^\times$, and then uses a version of Dirichlet's argument, but applied to Hecke characters χ that extend the relevant θ.)

15.3.6 Summary

If one unpacks Tate's proof of Hecke's functional equation in sufficient detail, it does not differ so much in its technical details from Hecke's original proof. (As one example, the step in Hecke's argument, going back to Dirichlet's original analysis of his L-functions, in which one decomposes the L-function into a sum of separate series, each indexed by ideals lying in a given ideal class, is related to the proof of the compactness of $\mathbb{A}_K^1/K^\times$, in which (as we noted above) the finiteness of the class group plays a key role.) But the difference in presentation is enormous! Tate's integral formulas lead directly to the presentation of the L-functions as an Euler product, his local calculations explicate the structure of the global root number by writing it as a product of local factors, and the class number formula falls out as another manifestation of the basic structure of $\mathbb{A}_K^1/K^\times$.

We close this discussion by a recapitulation of the role of the adèles in this argument, and, more generally, in contemporary number theory. Like many mathematical innovations, the use of the adèles at first seems to be a merely technical device, which one becomes used to over time; and which one might eventually regard as natural and even intuitive, just through habit alone. There is certainly an aspect of this with regard to the use of the adèles, and I believe that it reflects the attitude of many, perhaps even most, students of number theory upon encountering them for the first time. And the adèles do have many advantages, which appear purely technical at first; e.g. as we already indicated (in footnote 16), one such advantage is that we get to replace the ring $\mathbb{Z}$ with the field $\mathbb{Q}$. It seems, then, worthwhile to reflect on what fundamental number theoretic ideas and intuitions might be being captured in the concept of the adèles, beyond their purely technical advantages. We conclude with a short series of remarks in this spirit.

The product structure of the adèles, a *structural* product in the sense of modern algebra, reflects in a subtle but important way the *literal, arithmetic* product that appears in the statement of the fundamental theorem of arithmetic (that any natural number factors uniquely as a product of prime powers). The *restricted* nature of the product reflects the boundedness of the denominator of any rational number. And even though $\mathbb{Q}$ is a field, and so in some sense has no intrinsic arithmetic (unlike the ring $\mathbb{Z}$), the primes are still present and making their influence felt, once we consider the embedding $\mathbb{Q} \hookrightarrow \mathbb{A}$. One might summarize these points by saying that the embedding $\mathbb{Q} \hookrightarrow \mathbb{A}$ "externalizes" the traditional arithmetic concepts of prime number, unique factorization, and boundedness of denominators, and reinterprets them in a framework that more comfortably accommodates the mores of modern algebra, which require one to manipulate rings and fields themselves, rather than their individual elements.[24]

15.4 Class Field Theory

We now begin our discussion of *reciprocity*, by treating the so-called *abelian* case, which is encapsulated in the statements of class field theory. The history of class field theory is such a vast topic that we are unable to do it any justice here.[25] It has its origins in a multitude of topics from nineteenth-century algebraic number theory, such as Gauss's theory of quadratic forms, the theory of quadratic and higher reciprocity laws (studied first by Gauss, then by Eisenstein, Jacobi, and Kummer, among others), much of the theory of cyclotomic fields that Kummer developed in his study of higher reciprocity, the theory of complex multiplication elliptic curves as studied by Kronecker

---

[24] One might regard the passage from elements of sets to sets themselves, in one's view of what is an "atomic" mathematical object, as a first step in the process of *categorification*, in which the passage from sets or other mathematical objects to the categories of which they are constituents, then to higher categories, and so on, make up the subsequent steps. The transition from Hecke's viewpoint to that of Tate's thesis can then be regarded as just one step in the ongoing evolution of number theory in general and the Langlands program in particular. Nevertheless, however much the ideas and problems apparently transform, the connection to the concrete mathematics, not just of Hecke, but of Euler, Riemann, and Dirichlet, persists, since the inner nature of the number theory remains the same.

[25] We refer the reader interested in the history to the survey of Hasse [Has67].

(among others); and also the relationship between algebraic number theory and
L-functions brought to light by Dirichlet's proof of his theorem on primes in
arithmetic progressions. This last topic will be the starting point for our own
discussion of the theory.

15.4.1 Comparing Dirichlet and Dedekind

As is well known, and as we described above, the crux of Dirichlet's proof
of his theorem on primes in arithmetic progression is the nonvanishing of the
special values $L(1, \chi)$ when χ is a nontrivial character mod N, and the difficult
case of this result occurs when χ is a primitive quadratic (i.e. real-valued, i.e.
± 1-valued) character.

There is a bijection between nontrivial primitive quadratic characters
and quadratic extensions K of $\mathbb{Q}$, given by associating to χ the field
$K = \mathbb{Q}(\sqrt{\chi(-1)N})$. What Dirichlet shows is that

$$L(1, \chi) = \begin{cases} \dfrac{2Rh}{\sqrt{|D|}} & \text{if } \chi(-1) = 1 \\[2ex] \dfrac{2\pi h}{w\sqrt{|D|}} & \text{if } \chi(-1) = -1, \end{cases} \tag{15.24}$$

where h, etc., are the usual invariants attached to K, as in our discussion
of (15.19). Indeed, a moment's consideration shows that the right-hand
expression in this formula *is* the value on the right-hand side of the class
number formula (15.19) (for the particular quadratic field K at hand). Clearly,
there is a relationship between the two formulas!

Let us consider $\zeta_K(s)$. To reduce the number of symbols on the page, and
remembering that $\chi(-1)$ is simply a sign, we write $\pm N$ in place of $\chi(-1)N$
(the particular choice of sign of course being dictated by the value $\chi(-1)$).
The field $K = \mathbb{Q}(\sqrt{\pm N})$ is Galois over $\mathbb{Q}$, with Galois group of order 2; we
identify its Galois group with $\{\pm 1\}$. For each unramified prime in K, i.e. for
each $p \nmid N$, there is a Frobenius element $\sigma(p) \in \{\pm 1\}$, which determines
the splitting behavior of p in K: it equals $+1$ (respectively -1) if p splits
completely (respectively remains inert) in K. If p is odd, then this splitting
behavior depends just on the question of whether or not $\pm N$ is a square mod
p, and so we find that

$$\sigma(p) = \left(\frac{\pm N}{p} \right)$$

(the right-hand side denoting the Legendre symbol). We can now compute
$\zeta_K(s)$ via its Euler product (15.18); if p is ramified, i.e. if $p|N$, then there
is one prime of K above p, having norm p; if p is unramified and split

there are two such; and if p is unramified and inert then there is one prime above p of norm p^2. The corresponding factors in (15.18) multiply out to either

$$\left(1-\frac{1}{p^s}\right)^{-1}, \quad \left(1-\frac{1}{p^s}\right)^{-2}, \quad \text{or} \quad \left(1-\frac{1}{p^{2s}}\right)^{-1}$$

$$= \left(1-\frac{1}{p^s}\right)^{-1}\left(1+\frac{1}{p^s}\right)^{-1},$$

depending on which case we are in. The latter two cases admit the uniform description

$$\left(1-\frac{1}{p^s}\right)^{-1}\left(1-\frac{\sigma(p)}{p^s}\right)^{-1},$$

and so we obtain the formula

$$\zeta_K(s) = \prod_{p \text{ prime}} \left(1-\frac{1}{p^s}\right)^{-1} \times \prod_{\substack{p \text{ prime} \\ p\nmid N}} \left(1-\frac{\sigma(p)}{p^s}\right)^{-1}$$

$$= \zeta_{\mathbb{Q}}(s) \times \prod_{\substack{p \text{ prime} \\ p\nmid N}} \left(1-\frac{\sigma(p)}{p^s}\right)^{-1}, \tag{15.25}$$

where we have used (15.5) to rewrite the first of the two products over primes. If we recall that $\zeta_{\mathbb{Q}}(s)$ has a simple pole with residue 1 at $s = 1$ (the specialization of (15.19) to the case of the number field $\mathbb{Q}$), then we may rewrite the class number formula for K in the form

$$\lim_{s \to 1} \prod_{\substack{p \text{ prime} \\ p\nmid N}} \left(1-\frac{\sigma(p)}{p^s}\right)^{-1} = \begin{cases} \dfrac{2Rh}{\sqrt{|D|}} & \text{if } \chi(-1) = 1 \\[2ex] \dfrac{2\pi h}{w\sqrt{|D|}} & \text{if } \chi(-1) = -1. \end{cases} \tag{15.26}$$

Most readers who have made it through to this point no doubt know where we are heading, and those few who don't can surely guess: the key to Dirichlet's formula (15.24) is to identify the Euler product appearing in (15.26) with $L(s,\chi)$. But, taking into account the Euler product formula of (15.15), this amounts to verifying an identity of Euler products

$$\prod_{\substack{p \text{ prime} \\ p\nmid N}} \left(1-\frac{\chi(p)}{p^s}\right)^{-1} = \prod_{\substack{p \text{ prime} \\ p\nmid N}} \left(1-\frac{\sigma(p)}{p^s}\right)^{-1}, \tag{15.27}$$

which in turn amounts to verifying an identity of individual Euler factors, which itself comes down to establishing the formula[26]

$$\chi(p) = \sigma(p) = \left(\frac{\pm N}{p}\right); \qquad (15.28)$$

but if we recall that χ is a primitive quadratic character modulo N, then this formula reduces to the law of quadratic reciprocity (together with its supplementary laws, if N is even). We sketch a proof of this law in Section 15.4.3 below.

This is the crux of reciprocity: a Diophantine quantity, such as $\sigma(p)$, which we may think of as being defined by the condition

$$1 + \sigma(p) = \text{ the number of solutions to the equation } x^2 = \pm N \bmod p,$$

is related to an "automorphic" quantity, in this case the character χ. ("Automorphic" because it can be interpreted as an idèle class character, i.e. as a character of $GL_1(\mathbb{Q}) \setminus GL_1(\mathbb{A}_{\mathbb{Q}})$.)

And so we obtain an interpretation of the notion of reciprocity that goes well beyond the evident "switching p and N" meaning, which is at the origin of its use in the term "quadratic reciprocity": reciprocity is an identification of one kind of L-function, the *automorphic* kind with which we already have some familiarity, with a quite different kind of L-function, defined by Euler products such as the one appearing in (15.26) above, which are defined not by mod N characters or the like, but in terms of Frobenius elements of Galois groups!

In the subsequent sections we turn to a detailed discussion of these new kinds of L-functions (one might reasonably call them "Diophantine," but in fact the traditional label is "motivic"); but before doing that, we explain how the preceding example can be generalized; this is the content of class field theory.

15.4.2 Class Field Theory (I)

If we take into account the identification of Euler products (15.27), then we may rewrite the factorization (15.25) in the simple form

$$\zeta_K(s) = \zeta_{\mathbb{Q}}(s)L(s, \chi). \qquad (15.29)$$

As we have seen, it is this factorization that yields Dirichlet's class number formula (15.24).[27] Now the Dedekind ζ-function $\zeta_K(s)$ may be thought of

---

[26] For odd primes; we ignore the details related to the special case $p = 2$ when N is odd.

[27] We note that Dirichlet's theorem $L(1, \chi) \neq 0$ follows immediately from this factorization, since we know that each of the two ζ factors have a simple pole at $s = 1$. It is also interesting to compare the functional equation for $L(s, \chi)$ that one obtains from this factorization of $\zeta_K(s)$ with the functional equation that one obtains directly from Hecke's theory. Indeed, since

as *either* automorphic or motivic, from the perspective of the field K (for any number field K): it is the Hecke L-function for the trivial character on $K^\times \backslash \mathbb{A}_K^\times$ (and so automorphic) but it admits the Euler product

$$\prod_{\mathfrak{p}} \left(1 - \frac{1}{N(\mathfrak{p})^s} \right)^{-1},$$

and the coefficient 1 that we have emphasized (in order to make the analogy with the coefficients $\sigma(p)$ considered above) counts the number of solutions to the (extremely trivial!) equation $x = 0$ in the residue field at $\mathfrak{p}$. (Just to be clear, this equation has 1 (!) solution in any field.) This makes $\zeta_K(s)$ motivic.

However, from the perspective of $\mathbb{Q}$, the L-function $\zeta_K(s)$ is inexorably *motivic* – as we already saw in (15.25), its Euler factors (when we index them by rational primes p, rather than by the primes $\mathfrak{p}$ of K) relate to counting the number of solutions to a quadratic equation mod p. What (15.29) shows is that $\zeta_K(s)$ is nevertheless also automorphic from the perspective of $\mathbb{Q}$: it can be written as a product of Hecke L-functions *for characters of* $\mathbb{Q} \backslash \mathbb{A}_{\mathbb{Q}}^\times$.

It is natural to ask: for which other extensions K of $\mathbb{Q}$ does $\zeta_K(s)$ have this property? More generally, we can ask: for which extensions K/F of number fields can $\zeta_K(s)$ be written as a product of Hecke L-functions with respect to idèle class characters of F? Class field theory answers these questions.

Theorem 15.4.2.1 (Main theorem of class field theory, in terms of L-functions) *If K/F is an extension of number fields, then $\zeta_K(s)$ may be written as a product of Hecke L-functions with respect to idèle class characters of F if and only if K/F is Galois with abelian Galois group. Furthermore, the L-functions that appear in the factorization are ray class L-functions, and any ray class L-function over F appears in such a factorization (for an appropriately chosen abelian extension K of F).*

We elaborate on the statement of this theorem. Suppose to begin with that K/F is an abelian Galois extension, with Galois group G. We begin by generalizing the factorization (15.25). If $\chi : G \to \mathbb{C}^\times$ is *any* character, let $H_\chi \subseteq G$ denote its kernel, and let E_χ denote the subfield of K fixed by H_χ (so, by Galois theory, χ induces an embedding $\mathrm{Gal}(E_\chi/F) \hookrightarrow \mathbb{C}^\times$). Let S_χ denote the set of primes in F which ramify in E_χ. Given a prime $\mathfrak{p} \notin S_\chi$, we

---

each of $\zeta_{\mathbb{Q}}(s)$ and $\zeta_K(s)$ have a sign of $+1$ in their functional equations, we find that also $\Lambda(s, \chi) = \Lambda(1 - s, \chi)$ for the quadratic character χ; i.e. the global root number for $L(s, \chi)$ is also equal to $+1$. If one compares this with the formula (15.23) and computes the product of the local root numbers explicitly, one recovers Gauss's celebrated theorem on the sign of the quadratic Gauss sum (another example of unity and unification!). Indeed, the theory of ϵ factors is a modern descendent of that classical line of enquiry by Gauss. See Section 15.5.2 for a discussion of one aspect of this theory.

have a Frobenius element[28] $\mathrm{Frob}_{\mathfrak{p}} \in \mathrm{Gal}(E_\chi/F)$, which is taken, via χ, to an element $\chi(\mathrm{Frob}_{\mathfrak{p}}) \in \mathbb{C}^\times$. It is now an exercise, generalizing the computations leading to (15.25), to show that

$$\zeta_K(s) = \prod_{\chi} \prod_{\substack{\mathfrak{p} \text{ a prime of } F \\ \mathfrak{p} \notin S_\chi}} \left(1 - \frac{\chi(\mathrm{Frob}_{\mathfrak{p}})}{N(\mathfrak{p})^s}\right)^{-1}. \qquad (15.30)$$

For example, the factor corresponding to the trivial character is precisely $\zeta_F(s)$. If K/F is quadratic, then there is a unique nontrivial character $\chi : G \xrightarrow{\sim} \{\pm 1\}$, so (15.30) has just two factors. If we further assume that $F = \mathbb{Q}$, then (15.30) reduces to (15.25). (The quantity denoted there by $\sigma(p)$ coincides with the quantity $\chi(\mathrm{Frob}_p)$ of the present discussion.)

The first statement in Theorem 15.4.2.1 is then proved by establishing a generalization of the equality (15.27) – one proves that each of the various Euler products appearing in (15.30), labelled by the various characters χ of the Galois group, is in fact a Hecke L-function for some ray class character of F. In fact, one proves a more structural statement – one shows that, in an appropriate sense, the character χ *is* a ray class character, in such a way that $\chi(\mathrm{Frob}_{\mathfrak{p}})$ just becomes $\chi(\mathfrak{p})$. The Euler products in (15.30) will then literally be Euler products for ray class L-functions, and the factorization of $\zeta_K(s)$ as a product of such L-functions will be established.

15.4.3 Cyclotomic Fields

We begin by explaining a special case of the preceding discussion. Let $N > 0$ be a natural number, and set $K = \mathbb{Q}(\zeta_N)$. Then the irreducibility of the Nth cyclotomic polynomial implies that there is a canonical isomorphism $\mathrm{Gal}(K/\mathbb{Q}) \xrightarrow{\sim} (\mathbb{Z}/N\mathbb{Z})^\times$, given as follows: an element $a \in (\mathbb{Z}/N\mathbb{Z})^\times$ is identified with the Galois automorphism that maps ζ_N to ζ_N^a. If $p \nmid N$, so that K is unramified above p, then one sees that Frob_p corresponds to the the residue class of p itself. The characters of $\mathrm{Gal}(K/\mathbb{Q})$ are thus identified with the characters of $(\mathbb{Z}/N\mathbb{Z})^\times$.

A character χ has conductor M (i.e. is a primitive character mod M, for some divisor M of N) if and only if E_χ is contained in $\mathbb{Q}(\zeta_M)$, but in no smaller cyclotomic subfield of K. One then sees that E_χ is ramified at precisely

---

[28] Typically, the Frobenius elements $\mathrm{Frob}_{\mathfrak{q}} \in \mathrm{Gal}(E/F)$, for a Galois extension of number fields, are associated to unramified primes $\mathfrak{q}$ of the extension field E. Any two $\mathfrak{q}$ lying over a given unramified prime $\mathfrak{p}$ of F are conjugate by some element of $\mathrm{Gal}(E/F)$, implying that the corresponding Frobenius elements in $\mathrm{Gal}(E/F)$ are conjugate (in the sense of group theory); we write $\mathrm{Frob}_{\mathfrak{p}}$ to denote this conjugacy class (or, sometimes, some choice of element in it). If $\mathrm{Gal}(E/F)$ is *abelian* – which is the case for $E = E_\chi$ – then conjugacy classes in $\mathrm{Gal}(E/F)$ are singletons, so that $\mathrm{Frob}_{\mathfrak{p}}$ is indeed a well-defined element.

the primes dividing M, and so, identifying the Galois character χ with the corresponding character of $(\mathbb{Z}/M\mathbb{Z})^\times$, we find that

$$\prod_{p \notin S_\chi} \left(1 - \frac{\chi(\mathrm{Frob}_p)}{p^s}\right)^{-1} = \prod_{p \nmid M} \left(1 - \frac{\chi(p)}{p^s}\right)^{-1} = L(s, \chi).$$

Combining this identification of Euler products with (15.30), we find that

$$\zeta_K(s) = \prod_{\chi \text{ a char. mod } N} L(s, \chi),$$

establishing the claim of Theorem 15.4.2.1 for the cyclotomic extensions $K = \mathbb{Q}(\zeta_N)$ of $\mathbb{Q}$. Additionally, since Dirichlet's L-functions are precisely the ray class L-functions for $\mathbb{Q}$, we have established the second ("Furthermore, ...") statement of Theorem 15.4.2.1 in the case $F = \mathbb{Q}$.

Theorem 15.4.2.1 in general, for $F = \mathbb{Q}$, then comes down to the following celebrated theorem of Kronecker and Weber.

Theorem 15.4.3.1 *Any abelian Galois extension $K/\mathbb{Q}$ is contained in $\mathbb{Q}(\zeta_N)$ for some N.*

Let us sketch a proof of this result in the case of *quadratic* extensions $K/\mathbb{Q}$. Firstly, we note that if $\chi : \mathrm{Gal}\big(\mathbb{Q}(\zeta_N)/\mathbb{Q}\big) = (\mathbb{Z}/N\mathbb{Z})^\times \to \{\pm 1\}$ is a primitive quadratic character, then E_χ is a quadratic subextension of $\mathbb{Q}(\zeta_N)$ ramified precisely at the primes dividing N. Furthermore, it is real or imaginary depending on whether $\chi(-1) = 1$ or -1 (since $-1 \in (\mathbb{Z}/N\mathbb{Z})^\times$ corresponds to complex conjugation in $\mathrm{Gal}\big(\mathbb{Q}(\zeta_N)/\mathbb{Q}\big)$). From the explicit classification of quadratic extensions of $\mathbb{Q}$, this information uniquely determines E_χ, and indeed shows that

$$E_\chi = \mathbb{Q}\big(\sqrt{\chi(-1)N}\big). \tag{15.31}$$

As already noted, as χ runs over all primitive quadratic characters, the fields $\mathbb{Q}\big(\sqrt{\chi(-1)N}\big)$ run over all quadratic extensions of $\mathbb{Q}$. Thus we find that, indeed, every quadratic extension of $\mathbb{Q}$ is contained in some $\mathbb{Q}(\zeta_N)$.[29] Furthermore, from (15.31), and the fact that $\mathrm{Frob}_p \in \mathrm{Gal}\big(\mathbb{Q}(\zeta_N)/\mathbb{Q}\big)$ corresponds to $p \in (\mathbb{Z}/N\mathbb{Z})^\times$, we find that $\mathrm{Frob}_p \in \mathrm{Gal}\big(\mathbb{Q}(\sqrt{\chi(-1)N})/\mathbb{Q}\big) = \{\pm 1\}$ is equal to $\chi(p)$. This establishes the law of quadratic reciprocity (15.28), and illustrates how the apparently more abstract analysis of the Galois theory of abelian extensions can be used to deduce concrete reciprocity laws.

Remarks 15.4.3.2 Certainly $\mathbb{Q}$ admits other abelian extensions than quadratic ones. For example, since there are infinitely many primes $p \equiv 1 \bmod 3$,

---

[29] A more traditional proof of this result uses quadratic Gauss sums to express $\sqrt{\chi(-1)N}$ explicitly as an element of $\mathbb{Q}(\zeta_N)$.

there are infinitely many extensions $\mathbb{Q}(\zeta_p)$ that contain a degree-3 subextension. One might wonder, then, why there is not a more classical reciprocity law which relates to (say) the degree-3 case of Theorem 15.4.3.1 in the same way that the quadratic case is related to quadratic reciprocity.

The reason is that there is no simple family of Diophantine equations that gives rise to a corresponding family of degree-3 abelian extensions of $\mathbb{Q}$ in the same way that the family of equations $X^2 = D$ gives rise to the family of quadratic extensions. Indeed, while we can write down the family of cubic equations $x^3 + ax + b = 0$, such a cubic will typically have Galois group S_3 rather than C_3. In order to have abelian Galois group, we need to impose the additional condition that $-4a^3 - 27b^2$ be a square. Thus parameterizing such cubics amounts to finding rational points on the surface $4a^3 + 27b^2 + c^2 = 0$ – which is not a particularly simple parameter space. In practice, the simplest way to describe abelian extensions of $\mathbb{Q}$ of degree > 2 is as subfields of appropriate cyclotomic fields.

15.4.4 Class Field Theory (II)

In its more Galois-theoretic formulation, class field theory provides an analogue of Theorem 15.4.3.1 for general number fields. More precisely, if F is any number field, and $\mathfrak{m}$ is any conductor in F, then one introduces the notion of a *ray class field* $K_\mathfrak{m}$ of F. This is a finite Galois extension of F satisfying the following properties.

1. $\mathrm{Gal}(K_\mathfrak{m}/F)$ is isomorphic to the ray class group of K of conductor $\mathfrak{m}$;
2. $K_\mathfrak{m}$ is unramified[30] at places of F not dividing $\mathfrak{m}$. Further, if $\mathfrak{p} \nmid \mathfrak{m}$, then the Frobenius element $\mathrm{Frob}_\mathfrak{p} \in \mathrm{Gal}(K_\mathfrak{m}/F)$ is identified with the class of $\mathfrak{p}$ in the ray class group (for an appropriate choice of the isomorphism in (1), which is then pinned down uniquely by this requirement).

Example 15.4.4.1   The cyclotomic field $\mathbb{Q}(\zeta_N)$ is the ray class field of $\mathbb{Q}$ of conductor $N\infty$.

One then proves the following theorem.

Theorem 15.4.4.2 (Main theorem of class field theory, in terms of abelian extensions) *For any choice of conductor $\mathfrak{m}$ in F, a corresponding ray class field $K_\mathfrak{m}$ exists (and is unique up to isomorphism). Further, if K/F is an abelian Galois extension, then K is contained in some ray class field of F.*

---

[30] Here one has to extend the notion of ramification to the infinite places: if a real place of F lifts to a complex place of $K_\mathfrak{m}$, we say it is ramified; otherwise we say it is unramified. Complex places are always unramified.

If we take into account Example 15.4.4.1, then this result generalizes Theorem 15.4.3.1 in an evident sense. We also see (applying (15.30)) that

$$\zeta_{K_{\mathfrak{m}}}(s) = \prod_{\chi} L(s, \chi),$$

where the product is taken over all ray class characters of conductor dividing $\mathfrak{m}$. More generally, if K/F is a subextension of $K_{\mathfrak{m}}$, and if we write $H = \mathrm{Gal}(K_{\mathfrak{m}}/K) \subseteq \mathrm{Gal}(K_{\mathfrak{m}}/F)$, so that H is a subgroup of the ray class group of conductor $\mathfrak{m}$, then

$$\zeta_K(s) = \prod_{\chi \text{ which are trivial on } H} L(s, \chi). \tag{15.32}$$

Thus Theorem 15.4.2.1 follows from Theorem 15.4.4.2.

As the conductor $\mathfrak{m}$ varies, the ray class fields $K_{\mathfrak{m}}$ form a tower of field extensions (if $\mathfrak{m}|\mathfrak{m}'$ then $K_{\mathfrak{m}} \subseteq K_{\mathfrak{m}'}$), and so we may form their union. Theorem 15.4.4.2 shows that this is precisely the maximal abelian Galois extension F^{ab} of F. Thus

$$\mathrm{Gal}(F^{\mathrm{ab}}/F) = \varprojlim_{\mathfrak{m}} \mathrm{Gal}(K_{\mathfrak{m}}/F) = \varprojlim_{\mathfrak{m}} (\text{ray class group of conductor } \mathfrak{m}).$$

Recalling that this last inverse limit is precisely the group of connected components of the idèle class group, we obtain the following reformulation of the main theorem of class field theory.

Theorem 15.4.4.3 (Main theorem of class field theory, in terms of idèles) *There is a canonical isomorphism* $\mathrm{Gal}(F^{\mathrm{ab}}/F) \xrightarrow{\sim} \pi_0(F^{\times} \backslash \mathbb{A}_F^{\times})$.

15.4.4.4 Attributions Class field theory was first proved in general by Takagi [Tak33], although the characterization of ray class fields that he used was different. Class fields were first introduced by Weber, who defined them in terms of the splitting behavior of primes in $K_{\mathfrak{m}}/F$ (but could not prove their existence in general).[31] In his work, Takagi begins with a different definition, related to Weber's as well as to norms;[32] we refer the reader to Hasse's excellent historical survey [Has67] for more details on Takagi's point-of-view

---

[31] Suppose that K is an extension of F in which the primes that split completely are precisely the primes that are trivial in a given ray class group. Then the result described in footnote 12, together with Hecke's density result discussed in Section 15.3.5, shows that $[K : F]$ necessarily equals the order of the ray class group. This argument sketch may give the reader a sense of how one can even begin to relate the splitting behavior in a field to other of its properties.

[32] The role of norms from an extension K down to F is fundamental in the study of class field theory, although since it is not so relevant for our purposes, we have omitted any discussion of it here.

(and on all of the history that we are summarizing here). In our formulation they are characterized by the description of Frobenius elements in the Galois group (the splitting behavior of primes then being a consequence of this description).

The emphasis on Frobenius elements is due to Artin [Art27], who reformulated the theory in terms of his famous reciprocity law. Although none of our various formulations of the main theorems of class field theory states Artin's reciprocity law explicitly, it follows from condition (2) in the definition of the ray class field K_m, together with the statement that any abelian extension of a given ground field F is contained in a ray class field.

The idèlic reformulation is due to Chevalley [Che40].

15.4.5 Algebraic Hecke Characters

Since the ray class characters of a number field K are precisely the finite order idèle class characters, they may also be regarded as the characters of $\pi_0(K^\times \backslash \mathbb{A}_K^\times)$. Theorem 15.4.4.3 then gives a very direct sense to the notion that characters of Galois groups are the same as ray class characters.

What is less obvious from Theorem 15.4.4.3 is that certain *infinite-order* idèle class characters are also related to characters of Galois groups, in a manner which we now explain.

Firstly, among all the characters

$$\theta : (\mathbb{R} \otimes_\mathbb{Q} K)^\times = (\mathbb{R}^\times)^{r_1} \times (\mathbb{C}^\times)^{r_2} \to \mathbb{C}^\times,$$

let us single out those that are *algebraic*, in the sense that they are given by a formula (using obvious notation)

$$(x_1, \ldots, x_{r_1}, z_1, \ldots, z_{r_2}) \mapsto x_1^{n_1} \ldots x_{r_1}^{n_{r_1}} z_1^{p_1} \overline{z}_1^{q_1} \cdots z_{r_2}^{p_{r_2}} \overline{z}_{r_2}^{q_{r_2}},$$

for some integers $n_1, \ldots, n_{r_2}, p_1, q_1, \ldots, p_{r_2}, q_{r_2}$. We then say that an idèle class character χ is *algebraic*[33] if the associated character θ is algebraic in the above sense.

Now let ℓ be a prime,[34] and choose an isomorphism[35] $\iota : \mathbb{C} \xrightarrow{\sim} \overline{\mathbb{Q}}_\ell$.

Let ψ be an algebraic Hecke character, with associated algebraic character $\theta : (\mathbb{R}^\times \otimes_\mathbb{Q} K)^\times \to \mathbb{C}^\times$. Note that $(\mathbb{C} \otimes_\mathbb{Q} K)^\times = (\mathbb{C}^\times)^{r_1} \times (\mathbb{C}^\times \times \mathbb{C}^\times)^{r_2}$,

---

[33] In the original terminology of Weil [Wei56b], such a character is said to be of *type* (A_0).

[34] The use of ℓ, rather than p, in this context is traditional.

[35] Such an isomorphism exists, by the axiom of choice. Its precise nature is not terribly important, and making a choice of ι is largely an expedient. Its main purpose is to allow us to match the collection of embeddings $K \hookrightarrow \mathbb{C}$ with the collection of embeddings $\iota : K \hookrightarrow \overline{\mathbb{Q}}_\ell$, and to do this it suffices to identify the algebraic closures of $\mathbb{Q}$ inside $\mathbb{C}$ with the algebraic closure of $\mathbb{Q}$ inside $\overline{\mathbb{Q}}_\ell$. Such a choice of identification *is* provided by ι, but is a considerably tamer piece of data than ι itself.

and so θ extends to a character

$$(\mathbb{C} \otimes_{\mathbb{Q}} K)^{\times} \to \mathbb{C}^{\times}$$

via the formula (again using obvious notation)

$$(x_1, \ldots, x_{r_1}, z_1, w_1, \ldots, z_{r_2}, w_{r_2}) \mapsto x_1^{n_1} \ldots x_{r_1}^{n_{r_1}} z_1^{p_1} w_1^{q_1} \cdots z_{r_2}^{p_{r_2}} w_{r_2}^{q_{r_2}}.$$

We continue to denote this extension via θ.

The isomorphism ı allows us to regard θ as a character

$$(\overline{\mathbb{Q}}_{\ell} \otimes_{\mathbb{Q}} K)^{\times} \to \overline{\mathbb{Q}}_{\ell}^{\times}.$$

Composing this with the embedding $(\mathbb{Q}_{\ell} \otimes_{\mathbb{Q}} K)^{\times} \hookrightarrow (\overline{\mathbb{Q}}_{\ell} \otimes_{\mathbb{Q}} K)^{\times}$, and remembering that $\mathbb{Q}_{\ell} \otimes_{\mathbb{Q}} K = \prod_{\lambda | \ell} K_{\lambda}$, we obtain a character

$$\theta_{\ell} : \prod_{\lambda | \ell} K_{\lambda}^{\times} \to \overline{\mathbb{Q}}_{\ell}.$$

Furthermore, this character is again *continuous*.[36]

Now define a character $\mathbb{A}_K^{\times} \to \overline{\mathbb{Q}}_{\ell}^{\times}$ via the formula

$$x \mapsto \imath\big(\psi(x)\theta(x)^{-1}\big)\theta_{\ell}(x); \tag{15.33}$$

here $\theta(x)$ is computed by projecting x to its archimedean components, while $\theta_{\ell}(x)$ is computed by projecting x to its components at primes $\lambda | \ell$. This a continuous character, which by its definition and the construction of θ_{ℓ} is trivial on $K^{\times}$. Furthermore, since, in the formula (15.33), we have shifted the "infinite order" aspects contributed by θ from the archimedean primes to the ℓ-adic primes of K, we see that (15.33) is also trivial on the connected part of $\mathbb{A}_K^{\times}$ (as indeed it must be, since it is continuous and takes values in the totally disconnected group $\overline{\mathbb{Q}}_{\ell}^{\times}$). Thus (15.33) defines a continuous character[37]

$$\widetilde{\psi}_{\ell} : \pi_0(K^{\times} \backslash \mathbb{A}_K^{\times}) \to \overline{\mathbb{Q}}_{\ell}^{\times},$$

which by Theorem 15.4.4.3 we may equally well regard as a continuous character

$$\widetilde{\psi}_{\ell} : \mathrm{Gal}(\overline{\mathbb{Q}}/F) \to \overline{\mathbb{Q}}_{\ell}^{\times}. \tag{15.34}$$

Thus the algebraic Hecke character ψ gives rise to a *family*[38] of ℓ-adic characters (15.34). If ψ has conductor $\mathfrak{m}$, then one sees from the construction

---

[36] It is here that we use the assumption that θ is algebraic. Indeed, the isomorphism ı is certainly *not* continuous, since $\mathbb{C}$ and $\overline{\mathbb{Q}}_{\ell}$ have utterly different topological natures. But the character θ is given simply by raising to various integer powers, and such homomorphisms are continuous on *any* topological abelian group.

[37] We use the notation $\widetilde{\psi}_{\ell}$ just to avoid conflicting with our already-established convention, in the particular case when $K = \mathbb{Q}$, of writing ψ_{ℓ} for the local component of ψ at ℓ.

[38] "Family," because we may vary the prime ℓ.

that $\widetilde{\psi}_\ell$ is unramified at primes $\mathfrak{p} \nmid \mathfrak{m}\ell$, and for such a prime $\mathfrak{p}$, one computes[39]

$$\widetilde{\psi}_\ell(\mathrm{Frob}_\mathfrak{p}) = \psi_\mathfrak{p}(\pi_\mathfrak{p}^{-1}), \tag{15.35}$$

where $\psi_\mathfrak{p}$ denotes the local factor of ψ at $\mathfrak{p}$, and $\pi_\mathfrak{p}$ denotes some uniformizer at $\mathfrak{p}$ (the indicated expression being independent of this choice). This value is *independent* of ℓ (it depends just on the original character ψ), and gives meaning to the idea that the $\widetilde{\psi}_\ell$ are a *compatible family* of ℓ-adic Galois characters.

Remarks 15.4.5.1 If ψ is in fact a ray class character, then θ is trivial, the character ψ takes values in $\overline{\mathbb{Q}}^\times$, and $\widetilde{\psi}_\ell$ coincides with ψ (just using $\imath$ to relocate $\overline{\mathbb{Q}}$ from being a subfield of $\mathbb{C}$ to being a subfield of $\overline{\mathbb{Q}}_\ell$). So in this case the family $\widetilde{\psi}_\ell$ is really just the single character $\widetilde{\psi}$, reinterpreted as a finite-order Galois character via Theorem 15.4.4.3.

If, however, θ is nontrivial, then the $\widetilde{\psi}_\ell$ are truly distinct characters as ℓ varies (for example, $\widetilde{\psi}_\ell$ will be infinitely ramified above ℓ, but not above any other rational prime), and the sense in which they are related is a subtle one, mediated by the compatibility condition (15.35).

The notion of compatible families of ℓ-adic Galois representations was introduced by Taniyama [Tan57], and further developed by Serre [Ser68]. It has evolved into a fundamental notion in number theory, and as we will see in what follows, it is the notion that underlies the definition of *motivic* L-functions. The preceding construction allows us to produce compatible families of ℓ-adic characters whose associated L-function is just the Hecke L-function $L(\psi, s)$; thus these are compatible families for which reciprocity is known to hold.

Example 15.4.5.2 The absolute value character $\psi = |\;| : \mathbb{Q}^\times \backslash \mathbb{A}_\mathbb{Q}^\times \to \mathbb{R}_{>0}^\times$ is algebraic, and the associated family of ℓ-adic characters is just the family of ℓ-adic cyclotomic characters $\chi_\ell : \mathrm{Gal}(\overline{\mathbb{Q}}/\mathbb{Q}) \to \mathbb{Z}_\ell^\times$, defined by $\sigma(\zeta) = \zeta^{\chi_\ell(\sigma)}$ for any Galois automorphism σ and any ℓ-power root of unity ζ.

15.4.6 Brauer Groups and Weil Groups

Another aspect of class field theory that we have so far not mentioned is the role of the theory of Brauer groups (over both local and global fields). The reader can consult [Has67] for some indication of the historical importance of Brauer groups in the development of the subject. From the modern cohomological

---

[39] Whether or not one includes the power -1 in the following formula depends on the normalization of the isomorphism in Theorem 15.4.4.3; it is adapted to our (admittedly implicit) choice of normalization. See also footnote 19.

perspective, the construction of the fundamental class in $H^2(\mathrm{Gal}(K/F), K^\times \backslash \mathbb{A}_K^\times)$ is one of the pivotal steps in the entire development of the theory. We may interpret this class as an extension of $\mathrm{Gal}(K/F)$ by $K^\times \backslash \mathbb{A}_K^\times$; we call this extension the *Weil group* $W_{K/F}$. (See [Wei51] for Weil's original construction, and [Tat79] for a presentation along more modern lines.)

Just as $K^\times \backslash \mathbb{A}_K^\times$ relates to $\mathrm{Gal}(K^{\mathrm{ab}}/K)$, the group $W_{K/F}$ also relates to a Galois-theoretic object. Namely, if we let $\mathrm{Gal}(\overline{\mathbb{Q}}/K)^c$ denote the closure of the commutator subgroup of $\mathrm{Gal}(\overline{\mathbb{Q}}/K)$, so that $\mathrm{Gal}(K^{\mathrm{ab}}/K) = \mathrm{Gal}(\overline{\mathbb{Q}}/K)/\mathrm{Gal}(\overline{\mathbb{Q}}/K)^c$, then we have a canonical morphism of short exact sequences of groups

$$
\begin{array}{ccccccccc}
1 & \longrightarrow & K^\times \backslash \mathbb{A}_K^\times & \longrightarrow & W_{K/F} & \longrightarrow & \mathrm{Gal}(K/F) & \longrightarrow & 1 \\
& & \downarrow & & \downarrow & & \| & & \\
1 & \longrightarrow & \mathrm{Gal}(K^{\mathrm{ab}}/K) & \longrightarrow & \mathrm{Gal}(\overline{\mathbb{Q}}/F)/\mathrm{Gal}(\overline{\mathbb{Q}}/K)^c & \longrightarrow & \mathrm{Gal}(K/F) & \longrightarrow & 1
\end{array}
$$

(15.36)

where the left vertical arrow is induced by (the inverse to) the isomorphism of Theorem 15.4.4.3.

If ψ denotes an algebraic Hecke character on $K^\times \backslash \mathbb{A}_K^\times$, then we may induce ψ to obtain a representation $\rho : W_{K/F} \to \mathrm{GL}_n(\mathbb{C})$, if $n = [K : F]$. Correspondingly, we may induce each of the ℓ-adic characters $\widetilde{\psi}_\ell$ of $\mathrm{Gal}(K^{\mathrm{ab}}/K)$ to obtain a representation $\widetilde{\rho}_\ell : \mathrm{Gal}(\overline{\mathbb{Q}}/F) \to \mathrm{GL}_n(\overline{\mathbb{Q}_\ell})$ (which is trivial on $\mathrm{Gal}(\overline{\mathbb{Q}}/K)^c$). The $\widetilde{\rho}_\ell$ give an example of a compatible family of n-dimensional ℓ-adic Galois representations.

We finish this discussion by recalling how to construct the *absolute* Weil group of F. First, we recall that if $E \subseteq K$ with E again Galois over F, then there is a natural surjection $W_{K/F} \to W_{E/F}$, inducing the natural surjection on Galois groups, and the norm map on idèle class groups. In particular, taking $E = F$, and noting that by construction $W_{F/F} = F^\times \backslash \mathbb{A}_F^\times$, we obtain a surjection $W_{K/F} \to F^\times \backslash \mathbb{A}_F^\times$ that identifies the target with the abelianization of its source. We then find that the kernel $W^1_{K/F}$ of

$$
W_{K/F} \to W^{\mathrm{ab}}_{K/F} = F^\times \backslash \mathbb{A}_F^\times \xrightarrow{|\ |} \mathbb{R}_{>0}^\times
$$

sits in a short exact sequence

$$
1 \to K^\times \backslash \mathbb{A}_K^1 \to W^1_{K/F} \to \mathrm{Gal}(K/F) \to 1,
$$

and since $K^\times \backslash \mathbb{A}_K^1$ is compact, so is $W^1_{K/F}$. Thus $W_{K/F}$ is an extension of $\mathbb{R}_{>0}^\times$ by a compact group.

Now, just as we may take an inverse limit to form the profinite group $\text{Gal}(\overline{\mathbb{Q}}/F) = \varprojlim_K \text{Gal}(K/F)$, we may form an inverse limit $W_F = \varprojlim_K W_{K/F}$. This is a locally compact group (in fact, it is an extension of $\mathbb{R}^\times_{>0}$ by $\varprojlim_K W^1_{K/F}$, and the latter group is compact, being an inverse limit of compact groups), which admits a continuous morphism $W_F \to \text{Gal}(\overline{\mathbb{Q}}/F)$. The kernel of this morphism is *connected*; it is an inverse limit, under the norm maps, of the connected parts of all the various $K^\times \setminus \mathbb{A}^\times_K$, and so is an extension of $\mathbb{R}^\times$ by a very complicated solenoid. (See Example 15.3.3.3.)

15.5 Motivic *L*-Functions

Our discussion up to this point essentially completes the story of reciprocity in the *abelian* case (in summary form, to be sure!). As far as chronology is concerned, it takes us up to more-or-less the middle of the twentieth century.

We must now turn to the discussion of *nonabelian* reciprocity, which might also be called *nonabelian* class field theory. The search for such a theory began already in the first half of the twentieth century, but even the conjectural statements and frameworks remained entirely elusive outside of certain special cases, and it was only with Langlands' letter to Weil [Lana], and his elaborations [Lan70] and [Lan71] on this letter, that a general framework for nonabelian reciprocity was proposed. There is no doubt that this is indeed the correct framework within which to consider the problem, although much remains to be proved.

To begin with, we will describe nonabelian *L*-functions on the motivic side, since it is these *L*-functions that were the first to be discovered in the nonabelian context. Indeed, while these *L*-functions exist in abundance, the absolutely fundamental difficulty in making even a conjectural statement of a general nonabelian reciprocity law, prior to Langlands' letter to Weil, was the seeming lack of any kind of *L*-functions on the automorphic side that remotely resembled the creatures littering the zoo of nonabelian motivic *L*-functions.

15.5.1 Artin

Suppose that K/F is a finite Galois extension of number fields, and write $G = \text{Gal}(K/F)$. Artin [Art31] was the first to observe that even when G is nonabelian, the Dedekind ζ-function $\zeta_K(s)$ admits a factorization analogous to (15.30).

Let $\rho : G \to \text{GL}_n(\mathbb{C})$ be a representation. If $\mathfrak{p}$ is a prime of F unramified in K, then the conjugacy class in G of $\text{Frob}_\mathfrak{p}$ is well-defined, and so $\rho(\text{Frob}_\mathfrak{p})$ is a well-defined as a matrix up to conjugation. In particular, its characteristic polynomial is well-defined. Actually, we are more interested in its reciprocal characteristic polynomial $\det\big(\mathrm{Id}_{n\times n} - T\rho(\mathrm{Frob}_{\mathfrak{p}})\big)$, which is a degree-$n$ polynomial in T with constant term 1. We use this latter polynomial to construct an Euler factor

$$\det\big(\mathrm{Id}_{n\times n} - N(\mathfrak{p})^{-s}\rho(\mathrm{Frob}_{\mathfrak{p}})\big)^{-1}; \qquad (15.37)$$

this expression is now the reciprocal of a degree-n polynomial in the quantity $N(\mathfrak{p})^{-s}$. In fact, if $\mathfrak{p}$ is unramified in the splitting field of ρ (i.e. the fixed field of the kernel of ρ) then this same definition makes sense. Note that if ρ is one-dimensional, i.e. just a character χ, then we recover the definition of the Euler factors appearing in the Euler product labelled by χ that appears in the factorization (15.30).

If $\mathfrak{p}$ is ramified in the splitting field of ρ, then the definition of the Euler factor $\mathfrak{p}$ is a little more involved. It will be of degree $< n$, but (since $n > 1$ if ρ is not a character) will typically not be trivial (whereas in the abelian contexts we've considered up till now, the Euler factors at ramified primes have always been trivial.) To define it, we let V denote the vector space $\mathbb{C}^n$ on which G acts via ρ. We also make a choice $I_{\mathfrak{p}} \subset D_{\mathfrak{p}} \subset G$ of inertia and decomposition group at $\mathfrak{p}$ (again well-defined up to conjugation). If $\mathrm{Frob}_{\mathfrak{p}} \in D_{\mathfrak{p}}$ denotes a choice of element lifting the Frobenius element of $D_{\mathfrak{p}}/I_{\mathfrak{p}}$, then the reciprocal characteristic polynomial of $\rho(\mathrm{Frob}_{\mathfrak{p}})$ acting on the invariant subspace $V^{I_{\mathfrak{p}}}$ is well-defined independent of choices, and we form the Euler factor

$$\det\big(\mathrm{Id}_{V^{I_{\mathfrak{p}}}} - N(\mathfrak{p})^{-s}\rho(\mathrm{Frob}_{\mathfrak{p}})_{|V^{I_{\mathfrak{p}}}}\big)^{-1}. \qquad (15.38)$$

Of course, if $\mathfrak{p}$ is unramified in K (or, more generally, in the splitting field of ρ), then $V^{I_{\mathfrak{p}}} = V = \mathbb{C}^n$, and so the two Euler factors (15.38) and (15.37) coincide. In general, the quantity in (15.38) is the reciprocal of a polynomial in $N(\mathfrak{p})^{-s}$ of degree equal to $\dim V^{I_{\mathfrak{p}}}$.

We now define the Artin L-function of ρ as the product over all $\mathfrak{p}$ of these Euler factors, i.e.

$$L(s,\rho) = \prod_{\mathfrak{p}} \det\big(\mathrm{Id}_{V^{I_{\mathfrak{p}}}} - N(\mathfrak{p})^{-s}\rho(\mathrm{Frob}_{\mathfrak{p}})_{|V^{I_{\mathfrak{p}}}}\big)^{-1}. \qquad (15.39)$$

These L-functions are perhaps the first to appear in the literature that are truly *defined* as an Euler product, rather than as a Dirichlet series which is subsequently factored into a product over primes. The usual comparisons with $\zeta(s)$ show that the product defining $L(s,\rho)$ converges to a holomorphic function on the half-plane $\Re s > 1$; and this all one can say about them to begin with.

In the case of a character $\chi : G \to \mathbb{C}^{\times}$, class field theory allows us to identify χ with a ray class character, so that $L(s,\chi)$ (as defined above) coincides with the Hecke L-function $L(s,\chi)$ (as the formula (15.35) shows). In particular, in this case, $L(s,\chi)$ has an analytic continuation and functional equation.

One aspect of defining L-functions for general representations of G, rather than just one-dimensional representations, is that the set of all representations of G has more structure: for example, we can take direct sums of representations, or induce representations from subgroups. (Note that both of these processes tend to output representations of dimension > 1, and so were not visible in the context of the L-functions of characters that we've studied up till now.)

One rather easily verifies the following properties of Artin L-functions:

$$L(s, \rho_1 \oplus \rho_2) = L(s, \rho_1) L(s, \rho_2), \tag{15.40}$$

and, if $H \subset G$ is a subgroup and θ is a representation of H, then

$$L(s, \operatorname{Ind}_H^G \theta) = L(s, \theta). \tag{15.41}$$

This second formula looks a bit paradoxical at first, since the dimensions of the two representations $\operatorname{Ind}_H^G \theta$ and θ itself are different (assuming H is a proper subgroup of G) – the dimension of the former is $[G : H]$ times the dimension of the latter, and so the Euler factors in the Euler product (at least all the unramified primes) are of different degrees! But the Euler products are taken over different collections of primes! The representation $\operatorname{Ind}_H^G \theta$ is a representation of G, and so $L(s, \operatorname{Ind}_H^G)$ is defined as an Euler product over primes of F. On the other hand, θ is a representation of $H = \operatorname{Gal}(K/E)$, if E denotes the fixed field of H, and so $L(s, \theta)$ is defined as an Euler product over the primes of E. Now $[E : F] = [G : H]$, and so degree-d Euler factors indexed by primes in E, if we write them in terms of primes of F, will indeed become of degree $[G : H]d$. Thus we see that (15.41) is related to, indeed formalizes and generalizes, the computations we made to derive the factorization (15.25).

Indeed, let us apply these formulas in the case when $H = \{1\} \subset G$, with θ being the (necessarily) trivial character of H. Then the fixed field of H is just K itself, and the Artin L-function of the trivial character equals the Hecke L-function of the trivial character equals $\zeta_K(s)$. On the other hand, the induction of the trivial character from $\{1\}$ to G is just the regular representation $\mathbb{C}[G]$ of G, and we have the well-known decomposition

$$\mathbb{C}[G] = \bigoplus_\rho \rho^{\oplus \dim \rho},$$

where ρ runs over (a set of isomorphism class representatives of) all the irreducible representations of G. Thus, applying first (15.41) and then (15.40), we find that

$$\zeta_K(s) = \prod_\rho L(s, \rho)^{\dim \rho}, \tag{15.42}$$

where the product is taken over the all the irreducible representations ρ of G.

If G is abelian, then the irreducible representations of G are exactly the various characters of G, and so (15.42) reduces to (15.30), which we now recognize as a factorization of $\zeta_K(s)$ into a product of Artin L-functions. Class field theory then allows us to reinterpret this as expressing $\zeta_K(s)$ (for the abelian extension K of F) as a product of Hecke L-functions for F. The factorization (15.29), essentially due to Dirichlet, gives one of the simplest illustrations of this phenomenon, and already has far-reaching consequences (being the crux of Dirichlet's proof of this theorem, in so far as it leads to the nonvanishing of $L(1,\chi)$).

What are the number-theoretic consequences of the more general factorizations (15.42)? Is there a nonabelian class field theory that gives us an alternative interpretation of the L-functions $L(s,\rho)$ that appear in it? The answers to these questions (some known, most still conjectural) are provided by Langlands's notions of reciprocity and functoriality, to be discussed in the following section.

For now, we state the following celebrated conjecture of Artin.

Conjecture 15.5.1.1 (Artin) *If ρ is an irreducible representation of G distinct from the trivial character, then $L(\rho,s)$ admits a holomorphic continuation to the complex plane and satisfies a functional equation relating (a suitably completed form of) $L(\rho,1-s)$ and (a suitably completed form of) $L(\rho^\vee,s)$. (Here $\rho^\vee$ denotes the contragredient to ρ.)*

Example 15.5.1.2 We explain one concrete consequence of Artin's conjecture. If E/F is an arbitrary finite extension, let K denote the Galois closure of E, and write $H = \mathrm{Gal}(K/E) \subseteq \mathrm{Gal}(K/F) = G$. If θ denotes the trivial character of H, then $\mathrm{Ind}_H^G \theta$ is a transitive permutation representation, and so contains exactly one copy of the trivial character of G as direct summand. Thus we find that

$$\zeta_E(s) = L(s,\theta) = L(s, \mathrm{Ind}_H^G \theta) = \zeta_F(s) \prod_\rho L(\rho,s), \qquad (15.43)$$

where ρ ranges over the collection of nontrivial irreducible direct summands of $\mathrm{Ind}_H^G \theta$. Thus, if Artin's conjecture is true, we find that $\zeta_E(s)/\zeta_F(s)$ is an entire function, or (more expressively), that $\zeta_F(s)$ *divides* $\zeta_E(s)$.

One approach to studying Artin's conjecture, initiated by Artin himself and further pursued by Brauer, is to attempt to express representations of G in terms of inductions of characters (i.e. one-dimensional representations) of subgroups of G. This becomes a group-theoretic problem, via the study of which Brauer was led to prove [Bra47a] the following result.

Theorem 15.5.1.3 (Brauer) *If ρ is a nontrivial irreducible representation of $G = \mathrm{Gal}(K/F)$, then $L(\rho,s)$ is a product of integral powers of Hecke L-functions (attached to Hecke characters of various fields intermediate*

between F and K). In particular, L(ρ, s) has a meromorphic *continuation to
the entire complex plane, and satisfies a functional equation of the expected
form.*

Artin had already proved the corresponding statement with rational rather
than integer powers, thus obtaining an a priori multivalued meromorphic
continuation. The integral powers appearing in Brauer's theorem can be
negative in general, and so it does not suffice to prove the *holomorphic*
continuation that Artin conjectured. For particular (not necessarily irreducible)
ρ, variants are possible; for example, in some contexts one can obtain powers
that are positive rational numbers. Combined with Brauer's general theorem,
this suffices to prove that $L(s, \rho)$ is entire. For example, in this way Brauer
proved [Bra47b] the divisibility discussed in Example 15.5.1.2 when E/F is
already Galois.

15.5.2 Epsilon Factors

There is a subtle but important point that arises with regard to the functional
equation for $L(s, \rho)$ that comes out of Brauer's theorem. It indeed has the
"expected" form (as formulated by Artin – it is a natural generalization of
the form of the functional equation in the abelian case, involving Γ-factors for
the archimedean places, and a power of the discriminant of K and also of the
conductor of ρ; the notion of conductor having been formulated by Artin with
exactly this purpose in mind). But the root number in the equation comes as a
single quantity; it does not appear as a product of local factors as abelian root
numbers do (via (15.23)).

It is natural to hope that such a factorization of the nonabelian root numbers
should exist, especially if one anticipates that Artin L-functions, and Artin's
conjecture, are to be explained by a nonabelian class field theory. Versions of
this problem were studied by Hasse [Has54] and by Dwork [Dwo56], among
others; and Dwork succeeded in constructing a local root number "up to a
sign," in an appropriate sense.

The difficulties in studying this problem are two-fold. Firstly, there is the
problem of characterizing the local ϵ factors. One imagines that they should
satisfy analogues of the properties (15.40) and (15.41); given this, one can follow Brauer's proof of Theorem 15.5.1.3 and show that these properties allow
one to uniquely compute the local root numbers, assuming they exist. The
second, and seemingly more serious, difficulty is that the product appearing in
Brauer's theorem is not at all unique, and so correspondingly, the local root
numbers whose computation we have just described are not at all obviously
well-defined. In fact, if one adapts (15.41) in the most naive manner, it turns out
the local root numbers can *only* be well-defined up to a sign, and this explains
the limitation on Dwork's result.

The problem was subsequently taken up by Langlands, who realized that the key to finding the correct construction of the local root numbers was to adapt (15.41) in a more subtle manner.[40] With the correct characterization of the local ϵ factors in hand, Langlands proceeded, in his manuscript [Lanc], to show that they are well-defined. Again, this required dealing with the nonuniqueness in Brauer's theorem; the necessary calculations are famously elaborate, and make [Lanc] very long. By employing Langlands's characterization, Deligne soon after found a more indirect, but shorter, local–global argument that establishes the well-definedness [Del73].

15.5.3 Tschebotareff

We recall an important consequence of Brauer's Theorem 15.5.1.3. Namely, an appropriate generalization of Hadamard and de la Vallée Poussin's work shows that Hecke L-functions are zero-free on the closed half-plane $\Re s \geq 1$. Thus the ratio of Hecke L-functions appearing in Brauer's theorem is holomorphic in the same half-plane. This is enough to deduce an analogue of the Prime Number Theorem, which in this context proves that as $\mathfrak{p}$ ranges over the primes in F (excluding the finitely many primes that ramify in K), the Frobenius elements $\mathrm{Frob}_\mathfrak{p}$ are equidistributed in $\mathrm{Gal}(K/F)$. More precisely, since $\mathrm{Frob}_\mathfrak{p}$ is only well-defined up to conjugacy, the asymptotic proportion of $\mathfrak{p}$ for which $\mathrm{Frob}_\mathfrak{p}$ lies in a given conjugacy class is proportional to the size of the class; in particular, there are an infinitude of $\mathfrak{p}$ for which $\mathrm{Frob}_\mathfrak{p}$ lies in a given conjugacy class. Note that if K/F is abelian, and thus contained in a ray class field, this reduces to Hecke's theorem regarding primes lying in a given ray class. (And the proof becomes Hecke's proof.)

This result was first proved by Tschebotareff [Tsc26], using a different argument, building on earlier partial results of Frobenius (see, for example, the result mentioned in footnote 12 above for the simplest such precursor). Tschebotareff's argument was the inspiration for Artin's proof of his reciprocity law.

15.5.4 Motivic ζ-Functions

If we recall the definition of the Dedekind ζ-function of a number field K as a Dirichlet series $\zeta_K(s) = \sum_\mathfrak{a} \frac{1}{N(\mathfrak{a})^s}$, where $\mathfrak{a}$ runs over all nonzero ideals of $\mathcal{O}_K$, an obvious generalization suggests itself. Namely, if A is any ring (commutative, and with a unit), one could attempt to imitate this definition so as to obtain a ζ-function attached to A, by taking a sum over all *cofinite* ideals in A (i.e. ideals $\mathfrak{a}$ for which $A/\mathfrak{a}$ is finite; the point being that, in

---

[40] We don't go into the details here, but the key point is that local ϵ factors are literally induction invariant only for the induction of virtual representations of degree 0.

the number field case, $N(\mathfrak{a}) = |\mathcal{O}_K/\mathfrak{a}|$). However, there is a serious defect with this definition: ignoring for now the question of convergence of such a series (which would be related to bounding the number of ideals $\mathfrak{a}$ in A with $|A/\mathfrak{a}| = n$, as n grows), we see that, since there is no unique factorization of ideals into prime ideals in a general ring,[41] there is no reason in general for a ζ-function, so defined, to have an Euler product; and hence there is no reason (at least for our purposes) to imagine that this proposed definition is useful!

Indeed, we immediately abandon the momentary proposal of the preceding paragraph. Instead, we will insist that ζ-function of A (for suitable A) has an Euler product, and we will enforce this by definition! That is, we will imitate not the description of $\zeta_K(s)$ as a series, but its description as an Euler product. Any cofinite prime ideal in A is certainly maximal, and the natural condition to place on A (which will also lets us control the counting problem related to convergence) is that it be finitely generated as a $\mathbb{Z}$-algebra. Indeed, the Nullstellensatz then implies that any maximal ideal of A is cofinite, and so we define

$$\zeta_A(s) = \prod_{\substack{\mathfrak{p} \text{ maximal in } A}} \left(1 - \frac{1}{|A/\mathfrak{p}|^s}\right)^{-1}. \qquad (15.44)$$

Any cofinite ideal of A has a positive residue characteristic, and so (assuming some half-plane of convergence) we may rearrange the product in (15.44) to write

$$\zeta_A(s) = \prod_{p \text{ prime}} \prod_{\substack{\mathfrak{p} \text{ max. in } A \\ \text{of res. char. } p}} \left(1 - \frac{1}{|A/\mathfrak{p}|^s}\right)^{-1}. \qquad (15.45)$$

Each Euler factor in the product labeled by the prime number p is of the form $\left(1 - \frac{1}{p^{fs}}\right)^{-1}$, if $A/\mathfrak{p}$ is a finite field of order p^f. The difference between this general setting and the case when $A = \mathcal{O}_K$ is that there might be infinitely many such factors for each prime p!

There is an important geometric perspective on this construction: maximal ideals in A are the same as closed points in Spec A. In this optic, we can generalize the preceding construction to any finite-type scheme X over Spec $\mathbb{Z}$.

---

[41] Essentially the only interesting and general context in which we have unique factorization of cofinite ideals into cofinite prime ideals is when A is a product of finitely many Dedekind domains which are of finite type over $\mathbb{Z}$. Working on each Dedekind domain factor separately, and so assuming that A *is* a Dedekind domain, we then find that either $A = \mathcal{O}_K[1/a]$ for some number field K and some $a \in K$, or that Spec A is a smooth curve over a finite field; we will discuss this latter case in Section 15.5.5.

If X is such a scheme, write $\kappa(x)$ for the residue field of a point $x \in X$; then we can define[42]

$$\zeta_X(s) = \prod_{x \in X \text{ closed}} \left(1 - \frac{1}{|\kappa(x)|^s}\right)^{-1}. \tag{15.46}$$

Since the underlying topological space of X, and the residue fields at its closed points, are insensitive to any nilpotent elements in the structure sheaf of X, we may certainly assume that X is reduced when we analyze $\zeta_X(s)$. (In the affine case when $X = \operatorname{Spec} A$, this amounts to quotienting out A by its nilradical, and so supposing that A contains no nonzero nilpotents.) Also, if we write $X = Y \coprod Z$ as the disjoint union of two subschemes, then clearly

$$\zeta_X(s) = \zeta_Y(s)\zeta_Z(s). \tag{15.47}$$

Since any reduced scheme of finite type over $\operatorname{Spec} \mathbb{Z}$ may be stratified into a disjoint union of integral affine locally closed subschemes, in analyzing some of the coarser properties of $\zeta_X(s)$, e.g. its convergence properties, we may assume that $X = \operatorname{Spec} A$ with A an integral domain.

Momentarily suppose, then, that A is an integral domain of finite type over $\mathbb{Z}$. Either A is torsion free, in which case A contains maximal ideals of residue characteristic p for all but finitely many primes p, or A is of characteristic p for some fixed prime p. In the second case, a rough count of closed points on $\operatorname{Spec} A$ (e.g. via a Noether normalization to compare $\operatorname{Spec} A$ with an affine space of the same dimension) shows that $\zeta_A(s)$ converges if $\Re s > \dim A$. In the first case, we find that $\operatorname{Spec} A/p$ is either empty (if p is invertible in A, which happens only for a finite number of p) or of dimension equal to $\dim A - 1$. Inputting the estimates from our analysis of the second (i.e. positive characteristic) case into the product over (all but finitely many) p, we again find that $\zeta_A(s)$ converges if $\Re s > \dim A$. Working backwards through the stratification argument, we then find, for any scheme X of finite type over $\mathbb{Z}$, that $\zeta_X(s)$ converges if $\Re s > \dim X$.

Note that the preceding argument provides an important geometric interpretation of the first product (the product over p) in (15.45). If we consider the more general case of a finite type $\mathbb{Z}$-scheme X, then we may break up the product in (15.46) into a product over primes p of a product over all closed

---

[42] The author first learnt this definition from the note [Ser65] of Serre. The author's impression is that this definition developed via a complex and somewhat organic process over several decades, beginning with the function field analogue of Dedekind ζ-functions of quadratic field in Artin's thesis [Art24], progressing through the work of several others, especially Hasse and Weil, until by the 1960s, this emerged as the evidently natural definition. We strongly recommend [Ser65] to the reader, as well as Tate's article [Tat65], which immediately follows it in the same volume; our discussion here owes much to both papers.

points x of residue characteristic p. Since these latter points coincide with the closed points in the base-change $X_{/\mathbb{F}_p}$, we find that

$$\zeta_X(s) = \prod_p \prod_{\substack{x \in X_{/\mathbb{F}_p} \\ \text{closed}}} \left(1 - \frac{1}{|\kappa(x)|^s}\right)^{-1} = \prod_p \zeta_{X_{/\mathbb{F}_p}}(s). \qquad (15.48)$$

This suggests that in order to analyze $\zeta_X(s)$ for general X, we should first consider the case when X lies over a particular finite field, say $\mathbb{F}_p$.

Remarks 15.5.4.1 How do these considerations connect to number theory? Well, a finite type $\mathbb{Z}$-algebra A is just a quotient

$$A = \mathbb{Z}[x_1, \ldots, x_n]/(f_1, \ldots, f_r)$$

for some variables $x_1, \ldots, x_n$, and some polynomials f_i (with integer coefficients) in these variables. Maximal ideals of A, or equivalently, closed points of $X = \operatorname{Spec} A$, correspond (more or less – we will see the precise relationship soon) to solutions to the equations

$$f_1 = \cdots = f_n = 0 \qquad (15.49)$$

over some finite field $\mathbb{F}_q$. Maximal ideals of fixed residue characteristic p then correspond to such solutions with q taken to run through the powers of some fixed p. So $\zeta_{X_{/\mathbb{F}_p}}(s)$ encodes information about solutions to the system of simultaneous Diophantine equations (15.49) modulo the given prime p, while $\zeta_X(s)$ then packages this information together as the prime p varies.

15.5.5 ζ-Functions for Varieties over Finite Fields

We very briefly recall the theory of ζ-functions of varieties over finite fields. This is an enormous topic in its own right, which is the subject of Weil's celebrated series of conjectures [Wei49] – now theorems of Dwork, Grothendieck (and his collaborators), and Deligne.

Suppose that X is of finite type over $\mathbb{F}_p$. We first recall the precise relationship between $\zeta_X(s)$ and counting solutions to Diophantine equations. If X is cut out by some equations, then giving an $\mathbb{F}_{p^n}$-valued solution to the equations cutting out X is the same as giving a morphism $\operatorname{Spec} \mathbb{F}_{p^n} \to X$, i.e. an $\mathbb{F}_{p^n}$-valued point of X – as usual, we denote this set of solutions, or equivalently, morphisms, by $X(\mathbb{F}_{p^n})$. Such a morphism has a closed image, say $x \in X$, and so determines, and is determined by, an embedding $k(x) \hookrightarrow \mathbb{F}_{p^n}$.

Such an embedding exists if and only if $k(x) = \mathbb{F}_{p^m}$ for some $m \mid n$, and the number of such embeddings is then equal to m (the order of the automorphism group of $\mathbb{F}_{p^m}$). From these remarks, we compute that

$$\log \zeta_X(s) = \sum_{n=1}^{\infty} \frac{|X(\mathbb{F}_{p^n})|}{np^{ns}}. \tag{15.50}$$

Example 15.5.5.1   If $X = \mathbb{A}^d_{/\mathbb{F}_p}$ (affine d-space over $\mathbb{F}_p$), then (15.50) shows that

$$\zeta_{\mathbb{A}^d_{/\mathbb{F}_p}}(s) = \frac{1}{(1 - p^{d-s})}.$$

If $X = \mathbb{P}^d_{/\mathbb{F}_p}$ (projective d-space of $\mathbb{F}_p$), then writing $\mathbb{P}^d_{/\mathbb{F}_1}$ as a union of $\mathbb{A}^d_{/\mathbb{F}_p}$ and a copy of $\mathbb{P}^{d-1}_{/\mathbb{F}_p}$ at infinity in the usual way, taking into account (15.47), and arguing inductively, we find that

$$\zeta_{\mathbb{P}^d_{/\mathbb{F}_p}}(s) = \frac{1}{(1 - p^{-s})(1 - p^{1-s}) \cdots (1 - p^{d-s})}.$$

As we noted in footnote 41 above, if A is the ring of regular functions on a smooth affine curve over $\mathbb{F}_p$, then A is a Dedekind domain, and in this case the Euler product defining ζ_A can be expanded as a sum over nonzero ideals in A, and so it is in this case that ζ_A is most similar to a Dedekind ζ-function. In fact, it is more natural to consider $\zeta_X(s)$ for X a *complete* curve over $\mathbb{F}_p$ (this is analogous to forming the completed ζ-function by adding in the Γ-factors, etc., in the number field case). In the number field case, the completed ζ-function conjecturally has all its zeros along the line $\Re s = 1/2$ (the Γ-factors absorb the trivial zeros), and Artin [Art24] proposed the analogous conjecture for a complete curve X.

The case of elliptic curves was proved by Hasse [Has36] and the general case was proved by Weil.[43] In fact, Hasse (in the elliptic curve case), and Weil (for general curves), proved a much more precise statement, which we illustrate by recalling the elliptic curve case.

Example 15.5.5.2   If E is an elliptic curve over $\mathbb{F}_p$, then there are two algebraic numbers α and β both of absolute value $\sqrt{p}$, satisfying $\alpha\beta = p$, such that

$$|E(\mathbb{F}_{p^n})| = 1 + p^n - \alpha^n - \beta^n,$$

---

[43] The precise history of and citations for Weil's multiple proofs are involved, and rather than giving them here, we refer the reader to Milne's recent excellent survey [Mil16] for a thorough discussion and references.

for any n. Thus

$$\zeta_E(s) = \frac{(1 - \alpha p^{-s})(1 - \beta p^{-s})}{(1 - p^{-s})(1 - p^{1-s})}.$$

Note that the Riemann Hypothesis for $\zeta_E(s)$ is encapsulated in the statement about the absolute values of α and β.

By examining the precise shape of $\zeta_X(s)$ in a range of examples for which it could be computed (such as each of the previous examples, the case of general curves, and the case of diagonal hypersurfaces – this last case being discussed in a very readable form in [Wei49]), Weil was led to make his eponymous conjectures on $\zeta_X(s)$ for a general smooth projective variety X over $\mathbb{F}_p$. These conjectures suggested the existence of a robust cohomology theory of such varieties, compatible in a suitable sense with the usual (that is, singular) cohomology of (the spaces consisting of the complex points of) varieties over $\mathbb{C}$. Such a cohomology theory was then constructed, in the form of Grothendieck's ℓ-adic cohomology, and, using it as a tool, Weil's conjectures were proved.[44] Rather than recapitulating Weil's well-known conjectures here,[45] we proceed straight to the upshot, by recalling the description of $\zeta_X(s)$ that comes out of the ℓ-adic theory.

If X is a variety over a separably closed field k, then we can define its ℓ-adic cohomology groups $H^i(X, \mathbb{Q}_\ell)$ for any prime ℓ different from the characteristic of k (i.e. if we are in positive characteristic p, we require $\ell \neq p$). When $k = \mathbb{C}$, we recover singular cohomology with $\mathbb{Q}_\ell$ coefficients. If X is defined over a not-necessarily separably closed field k, and we let $\overline{X}$ denote the base-change of X to some separably closed field Ω containing k, then we may form the cohomology groups $H^i(\overline{X}, \mathbb{Q}_\ell)$. These are in fact independent of the choice of Ω, and so it is no loss of generality to take[46] $\Omega = \overline{k}$, a separable closure of k. The action of $\mathrm{Gal}(\overline{k}/k)$ on $\overline{X}$ then induces, by an appropriate functoriality, a continuous action of $\mathrm{Gal}(\overline{k}/k)$ on $H^i(\overline{X}, \mathbb{Q}_\ell)$.

Now if $k = \mathbb{F}_p$, then Frob_p generates $\mathrm{Gal}(\overline{\mathbb{F}}_p/\mathbb{F}_p)$, and so if X is a variety over $\mathbb{Q}$, we obtain an action of Frob_p on each $H^i(\overline{X}, \mathbb{Q}_\ell)$. Now the $\mathbb{F}_{p^n}$ points of X are precisely the Frob_p^n-fixed points of $\overline{X}$, and so if X is projective, we

---

[44] The *rationality* part of the conjectures was in fact proved first by Dwork [Dwo60], using p-adic techniques. It was reproved by Grothendieck [AGV72] using ℓ-adic cohomology, as was the functional equation of $\zeta_X(s)$. Finally, the general Riemann Hypothesis was proved by Deligne [Del74].

[45] We again refer the reader to [Mil16] for more details. Appendix C of Hartshorne's textbook [Har77] also gives a short readable summary.

[46] The point of allowing more general Ω is that if, e.g., X is a variety over $\mathbb{Q}$, then we find that the cohomology of X base-changed to $\overline{\mathbb{Q}}$, which inherits a $\mathrm{Gal}(\overline{\mathbb{Q}}/\mathbb{Q})$-action, coincides with the cohomology of X base-changed to $\mathbb{C}$, which can then be interpreted as singular cohomology.

may compute the fixed points of Frob_p^n via the ℓ-adic form of the Lefschetz fixed-point formula.[47] The upshot is that[48]

$$\zeta_X(s) = \prod_{i=1}^{2d} \det\big((\mathrm{Id} - p^{-s}\mathrm{Frob}_p^{-1}) \text{ acting on } H^i(\overline{X}, \mathbb{Q}_\ell)\big)^{(-1)^{i-1}}; \quad (15.51)$$

here d denotes the dimension of X (or, equivalently, of $\overline{X}$, so that the cohomology lies in degrees between 0 and $2d$). Each factor in this product is a polynomial in p^{-s} (raised to a power ± 1), of degree equal to the dimension of the corresponding cohomology group. Thus this expression exhibits, for general X, the structure of $\zeta_X(s)$ that can be observed in the simple examples described above. In particular, from this formula one immediately obtains the meromorphic continuation of $\zeta_X(s)$.

If X is smooth as well as projective, then the cohomology of $\overline{X}$ satisfies Poincaré duality, and this gives rise to a functional equation for $\zeta_X(s)$. Finally, the *Riemann Hypothesis* in this context (i.e. for X smooth and projective over $\mathbb{F}_p$) states that the eigenvalues of Frob_p^{-1} on H^i are algebraic numbers all of whose conjugates have absolute value $p^{i/2}$. As Serre explained in his letter [Ser60] to Weil, this also admits a cohomological "explanation," by analogy with the Hodge theory of smooth projective complex varieties (the particular focus of the analogy being the definiteness of the polarization pairings on primitive cohomology that comes out of the Hodge–Riemann bilinear relations). Grothendieck proposed an approach to proving the Riemann Hypothesis in the present context via a series of so-called "standard conjectures" [Kle94], which would allow one to make the same kind of definiteness arguments as in the context of complex varieties. (Such arguments were already employed by Hasse and Weil in their proofs the Riemann Hypothesis in the case when

---

[47] If X is not necessarily projective, then this still works, provided that we use compactly supported ℓ-adic cohomology. We also remark that Weil proposed this "Lefschetz formula" interpretation of his conjectures in his ICM address [Wei56a].

[48] The reason that Frob_p^{-1}, rather than Frob_p, appears in the formula is slightly technical: in order to apply the Lefschetz formula, we have to work with a morphism of varieties, *not* a Galois automorphism. Thus we work with the Frobenius endomorphism $F_{\overline{X}}$ of $\overline{X}$, defined by raising the coordinates of the a point to the pth power. If we compose this endomorphism with the Galois automorphism Frob_p of $\overline{X}$, we obtain the *absolute Frobenius* of $\overline{X}$ (defined simply by the map $f \mapsto f^p$ on the structure sheaf $\mathcal{O}_{\overline{X}}$). This latter morphism induces the identity on the underlying topological space of $\overline{X}$, and consequently acts via the identity on the cohomology of $\overline{X}$. Thus the traces of powers Frob_p^{-1} on cohomology coincide with the traces of powers of $F_{\overline{X}}$, and so it is Frob_p^{-1} that appears in the Lefschetz formula for $\zeta_X(s)$.

Motivated by this relationship between Frob_p^{-1} and $F_{\overline{X}}$, the Galois element Frob_p^{-1} is referred to as the *geometric* Frobenius automorphism, in contrast to the *arithmetic* Frobenius Frob_p itself.

X is a curve.) Unfortunately, these conjectures remain unproved in general, and Deligne [Del74] found a different approach to proving the Riemann Hypothesis.[49]

We close this discussion by pointing out that if X is smooth and projective, then the various factors

$$\det\big((\mathrm{Id} - p^{-s}\mathrm{Frob}_p^{-1}) \text{ acting on } H^i(\overline{X}, \mathbb{Q}_\ell)\big) \tag{15.52}$$

in (15.51) are mutually coprime as i varies, since the Riemann Hypothesis ensures that they have no roots in common. This implies that the factorization of $\zeta_X(s)$ given by the right-hand side of (15.51) is *intrinsic* to $\zeta_X(s)$, and can be constructed independently of the choice of $\ell \neq p$. In particular, the various determinants (15.52) have integer coefficients (as polynomials in p^{-s}), and are *independent* of the choice of ℓ!

15.5.6 Hasse–Weil ζ- and L-Functions

Let us return now to the case of a finite-type scheme over $\mathbb{Z}$, which we now assume to be flat over $\mathbb{Z}$, so that it does not lie in any particular residue characteristic p; the reader should have in mind an integral model of a variety over $\mathbb{Q}$ or some other number field. If we combine (15.48) with (15.51), we obtain the formula (in which d denotes the relative dimension of X over $\mathbb{Z}$)

$$\zeta_X(s) = \prod_p \prod_{i=1}^{2d} \det\big((\mathrm{Id} - p^{-s}\mathrm{Frob}_p^{-1}) \text{ acting on } H^i(\overline{X}, \mathbb{Q}_\ell)\big)^{(-1)^{i-1}},$$
$$\tag{15.53}$$

which expresses $\zeta_X(s)$ as an Euler product, where the Euler factors are now determined by the action of (geometric) Frobenius on the ℓ-adic cohomology of the various base-changes $X_{/\mathbb{F}_p}$. But this looks very similar[50] to Artin's definition (15.39) of his L-functions! It is also slightly inconsistent: for any particular p, we are free to choose any $\ell \neq p$ to compute the Euler factor at p; but since ℓ *is* a prime, and so appears as one of the indices in the product over p, there is no single choice of ℓ which works for every factor in the product (15.53)!

[49] There is still a certain "definiteness," or "positivity," that goes into Deligne's proof, but it is more elementary – it is, *grosso modo*, just the fact that the point counts that are encoded by $\zeta_X(s)$ are necessarily nonnegative.
[50] The use of geometric Frobenius rather than arithmetic Frobenius can be thought of as simply a certain choice of normalization.

To further develop the observations just made, we now change notation. Namely, we now let X be a $\mathbb{Q}$-scheme, assumed to be smooth and projective, and we let $\mathcal{X}$ be a model of X over $\mathbb{Z}$, i.e. a projective flat $\mathbb{Z}$-scheme whose fiber over Spec $\mathbb{Q}$ equals X. (We can obtain $\mathcal{X}$ just by suitably clearing denominators in some system of equations defining X; and $\mathcal{X}$ will now be the $\mathbb{Z}$-scheme that was previously denoted by X.) There is a finite set S of primes (the set of primes of *bad reduction*) such that $\mathcal{X}_{/\mathbb{F}_p}$ is smooth if $p \notin S$.

Now write $\overline{X}$ to denote the base-change of X to $\overline{\mathbb{Q}}$. If we choose a prime ℓ, then we may form the ℓ-adic cohomology $H^i(\overline{X}, \mathbb{Q}_\ell)$, with its natural $\mathrm{Gal}(\overline{\mathbb{Q}}/\mathbb{Q})$-action; we are in characteristic zero, so no primes ℓ are excluded. General principles of ℓ-adic cohomology ("local acyclicity") imply that if $p \notin S$, and if $\ell \neq p$, then

$$H^i(\overline{X}, \mathbb{Q}_\ell) \cong H^i(\overline{\mathcal{X}_{/\mathbb{F}_p}}, \mathbb{Q}_\ell) \tag{15.54}$$

(where, as above, $\overline{\mathcal{X}_{/\mathbb{F}_p}}$ denotes the base-change of $\mathcal{X}_{/\mathbb{F}_p}$, the reduction mod p of $\mathcal{X}$, to $\overline{\mathbb{F}}_p$.) To make this identification canonical, we should choose a place of $\overline{\mathbb{Q}}$ above p; or equivalently, fix an embedding $\overline{\mathbb{Q}} \hookrightarrow \overline{\mathbb{Q}}_p$, up to the action of $\mathrm{Gal}(\overline{\mathbb{Q}}_p/\mathbb{Q}_p)$ on the target; or, again equivalently, fix a choice of decomposition group D_p at p in $\mathrm{Gal}(\overline{\mathbb{Q}}/\mathbb{Q})$; or, yet again equivalently, fix a choice of surjection from the integral closure of $\mathbb{Z}$ in $\overline{\mathbb{Q}}$ onto $\overline{\mathbb{F}}_p$. With this choice made, the isomorphism (15.54) is functorial, and so is equivariant for the action of the chosen decomposition group D_p, which acts on the left-hand side via restricting the $\mathrm{Gal}(\overline{\mathbb{Q}}/\mathbb{Q})$-action to D_p, and acts on the right-hand side via the surjection $D_p \to D_p/I_p = \mathrm{Gal}(\overline{\mathbb{F}}_p/\mathbb{F}_p)$, and the action of $\mathrm{Gal}(\overline{\mathbb{F}}_p/\mathbb{F}_p)$ on the right-hand side.

In particular, we find that the $\mathrm{Gal}(\overline{\mathbb{Q}}/\mathbb{Q})$-action on $H^i(\overline{X}, \mathbb{Q}_\ell)$ is unramified at $p \notin S \cup \{\ell\}$, and (for such p) we have

$$\det\big((\mathrm{Id} - p^{-s}\mathrm{Frob}_p^{-1}) \text{ acting on } H^i(\overline{X}, \mathbb{Q}_\ell)\big)^{(-1)^{i-1}}$$

$$= \det\big((\mathrm{Id} - p^{-s}\mathrm{Frob}_p^{-1}) \text{ acting on } H^i(\overline{\mathcal{X}_{/\mathbb{F}_p}}, \mathbb{Q}_\ell)\big)^{(-1)^{i-1}}. \tag{15.55}$$

Finally, the "independence of ℓ" noted above shows that the determinant (15.55) is independent of ℓ, as long as we assume that $p \notin S$ and $\ell \neq p$.

The conclusion is that (for each i) we have a family of ℓ-adic representations of $\mathrm{Gal}(\overline{\mathbb{Q}}/\mathbb{Q})$, namely the representations $H^i(\overline{X}, \mathbb{Q}_\ell)$, each of which is unramified outside $S \cup \{\ell\}$ (where S is fixed independently of ℓ), and for which the characteristic polynomial of Frobenius at p, and hence the Euler factor (15.55), is independent of the choice of $\ell \neq p$, for all $p \notin S$. We say that the $H^i(\overline{X}, \mathbb{Q}_\ell)$ form a *compatible family*, and we define the associated (partial) *Hasse–Weil* ζ-function via Artin's formula[51]

$$\zeta_X^S(s) = \prod_{p \notin S} \prod_{i=0}^{2d} \det\big((\text{Id} - p^{-s}\text{Frob}_p^{-1}) \text{ acting on } H^i(\overline{X}, \mathbb{Q}_\ell)\big)^{(-1)^{i-1}},$$

(15.56)

where it is understood that to compute the factor at a given prime p, we choose any $\ell \neq p$. The superscript S indicates that we are forming an imprimitive ζ-function, in which we have omitted Euler factors at the finitely many bad primes in S. (We will return to this point below.)

The formula defining $\zeta_X^S(s)$ then suggests that we should follow Artin in defining an L-function for each of the individual compatible families of representations $H^i(\overline{X}, \mathbb{Q}_\ell)$, and indeed we can do just that, setting

$$L^S\big(s, H^i(X)\big) = \prod_{p \notin S} \det\big((\text{Id} - p^{-s}\text{Frob}_p^{-1}) \text{ acting on } H^i(\overline{X}, \mathbb{Q}_\ell)\big)^{-1}$$

with the same convention as above, namely that in computing the pth factor we choose any $\ell \neq p$. We then have

$$\zeta_X^S(s) = \prod_{i=0}^{2d} L^S\big(s, H^i(X)\big)^{(-1)^i}.$$

The Riemann Hypothesis gives an improvement on the half-plane of convergence for the individual L-functions $L^S\big(s, H^i(X)\big)$; namely, this L-function converges for $\Re s > 1 + \frac{i}{2}$.

How should we eliminate the imprimitivity of the L-functions $L^S\big(s, H^i(X)\big)$, or the ζ-function $\zeta_X^S(s)$? That is, what Euler factors should we include for the missing primes $p \in S$? One would like to use Artin's recipe (forming the reciprocal characteristic polynomial of geometric Frobenius on the I_p-invariants). Unfortunately, the problem now arises as to whether or not this is independent of ℓ. This problem is related to the following one: our definition of $\zeta_X^S(s)$, and of the $L^S\big(s, H^i(X)\big)$, no longer uses the integral model $\mathcal{X}$! For example, it could be that for some primes in S, the action of $\text{Gal}(\overline{\mathbb{Q}}/\mathbb{Q})$ on $H^i(\overline{X}, \mathbb{Q}_\ell)$ happens to be unramified for some choice of ℓ; does this imply the same for other choices of ℓ? Even if it does, are the resulting characteristic polynomials of Frobenius independent of ℓ? One way this could happen is if X admitted a different integral model $\mathcal{X}'$ which *did* have good reduction at S. Then the answers to these questions would be "yes." But in general the answers to these questions aren't known (although the answers *are* still expected to be "yes").

---

[51] Up to working with geometric rather than arithmetic Frobenius.

The preceding question is closely related to the problem of finding "good" (e.g. semistable) integral models of smooth projective schemes over $\mathbb{Q}$, or over $\mathbb{Q}_p$ (since the ramification behavior at p can be investigated locally at p). And this is related to problems such as resolution of singularities in positive characteristic, or in mixed characteristic (i.e. over $\mathbb{Z}_p$) – problems that also remain open. For example, the isomorphism (15.54) typically won't hold at primes of bad reduction, and so the question arises as to what the true relationship between the two sides is in that case; the theory of nearby and vanishing cycles is aimed at answering this question. But even for varieties over $\mathbb{F}_p$, if they are singular then we don't know independence of ℓ for their ℓ-adic cohomology in general, and without knowing this it doesn't seem reasonable to imagine we could prove an independence of ℓ result at a prime of bad reduction for a general smooth projective variety over $\mathbb{Q}$.

All that being said, one *does expect* that the compatible families of ℓ-adic representations $H^i(\overline{X}, \mathbb{Q}_\ell)$ *are* independent of $\ell \neq p$ locally at every prime p in a suitable sense, and in particular, that if we apply Artin's formula, we obtain a well-defined L-function

$$L(s, H^i(X)) = \prod_p \det\big((\mathrm{Id} - p^{-s}\mathrm{Frob}_p^{-1})\text{ acting on } H^i(\overline{X}, \mathbb{Q}_\ell)^{I_p}\big)^{-1},$$

as well as a ζ-function

$$\zeta_X(s) = \prod_{i=0}^{2d} L(s, H^i(X))^{(-1)^i}.$$

Independence of ℓ is known in many interesting cases, including the case when X is a curve, or an abelian variety. We refer to Serre's lecture [Ser70] for a more-detailed discussion, as well as for a discussion of how to complete these functions at the archimedean place by adding an appropriate Γ-factor.

Once we have added the appropriate factors, we conjecture that $\zeta_X(s)$ will have an analytic continuation (this doesn't actually depend on the manner in which $\zeta_X(s)$ is completed at the $p \in S$ or at ∞, since the Euler factors that we add are themselves meromorphic functions) and a functional equation (and having a functional equation is very much dependent on completing in the correct way!).

More precisely, the functional equation will relate $\zeta_X(s)$ to $\zeta_X(d + 1 - s)$; in terms of the individual L-factors, it will relate $L(s, H^i(X))$ to $L(1 + d - s, H^{2d-i})$; if one takes into account the cup product pairing of H^i and H^{2d-i} into H^{2d}, and the Galois action on H^{2d}, this in fact becomes a functional equation relating $L(s, H^i(X))$ to $L(1 - s, H^i(X)^\vee)$, just as in the formalism of Artin L-functions. And just as is the case for Artin's L-functions, we anticipate that the global root number will be a product of local root numbers, which can be determined using the formalism of Langlands–Deligne discussed above.

15.5.7 Motives

When we write $L\big(s, H^i(X)\big)$, what "is" $H^i(X)$? For now, it is just a symbol standing for the compatible family of ℓ-adic $\mathrm{Gal}(\overline{\mathbb{Q}}/\mathbb{Q})$-representations

$$\{H^i\big(\overline{X}, \mathbb{Q}_\ell\big)\}.$$

Grothendieck's theory of motives is intended to supply a more substantive answer to this question.

Grothendieck's proposal is based on the fact that although ℓ-adic cohomology is a functor on the category of varieties (e.g. over $\mathbb{Q}$, or $\mathbb{F}_p$), it is functorial with respect to a larger class of morphisms then just the morphisms of varieties. Namely, because cohomology is valued in linear objects (vector spaces), multivalued functions between varieties can induce single-valued functions between their cohomology groups (by simply adding up the multiple values!). Now multivalued functions between varieties are actually very natural objects to consider – they are just the correspondences. (The reader might think of the Hecke correspondences on modular curves, or if $f : C \to C'$ is a surjective morphism of curves of degree > 1, one can think of the "reflected" graph of f in $C \times C' \xrightarrow{\sim} C' \times C$ as providing a multivalued morphism $f^{-1} : C' \to C$.) So cohomology extends to a functor on the category of varieties, with suitable correspondences as morphisms. But we can also do spectral theory of operators on cohomology groups – decomposing them into eigenspaces. So we should also allow ourselves to decompose varieties into "eigenspaces" for the action of correspondences. Roughly speaking, this gives us the category of motives. But many technicalities arise – for example, two correspondences that can be algebraically deformed, one into the other, will induce the same morphism on cohomology, and so we might want to identify them. Very quickly, in trying to make a precise definition and investigate its properties, one gets into thorny questions of equivalence relations on cycles. Grothendieck's standard conjectures [Kle94] are intended in part to resolve some of these questions, and thus to define a category – the category of so-called *Grothendieck motives*. The objects $H^i(X)$ are then supposed to be motives. (How would we cut them out as eigenspaces of correspondences? We saw the answer above, at least over $\mathbb{F}_p$, in the course of proving independence of ℓ: the characteristic polynomials of Frobenius on the different H^i are coprime polynomials, and so we can use the Frobenius endomorphism of X to cut X up into the various $H^i(X)$.)

Let us return momentarily to the context of Artin L-functions. If K is a number field, then $\mathrm{Spec}\, K$ is a zero-dimensional smooth $\mathbb{Q}$-scheme. If $F \subset K$ is a subfield for which K/F is Galois, then $\mathrm{Gal}(K/F)$ acts as automorphism of $\mathrm{Spec}\, K$, and elements in the group ring of $\mathrm{Gal}(K/F)$ induces correspondences on $\mathrm{Spec}\, K$. "Cutting up" $\mathrm{Spec}\, K$ using these correspondences amounts to decomposing $\mathrm{Spec}\, K$ in the category of motives. Thus the Galois representations appearing in the theory of Artin L-functions are examples of motives (zero-dimensional motives). Note that there was no ℓ in that story; this is because H^0 (the only degree of cohomology involved in the zero-dimensional context) can be perfectly well defined with $\mathbb{Q}$-coefficients, so "independence of ℓ" is elementary for H^0.

Note, though, that not all $\mathrm{Gal}(K/F)$-representations will necessarily be defined over $\mathbb{Q}$. In the theory of Artin L-functions we used representations on $\mathbb{C}$-vector spaces, but since $\mathrm{Gal}(K/F)$ is a finite group, we could just as well have considered representations on $\overline{\mathbb{Q}}$-vector spaces. For higher-degree cohomology, when we have to allow ℓ to vary, we should take cohomology with $\overline{\mathbb{Q}}_\ell$-coefficients, so as to allow finer motivic decompositions than can be obtained with $\mathbb{Q}_\ell$-coefficients. (One way to express this passage from $\mathbb{Q}_\ell$ to $\overline{\mathbb{Q}}_\ell$-coefficients is to say that we are allowing motives "with coefficients.")

Whatever a motive M is, then, it should have cohomology (however we cut up our original scheme X, we can correspondingly cut up its cohomology into the appropriate eigenspaces), and this cohomology will have a $\mathrm{Gal}(\overline{\mathbb{Q}}/\mathbb{Q})$-action (as long as we are using correspondences defined over $\mathbb{Q}$), or at least a $\mathrm{Gal}(\overline{\mathbb{Q}}/F)$-action (if we use correspondences defined over some extension F of $\mathbb{Q}$, as we did above when we decomposed Spec K into its various irreducible Artin representations). We anticipate that, as ℓ varies, the cohomology of M will give compatible families of Galois representations.[52] Each motive will then have an associated L-function $L(s, M)$, which we again anticipate will admit an analytic continuation and functional equation (the latter relating $L(s, M)$ to $L(1 - s, M^\vee)$ for a suitably understood dual[53] motive $M^\vee$, with a root number determined by the Langlands–Deligne theory).

We have written "whatever a motive is" above, because Grothendieck's standard conjectures remain unproved, and so the theory of motives is not complete. There are other standard conjectures relating cohomology and cycles (not in the sense of being on Grothendieck's list, but in the usual meaning of "standard"), such as the Hodge conjecture and the Tate conjecture [Tat65], which would also have to be resolved to give anything like a complete theory of motives. (For example, one conjectures[54] that if X is smooth and projective over $\mathbb{Q}$, then the $\mathrm{Gal}(\overline{\mathbb{Q}}/\mathbb{Q})$-action on each $H^i(\overline{X}, \mathbb{Q}_\ell)$ is semisimple, and that

---

[52] Meaning that there is a number field E independent of ℓ such that, for $\ell \neq p$, the characteristic polynomial of Frob$_p$ has coefficients in E (thought of as lying inside $\overline{\mathbb{Q}}_\ell$), and is independent of ℓ.

[53] To obtain duals, there is an additional step that we should perform in the construction of the category of motives, namely "invert the Lefschetz motive," i.e. the motive $H^2(\mathbb{P}^1)$. If we do this, then in fact $H^{2d}(X)$ becomes an invertible motive for any smooth projective geometrically connected X of dimension d, and so the cup product pairings on cohomology can be used to construct dual motives.

[54] This conjecture is usually included under the general umbrella of "the Tate conjecture," although in [Tat65], Tate attributes it to Grothendieck and Serre. For a discussion of this conjecture and its relation to other conjectures regarding the Galois action on ℓ-adic cohomology, one can consult [Tat94].

furthermore each simple constituent is cut out by a suitable correspondence, i.e. is *motivic*. This would make the analogy between our general context and the original context of Artin L-functions particularly tight and compelling; but this semisimplicity conjecture is proved in very few cases.)

15.5.8 Compatible Families of Galois Representations

Number theorists often circumvent the difficulties of rigorously defining/constructing motives by working directly with compatible families of ℓ-adic representations. Since this notion was introduced by Taniyama [Tan57], it has evolved into a fundamental concept, which provides a practical and flexible replacement for the notion of motive. Being inherently linear, Galois representations are easier to manipulate than varieties!

One technical difficulty in working with compatible families, such as the families $\{H^i(\overline{X}, \mathbb{Q}_\ell)\}$ considered above, is ensuring compatibility at *all* primes p (i.e. including the bad ones). Although, as already noted, arithmetic geometry is not currently advanced enough to prove this directly in all cases of interest, in practice, there are many compatible families for which one *does* have control at all primes, and hence for which one can define complete L-functions unambiguously. As we already mentioned, the families $\{H^1(\overline{X}, \mathbb{Q}_\ell\}$ for a smooth projective curve X over a number field give examples of such families.

In Section 15.4.6 we gave examples of compatible families of ℓ-adic representations $\{\widetilde{\rho}_\ell\}$ of $\mathrm{Gal}(\overline{\mathbb{Q}}/F)$ (for some number field F), constructed from algebraic Hecke characters ψ on some extension K of F. These particular compatible families are in fact motivic. Indeed, the compatible families of characters $\{\widetilde{\psi}_\ell\}$ from Sections 15.4.5 and 15.4.6 are associated to motives over K that can be constructed from abelian varieties with complex multiplication. (This goes back to Taniyama [Tan57] and [Ser68].) The group-theoretic induction that constructs $\{\widetilde{\rho}_\ell\}$ from $\{\widetilde{\psi}_\ell\}$ then corresponds in the world of varieties and motives to restriction of scalars from K to F. Thus the L-functions $L(s, \{\widetilde{\rho}_\ell\})$ are motivic L-functions. In this case, the induction formalism for L-functions (15.41), together with (15.35), shows that

$$L(s, \{\widetilde{\rho}_\ell\}) = L(s, \psi), \tag{15.57}$$

and so the L-functions $L(s, \{\widetilde{\rho}_\ell\})$ are also automorphic[55] (in this particular instance they are Hecke L-functions). In particular, these L-functions *do* admit analytic continuations and satisfy functional equations.

---

[55] Automorphic *over* K, since we obtain Hecke L-functions for K. The theory of *automorphic induction*, briefly recalled in Section 15.7.3, shows that they are also automorphic over F.

15.5.9 The Sato–Tate Conjecture, Unitary Normalization, and the Langlands Group

We begin by considering a specific, historically important, class of examples: elliptic curves E over $\mathbb{Q}$.

Suppose that E has good reduction outside the finite set S of primes; we abuse notation by writing simply $E_{/\mathbb{F}_p}$ for the reduction modulo $p \notin S$. For each $p \notin S$, we see from Example 15.5.5.2 that

$$\zeta_{E_{/\mathbb{F}_p}}(s) = \frac{(1 - \alpha_p p^{-s})(1 - \beta_p p^{-s})}{(1 - p^{-s})(1 - p^{1-s})},$$

where $|\alpha_p| = |\beta_p| = \sqrt{p}$, and $\alpha_p \beta_p = p$. The numerator arises from H^1, and the factors in the denominator from H^0 and H^2. We deduce that

$$L(s, H^0(E)) = \zeta(s) \quad \text{and } L(s, H^2(E)) = \zeta(s-1)$$

(strictly speaking we haven't computed the missing Euler factors at $p \in S$, but completing to the Riemann ζ-function and its translate in this way are the only possibilities compatible with a functional equation), while

$$L^S(s, H^1(E)) = \prod_{p \notin S} \frac{1}{(1 - \alpha_p p^{-s})(1 - \beta_p p^{-s})} = \prod_{p \notin S} \frac{1}{1 - a_p p^{-s} + p^{1-2s}},$$

where

$$a_p = \alpha_p + \beta_p = 1 + p - |E_{/\mathbb{F}_p}(\mathbb{F}_p)|.$$

The condition $|\alpha_p| = |\beta_p| = \sqrt{p}$ implies (indeed, is equivalent to) the condition

$$a_p \in [-2\sqrt{p}, 2\sqrt{p}]; \tag{15.58}$$

and both conditions are equivalent to the characteristic polynomial of Frob_p^{-1} on $H^1(\overline{E}, \mathbb{Q}_\ell)$ – which equals $X^2 - a_p X + p$ – having nonpositive discriminant.[56]

Our next immediate goal is to discuss the *Sato–Tate* conjecture, which is a conjecture describing the distribution of the quantities a_p as p varies. But we also intend to explain how it relates to a more far-reaching idea, that of the *Langlands group*.

To begin with, a consideration of (15.58) suggests that we rescale the a_p, and consider (for each $p \notin S$) the quantity

$$A_p = \frac{a_p}{\sqrt{p}} \in [-2, 2];$$

---

[56] The ℓ-adic H^1 of E is dual to the ℓ-adic Tate module of E, and so this polynomial is also the characteristic polynomial of Frob_p on the ℓ-adic Tate module of E; which is how it is often described in discussions of elliptic curves.

the quantities A_p now lie in the same interval, as p varies, and so it makes sense to discuss their distribution. It turns out to be convenient to write $A_p = 2\cos\theta_p$, where $\theta_p \in [0, \pi]$. (Then, up to switching the labelling, we have $\alpha_p = \sqrt{p}e^{i\theta_p}$ and $\beta_p = \sqrt{p}e^{-i\theta_p}$.)

If E admits CM by an imaginary quadratic field K, then it is well known (and easily seen) that $a_p = 0$ if p is inert in K (which happens half the time, by Dirichlet), so that $\theta_p = \pi/2$ for all such p; while for the p that are split in K, one finds that θ_p is distributed uniformly throughout $[0, \pi]$.

If E does not admit CM, then the Sato–Tate conjecture[57] states that θ_p is distributed over $[0, \pi]$ according to the measure $\frac{2}{\pi}\sin^2\theta d\theta$.

Before we try to explain the meaning of these results, we begin by describing a more structural interpretation of the above rescaling by $p^{-1/2}$. If we let $\{\chi_\ell\}$ denote the compatible family of cyclotomic characters (see Example 15.4.5.2), then $\chi(\text{Frob}_p^{-1}) = p^{-1}$. Suppose for a moment that we could form a compatible family $\{\chi_\ell^{1/2}\}$, with the property that $\chi_\ell^{1/2}(\text{Frob}_p^{-1}) = p^{-1/2}$ for each $p \neq \ell$. Then we could form the twisted compatible family $H^1(E) \otimes \chi^{1/2}$, and for a given $p \notin S$ (working with $\ell \neq p$) we would find that the characteristic polynomial of Frob_p^{-1} on this twist is

$$X^2 - A_p X + 1. \tag{15.59}$$

Now unfortunately, the characters χ_ℓ don't admit square roots. (For example, complex conjugation is an element $c \in \text{Gal}(\overline{\mathbb{Q}}/\mathbb{Q})$ of order 2 for which $\chi_\ell(c) = -1$.) On the other hand, the compatible family $\{\chi_\ell\}$ arises from the absolute value character $|\ |$ on $\mathbb{Q}^\times \setminus \mathbb{A}_\mathbb{Q}^\times$, and this character evidently admits a square root, namely $|\ |^{1/2}$.

This prompts the question: is there a process via which we can "unravel" the various two-dimensional ℓ-adic representations $H^1(\overline{E}, \mathbb{Q}_\ell)$, so as to obtain a single two-dimensional representation, over the complex numbers, of some group – in the way that we can "undo" the passage from $|\ |$ to $\{\chi_\ell\}$, at the cost of passing from $\text{Gal}(\mathbb{Q}^{ab}/\mathbb{Q}) = \pi_0(\mathbb{Q}^\times \setminus \mathbb{A}_\mathbb{Q}^\times)$ to the group $\mathbb{Q}^\times \setminus \mathbb{A}_\mathbb{Q}^\times$ itself? If so, then we could twist this unraveled representation by $|\ |^{1/2}$, and obtain a representation which (we might hope) would give a conceptual sense to the rescaled characteristic polynomials (15.59).

---

[57]   Now a theorem for elliptic curves over any totally real of CM number field. Over totally real fields, the proof is due to a large number of people, including Clozel–Harris–Taylor [CHT08], Harris–Shepherd-Barron–Taylor [HSBT06], Taylor [Tay08], Shin [Shi11], Clozel–Harris–Labesse–Ngô [CHLN11], and Barnet-Lamb–Geraghty–Harris–Taylor [BLGHT11]. The recent extension [ACC+18] to the case of CM fields is discussed in Section 15.7.3.

Example 15.5.9.1 Consider the elliptic curve E cut out by the Weierstrass equation

$$y^2 = x^3 - x.$$

This is an elliptic curve with complex multiplication by $\mathbb{Z}[i]$, defined over $\mathbb{Q}(i)$, via the formula

$$[i](x, y) = (-x, iy).$$

This action makes each $H^1(\overline{E}, \mathbb{Q}_\ell)$ free of rank one over $\mathbb{Q}_\ell \otimes_{\mathbb{Q}} \mathbb{Q}(i)$, so that the $\mathrm{Gal}(\overline{\mathbb{Q}}/\mathbb{Q}(i))$-action on $H^1(\overline{E}, \mathbb{Q}_\ell)$ (which commutes with the $\mathbb{Q}_\ell \otimes_{\mathbb{Q}} \mathbb{Q}(i)$-action) is abelian. In fact, this action is given by the character $\widetilde{\psi}_\ell$, where $\widetilde{\psi}_\ell$ is constructed from the (algebraic!) Hecke character ψ which is *inverse* to the character of Examples 15.3.2.2 and 15.3.3.2. It is then easy to see that the $\mathrm{Gal}(\overline{\mathbb{Q}}/\mathbb{Q})$-action on $H^1(\overline{E}, \mathbb{Q}_\ell)$ is given by the two-dimensional representation $\widetilde{\rho}_\ell$ obtained by inducing $\widetilde{\psi}_\ell$.

In this case, an "unraveling" of the desired type is possible: we have the Weil-group representation $\rho : W_{\mathbb{Q}(i)/\mathbb{Q}} \to \mathrm{GL}_2(\mathbb{C})$ obtained by inducing ψ, as in Section 15.4.6, which we can then twist to form

$$\rho \otimes |\ |^{1/2} = \mathrm{Ind}^{W_{\mathbb{Q}(i)/\mathbb{Q}}}_{\mathbb{Q}(i)^\times \backslash \mathbb{A}^\times_{\mathbb{Q}(i)}} (\psi \otimes |\ |^{1/2}).$$

Thus $\rho \otimes |\ |^{1/2}$ is also obtained by the induction construction of Section 15.4.6, but now applied to $\psi \otimes |\ |^{1/2}$, which is the unitarization of ψ (see Example 15.3.2.2). Thus $\rho \otimes |\ |^{1/2}$ is a *unitary* representation of $W_{\mathbb{Q}(i)/\mathbb{Q}}$, and we see that the polynomials (15.59) are the characteristic polynomials of geometric Frobenius elements[58] in $W_{\mathbb{Q}(i)/\mathbb{Q}}$. Thus they are the characteristic polynomials of *unitary* 2×2-matrices – which explains why their roots have the form $e^{\pm i\theta_p}$!

The result about the distribution of the A_p is then seen to be a result about the distribution of the images $\rho(\mathrm{Frob}_p^{-1})$, and so is a certain kind of generalization of Tschebotareff density.

On the other hand, if E is *not* a CM elliptic curve, then Serre showed in [Ser68] that the action of $\mathrm{Gal}(\overline{\mathbb{Q}}/\mathbb{Q})$ on $H^1(\overline{E}, \mathbb{Q}_\ell)$ is irreducible, and remains so after restriction to any open subgroup $\mathrm{Gal}(\overline{\mathbb{Q}}/K)$ of $\mathrm{Gal}(\overline{\mathbb{Q}}/\mathbb{Q})$. Thus ρ_ℓ is *not* induced from a character, and so we cannot describe the compatible family $\{\rho_\ell\}$ in terms of a Weil group representation ρ.

However, we have the following conjecture. (We remind the reader that we introduced the absolute Weil group W_F of F in Section 15.4.6.)

---

[58] We don't recall the details of the construction of these elements right now; it is part of the local–global compatibility theory for Weil groups, which is discussed in Section 15.5.10.

Conjecture 15.5.9.2 (Existence of the Langlands group) *For any number field F, there is a locally compact topological group L_F, the* Langlands group *of F, admitting a surjection $L_F \rightarrow W_F$ with compact kernel and for which the induced morphism $L_F^{\mathrm{ab}} \rightarrow W_F^{\mathrm{ab}} = F^\times \backslash \mathbb{A}_F^\times$ is an isomorphism, and such that compatible families of n-dimensional ℓ-adic Galois representations $\{\rho_\ell\}$ arising from motives arise (in a suitable sense) from certain ("algebraic") complex representations $L_F \rightarrow \mathrm{GL}_n(\mathbb{C})$.*

This conjecture is somewhat informal as stated, but we will make it more precise in the following subsection. For now, we just remark that the notion of "algebraic," for representations of L_F, should be analogous to the corresponding notion for Hecke characters, and that the mechanism which produces the compatible family $\{\rho_\ell\}$ from ρ should be (at least formally) analogous to the mechanism which produces the compatible family $\{\widetilde{\psi}_\ell\}$ from an algebraic Hecke character ψ.

If we admit this conjecture, then we find, for any elliptic curve, that the family of representations $H^1(\overline{E}, \mathbb{Q}_\ell)$ arises from a representation $L_{\mathbb{Q}} \rightarrow \mathrm{GL}_2(\mathbb{C})$. If we let ρ denote the twist of this latter representation by $|\ |^{1/2}$ (this twist now makes sense!), then ρ in fact factors through $SU(2) \subset \mathrm{GL}_2(\mathbb{C})$. Indeed its determinant is trivial, as each characteristic polynomial (15.59) has constant term 1, and since L_F is compact outside of its $\mathbb{R}_{>0}^\times$ part coming from W_F, we see that the image is compact with trivial determinant, and so can be factored through $SU(2)$ after an appropriate choice of basis. Now the image is a compact subgroup, and there aren't many possibilities for it: either a copy of $U(1)$; the normalizer of a copy of $U(1)$ in $SU(2)$ (which is a non-split extension of a group of order 2 by $U(1)$, with the non-trivial element in the group of order 2 acting on $U(1)$ via inversion); or $SU(2)$ itself. The first possibility can't occur, since the representations of $\mathrm{Gal}(\overline{\mathbb{Q}}/\mathbb{Q})$ on $H^1(\overline{E}, \mathbb{Q}_\ell)$ aren't abelian, and so one of the latter two possibilities must occur.

In the second case, restricting to an index-two open subgroup of $L_{\mathbb{Q}}$, i.e. to L_K for some quadratic extension of K, yields an abelian representation. This is evidently the CM case. Otherwise, we must be in the third case. And now we see the meaning of the results on the distributions of the θ_p: what they are saying is that $\rho(\mathrm{Frob}_p^{-1})$ is equidistributed in the image of ρ, as p ranges over all primes $\notin S$. Just as in the case of the Tschebotareff's theorem, we must remember that Frob_p^{-1} is only defined up to conjugacy, so we have to understand this equidistribution in the framework of conjugacy classes.

The normalizer of $U(1)$ has two connected components. The nonidentity component is a single conjugacy class. The other component is just $U(1)$ itself, and the conjugacy classes are just the pairs $\{z, \overline{z}\}$ for $z \in U(1)$, which can be parameterized by the elements $z = e^{i\theta}$, with $\theta \in [0, \pi]$. Thus, when E has CM, we see that half the $\rho(\mathrm{Frob}_p^{-1})$ land in the nonidentity component – these are the half of the primes for which $a_p = A_p = 0$. And the other half of the primes satisfy $\mathrm{Frob}_p^{-1} \in U(1)$, and θ_p is equidistributed in $[0, \pi]$.

The conjugacy classes in $SU(2)$ are again parameterized by their pairs of eigenvalues $\{z, \bar{z}\}$, and hence by $z = e^{i\theta}$ with $\theta \in [0, \pi]$. But the uniform (i.e. Haar) measure on $SU(2)$ doesn't induce the uniform measure on $[0, \pi]$. Rather, it induces the Sato–Tate measure $\frac{2}{\pi} \sin^2 \theta d\theta$. And the Sato–Tate conjecture affirms that if E does not have CM, then the θ_p are indeed equidistributed according to this measure.

How does one *prove* this equidistribution? It is a density theorem, and the proof proceeds like the proof of all such theorems, via an appropriate Fourier analysis facilitated by a result on the nonvanishing of appropriate L-functions. Consider first Example 15.5.9.1. For primes p that are split in $\mathbb{Q}(i)$, say $p = \pi\bar{\pi}$, with $\pi \equiv 1 \bmod (1+i)^3$, the corresponding conjugacy class in $U(1)$ is simply $(\pi/\sqrt{p}, \bar{\pi}/\sqrt{p})$ (see the definition of ψ in Example 15.3.2.2), and so the claimed equidistribution amounts to the equidistribution of Gaussian primes in angular sectors that was already mentioned in Section 15.3.5.

What about the non-CM case? Then we have to perform a Fourier analysis of the conjugacy classes in $SU(2)$. To expand the indicator function of a conjugacy class in terms of traces of representations, we have to use not just the character of the standard two-dimensional representation of $SU(2)$, but the characters of *all* of its irreducible representations. These are obtained by taking the symmetric powers $\mathrm{Sym}^n \rho$ of the two-dimensional representation ρ. Now $\mathrm{Sym}^n \rho$ is a obtained as a direct summand of $\rho^{\otimes n}$ by writing it as the image of a certain symmetrization projector, and since ρ arises from $H^1(E)$ by a twist, the Künneth theorem assures us that $\rho^{\otimes n}$, and hence also $\mathrm{Sym}^n \rho$, will be cut out of (some twist of) $H^n(E^n)$ by some correspondence. Thus $\mathrm{Sym}^n \rho$ is motivic. So what we need to know, to prove the Sato–Tate conjecture, is that each of the motivic L-functions $L(s, \mathrm{Sym}^n \rho)$ has an analytic continuation up to $\Re s \geq 1$, with no zeros or poles in this region.

Now, although the representation ρ is only hypothetical, its L-function is not. Indeed, since hypothetically $\rho = H^1(E) \otimes |\ |^{1/2}$, we see that (again, hypothetically)[59] $\mathrm{Sym}^n \rho = \mathrm{Sym}^n H^1(E) \otimes |\ |^{n/2}$. Thus

$$L(s, \mathrm{Sym}^n \rho) = L\big(s, \mathrm{Sym}^n H^1(E) \otimes |\ |^{n/2}\big) = L\left(s + \frac{n}{2}, \mathrm{Sym}^n H^1(E)\right),$$

and this equation gives the left-hand L-function a meaning, even if ρ is only hypothetical. Furthermore, the Riemann Hypothesis ensures that the Euler product defining $L(s, \mathrm{Sym}^n H^1(E))$ converges if $\Re s > 1 + \frac{n}{2}$, and so we

---

[59] Now $H^1(E)$ just stands for the compatible family $\{H^1(\bar{E}, \mathbb{Q}_\ell)\}$, and we can simply form its symmetric power to obtain a compatible family $\mathrm{Sym}^n H^1(E)$, given by the representations $\{\mathrm{Sym}^n H^1(\bar{E}, \mathbb{Q}_\ell)\}$.

see that the Euler product defining $\mathcal{L}(s, \mathrm{Sym}^n \rho)$ converges if $\Re s > 1$ – so that L-functions of the unitary Langlands group representation $\mathrm{Sym}^n \rho$ behave formally just like the L-functions of unitary Weil group representations, or unitary Hecke characters: the Euler products converge if $\Re s > 1$ – and our problem is to analytically continue them to the line $\Re s = 1$.

For Artin L-functions, this was achieved by Brauer's Theorem 15.5.1.3. We will briefly recall in Section 15.7.3 below that Brauer's theorem plays a role in achieving the analogous result here as well.

We remark that in general, for any smooth projective $\mathbb{Q}$-scheme X (or, more generally, any motive), we may hypothetically form $H^i(X) \otimes |\ |^{i/2}$ as a representation of the Langlands group $L_{\mathbb{Q}}$, and the Riemann Hypothesis will ensure that this representation, if it existed, would be unitary. The Riemann Hypothesis also ensures that

$$L\big(s, H^i(X) \otimes |\ |^{i/2}\big) = L\big(s + \frac{i}{2}, H^i(X)\big)$$

converges if $s > 1$. The left-hand L-function is sometimes said to be *unitarily* normalized. This can be thought of[60] as referring to the hypothetical unitary representation $H^i(X) \otimes |\ |^{i/2}$.

15.5.10 The Conjectural Langlands Group – an Elaboration

We first recall the definition of the local Weil groups. They come in two flavors, archimedean and nonarchimedean. The archimedean local Weil groups are the easiest to describe: $W_{\mathbb{C}} = \mathbb{C}^\times$, while $W_{\mathbb{R}}$ is the unique nonsplit extension of $\mathrm{Gal}(\mathbb{C}/\mathbb{R})$ by $\mathbb{C}^\times$ (with $\mathrm{Gal}(\mathbb{C}/\mathbb{R})$ acting on $\mathbb{C}^\times$ via complex conjugation).[61]

In analogy with the short exact sequence (15.36) (and thinking of $W_{\mathbb{C}}$ as $W_{\mathbb{C}/\mathbb{C}}$ – since $\mathbb{C}$ has no nontrivial finite extensions – and of $W_{\mathbb{R}}$ as $W_{\mathbb{C}/\mathbb{R}}$ – since $\mathbb{C}$ is the algebraic closure of $\mathbb{R}$), we may place these Weil groups in short exact sequences

$$1 \longrightarrow \mathbb{C}^\times \longrightarrow W_{\mathbb{C}} \longrightarrow \mathrm{Gal}(\mathbb{C}/\mathbb{C}) = 1 \tag{15.60}$$

and

$$1 \longrightarrow \mathbb{C}^\times \longrightarrow W_{\mathbb{R}} \longrightarrow \mathrm{Gal}(\mathbb{C}/\mathbb{R}) \longrightarrow 1. \tag{15.61}$$

---

[60] Although it might be historically more accurate to say that this name is derived from its meaning on the automorphic side of reciprocity, where the corresponding process amounts to twisting so as to obtain a unitary central character.

[61] If we regard $\mathbb{R}_{>0}^\times$ as a subgroup of scalar matrices in $\mathrm{GL}_2(\mathbb{C})$, then $\mathbb{C}^\times = \mathbb{R}_{>0}^\times U(1)$ embeds into $\mathbb{R}_{>0}^\times SU(2)$, and $W_{\mathbb{R}}$ can also be described as the normalizer of $\mathbb{C}^\times$ in $\mathbb{R}_{>0}^\times SU(2)$.

Of course, $W_{\mathbb{C}}^{ab} = W_{\mathbb{C}/\mathbb{C}} = W_{\mathbb{C}} = \mathbb{C}^{\times}$, while $W_{\mathbb{R}}^{ab} = W_{\mathbb{R}/\mathbb{R}} = \mathbb{R}^{\times}$, the identification with $\mathbb{R}^{\times}$ being uniquely determined by the requirement that it induces the morphism $z \mapsto z\bar{z}$ when pulled back to $\mathbb{C}^{\times} \subset W_{\mathbb{R}}$.

If F_v is a nonarchimedean local field (which we denote in this fashion in anticipation of the fact that it will arise by completing a global field F at some nonarchimedean place v), with residue field $\mathbb{F}_q$, then there there is a short exact sequence of groups

$$1 \to I_v \to \mathrm{Gal}(\overline{F}_v/F_v) \to \mathrm{Gal}(\overline{\mathbb{F}}_q/\mathbb{F}_q) \to 1,$$

where I_v denotes the inertia subgroup of $\mathrm{Gal}(\overline{F}_v/F_v)$. The Galois theory of finite fields shows that the group $\mathrm{Gal}(\overline{\mathbb{F}}_q/\mathbb{F}_q)$ is isomorphic to $\widehat{\mathbb{Z}}$, with topological generator the Frobenius element $\mathrm{Frob}_v = \mathrm{Frob}_q$ (the automorphism of $\overline{\mathbb{F}}_q$ defined by $x \mapsto x^q$). Inside this Galois group we may consider the cyclic subgroup (just a copy of $\mathbb{Z}$) that is literally generated by Frob_v. Then the Weil group W_{F_v} is the preimage of this cyclic group in $\mathrm{Gal}(\overline{F}_v/F_v)$, so that W_{F_v} sits in a short exact sequence

$$1 \to I_v \to W_{F_v} \to \langle \mathrm{Frob}_v \rangle \to 1.$$

We recall that if K_v is a finite extension of F_v, then $\mathrm{Gal}(\overline{F}_v/K_v)^c$ (the closure of the commutator subgroup of $\mathrm{Gal}(\overline{F}_v/K_v)$) is contained in I_v, and so we may set $W_{K_v/F_v} = W_{F_v}/\mathrm{Gal}(\overline{F}_v/K_v)^c$. Local class field theory then yields a morphism of exact sequences analogous to (15.36)

$$\begin{array}{ccccccccc}
1 & \longrightarrow & K_v^{\times} & \longrightarrow & W_{K_v/F_v} & \longrightarrow & \mathrm{Gal}(K_v/F_v) & \longrightarrow & 1 \\
 & & \downarrow & & \downarrow & & \| & & \\
1 & \longrightarrow & \mathrm{Gal}(K_v^{ab}/K_v) & \longrightarrow & \mathrm{Gal}(\overline{F}_v/F_v)/\mathrm{Gal}(\overline{F}_v/K_v)^c & \longrightarrow & \mathrm{Gal}(K_v/F_v) & \longrightarrow & 1
\end{array}$$
$$(15.62)$$

and of course $W_{F_v} = \varprojlim_{K_v} W_{K_v/F_v}$.

If F_v is a completion of the number field F at a place v, then there is a morphism[62] $W_{F_v} \to W_F$ compatible with the short exact sequences (15.36) and whichever choice of (15.60), (15.61), or (15.62) is appropriate to v.

We may also define local versions of the Langlands group. If v is archimedean, then $L_{F_v} = W_{F_v}$. If v is nonarchimedean, then $L_{F_v} = W_{F_v} \times SU(2)$. (From the viewpoint of Conjecture 15.5.9.2, relating the hypothetical global Langlands group to compatible families of ℓ-adic Galois representations, the role of the $SU(2)$-factor is to account for the potentially

---

[62] To determine this unambiguously, we should extend the embedding $F \hookrightarrow F_v$ to an embedding $\overline{\mathbb{Q}} = \overline{F} \hookrightarrow \overline{F}_v$.

infinite-order action of tame inertia on ℓ-adic representations.) In either case, we have a surjection $L_{F_v} \to W_{F_v}$.

We now elaborate on Conjecture 15.5.9.2. We conjecture the existence of a locally compact group L_F, the Langlands group of F, equipped with a surjection $L_F \to W_F$ whose kernel is compact and which induces an isomorphism

$$L_F^{\mathrm{ab}} \to W_F^{\mathrm{ab}} = W_{F/F} = F^\times \backslash \mathbb{A}_F^\times,$$

and also equipped with morphisms $L_{F_v} \to L_F$ for each place v that are compatible with the maps to and between local and global Weil groups. Granting this, if $\rho : L_F \to \mathrm{GL}_n(\mathbb{C})$ is a continuous representation, then we may pull-back ρ to a morphism $\rho_v : W_{F_v} = L_{F_v} \to \mathrm{GL}_n(\mathbb{C})$ for each archimedean place v of F. We say that ρ is *algebraic* if each of these pull-backs ρ_v becomes algebraic when restricted to the copy of $\mathbb{C}^\times \subset W_{F_v}$; i.e. if the induced morphism $\mathbb{C}^\times \to \mathrm{GL}_n(\mathbb{C})$ is a product of n algebraic characters of $\mathbb{C}^\times$, i.e. characters of the form $z \mapsto z^p \bar{z}^q$ for some $p, q \in \mathbb{Z}$.

We now conjecture that algebraic representations of L_F correspond to motivic compatible families of representations of $\mathrm{Gal}(\overline{\mathbb{Q}}/F)$. To make this precise, we follow Langlands [Lan79] in using the language of Tannakian categories.[63] We can consider the category of algebraic representations of L_F; this will be a Tannakian category, and we can form its Tannakian Galois group, a proalgebraic group over $\mathbb{C}$, which we denote by L_F^{alg}. Assuming that Grothendieck's category of motives exists, we may also form its Tannakian group G_F^{mot}; this is a proalgebraic group over $\mathbb{Q}$, or over $\overline{\mathbb{Q}}$ if we allow our motives to have coefficients. The existence of compatible families of Galois representations can be expressed in terms of the existence of morphisms $\mathrm{Gal}(\overline{\mathbb{Q}}/F) \to G_F^{\mathrm{mot}}(\overline{\mathbb{Q}}_\ell)$ (one for each ℓ and each choice of embedding $\overline{\mathbb{Q}} \to \overline{\mathbb{Q}}_\ell$) having certain properties. A more precise formulation of Conjecture 15.5.9.2, then, is that there is an isomorphism of proalgebraic groups over $\mathbb{C}$

$$L_F^{\mathrm{alg}} \xrightarrow{\sim} (G_F^{\mathrm{mot}})_{/\mathbb{C}}.$$

This isomorphism should again satisfy various compatibilities, whose details we omit. (For example, it should be compatible with the construction of Section 15.4.5.)

This more precise form of Conjecture 15.5.9.2 is the top layer of a rather tall stack of conjectures, since its formulation depends on the existence of a category of motives satisfying a range of desirable, but so far unproved, properties. Nevertheless, in his work (e.g. [Lan79] and [LR87]) Langlands has consistently shown the value of substantively engaging with this circle of ideas, and of incorporating them into one's thinking. For this reason, we believe it is worthwhile, and in keeping with Langlands's own point-of-view, to place Conjecture 15.5.9.2 at the centre of our discussion of reciprocity.

[63] For abelian representations, this viewpoint was already introduced in [Ser68].

15.6 Automorphic L-Functions and Reciprocity

15.6.1 Automorphic Representations

Following the definition in Hecke [Hec27] (or Serre [Ser77] for a more contemporary reference), we may regard a modular form of weight k and level 1 as a function

$$f : \{ \Lambda \mid \Lambda \subset \mathbb{C} \text{ a lattice } \} \to \mathbb{C},$$

which satisfies the condition $f(\lambda \Lambda) = \lambda^{-k} f(\Lambda)$ for any $\lambda \in \mathbb{C}^{\times}$, which is holomorphic as a function of those τ for which $\Im \tau > 0$, if we associate to τ the lattice $\mathbb{Z} + \mathbb{Z}\tau$, and which satisfies a moderate growth condition as $\tau \to i\infty$.

Now choosing a *based* lattice in $\mathbb{C} \cong \mathbb{R}^2$ is the same as choosing a basis of $\mathbb{R}^2$, i.e. an element of $\mathrm{GL}_2(\mathbb{R})$. Changing basis amounts to applying an element of $\mathrm{GL}_2(\mathbb{Z})$. So modular forms of weight k and level 1 are functions on the quotient

$$\mathrm{GL}_2(\mathbb{Z}) \backslash \mathrm{GL}_2(\mathbb{R}) \tag{15.63}$$

satisfying certain additional conditions. We now follow the adèlic philosophy of replacing $\mathbb{Z} \subseteq \mathbb{R}$ by $\mathbb{Q} \subseteq \mathbb{A}_{\mathbb{Q}}$, and observe that we can rewrite (15.63) as

$$\mathrm{GL}_2(\mathbb{Q}) \backslash \mathrm{GL}_2(\mathbb{A}_{\mathbb{Q}}) / \mathrm{GL}_2(\widehat{\mathbb{Z}}),$$

so that modular forms of weight k and level 1 are functions on the quotient $\mathrm{GL}_2(\mathbb{Q}) \backslash \mathrm{GL}_2(\mathbb{A}_{\mathbb{Q}})$ satisfying certain conditions. Being invariant under the right translation action of $\mathrm{GL}_2(\widehat{\mathbb{Z}})$ is one of those conditions, and it is natural to omit it, thus incorporating modular forms of higher level into our picture. Relaxing the archimedean conditions (holomorphicity and the precise weight k condition) is also a sensible step; this incorporates Maass forms and nonholomorphic Eisenstein series into our picture.

Thus we are led to define the space of automorphic forms $\mathcal{A}(\mathrm{GL}_2(\mathbb{Q}) \backslash \mathrm{GL}_2(\mathbb{A}))$; it is the space of functions $f : \mathrm{GL}_2(\mathbb{Q}) \backslash \mathrm{GL}_2(\mathbb{A}) \to \mathbb{C}$, which are smooth in the archimedean variables and of moderate growth as these variables tend to ∞; invariant under some open subgroup of $\mathrm{GL}_2(\mathbb{A}_{\mathbb{Q}}^{\infty})$; and which are generalized eigenvectors for the centre of the enveloping algebra of the Lie algebra $\mathfrak{gl}_2$ of $\mathrm{GL}_2(\mathbb{R})$ (so essentially we are asking them to be eigenvectors for the Casimir operator, which can be interpreted as the hyperbolic Laplacian in the classical picture). The space $\mathcal{A}(\mathrm{GL}_2(\mathbb{Q}) \backslash \mathrm{GL}_2(\mathbb{A}))$ is a complete locally convex topological vector space,[64] which becomes an *admissible* representation of $\mathrm{GL}_2(\mathbb{A})$ under the action by right translation. An automorphic representation of $\mathrm{GL}_2(\mathbb{A})$ is then, by definition, an irreducible

---

[64] We work with topological vector spaces so as to avoid having to take $O(2)$-finite vectors; the equivalence between this viewpoint and the $O(2)$-finite vector viewpoint follows from Harish-Chandra's theory of admissible representations of real reductive groups.

subquotient (in the topological sense, i.e. the quotient of one closed invariant subspace by another that it contains, which is irreducible in the topological sense of containing no proper closed invariant subspace) π of $\mathcal{A}\big(\mathrm{GL}_2(\mathbb{Q}) \setminus \mathrm{GL}_2(\mathbb{A})\big)$.

These notions extend directly to any connected reductive affine algebraic group G over a number field F – we may define the space $\mathcal{A}\big(G(F) \setminus G(\mathbb{A}_F)\big)$ of automorphic forms, which is an admissible representation of $G(\mathbb{A}_F)$ via right translation, and then define an *automorphic representation* π of $G(\mathbb{A}_F)$ to be a topologically irreducible subquotient of this space. Since $G(\mathbb{A}_F)$ is the restricted direct product of the groups $G(F_v)$ (as v runs over all places of F), we may factor the irreducible representation π as a restricted tensor product (suitably defined)

$$\pi = \bigotimes{}' \pi_v, \tag{15.64}$$

where each π_v is an irreducible admissible representation of $G(F_v)$ [Fla79]. (This is the analogue in the nonabelian setting of the factorization (15.20) of an idèle class character.)

The space $\mathcal{A}\big(G(F) \setminus G(\mathbb{A}_F)\big)$ contains a subspace $\mathcal{A}^\circ\big(G(F) \setminus G(\mathbb{A}_F)\big)$ of cuspforms, whose elements are characterized by the vanishing of certain integrals that compute their "constant terms" at the boundary of $G(F)\setminus G(\mathbb{A}_F)$. The space $\mathcal{A}^\circ\big(G(F) \setminus G(\mathbb{A}_F)\big)$ has a positive definite $G(\mathbb{A}_F)$-equivariant inner product (the L^2-inner product, also called the Petersson inner product), and decomposes as a direct sum of irreducible representations. These are the *cuspidal* automorphic representations.

We refer the reader to [BJ79] for more details regarding these definitions.

Returning to the case of GL_2 over $\mathbb{Q}$ and classical modular forms: if f is a classical cuspform, then we have seen that we can interpret f as an automorphic form, i.e. as an element $f \in \mathcal{A}^0(\mathrm{GL}_2(\mathbb{Q}) \setminus \mathrm{GL}_2(\mathbb{A}))$, and we may then consider the $\mathrm{GL}_2(\mathbb{A}_\mathbb{Q})$-subrepresentation that it generates. It is a theorem, essentially due to Atkin and Lehner [AL70], that f (thought of as an automorphic form) generates an irreducible subrepresentation (i.e. an automorphic representation) if and only if f (thought of as a modular form of some weight and level) is an eigenform for the Hecke operator T_p for all but finitely many primes p. Thus automorphic representations can be regarded as a representation-theoretic generalization of the notion of Hecke eigenform, and the list of the factors π_v occurring in the the decomposition (15.64) as a refinement of the classical data consisting of the list of Hecke eigenvalues of f.

In particular, returning to the general case, if π_v is a constituent of an automorphic form that is *unramified*, then it gives rise to a Frobenius–Hecke conjugacy class $c_v \in \widehat{G} \times \mathrm{Frob}_v^{-1}$ (as explained by Langlands in [Lan70], building nontrivially on Satake's work [Sat63]; see also Shahidi's chapter in this volume [Sha] and the discussion of Section 15.6.3 below); this class is an analogue of the classical pth Hecke polynomial $X^2 - a_p X + \varepsilon(p)p^{k-1}$ associated to a modular Hecke eigenform. (Up to a possible issue of normalization, this polynomial is the characteristic polynomial of the Frobenius–Hecke conjugacy class that Langlands associates to the factor π_p of the automorphic representation π generated by f.)

15.6.2 Hecke

In his paper [Hec36], Hecke generalized Riemann's second proof of the analytic continuation and functional equation of $\zeta(s)$ to show that the Mellin transform $\Lambda(f,s)$ of a cuspform f of level 1 is the product of a Dirichlet series $L(f,s)$ with an appropriate Γ-factor, which admits an analytic continuation to an entire function and satisfies a functional equation. He also proved the converse: any entire function which satisfies a functional equation of the appropriate form, and which satisfies an appropriate growth condition, arises from a modular form of level 1.

In the papers [Hec37], he introduced the operators that now bear his name, and proved that f is a Hecke eigenform if and only if $L(f,s)$ admits an Euler product – in which case the Euler factors are of the form $(1 - a_p p^{-s} + p^{k-2s})^{-1}$, where a_p is the pth Hecke eigenvalue, and k is the weight of f. Although he extended the theory of Hecke operators to higher-level modular forms (and the theory, at least from a classical – i.e. nonadèlic – viewpoint was essentially completed by Atkin and Lehner [AL70]), the converse theorem for levels > 1 is not as obvious. It was eventually proved by Weil [Wei67]. Jacquet and Langlands [JL70] proved another form of the converse theorem, this time in the context of automorphic representations of $\mathrm{GL}_2(\mathbb{A}_F)$ for arbitrary number fields F. A key feature of their converse theorem is that the hypothesized functional equations must have the correct ϵ-factors. This provides one of the motivations for developing the theory discussed in Section 15.5.2.

We close this discussion by emphasizing that, just as in Riemann's work, Hecke's description of his L-functions associated to automorphic forms via an integral formula, and his analysis of their Euler product factorization, are completely independent of one another.

15.6.3 Langlands

We sketch Langlands's definition of automorphic L-functions, referring to Shahidi's chapter in this volume [Sha] for more details.

Recall that, if G is a connected reductive affine algebraic group over the number field F, then we may form first the dual group $\widehat{G}$, which is a group over $\mathbb{C}$ with dual root datum to that of G (the construction of $\widehat{G}$ depends only on G over $\overline{\mathbb{Q}}$), and then construct an action of the Galois group $\mathrm{Gal}(K/F)$ on $\widehat{G}$, so as to form the L-group

$$^L G = \widehat{G} \rtimes \mathrm{Gal}(K/F)$$

of G. The action of $\mathrm{Gal}(K/F)$ on $\widehat{G}$, and hence the L-group of G, depends only on the inner class of G over F.

Note that the L-group isn't entirely well-defined, since we are free to enlarge K, so it is better to speak of *an* L-group of G. This is actually an advantage in the theory. We are not even restricted to taking the semidirect product with a Galois group. Any group that surjects onto $\mathrm{Gal}(K/F)$ (and hence can be made to act on $\widehat{G}$ through the original $\mathrm{Gal}(K/F)$-action) will do. Thus we can form the L-group in its "Weil form," as $^L G = \widehat{G} \rtimes W_{K/F}$, or $^L G = \widehat{G} \rtimes W_F$. If we grant ourselves the existence of the Langlands group, we could even form $^L G = \widehat{G} \rtimes L_F$.

In general, suppose that Γ is a group that surjects onto $\mathrm{Gal}(K/F)$, and that we form $^L G = \widehat{G} \rtimes \Gamma$. Suppose now that H is any group equipped with a homomorphism to Γ (and so also with a homomorphism to $\mathrm{Gal}(K/F)$, by composition). Then giving an L-*homomorphism* $H \to {}^L G = \widehat{G} \rtimes \Gamma$, i.e. a homomorphism that induces the given map upon projection to Γ, is the same as giving an L-*homomorphism* $H \to {}^L G = \widehat{G} \rtimes \mathrm{Gal}(K/F)$. (This just follows from description of $\widehat{G} \rtimes \Gamma$ as the fiber product $(\widehat{G} \rtimes \mathrm{Gal}(K/F)) \times_{\mathrm{Gal}(K/F)} \Gamma$.) Thus the notion of L-homomorphism is suitably compatible with the flexibility in the formation of the L-group of G.

If $\mathfrak{p}$ is a prime F, and $K_{\mathfrak{p}}$ is an extension of $F_{\mathfrak{p}}$ that splits G, then we may perform a similar local construction, obtaining a semidirect product $\widehat{G} \rtimes \mathrm{Gal}(K_{\mathfrak{p}}/F_{\mathfrak{p}})$, and similar considerations apply. In particular, if G is unramified at $\mathfrak{p}$ (i.e. quasisplit and split over an unramified extension), then we may choose $K_{\mathfrak{p}}/F_{\mathfrak{p}}$ to be unramified, so that $\mathrm{Gal}(K_{\mathfrak{p}}/F_{\mathfrak{p}}) = \langle \mathrm{Frob}_{\mathfrak{p}} \rangle$. Then $\widehat{G}$-conjugacy classes of L-homomorphisms $\varphi : L_{F_{\mathfrak{p}}} \to {}^L G$ (for any choice of L-group; and remember that $L_{F_{\mathfrak{p}}} = W_{F_{\mathfrak{p}}} \times SU(2)$ is the Langlands group of $F_{\mathfrak{p}}$) correspond to $\widehat{G}$-conjugacy class of L-homomorphisms $L_{F_{\mathfrak{p}}} \to \widehat{G} \rtimes \langle \mathrm{Frob}_{\mathfrak{p}} \rangle$. We define one of the latter (and hence also one of the former) homomorphisms to be unramified if it is trivial on $I_v \times SU(2)$; giving an unramified φ then amounts to giving an element $c \in \widehat{G} \times \mathrm{Frob}_p^{-1}$, up to $\widehat{G}$-conjugacy, the *Frobenius–Hecke* conjugacy class associated to the unramified L-homomorphism φ. The class c and the unramified L-homomorphism φ determine one another.

We now consider a representation $\rho : {}^L G \to \mathrm{GL}_n$, required to be algebraic when restricted to $\widehat{G}$, and continuous on the Galois/Weil factor in $^L G$. Thus ρ will be unramified at all but finitely many primes $\mathfrak{p}$ of F. In other words, there is a finite set of primes S such that for $\mathfrak{p} \notin S$, when we restrict ρ to $\widehat{G} \rtimes W_{F_{\mathfrak{p}}}$ for $\mathfrak{p} \notin S$, the action of $W_{F_{\mathfrak{p}}}$ on $\widehat{G}$ factors through the quotient $W_{F_{\mathfrak{p}}}/I_{\mathfrak{p}} = \langle \mathrm{Frob}_{\mathfrak{p}} \rangle$, and the restriction of ρ to $\widehat{G} \rtimes W_{F_{\mathfrak{p}}}$ factors through $\widehat{G} \rtimes \langle \mathrm{Frob}_{\mathfrak{p}} \rangle$.

Suppose now that π is an automorphic representation of $G(\mathbb{A}_F)$. Then we may form the restricted tensor factorization $\pi = \bigotimes'_v \pi_v$. The group G is unramified at all but finitely many primes $\mathfrak{p}$, and for all but finitely many of these primes, the local factor $\pi_{\mathfrak{p}}$ will be unramified. As recalled in Section 15.6.1 above, for each such prime p, we obtain a corresponding

Frobenius–Hecke conjugacy class, i.e. an element $c_{\mathfrak{p}} \in \widehat{G} \rtimes \mathrm{Frob}_{\mathfrak{p}}^{-1}$, well-defined up to $\widehat{G}$-conjugacy.

Now enlarge the finite set of places S so that it includes all the bad primes for ρ, all the bad primes for π (i.e. the primes for which $c_{\mathfrak{p}}$ is not defined), and also all the infinite places. Langlands then defines the (imprimitive) automorphic L-function associated to π and a representation $\rho : {}^{L}G \to \mathrm{GL}_n$ *as an Euler product*, via the formula

$$L^{S}(s,\pi,\rho) = \prod_{\mathfrak{p} \notin S} \frac{1}{\det\big(\mathrm{Id}_{n \times n} - N(\mathfrak{p})^{-s}\rho(c_{\mathfrak{p}})\big)} \qquad (15.65)$$

Note the similarity to Artin's definition (15.39) of his L-functions!

The problem of completing the L-function (15.65) by adding Euler factors at the primes in S is a difficult one. We first recast the Frobenius–Hecke classes $c_{\mathfrak{p}}$ as unramified L-homomorphisms $\varphi_{\mathfrak{p}} : L_{F_{\mathfrak{p}}} \to {}^{L}G$, in the manner described above. We would then like to use the remaining $\pi_{\mathfrak{p}}$ to produce L-homomorphisms $\varphi_v : L_{F_v} \to {}^{L}G$ which are no longer necessarily unramified. The *local Langlands conjecture* for G states that this is possible. It is proved in a number of cases (we can vary G, and also, having fixed G, impose various ramification conditions on the π), and in particular is proved in the archimedean case for arbitrary G by Langlands himself [Lan89], and is proved completely for GL_n by Harris–Taylor and Henniart [HT01, Hen00]; but it remains open in general.

Langlands proves that the Euler product (15.65) converges in some right half-plane. There is one particular case in which we can say more. Let $G = \mathrm{GL}_n$, so that $\widehat{G} = \mathrm{GL}_n$ also, and take ρ to be the *standard representation*, i.e. simply the identity map of GL_n to itself; the resulting L-functions are referred to as the *standard L-functions*, and we denote them $L(s,\pi,\mathrm{standard})$. The work of Godement–Jacquet [GJ72] establishes the analytic continuation and functional equation for these L-functions by methods generalizing those of Tate's thesis. In general, Langlands conjectures that, after completing the Euler product (15.65) by adding factors at the missing places, the resulting function will admit an analytic continuation to the entire plane, and satisfy an appropriate functional equation. But in fact, he conjectures more: as we recall below, his *functoriality* conjecture states that an arbitrary automorphic L-function $L(s,\pi,\rho)$ should in fact be a standard L-function (for some other π, to be sure), whose analytic continuation and functional equation is then assured by [GJ72].

15.6.4 The Reciprocity Conjecture

If π is an automorphic representation of $G(\mathbb{A}_F)$, then we obtain the unramified L-homomorphisms $\varphi_{\mathfrak{p}} : W_{F_{\mathfrak{p}}} \to {}^{L}G$, for $\mathfrak{p} \notin S$, and, if we grant the existence of the local Langlands correspondence, we in fact obtain L-homomorphisms $\varphi_v : L_{F_v} \to {}^L G$ for every place v. We could now ask: is there an L-homomorphism $\varphi : L_F \to {}^L G$ which induces the various φ_v. The answer is "no" in general,[65] even if $G = GL_n$. But we have the following conjecture.

Conjecture 15.6.4.1 (Langlands reciprocity) *There is a bijection between irreducible representations* $\varphi : L_F \to GL_n$ *and cuspidal automorphic forms* π *on* $GL_n(\mathbb{A}_F)$, *via which* φ *and* π *correspond if* $\varphi_{|F_v} = \varphi_v$ *(the local L-homomorphism associated to* π_v *via the local Langlands correspondence) for all places v.*

Why do we call this a reciprocity law? Because, according to Conjecture 15.5.9.2, each motive over F gives rise to a representation ρ of L_F. By the conjecture, each irreducible constituent of ρ then corresponds to a cuspidal automorphic representation π of some $GL_n(\mathbb{A}_F)$. In particular, $L(s, \rho)$ (which will coincide with the L-function of our original motive) is a product of standard L-functions $L(s, \pi, \text{standard})$, and so has the anticipated analytic continuation and functional equation.

The conjecture also has an implication in the reverse direction: if π is a cuspidal automorphic representation on $GL_n(\mathbb{A}_F)$, whose factors at the infinite places are algebraic,[66] then the conjecture produces an irreducible representation ρ of L_F, which is algebraic in the sense of Conjecture 15.5.9.2, and so should correspond to a compatible family of ℓ-adic Galois representations coming from a motive over F.

15.6.4.2 Reciprocity in Terms of L-Functions

A key point, which we have already intimated, is that the reciprocity conjecture can be phrased in terms of L-functions, and hence stated (and studied!) independently of the problem of the existence of the Langlands group, at least in the algebraic case. Thus, suppose that π is an algebraic cuspidal automorphic representation of $GL_n(\mathbb{A}_F)$, conjecturally corresponding to an

---

[65] For $G = GL_n$, this is due to *nonisobaric* automorphic representations, in the sense of Langlands' paper [Lan79]. For more general G, it is then related to the representations that Langlands (in the same paper) calls *anomolous*. Arthur's theory of nontempered endoscopy [Art89] aims to explain automorphic representations that appear in $L^2\big(G(F) \setminus G(\mathbb{A}_F)\big)$, and which are anomolous in the sense of [Lan79], in terms of L-homomorphisms $L_F \times SL_2 \to {}^L G$.

[66] Suppose that $\pi = \bigotimes' \pi_v$ is an automorphic representation of some $G(\mathbb{A}_F)$. At each archimedean place v, the factor π_v induces a morphism $\varphi_v : L_{F_v} \to {}^L G$ via the local Langlands correspondence [Lan89], which in turn restricts to a morphism $\mathbb{C}^\times \to \widehat{G}$. We can then ask that this morphism be algebraic, in the sense that it factor through a maximal torus $\widehat{T}$ of $\widehat{G}$ where it is then described by characters of the form $z \mapsto z^p z^q$. If this condition holds, we say that π_∞ is *algebraic*. This corresponds to the notion of L-algebraic, in the terminology of [BG14].

algebraic representation $\varphi : L_F \to GL_n(\mathbb{C})$. This algebraic representation in turn corresponds to a motive M over F, and the conjectured relationship between π and φ can be rephrased as an equality

$$L(s, \pi, \text{standard}) = L(s, \varphi) = L(s, M). \tag{15.66}$$

Thus, phrased in terms of L-functions this way, we can study the conjecture in either direction: we can begin with an algebraic cuspidal automorphic representation π and try to construct a motive M, or at least a compatible family of ℓ-adic Galois representations, satisfying (15.66). Conversely, we can begin with a (suitably irreducible) motive M and try to construct a cuspidal automorphic representation π satisfying (15.66).

15.6.4.3 Functoriality[67]

Langlands has made another fundamental conjecture related to L-groups and his L-functions, namely his *functoriality* conjecture. To describe it, we suppose that G and H are two connected reductive linear algebraic groups over F, and that $\rho : {}^L G \to {}^L H$ is an L-homomorphism of their L-groups. Suppose also that π is an automorphic representation of $G(\mathbb{A}_F)$, giving rise to the Frobenius–Hecke classes $c_\mathfrak{p}$ for good primes $\mathfrak{p}$. We may then consider the collection of classes $\rho(c_\mathfrak{p})$ in ${}^L H$, and Langlands conjectures that if H is furthermore quasi-split,[68] then there is an automorphic representation Π of $H(\mathbb{A}_F)$ whose associated Frobenius–Hecke classes are equal to the classes $r(c_\mathfrak{p})$ (for all but finitely many $\mathfrak{p}$). In short, automorphic representations are *functorial* in the L-group.

We can describe the conjectured relationship between π and Π in terms of their L-functions: if $r : {}^L H \to GL_n$, then a direct consideration of the definitions shows that (taking S to be sufficiently large)

$$L^S(s, \pi, r \circ \rho) = L^S(s, \Pi, r). \tag{15.67}$$

Example 15.6.4.4 One example of functoriality is the case when $H = GL_n$. Then ${}^L H$ is simply the direct product of GL_n with the Galois group, and so we may ignore the Galois factor, so that ρ is just a homomorphism ${}^L G \to GL_n$. In this case, taking r to be the standard representation, the formula (15.67) reduces to the equality

$$L^S(s, \pi, \rho) = L^S(s, \Pi, \text{standard}). \tag{15.68}$$

Thus functoriality implies that *all* automorphic L-functions are *standard* L-functions, and so in particular have analytic continuation and satisfy a functional equation.

---

[67] For an elaboration on this topic, see Arthur's chapter in this volume [Art].
[68] In Example 15.6.4.6 we explain the significance of this assumption.

Example 15.6.4.5 Consider now the case when G is the trivial group! As Langlands has emphasized, although G is trivial, its L-group is not. If we take the Galois form for L-groups, then can write its L-group as $\mathrm{Gal}(K/F)$ for some finite extension K, and $\rho : \mathrm{Gal}(K/F) \to \mathrm{GL}_n$ is simply a representation of the kind considered in Section 15.5.1. There is only one automorphic representation of the trivial group, namely the trivial representation, and its Frobenius–Hecke classes are just the Frobenius conjugacy classes $\mathrm{Frob}_{\mathfrak{p}} \in \mathrm{Gal}(K/F)$. Now apply functoriality, in particular (15.68). The L-function of the trivial representation is precisely the Artin L-function of ρ. Thus functoriality predicts the existence of an automorphic representation Π of $\mathrm{GL}_n(\mathbb{A})$ for which

$$L(s, \rho) = L(s, \Pi, \text{standard})). \qquad (15.69)$$

If ρ is irreducible then we anticipate that Π should be cuspidal, in which case Artin's Conjecture 15.5.1.1 would hold for ρ. In this way Artin's conjecture is situated in the much more general context of functoriality.

Of course, since the representation ρ is a particular example of a motive, the existence of Π satisfying (15.69) is also a special case of the reciprocity conjecture. In fact, assuming that the Langlands group L_F exists, we could use the L_F-form of the L-group. In this case, if G is the trivial group, then $^LG = L_F$, and giving an L-homomorphism ρ as above amounts to giving a morphism $\rho : L_F \to \mathrm{GL}_n$. Applying functoriality in this context, we obtain an automorphic representation Π of $\mathrm{GL}_n(\mathbb{A}_F)$ for which $L(s, \rho) = L(s, \Pi, \text{standard})$. Thus functoriality, when interpreted in a sufficiently broad manner, implies the "motivic to automorphic" direction of reciprocity!

Example 15.6.4.6 If G is any connected linear reductive group over F, we can take H to be the quasisplit inner form of G, and let ρ be the equality $^LG = {}^LH$. We thus expect to have a functorial transfer of automorphic representations from $G(\mathbb{A}_F)$ to $H(\mathbb{A}_F)$. It is typically *not* surjective – automorphic representations for the quasi-split group H can't always be "descended" to the inner form H. This is why, in the statement of functoriality, we assume that H is quasisplit.

In the case when $H = \mathrm{GL}_2$, this transfer was established by Jacquet and Langlands in their book [JL70]. They did this via an application of the trace formula, and Langlands has emphasized the importance of the trace formula as a tool to study functoriality in general.

Example 15.6.4.7 There are many other examples of functoriality that we could mention in this context, and which will play a role in the discussion of the next section. Among them are base change, automorphic induction, symmetric power functoriality (corresponding to the representation $\mathrm{Sym}^n : \mathrm{GL}_2 = \widehat{\mathrm{GL}}_2 \to \mathrm{GL}_{n+1} = \widehat{\mathrm{GL}}_{n+1}$), and endoscopic functoriality. Another important example occurs when G is a classical (i.e. special orthogonal or symplectic) group. In this case $\widehat{G}$ is again classical, and we may consider $\rho : {}^L G \rightarrow GL_n$ arising from the standard representation of $\widehat{G}$. This case of functoriality (subject to some splitness/quasisplitness assumptions on G) has been proved by Arthur [Art13], as part of his theory of nontempered endoscopy and stabilization of the trace formula [Art89].

15.7 Progress

The proof of class field theory – abelian reciprocity – can be thought of as consisting of two steps: (i) constructing the ray class fields, for which one has an explicit reciprocity law (i.e. their Galois groups, and the Frobenius elements inside them, are described in automorphic terms), and (ii) showing that any abelian extension is contained in a ray class field. In practice, in modern proofs of class field theory, this order of argumentation is not necessarily observed (e.g. one can prove Artin reciprocity for arbitrary abelian extensions before proving the existence theorem for ray class fields), but it provides a helpful framework for viewing the dominant contemporary approaches to studying reciprocity.

The analogue of ray class fields, then, will be certain varieties (or motives) that admit an explicit reciprocity law, i.e. for which the Galois action on their ℓ-adic cohomology can be described in automorphic terms. These are the *Shimura varieties*. Although not all motives over number fields can be constructed out of them, they play a fundamental role in the study of reciprocity.

15.7.1 Shimura Varieties

These are a class of varieties attached to certain reductive groups over $\mathbb{Q}$, originally studied by Shimura, and subsequently by Langlands (who coined their name) and a host of other researchers. The simplest examples arise from the groups GL_2 over $\mathbb{Q}$, for which the associated Shimura varieties are modular curves, and the groups $\mathrm{Res}_{\mathbb{Q}}^{E}\mathbb{G}_m$ for imaginary quadratic extensions E of $\mathbb{Q}$ (which are forms of the group $\mathbb{R}_{>0}^{\times}U(1) \subset GL_2(\mathbb{C})$), whose associated Shimura varieties are finite sets of "singular" j-invariants (i.e. j-invariants of CM elliptic curves).

The second example above is of course famously related to the class field theory of quadratic imaginary fields (Kronecker's *Jugendtraum*). The first example was also intensively studied throughout the nineteenth and twentieth centuries. In the 1950's, it came to be realized (by Eichler [Eic54] and Shimura [Shi58] particularly, and by Taniyama as well) that the ζ-functions of modular curves admitted a description in terms of the L-functions of weight-2 cuspforms. Exploiting the universal family of elliptic curves over the modular curve (or, in an alternative formulation, allowing nonconstant coefficients over the modular curves), Ihara [Iha67] found that higher weight cuspforms were also related to motivic ζ-functions.

Shimura introduced his eponymous varieties as generalizations of the modular curves, and (exploiting the twentieth-century outgrowths of Kronecker's *Jugendtraum*, i.e. class field theory and the theory of CM abelian varieties) he constructed their canonical models,[69] over an appropriate *reflex field*.[70] Langlands proposed the problem of studying their ζ-functions (with both constant and twisted coefficients), as a natural generalization of Kronecker's *Jugendtraum* [Lan76]. He further developed this suite of ideas in [Lan79] and [LR87].

There is an enormous amount of subsequent research on this problem, and it has proved to be one of the most stimulating directions of research in contemporary number theory and arithmetic geometry. (As just one example, Langlands's beginning investigations into this question led to his discovery of endoscopy.) We won't begin to attempt to describe the current state of the field. We do mention that Arthur has given a precise conjectural description of the ζ-function of the intersection cohomology of the minimal compactification of a Shimura variety in [Art89, §9]. These conjectures imply that these ζ-functions can be expressed in terms of automorphic L-functions. We also mention that these are typically not exhibited as standard L-functions, but as various L-functions attached to G and its endoscopic groups (if G is the group giving rise to the Shimura variety).[71]

We also mention Langlands's conjecture on conjugation of Shimura varieties [Lan79]. This conjecture proposes that the collection of Shimura varieties is stable under the action of $\mathrm{Gal}(\overline{\mathbb{Q}}/\mathbb{Q})$, even though any particular Shimura variety is defined only over its reflex field E (and so stable only by $\mathrm{Gal}(\overline{\mathbb{Q}}/E)$). In fact, if we think of Shimura's results (the so-called *Shimura reciprocity law*) as describing the action of $\mathrm{Gal}(\overline{\mathbb{Q}}/E)$ on H^0 of a particular Shimura variety, then Langlands's proposed extension describes the action of $\mathrm{Gal}(\overline{\mathbb{Q}}/E)$ on the collection of H^0s taken over all Shimura varieties. Langlands' reciprocity law was proved (in most cases) in [DMOS82]. As far as I know, no one has posited an extension of Langlands's conjecture to the general H^i of Shimura varieties,

---

[69] In many cases; the general case is treated in [Mil83].

[70] Shimura actually worked with what now called *connected* Shimura varieties, and showed that they are defined over ray class fields of the reflex field, the conductor of the field depending on the level. We follow Deligne's formulation of the theory [Del71b, Del79], and consider (disconnected, i.e. what are now called) Shimura varieties, which are defined over the reflex field, no matter what the level.

[71] The representation ρ of the L-group that appears in these L-functions is related to the datum originally used to define the Shimura variety. It first appeared in Langlands's letter to Lang [Lanb], and is also described in [Lan79].

although there should be such an extension, generalizing the conjecture of [Art89] for the action of $\mathrm{Gal}(\overline{\mathbb{Q}}/E)$ on the cohomology of a particular Shimura variety. The paper [Tay12] establishes a particular result of this type.

15.7.2 From Automorphic Representations to Galois Representations

Shimura varieties are endowed with large numbers of correspondences – "Hecke correspondences" – and so can be decomposed into motives according to the eigenvalues of the Hecke correspondences. Correspondingly, their ζ-functions will decompose into the product of the L-functions of these motives, which (following e.g. the general conjecture of [Art89]) one anticipates will be related to automorphic L-functions associated to G or to its endoscopic groups. Reversing the order of this discussion, one sees that if a particular automorphic representation "contributes" to the decomposition of the ζ-function, one can hope to find a motive which corresponds to this automorphic representation, in the sense that the L-function of the motive coincides with an L-function of the given automorphic representation.

The simplest case of this analysis is that of modular curves; in this case the upshot is the construction of a compatible family of ℓ-adic representations [Del71a] (and even a motive [Sch90]) associated to the automorphic representation of $\mathrm{GL}_2(\mathbb{A}_{\mathbb{Q}})$ generated by any cuspidal Hecke eigenform of weight $k \geq 2$. (The weight-2 case was treated first, by Shimura [Shi71].) This proves one direction of the reciprocity conjecture for such automorphic representations.

The preceding construction extends to the case of Hilbert modular forms, but in an indirect way: if F is totally real, then while $\mathrm{GL}_2(\mathbb{A}_F)$ gives rise to a Shimura variety (essentially a Hilbert modular variety), the reflex field is $\mathbb{Q}$, not F, and the cohomology of this Shimura variety won't produce the two-dimensional representations that ought to be attached to cuspidal automorphic representations on $\mathrm{GL}_2(\mathbb{A}_F)$. (Rather, it produces their "tensor inductions" from $\mathrm{Gal}(\overline{\mathbb{Q}}/F)$ to $\mathrm{Gal}(\overline{\mathbb{Q}}/\mathbb{Q})$.) Instead, one has to work with Shimura curves [Car86]; these are Shimura varieties attached to a nonsplit inner form of GL_2 over F, and so there are obstructions to functorially transferring automorphic representations from $\mathrm{GL}_2(\mathbb{A}_F)$ to these inner forms. The upshot is that one constructs compatible families of Galois representations for most, but not all, Hilbert modular eigenforms of weights $k_i \geq 2$.

In the case of modular eigenforms of weight $k = 1$, or Hilbert modular eigenforms with one or more of its weights $= 1$, there is no direct connection between these modular forms (or the automorphic representations that they generate) and the motives arising from modular curves or Shimura curves. One can still find associated families of ℓ-adic representations in these cases, but they are obtained by ℓ-adic limiting processes from the Galois representations associated to higher weight forms (using the theory of congruences of modular forms) [DS74, RT83, Oht84, Jar97]. In this case one works with a single ℓ at a time to construct the family; the compatibility comes at the end, by comparing the representations with varying ℓ to the fixed modular eigenform. There are no obvious motives giving rise to these compatible families. In the case of weight-1 modular forms, or parallel weight-1 Hilbert modular forms, one can prove a posteriori that the resulting Galois representations are of finite image, and hence *are* motivic. For partial weight-1 Hilbert modular forms, the problem of showing that these Galois representation are motivic remains open in general. The construction of compatible families associated to Hilbert modular forms of weights $k_i \geq 2$ that cannot be transferred to Shimura curves also proceeds by ℓ-adic limiting/congruence arguments. Again, the resulting compatible family is not evidently motivic. An alternative approach in this case is to find these compatible families in the cohomology of a higher-dimensional Shimura variety via endoscopy [BR93]; this approach produces a motive, at least if one grants the Tate conjecture.

The problems of the preceding paragraphs compound if one tries to treat $GL_n(\mathbb{A}_F)$ for $n > 2$. The group GL_n never gives rise to a Shimura variety if $n > 2$. If F is totally real or CM, one *does* have Shimura varieties associated to forms of $GU(n)$ defined over F. These groups are *outer* twists of GL_n. Roughly, the cuspidal automorphic representations of $GL_n(\mathbb{A}_F)$ that satisfy an appropriate self-duality arise as functorial transfers from automorphic representations on some form of $GU(n)$. One also has the happy fact that if the generalized unitary group under consideration has signature $(1, n - 1)$ at one infinite place, and definite signature at all others, than the most significant of the L-functions appearing in the description of its ζ-function are, in fact, standard L-functions.

Applying appropriate extensions of all the methods described in the GL_2-case (cohomology of Shimura varieties, endoscopy, ℓ-adic limiting arguments), as well as a passage from dimension n to dimension $2n$ to impose self-duality in the nonessentially self-dual case, leads to the following result.[72]

Theorem 15.7.2.1 *If F is totally real or CM, and if π is a cuspidal automorphic representation whose infinitesimal characters at all infinite places are regular, then there exists a compatible family of Galois representations $\{\varphi_\ell\}$ associated to π.*

In many of the cases when π is self-dual, these are known to be motivic (either unconditionally or subject to the Tate conjecture), since they are

---

[72] An enormous number of authors are involved in the complete proof of this result. In the essentially self-dual case, we should mention [Clo91], [Kot92], [HT01], [Shi11], and [CH13]. In the not-necessarily self-dual case, the references are [HLTT16] and [Sch15].

constructed in the cohomology of Shimura varieties. But not always, and not at all if π is not self-dual (because of the ℓ-adic limiting arguments that are required).

15.7.3 From Galois Representations to Automorphic Representations

We now discuss the general problem of passing from a motive, or its associated compatible family of Galois representations, to an associated automorphic representation. It is helpful to consider this problem through the lens of various particular examples of two-dimensional Galois representations.

To begin with, suppose that $K/\mathbb{Q}$ is a degree two extension, and that ψ is an algebraic Hecke character for K. The construction of Section 15.4.6 gives rise to a compatible family $\{\widetilde{\rho}_\ell\}$ of two-dimensional Galois representations, which, as we observed in Section 15.5.8, are motivic. As we also observed there, $L(s, \{\widetilde{\rho}_\ell\}) = L(s, \psi)$, and so in this case the associated L-function *does* satisfy the expected analytic continuation and functional equation. If K is imaginary, and in certain cases if K is real, one can then apply the Hecke–Weil converse theorem discussed in 15.6.2 to show that $L(s, \psi) = L(s, \pi, \text{standard})$ for some cuspidal automorphic representation π of $GL_2(\mathbb{A}_\mathbb{Q})$. In fact, using the relationship between quadratic fields and quadratic forms, one can also directly use arguments with θ-series (so, ultimately, the Poisson summation formula) to construct π (or, rather, to construct a modular eigenform f that generates π).[73] Maass [Maa49] made a similar construction in the remaining real quadratic cases (and introduced Maass forms to this end).

There is an enormous generalization of the preceding construction, known as *automorphic induction*. If K/F is a cyclic Galois extension of degree n, if ψ is a Hecke character, and if ρ is the induction of ψ to $W_{K/F}$, then we may find an automorphic representation π of $GL_n(\mathbb{A}_F)$ such that $L(s, \rho) = L(s, \pi)$. In the case when ψ is algebraic, so that ρ gives rise to the motivic family $\{\widetilde{\rho}_\ell\}$ (generalizing the context of the preceding paragraph), we then have $L(s, \{\widetilde{\rho}_\ell\}) = L(s, \pi)$, and we have established reciprocity for the family $\{\widetilde{\rho}_\ell\}$.

When $n = 2$, the converse theorem of [JL70] can be used to prove this result. For general n, a different argument is required. One considers the problem in the optic of Section 15.6.4.3, by working with the Weil form of the L-group. Automorphic induction is then proved by an argument using the trace formula [AC89, Hen12].

Another two-dimensional example comes from two-dimensional Artin representations. If these have solvable image in $GL_2(\mathbb{C})$, then by analyzing the structure of these representations (e.g. relating them to certain inductions),

---

[73] One can think of this as taking the θ-series arguments that go into proving the functional equation for $L(s, \psi)$, and using them directly to construct the required modular form.

and using automorphic induction, as well as cases of *base change* functoriality, they can be shown to satisfy reciprocity – this is the celebrated Langlands–Tunnell theorem [Lan80, Tun81]. In particular, this establishes the Artin conjecture (Conjecture 15.5.1.1) for these representations. In fact, in this case, the Artin conjecture is essentially equivalent to reciprocity, by the Jacquet–Langlands converse theorem for GL_2.[74]

Yet another two-dimensional example is given by $H^1(E)$ for an elliptic curve E over $\mathbb{Q}$. The conjecture on meromorphic continuation of $\zeta(s, E)$ is attributed to Hasse by Weil [Wei52], while the reciprocity conjecture for $H^1(E)$ seems to have been discovered by Shimura and Taniyama. Taking into account the converse theorems for GL_2, this reciprocity is seen to be equivalent to the holomorphic continuation and functional equation of $L(s, H^1(E))$; and indeed, Weil proved his converse theorem in part to establish this equivalence. In particular, if E has CM, then we have seen that $H^1(E)$ arises from a Weil group representation, and so we are in the case discussed at the beginning of this section – the converse theorem indeed can be applied to deduce reciprocity for $H^1(E)$. For elliptic curves without CM, the compatible family $H^1(E)$ does not arise from a Weil group character, and there is no evident way to directly prove this analytic continuation and functional equation. Indeed, the only known proof of this analytic continuation and functional equation is *via* reciprocity.

In this case, reciprocity was famously proved by Wiles [Wil95] and Taylor–Wiles [TW95] for semistable E, and then by Breuil–Conrad–Diamond–Taylor [BCDT01] for general E. We will say the briefest amount about the methods. They are ℓ-adic in nature, exploiting the theory of congruences of modular forms. Very roughly, one finds an automorphic representation π', which is *known* to have an associated ℓ-adic representation, and for which this ℓ-adic representation, when reduced mod ℓ, is *isomorphic* to the mod ℓ reduction of $H^1(E)$ (so, concretely, to the Galois representation on the group of ℓ-division points on E). An *automorphy lifting theorem* (here "lifting" refers to lifting back from mod ℓ to the original ℓ-adic representation $H^1(\overline{E}, \mathbb{Q}_\ell)$) then implies that $H^1(\overline{E}, \mathbb{Q}_\ell)$ is itself "automorphic," i.e. the ℓ-adic representation associated to an automorphic representation π. Now we are done: $L(s, \pi) = L(s, H^1(E))$ (since we can check this by working at a single ℓ).

As Wiles remarks in the introduction to [Wil95], this method of argument does not need the entire compatible family of ℓ-adic representations, other than that some ℓ may be easier to argue with than others. Having flexibility in choosing which ℓ to work with turns out to be essential, though, because in implementing the method, one must exhibit an existing automorphic

---

[74] Tate once wryly noted that between the wars, both Artin and Hecke were in Hamburg, and both were studying the problem of analytic continuation of L-functions. Nevertheless, it took 30-odd more years before Langlands made the connection between their respective viewpoints.

representation π' that will give rise to the desired automorphic representation π via automorphy lifting.

In the original argument of [Wil95], the choice of $\ell = 3$ is made; since $GL_2(\mathbb{F}_3)$ is solvable, one can use the Langlands–Tunnell result to construct a π'. For certain E, the argument at $\ell = 3$ doesn't work, and one shifts to $\ell = 5$ instead. For this one, has to find a π' matching with the 5-torsion of E. To this end, one finds *another* elliptic curve E' with $E'[5] = E[5]$ (as $\mathrm{Gal}(\overline{\mathbb{Q}}/\mathbb{Q})$-modules), but with better-behaved 3-torsion, so that the Langlands–Tunnell result applies. The previous argument shows that E' satisfies reciprocity, so now one has a π' from which to construct the desired π via 5-adic automorphy lifting. The argument of [BCDT01] uses further switching between the primes 3 and 5. These switches are possible because there are an abundance of auxiliary E' available with prescribed 3- or 5-torsion, since the modular curves of level 3 and 5 are *rational* curves.

Once one tries to generalize this argument, the moduli schemes from which one is "sampling" one's auxiliary objects have essentially uncontrollable geometry, and there is no guarantee that they have rational points. This necessitates having to pass from $\mathbb{Q}$, or whatever ground field one is working over, to an extension field, in order to find the auxiliary automorphic representation π' that will effectuate the automorphy lifting. One also needs a source of "automatic automorphy," since an analogue of Langlands–Tunnell is typically not available. Automorphic induction provides the solution to this – mod ℓ Galois representations which are monomial automatically arise from automorphic representations π'.

We should also mention that the automorphy lifting arguments depend very much on the "automorphic to Galois" direction of reciprocity. And of course there are many other technical constraints which limit the precise arguments that can be made and, consequently, the scope of the results that can be proved. Nevertheless, much progress has been made in the past 25 years. Here are some sample recent results.

Theorem 15.7.3.1 (Allen–Calegari–Caraiani–Gee–Helm–Le Hung–Newton–Scholze-Taylor–Thorne [ACC+18]) *The Sato–Tate conjecture holds for any elliptic curve over a CM field.*[75]

Theorem 15.7.3.2 (Boxer–Calegari–Gee–Pilloni [BCGP18]) *If C is a genus-two curve over a totally real field, then $\zeta_C(s)$ admits a meromorphic continuation to the entire complex plane, and satisfies the expected functional equation.*

---

[75] As indicated in footnote 57, the case of elliptic curves over totally real fields was proved earlier.

Both results are about meromorphic continuation – the second evidently, and the first because, as we explained above, the proof of the Sato–Tate conjecture requires us to continue each $L(s, \text{Sym}^n H^1(E))$ to the line $s \geq 1 + \frac{n}{2}$, and to prove that it is zero-free along the edge of this region. Both results are established by proving a *potential* version of reciprocity; i.e. by constructing the relevant automorphic representation π that controls the given L-function, not over F, but over some (uncontrollably large) extension K of F. (As explained above, this is necessary to find the required auxiliary points on some moduli scheme.) A variant of Brauer's argument (Theorem 15.5.1.3) then shows that each $L(s, \text{Sym}^m H^1(E))$, in the context of the Theorem 15.7.3.1, or $L(s, H^1(C))$, in the context of Theorem 15.7.3.2, has the required meromorphic continuation. We also remark that the the automorphy lifting argument used to prove Theorem 15.7.3.2 takes place in the context of automorphic representations of a (generalized) symplectic group, and the authors rely on Arthur's work on functoriality for classical groups [Art13] (mentioned in Example 15.6.4.7) to compare these automorphic representations with automorphic representations on the group GL_4 – for example, in order to manipulate the associated Galois representations, and in order to apply base-change arguments such as those used in the Brauer-type argument for meromorphic continuation.

In his Shaw Prize address [Lan11], Langlands observes that the Sato–Tate conjecture, analogous as it is to Tschebotareff's theorem, is closer to Brauer's Theorem 15.5.1.3 than to Artin's Conjecture 15.5.1.1; this is a valid point, and indeed, the potential reciprocity that is being applied here to deduce it is weaker than true reciprocity. On the other hand, from the point of view of current technique, reciprocity and potential reciprocity don't seem all that different; to establish one rather than the other, one just has to find a method for constructing the auxiliary automorphic representation π' without making an unwanted field extension. One recent instance of this is the paper [AKT19] of Allen–Khare–Thorne, in which the authors are able to improve on some of the results of [ACC$^+$18] by finding a large class of elliptic curves over CM fields for which reciprocity holds genuinely, not just potentially. We also mention the paper [CCG20], which employs the arguments of [BCGP18] to give examples of abelian surfaces over $\mathbb{Q}$ for which reciprocity holds.

In the same address, Langlands also discusses the significance of establishing results such as Sato–Tate, which are consequences of functoriality, by using the technique of reciprocity instead. The same ten authors who proved Theorem 15.7.3.1 also established entirely new cases of the Ramanujan conjecture for cuspidal automorphic representations on $\text{GL}_2(\mathbb{A}_F)$ (where F is a CM field). This is another application of reciprocity to establish a consequence of functoriality.

As we noted in Section 15.6.4.3, from a certain viewpoint reciprocity can be entirely subsumed into functoriality. In any event, the two problems seem to be just as, or even more, intertwined than ever before. Whether this is merely a contingent phenomenon, or the reflection of something deeper, remains to be seen.

References

[AC89]   James Arthur and Laurent Clozel, *Simple Algebras, Base Change, and the Advanced Theory of the Trace Formula*, Annals of Mathematics Studies, vol. 120, Princeton University Press, Princeton, NJ, 1989.

[ACC⁺18]   Patrick B. Allen, Frank Calegari, Ana Caraiani, et al., *Potential automorphy over CM fields*, 2018. arXiv:1812.09999.

[AGV72]   Michael Artin, Alexander Grothendieck, and Jean-Louis Verdier, *Théorie des topos et cohomologie étale des schémas. Tome 2*, Lecture Notes in Mathematics, Vol. 270, Springer-Verlag, Berlin-New York, 1972, Séminaire de Géométrie Algébrique du Bois-Marie 1963–1964 (SGA 4), Dirigé par M. Artin, A. Grothendieck et J. L. Verdier. Avec la collaboration de N. Bourbaki, P. Deligne et B. Saint-Donat.

[AKT19]   Patrick B. Allen, Chandrashekhar Khare, and J. A. Thorne, *Modularity of* $GL_2(\mathbb{F}_p)$-*representations over CM fields*, 2019.

[AL70]   A. O. L. Atkin and Joseph Lehner, Hecke operators on $\Gamma_0(m)$, *Math. Ann.* **185** (1970), 134–160.

[Art]   James Arthur, An Introduction to Langlands Functoriality, this volume.

[Art24]   Emil Artin, Quadratische Körper im Gebiete der höheren Kongruenzen. I, II, *Math. Z.* **19** (1924), no. 1, 153–206, 207–246.

[Art27]   ——, Beweis des allgemeinen Reziprozitätsgesetzes, *Abh. Math. Sem. Univ. Hamburg* **5** (1927), no. 1, 353–363.

[Art31]   ——, Zur Theorie der *L*-Reihen mit allgemeinen Gruppencharakteren, *Abh. Math. Sem. Univ. Hamburg* **8** (1931), no. 1, 292–306.

[Art89]   J. Arthur, Unipotent automorphic representations: conjectures, no. 171–172, 1989, Orbites unipotentes et représentations, II, *Astérisque*, 13–71.

[Art02]   ——, A note on the automorphic Langlands group, *Canad. Math. Bull.* **45** (2002), no. 4, 466–482, Dedicated to Robert V. Moody.

[Art13]   ——, *The endoscopic classification of representations*, American Mathematical Society Colloquium Publications, vol. 61, American Mathematical Society, Providence, RI, 2013, Orthogonal and symplectic groups.

[BCDT01]   Christophe Breuil, Brian Conrad, Fred Diamond, and Richard Taylor, On the modularity of elliptic curves over **Q**: wild 3-adic exercises, *J. Amer. Math. Soc.* **14** (2001), no. 4, 843–939 (electronic).

[BCGP18]   George Boxer, Frank Calegari, Toby Gee, and Vincent Pilloni, *Abelian surfaces over totally real fields are potentially modular*, 2018.

[BG14]   Kevin Buzzard and Toby Gee, *The conjectural connections between automorphic representations and Galois representations*, Automorphic forms and Galois representations. Vol. 1, London Math. Soc. Lecture

Note Ser., vol. 414, Cambridge University Press, Cambridge, 2014, pp. 135–187.

[BJ79] Armand Borel and Hervé Jacquet, *Automorphic forms and automorphic representations*, Automorphic forms, representations and L-functions (Proc. Sympos. Pure Math., Oregon State University, Corvallis, Oregon, 1977), Part 1, Proc. Sympos. Pure Math., XXXIII, Amer. Math. Soc., Providence, RI, 1979, With a supplement "On the notion of an automorphic representation" by R. P. Langlands, pp. 189–207.

[BLGHT11] Tom Barnet-Lamb, David Geraghty, Michael Harris, and Richard Taylor, *A family of Calabi–Yau varieties and potential automorphy II, Publ. Res. Inst. Math. Sci.* **47** (2011), no. 1, 29–98.

[BR93] Don Blasius and Jonathan D. Rogawski, Motives for Hilbert modular forms, *Invent. Math.* **114** (1993), no. 1, 55–87.

[Bra47a] Richard Brauer, On Artin's L-series with general group characters, *Ann. of Math.* (2) **48** (1947), 502–514.

[Bra47b] , On the zeta-functions of algebraic number fields, *Amer. J. Math.* **69** (1947), 243–250.

[Car86] Henri Carayol, Sur les représentations l-adiques associées aux formes modulaires de Hilbert, *Ann. Sci. École Norm. Sup.* (4) **19** (1986), no. 3, 409–468.

[CCG20] Frank Calegari, Shiva Chidambaram, and Alexandru Ghitza, Some modular abelian surfaces, *Math. Comp.* **89** (2020), no. 321, 387–394.

[CH13] Gaëtan Chenevier and Michael Harris, Construction of automorphic Galois representations, II, *Camb. J. Math.* **1** (2013), no. 1, 53–73.

[Che40] Claude Chevalley, La théorie du corps de classes, *Ann. of Math.* (2) **41** (1940), 394–418.

[CHLN11] Laurent Clozel, Michael Harris, Jean-Pierre Labesse, and Bao-Châu Ngô (eds.), *On the stabilization of the trace formula*, Stabilization of the Trace Formula, Shimura Varieties, and Arithmetic Applications, vol. 1, International Press, Somerville, MA, 2011.

[CHT08] Laurent Clozel, Michael Harris, and Richard Taylor, Automorphy for some l-adic lifts of automorphic mod l Galois representations, *Pub. Math. IHES* **108** (2008), 1–181.

[Clo91] Laurent Clozel, Représentations galoisiennes associées aux représentations automorphes autoduales de GL(n), *Inst. Hautes Études Sci. Publ. Math.* (1991), no. 73, 97–145.

[Ded] Richard Dedekind, *Supplement XI* to Vorlesungen über Zahlentheorie, by L. Dirichlet, 4th ed., 1894.

[Del71a] Pierre Deligne, *Formes modulaires et représentations l-adiques*, Séminaire Bourbaki, Exposés 347–363, Lecture Notes in Mathematics, vol. 179, Springer-Verlag, 1971, pp. 139–172.

[Del71b] , *Travaux de Shimura*, Séminaire Bourbaki, 23ème année (1970/71), Exp. No. 389, 1971, pp. 123–165. Lecture Notes in Math., Vol. 244.

[Del73] , *Les constantes des équations fonctionnelles des fonctions L*, Modular functions of one variable, II (Proc. Internat. Summer School, University of Antwerp, Antwerp, 1972), 1973, pp. 501–597. Lecture Notes in Math., Vol. 349.

[Del74]   , La conjecture de Weil. I, Inst. *Hautes Études Sci. Publ. Math.* (1974), no. 43, 273–307.

[Del79]   , *Variétés de Shimura: interprétation modulaire, et techniques de construction de modèles canoniques*, Automorphic forms, representations and *L*-functions (Proc. Sympos. Pure Math., Oregon State University, Corvallis, Oregon, 1977), Part 2, Proc. Sympos. Pure Math., XXXIII, Amer. Math. Soc., Providence, RI, 1979, pp. 247–289.

[Dir]   Lejeune Dirichlet, Recherches sur diverses applications de l'analyse infinitésimale a la théorie des nombres, *J. Reine Angew. Math.* **19** (1839), 324–369, **21** (1840), 1–12, 134–155.

[Dir37]   , Beweis des satzes, dass jede unbegrenzte arithmetische progression, deren erstes glied und differenz ganze zahlen ohne gemeinschaftlichen factor sing, unendlich viele primzahlen erhält, *Abhandlungen der Königlich Preussischen Akademie der Wissenschaften* (1837), 45–81.

[DMOS82] Pierre Deligne, James S. Milne, Arthur Ogus, and Kuang-yen Shih, *Hodge cycles, motives, and Shimura varieties*, Lecture Notes in Mathematics, vol. 900, Springer-Verlag, Berlin-New York, 1982.

[DS74]   Pierre Deligne and Jean-Pierre Serre, Formes modulaires de poids 1, *Ann. Sci. Ec. Norm. Sup.* **7** (1974), 507–530.

[Dwo56]   Bernard Dwork, On the Artin root number, *Amer. J. Math.* **78** (1956), 444–472.

[Dwo60]   , On the rationality of the zeta function of an algebraic variety, *Amer. J. Math.* **82** (1960), 631–648.

[Eic54]   Martin Eichler, Quaternäre quadratische Formen und die Riemannsche Vermutung für die Kongruenzzetafunktion, *Arch. Math.* **5** (1954), 355–366.

[Eula]   Leonhard Euler, *De seriebus quibusdam considerationes*, Opera Omnia: Series 1, Volume 14, pp. 407–462.

[Eulb]   , *De summa seriei ex numeris primis formatae* $1/3 - 1/5 + 1/7 + 1/11 - 1/13 - 1/17 + 1/19 + 1/23 - 1/29 + 1/31$ *etc. ubi numeri primi formae* $4n - 1$ *habent signum positivum, formae autem* $4n + 1$ *signum negativum*, Opera Omnia: Series 1, Volume 4, pp. 146–162.

[Eulc]   , *De summis serierum reciprocarum*, Opera Omnia: Series 1, Volume 14, pp. 73–89.

[Euld]   , *Introductio in analysin infinitorum, volume 1*, Opera Omnia: Series 1, Volume 8.

[Eule]   , *Remarques sur un beau rapport entre les series desepuissances tant directes que reciproques*, Opera Omnia: Series 1, Volume 15, pp. 70–90.

[Eulf]   , *Variae observationes circa series infinitas*, Opera Omnia: Series 1, Volume 14, pp. 217–244.

[Fla79]   Daniel Flath, *Decomposition of representations into tensor products*, Automorphic forms, representations and *L*-functions (Proc. Sympos. Pure Math., Oregon State University, Corvallis, Oregon, 1977), Part 1, Proc. Sympos. Pure Math., XXXIII, Amer. Math. Soc., Providence, RI, 1979, pp. 179–183.

[GJ72]   Roger Godement and Hervé Jacquet, *Zeta functions of simple algebras*, Lecture Notes in Mathematics, Vol. 260, Springer-Verlag, Berlin-New York, 1972.

[Har77] Robin Hartshorne, *Algebraic Geometry*, Springer-Verlag, New York-Heidelberg, 1977, Graduate Texts in Mathematics, No. 52.

[Has36] Helmut Hasse, Zur Theorie der abstrakten elliptischen Funktionenkörper I, II, III, *J. Reine Angew. Math.* **175** (1936), 55–62, 69–88, 193–208.

[Has54] , Artinsche Führer, Artinsche *L*-Funktionen und Gaussche Summen über endlich-algebraischen Zahlkörpern, *Acta Salmanticensia. Ciencias: Sec. Mat.* **1954** (1954), no. 4, viii+113.

[Has67] , *History of class field theory*, Algebraic Number Theory (Proc. Instructional Conf., Brighton, 1965), Thompson, Washington, D.C., 1967, pp. 266–279.

[Hec17] Erich Hecke, Über die *L*-funktionen und den Dirichletschen Primzahlsatz für einen beliebigen Zahlkörper, *Nachr. Ges. Wiss. Göttingen* (1917), 299–318.

[Hec20] , Eine neue Art von Zetafunktionen und ihre Beziehungen zur Verteilung der Primzahlen, *Math. Z.* **6** (1920), no. 1–2, 11–51.

[Hec27] , Zur Theorie der elliptischen Modulfunktionen, *Math. Ann.* **97** (1927), no. 1, 210–242.

[Hec36] , Über die Bestimmung Dirichletscher Reihen durch ihre Funktionalgleichung, *Math. Ann.* **112** (1936), no. 1, 664–699.

[Hec37] , Über Modulfunktionen und die Dirichletschen Reihen mit Eulerscher Produktentwicklung. I, II, *Math. Ann.* **114** (1937), no. 1, 1–28, 316–351.

[Hen00] Guy Henniart, Une preuve simple des conjectures de Langlands pour GL(*n*) sur un corps *p*-adique, *Invent. Math.* **139** (2000), no. 2, 439–455.

[Hen12] , Induction automorphe globale pour les corps de nombres, *Bull. Soc. Math. France* **140** (2012), no. 1, 1–17.

[HLTT16] Michael Harris, Kai-Wen Lan, Richard Taylor, and Jack Thorne, On the rigid cohomology of certain Shimura varieties, *Res. Math. Sci.* **3** (2016), Paper No. 37, 308.

[HSBT06] Michael Harris, Nick Shepherd-Barron, and Richard Taylor, *A family of Calabi–Yau varieties and potential automorphy*, Preprint, 2006.

[HT01]   Michael Harris and Richard Taylor, *The geometry and cohomology of some simple Shimura varieties*, Annals of Mathematics Studies, vol. 151, Princeton University Press, Princeton, NJ, 2001, With an appendix by Vladimir G. Berkovich.

[Iha67] Yasutaka Ihara, Hecke Polynomials as congruence ζ functions in elliptic modular case, *Ann. of Math.* (2) **85** (1967), 267–295.

[Jar97] Frazer Jarvis, On Galois representations associated to Hilbert modular forms, *J. Reine Angew. Math.* **491** (1997), 199–216.

[JL70]   Hervé Jacquet and Robert P. Langlands, *Automorphic forms on* GL(2), Lecture Notes in Mathematics, Vol. 114, Springer-Verlag, Berlin-New York, 1970.

[Kat75] Nicholas M. Katz, *p-adic L-functions via moduli of elliptic curves*, Algebraic geometry (Proc. Sympos. Pure Math., Vol. 29, Humboldt State University, Arcata, California, 1974), 1975, pp. 479–506.

[Kle94] Steven L. Kleiman, *The standard conjectures*, Motives (Seattle, WA, 1991), Proc. Sympos. Pure Math., vol. 55, Amer. Math. Soc., Providence, RI, 1994, pp. 3–20.

[Kot84] Robert E. Kottwitz, Stable trace formula: cuspidal tempered terms, *Duke Math. J.* **51** (1984), no. 3, 611–650.

[Kot92] , On the λ-adic representations associated to some simple Shimura varieties, *Invent. Math.* **108** (1992), no. 3, 653–665.

[Kud03] Stephen S. Kudla, *Tate's thesis*, An introduction to the Langlands program (Jerusalem, 2001), Birkhäuser Boston, Boston, MA, 2003, pp. 109–131.

[Lana] Robert P. Langlands, *Letter to André Weil, January 1967*, published in "Emil Artin and beyond – class field theory and *L*-functions", European Mathematical Society (EMS), Zürich, 2015.

[Lanb] , *Letter to Lang*, December 5, 1970.

[Lanc] , *On the functional equation of the Artin L-functions*.

[Lan03] Edmund Landau, Neuer Beweis des Primzahlsatzes und Beweis des Primidealsatzes, *Math. Ann.* **56** (1903), no. 4, 645–670.

[Lan70] Robert P. Langlands, *Problems in the theory of automorphic forms*, Lectures in modern analysis and applications, III, 1970, pp. 18–61. Lecture Notes in Math., Vol. 170.

[Lan71] , *Euler Products*, Yale University Press, New Haven, Conn.-London, 1971, A James K. Whittemore Lecture in Mathematics given at Yale University, 1967, Yale Mathematical Monographs, 1.

[Lan76] , *Some contemporary problems with origins in the Jugendtraum*, Mathematical developments arising from Hilbert problems (Proc. Sympos. Pure Math., Vol. XXVIII, Northern Illinois University, De Kalb, Illinois, 1974), 1976, pp. 401–418.

[Lan79] , *Automorphic representations, Shimura varieties, and motives. Ein Märchen*, Automorphic forms, representations and *L*-functions (Proc. Sympos. Pure Math., Oregon State University, Corvallis, Oregon, 1977), Part 2, Proc. Sympos. Pure Math., XXXIII, Amer. Math. Soc., Providence, RI, 1979, pp. 205–246.

[Lan80] , Base Change for GL(2), *Annals of Math. Studies* **96** (1980).

[Lan89] , *On the classification of irreducible representations of real algebraic groups*, Representation theory and harmonic analysis on semisimple Lie groups, Math. Surveys Monogr., vol. 31, Amer. Math. Soc., Providence, RI, 1989, pp. 101–170.

[Lan11] , *Reflexions on receiving the Shaw Prize*, On certain *L*-functions, Clay Math. Proc., vol. 13, Amer. Math. Soc., Providence, RI, 2011, pp. 297–308.

[Lem] Franz Lemmermeyer (https://mathoverflow.net/users/3503/franzlemmermeyer), *History of the analytic class number formula*,

MathOverflow, URL: https://mathoverflow.net/q/242603 (version: 2016-06-19).

[LR87] Robert P. Langlands and Michael Rapoport, Shimuravarietäten und Gerben, *J. Reine Angew. Math.* **378** (1987), 113–220.

[Maa49] Hans Maass, Über eine neue Art von nichtanalytischen automorphen Funktionen und die Bestimmung Dirichletscher Reihen durch Funktionalgleichungen, *Math. Ann.* **121** (1949), 141–183.

[Maz77] Barry Mazur, Book Review: Ernst Edward Kummer, Collected Papers, *Bull. Amer. Math. Soc.* **83** (1977), no. 5, 976–988.

[Mil83] James S. Milne, *The action of an automorphism of C on a Shimura variety and its special points*, Arithmetic and geometry, Vol. I, Progr. Math., vol. 35, Birkhäuser Boston, Boston, MA, 1983, pp. 239–265.

[Mil16] , *The Riemann hypothesis over finite fields: from Weil to the present day [Reprint of 3525903]*, ICCM Not. **4** (2016), no. 2, 14–52.

[Oht84] Masami Ohta, *Hilbert modular forms of weight one and Galois representations*, Automorphic forms of several variables (Katata, 1983), Progr. Math., vol. 46, Birkhäuser Boston, Boston, MA, 1984, pp. 333–352.

[Rie] Bernhard Riemann, *Über die Anzahl der Primzahlen unter einer gegebenen Grösse*, Monatsberichte der Berliner Akademie, November 1859.

[RT83] Jonathan D. Rogawski and Jerrold B. Tunnell, On Artin *L*-functions associated to Hilbert modular forms of weight one, *Invent. Math.* **74** (1983), no. 1, 1–42.

[Sat63] Ichirô Satake, Theory of spherical functions on reductive algebraic groups over p-adic fields, *Inst. Hautes Études Sci. Publ. Math.* (1963), no. 18, 5–69.

[Sch90] Anthony J. Scholl, Motives for modular forms, *Invent. Math.* **100** (1990), no. 2, 419–430.

[Sch15] Peter Scholze, On torsion in the cohomology of locally symmetric varieties, *Ann. of Math.* (2) **182** (2015), no. 3, 945–1066.

[Ser60] Jean-Pierre Serre, Analogues kählériens de certaines conjectures de Weil, *Ann. of Math.* (2) **71** (1960), 392–394.

[Ser65] , *Zeta and L functions*, Arithmetical Algebraic Geometry (Proc. Conf. Purdue University, 1963), Harper & Row, New York, 1965, pp. 82–92.

[Ser68] , *Abelian l-adic representations and elliptic curves*, McGill University lecture notes written with the collaboration of Willem Kuyk and John Labute, W. A. Benjamin, Inc., New York-Amsterdam, 1968.

[Ser70] , *Facteurs locaux des fonctions zêta des variétés algébriques (définitions et conjectures)*, Séminaire Delange-Pisot-Poitou. 11e année: 1969/70. Théorie des nombres. Fasc. 1: Exposés 1 à 15; Fasc. 2: Exposés 16 à 24, Secrétariat Math., Paris, 1970, p. 15.

[Ser77] , *Cours d'arithmétique*, Presses Universitaires de France, Paris, 1977, Deuxième édition revue et corrigée, Le Mathématicien, No. 2.

[Sha] Freydoon Shahidi, Automorphic *L*-Functions, this volume.

[Shi58] Goro Shimura, Correspondances modulaires et les fonctions ζ de courbes algébriques, *J. Math. Soc. Japan* **10** (1958), 1–28.

[Shi71]   , *Introduction to the Arithmetic Theory of Automorphic Functions*, Princeton University Press, 1971.

[Shi11]   Sug Woo Shin, Galois representations arising from some compact Shimura varieties, *Ann. of Math.* (2) **173** (2011), no. 3, 1645–1741.

[Tak33]   Teiji Takagi, Über eine theorie des relativ-abel'schen Zahlkörpers, *J. Coll. Sci. Imp. Univ. Tokyo* **41** (1920), no. 9, 1–133.

[Tan57]   Yutaka Taniyama, *L*-functions of number fields and zeta functions of abelian varieties, *J. Math. Soc. Japan* **9** (1957), 330–366.

[Tat65]   John T. Tate, *Algebraic cycles and poles of zeta functions*, Arithmetical Algebraic Geometry (Proc. Conf. Purdue University, 1963), Harper & Row, New York, 1965, pp. 93–110.

[Tat67]   , *Fourier analysis in number fields, and Hecke's zeta-functions*, Algebraic Number Theory (Proc. Instructional Conf., Brighton, 1965), Thompson, Washington, D.C., 1967, pp. 305–347.

[Tat79]   , *Number theoretic background*, Automorphic forms, representations and *L*-functions (Proc. Sympos. Pure Math., Oregon State Univ., Corvallis, Ore., 1977), Part 2, Proc. Sympos. Pure Math., XXXIII, Amer. Math. Soc., Providence, RI, 1979, pp. 3–26.

[Tat94]   , *Conjectures on algebraic cycles in l-adic cohomology*, Motives (Seattle, WA, 1991), Proc. Sympos. Pure Math., vol. 55, Amer. Math. Soc., Providence, RI, 1994, pp. 71–83.

[Tay08]   Richard Taylor, Automorphy for some *l*-adic lifts of automorphic mod *l* Galois representations. II, *Pub. Math. IHES* **108** (2008), 183–239.

[Tay12]   , The image of complex conjugation in *l*-adic representations associated to automorphic forms, *Algebra Number Theory* **6** (2012), no. 3, 405–435.

[Tsc26]   Nikolai Tschebotareff, Die Bestimmung der Dichtigkeit einer Menge von Primzahlen, welche zu einer gegebenen Substitutionsklasse gehören, *Math. Ann.* **95** (1926), no. 1, 191–228.

[Tun81]   Jerrold Tunnell, Artin's Conjecture for representations of octahedral type, *Bull. Amer. Math. Soc.* **5** (1981), 173–175.

[TW95]   Richard Taylor and Andrew Wiles, Ring-theoretic properties of certain Hecke algebras, *Ann. of Math.* **142** (1995), 553–572.

[Wei49]   André Weil, Numbers of solutions of equations in finite fields, *Bull. Amer. Math. Soc.* **55** (1949), 497–508.

[Wei51]   , Sur la théorie du corps de classes, *J. Math. Soc. Japan* **3** (1951), 1–35.

[Wei52]   , *Number-theory and algebraic geometry*, Proceedings of the International Congress of Mathematicians, Cambridge, Mass., 1950, vol. 2, Amer. Math. Soc., Providence, RI, 1952, pp. 90–100.

[Wei56a]   , *Abstract versus classical algebraic geometry*, Proceedings of the International Congress of Mathematicians, 1954, Amsterdam, vol. III, Erven P. Noordhoff N.V., Groningen; North-Holland Publishing Co., Amsterdam, 1956, pp. 550–558.

[Wei56b]   , *On a certain type of characters of the idèle-class group of an algebraic number-field*, Proceedings of the international symposium on algebraic number theory, Tokyo & Nikko, 1955, Science Council of Japan, Tokyo, 1956, pp. 1–7.

[Wei67]   , Über die Bestimmung Dirichletscher Reihen durch Funktionalgleichungen, *Math. Ann.* **168** (1967), 149–156.

[Wei95]   , *Fonction zêta et distributions*, Séminaire Bourbaki, Vol. 9, Soc. Math. France, Paris, 1995, pp. Exp. No. 312, 523–531.

[Wil95] Andrew Wiles, Modular elliptic curves and Fermat's last theorem, *Ann. of Math.* **142** (1995), 443–551.

[Wym72] B. F. Wyman, What is a reciprocity law?, *Amer. Math. Monthly* **79** (1972), 571–586; correction, ibid. **80** (1973), 281.

16

On Some Early Sources for the Notion of Transfer in Langlands Functoriality

Part I An Overview with Examples

Diana Shelstad

16.1 Introduction

The purpose of this paper is to explain a quite simple-minded way of looking at some of Langlands' vast and visionary program of conjectures. Part II of our project, more concerned with precise general statements and their proofs, will be presented elsewhere.

What does *transfer* mean? We may just as well ask: what does *Langlands functoriality* mean? The two notions, whatever they are or should be, are intricately intertwined with each other.

To get started, what are the objects we study? And then, what does it mean to transfer them? Where do functoriality principles come into play? After very limited remarks towards answers in some generality, we examine, also briefly, a concrete example where we do have quite simple explicit answers. We also include sources and hints for our approach, including remarks on a short expository gem from Harish-Chandra in 1966.

16.2 Settings

We limit our attention to fields that have characteristic zero, even when that is not necessary. Then we fix a **field** F that either is **local**, i.e. a finite extension of a completion of $\mathbb{Q}$, or is **global**, i.e. a finite extension of $\mathbb{Q}$ itself (a number field). We are interested in connected reductive algebraic groups defined over F. There are three types of problems: those in the local setting, those in the global setting, and those concerned with the relationships among objects in the two settings.

In the local setting, our objects of study fall into two types, geometric and spectral. The geometric objects are the so-called **orbital integrals** on $G(F)$ and the spectral objects are **irreducible representations** of $G(F)$. For us, geometric transfer (the transfer of orbital integrals) emphatically comes first.

16.3 Flavors for Transfer

Following Langlands' vision, transfer itself comes in two flavors. First there is endoscopic transfer, which involves a severely limited family of groups and must be viewed as a preliminary step for the second transfer, which fully embraces the notion of Langlands functoriality and is very different both in overview and in details. We label the second transfer as stable–stable transfer (for reasons that will become apparent). Again we stress that our goal in this paper is to explain how these principles may be built out of elementary considerations.

16.4 Endoscopic Transfer

First some more words about orbital integrals and endoscopic transfer. A deep analysis of orbital integrals played a central role in the monumental work of Harish-Chandra in the 1940s, 1950s and 1960s on representation theory of real reductive Lie groups. That analysis is our starting point, and we can't stress enough that we plan to do only simple things with it.

Thus let G be a connected reductive linear algebraic group defined over $\mathbb{R}$, the field of real numbers. This forces $G(\mathbb{C})$ to be a connected complex Lie group and $G(\mathbb{R})$ to be a real Lie group with finitely many connected components. What is an orbital integral (on $G(\mathbb{R})$)? By an orbital integral we mean the set of integrals of a nice function f_G on $G(\mathbb{R})$ along the various conjugacy classes in $G(\mathbb{R})$. The measure on each conjugacy class must be specified. For the purposes of actually defining endoscopic transfer we are able to limit our attention to the so-called strongly regular conjugacy classes. We stress that this will be enough to provide us with a transfer for *all* conjugacy classes.

16.5 Focus on the Geometric Side

The strongly regular classes are the conjugacy classes of the regular semisimple elements γ_G in $G(\mathbb{R})$ for which the centralizer $Cent(\gamma_G, G)$ of γ_G in G is connected as algebraic group and so coincides with the maximal torus T_{γ_G} of G containing γ_G. Here is how we choose a measure on the conjugacy class of such γ_G. Let dg and dt_{γ_G} be Haar measures on $G(\mathbb{R})$ and $T_{\gamma_G}(\mathbb{R})$ respectively (the choices will be of no consequence when we arrive at a careful statement of endoscopic transfer). Then $\frac{dg}{dt_{\gamma_G}}$ will be the quotient measure on the space $T_{\gamma_G}(\mathbb{R}) \backslash G(\mathbb{R})$. This quotient is diffeomorphic to the conjugacy class of γ_G, a closed subset of $G(\mathbb{R})$, via $T_{\gamma_G}(\mathbb{R})g \longmapsto g^{-1}\gamma_G g$. We define the orbital integral $O(\gamma_G, f_G)$ at γ_G of f_G to be $\int_{T_{\gamma_G}(\mathbb{R}) \backslash G(\mathbb{R})} f_G(g^{-1}\gamma_G g) \frac{dg}{dt_{\gamma_G}}$.

We organize the strongly regular classes using Harish-Chandra's F_f-transform, or more generally his $'F_f$-transform, defined for all regular semisimple conjugacy classes, as our inspiration. Harish-Chandra made two different definitions of his transforms and it is crucial to our considerations that we use the second (final) version [6].

16.6 Stable Conjugacy

In fact, what will work much better for our goals, is to work with the notion of stable conjugacy. The stable conjugacy class of a strongly regular element in $G(\mathbb{R})$ consists of all elements in $G(\mathbb{R})$ that are conjugate to that element by an element of $G(\mathbb{C})$. Langlands' general definition of stable conjugacy of two elements requires further conditions on the chosen elements of $G(\mathbb{C})$, but what we may show eventually is that in endoscopic transfer basic results for *all* classes follow a simple pattern heralded by the strongly regular case.

16.7 Algebraic Groups Foremost

For all that we do it is crucial that we work in the algebraic group setting. There is much apparently lucky cancellation in otherwise unwieldy formulas. Nevertheless, the resulting simple formulas hold deep information (here we will concern ourselves only with some fairly immediate examples of our evidence for this).

One thing to notice is that while we work with the results of classical theory of real reductive Lie groups, we do not fix a Cartan decomposition up front. That comes only after we have our algebraic setting in place. Again, that we do these things, with algebraic information at the forefront, is key to our agenda.

Another point, quite minor, is that once we can deal with the case $F = \mathbb{R}$, it takes comparatively little effort to talk in terms of the general archimedean setting. We will save that for elsewhere, noting that much of what we need is found in Langlands' paper [8] on real groups.

16.8 Working with Real Groups: One Algebraic Feature that is Harder

Thus we start with a connected reductive linear algebraic group G defined over $\mathbb{R}$. And we are looking for a well-defined notion of endoscopic transfer. An immediate stumbling block is that the stable conjugacy classes are not quite big enough... for example, for a nonanistropic unitary group G in three variables, there are three conjugacy classes in a stable conjugacy class of regular elliptic elements in $G(\mathbb{R})$, whereas a little work shows we might reasonably expect four conjugacy classes. Where do we find the missing class?

The idea for our answer is due essentially to Vogan, although he considered not conjugacy classes but dual objects, namely irreducible representations, and we further capitalize on a refinement due to Kottwitz; see [3]. We define an extended group over $\mathbb{R}$ to be a (necessarily finite) collection of connected reductive linear algebraic groups G_i, each defined over $\mathbb{R}$, together with a family ψ_{ij} of isomorphisms $G_i \longrightarrow G_j$ over $\mathbb{C}$ for which $\sigma(\psi_{ij})\psi_{ij}^{-1}$ is inner, i.e. each ψ_{ij} is an inner twist, subject to constraints we will come to later.

An extended group may include several copies of an individual group, but there can be at most one copy of a group that is quasisplit over $\mathbb{R}$. Every group appears in some extended group. We call an extended group quasisplit over $\mathbb{R}$ if it does include a group that is quasisplit over $\mathbb{R}$. We stress that not every group appears in an extended group quasisplit over $\mathbb{R}$.

16.9 Endoscopic Transfer: Difficulty Making a Well-Defined Notion

We emphasize again that we seek a well-defined notion of geometric endoscopic transfer, and this is a most delicate issue. The considerations of Adams and Johnson in [1] are not adequate, nor are those of Adams, Barbasch, and Vogan [2]. The notion of geometric transfer discussed in the Wikipedia article on the Fundamental Lemma is not well-defined. We simply cannot use an endoscopic group alone as primary datum.

16.10 Our Primary Datum

Instead, we look to embeddings of (Weil group versions of) L-groups. Actually we need a technical modification of no serious interest here so we ignore that. Our primary datum for an endoscopic transfer will then be a pair $(s, {}^L H \hookrightarrow {}^L G)$, where ${}^L G = G^\vee \rtimes W_\mathbb{R}$ is the L-group of the extended group $\{(G_i, \psi_{ij})\}$, s is a semisimple element of $G^\vee$, and ${}^L H = Cent(s, G^\vee)^0 \rtimes W_\mathbb{R}$.

Another point to stress is that we want a robust notion of transfer: our "nice functions" must form a large enough space defined independently of our problem, although we keep in mind that the larger the geometric transfer is, the more limited the dual spectral transfer must be. We will find that there is a remarkable balance between our geometric and spectral transfers.

16.11 More on Well-Defined Notions

What precisely do we mean by a *well-defined notion of transfer*? And what is its significance?

Our first observation is that any geometric transfer uniquely determines a dual transfer of distributions.

Is it clear that this dual transfer is an endoscopic transfer of characters of irreducible representations? The short answer is *no*. However, there is considerable progress.

Could we start with making a well-defined notion for the transfer of (some) characters and then get a uniquely determined transfer of orbital integrals? In principle, the answer may be *yes*. However, a deep understanding of the representation theory of our group would be needed, and so we insist on geometric transfer as the starting point.

We have introduced our *primary datum* for endoscopic transfer, certain L-group information. Does this determine a unique endoscopic transfer? Almost! In fact, we get a family of transfers with a simply transitive $\mathbb{C}^\times$-action on the family. A more technical analysis shows this is exactly what we want, and so it is our definition of a well-defined transfer.

16.12 Measures in Place of Functions

It turns out that certain related measures are simpler to work with than nice functions themselves. We deal with that now. By a nice measure on $G(\mathbb{R})$ we will mean a measure of the form $f\,dg$, where f is a nice function on $G(\mathbb{R})$ and dg is a Haar measure on $G(\mathbb{R})$. This is all expressed more elegantly in terms of tensor products: see, for example, [11]. However, our more concrete approach will serve us well enough here.

16.13 What is a Nice Function?

We have two answers. First, we define a nice function to be a smooth function on $G(\mathbb{R})$ that is rapidly decreasing in the sense of Harish-Chandra. The set of all such functions forms a complete topological vector space, the Harish-Chandra Schwartz space $\mathcal{C}(G(\mathbb{R}))$ via the well-known Harish-Chandra seminorms. We will label the corresponding nice measures as HCS-measures. Our second space of nice functions is $C_c^\infty(G(\mathbb{R}))$, the set of smooth compactly supported functions on $G(\mathbb{R})$ under the topology of uniform convergence on compact sets. We then label the attached measures as C_c^∞-measures. The natural embedding of $C_c^\infty(G(\mathbb{R}))$ in $\mathcal{C}(G(\mathbb{R}))$, being continuous, provides us with a compatibility demand for the two transfers that we define. That demand will be satisfied thanks to work of Bouaziz [4].

16.14 A Different and Simpler Problem

We pause to look at an evidently much less complicated problem, combinatorial in nature. The results will be critical not only for our work on endoscopic transfer but also for the second, and main, transfer we have not yet addressed. We start with a single connected reductive group G defined over $\mathbb{R}$.

Let T, T' be maximal tori in G, each defined over $\mathbb{R}$. Then we say that T, T' are stably conjugate if there is $g \in G(\mathbb{C})$ such that the restriction of $Int(g)$ to T is defined over $\mathbb{R}$ and carries $T(\mathbb{C})$ to $T'(\mathbb{C})$. In that case, $Int(g)$ is easily seen to carry $T(\mathbb{R})$ to $T'(\mathbb{R})$, and by an old theorem [9], g may be chosen in $G(\mathbb{R})$. Thus for the base field $F = \mathbb{R}$, stable conjugacy for maximal tori over F coincides with $G(F)$-conjugacy.

We write $t_{st}(G)$ for the set of (stable) conjugacy classes of maximal tori in G that are defined over $\mathbb{R}$. This finite set has a partial ordering: let T, T' be maximal tori over $\mathbb{R}$, and write $\{T\}$, $\{T'\}$ for their stable conjugacy classes. Then we define $\{T\} \preceq \{T'\}$ if the unique maximal $\mathbb{R}$-split torus S_T in T is $G(\mathbb{R})$-conjugate to an $\mathbb{R}$-split torus in T' or, equivalently, there is $g \in G(\mathbb{R})$ such that $Int(g)$ carries S_T into $S_{T'}$. This partial ordering makes $t_{st}(G)$ a lattice with a unique minimal element, namely the class of fundamental maximal tori over $\mathbb{R}$, and a unique maximal element, namely the class of those maximal tori over $\mathbb{R}$ containing a maximal $\mathbb{R}$-split torus in G. In our pictures of these lattices we place the minimal element at the top.

16.15 Examples

We consider a few low-dimensional cases.

$G = SL(2)$

    o
    $\downarrow$
    o

$G = SU(2,1)$

    o
    $\downarrow$
    o

$G = Sp_4$, the $\mathbb{R}$-split symplectic group in four variables

$G = Sp(2,2)$, an example of a hyperbolic symplectic group

    o
    $\downarrow$
    o

16.16 Concrete View in the General Case

We have a concrete description of the structure of the lattice $t_{st}(G)$ in terms of the structure of G as algebraic group. Define T to be *adjacent to* (or to *immediately precede*) T' and $\{T\}$ to be adjacent to $\{T'\}$ if $\{T\} \preceq \{T'\}$ and $\dim S_{T'} = 1 + \dim S_T$.

Adjacency is key to the structure of $t_{st}(G)$, and the *symmetric orbits* which come next are key to understanding adjacency.

First we describe adjacency in concrete terms using Harish-Chandra's classification of roots, but in purely algebraic terms. Let α be a root of T in G. Then α is a rational character on T, and so is the root $\sigma\alpha$, the image of α under the action of the nontrivial element σ in $Gal(C/R)$. We consider the Galois orbit $\mathcal{O}_\alpha = \{\alpha, \sigma\alpha\}$ in $X^*(T)$, the module of all rational characters on T. If $\mathcal{O}_\alpha = \mathcal{O}_{-\alpha}$, where we write $-\alpha$ for the root $t \mapsto \alpha(t)^{-1}$, then we call $\mathcal{O}_\alpha$ symmetric. An orbit $\mathcal{O}_\alpha$ that is not symmetric must have the property that $\mathcal{O}_\alpha$ and $\mathcal{O}_{-\alpha}$ are disjoint, and we then call $\mathcal{O}_\alpha$ antisymmetric. These orbits are important too, but not yet.

16.17 More Prep for this View

For $\mathcal{O}_\alpha$ symmetric, the roots $\pm\alpha$ are *imaginary* in the sense of Harish-Chandra or they are *real* in his sense. This is according as $\sigma\alpha = -\alpha$ or $\sigma\alpha = \alpha$. In contrast to the usual practice, we now make a purely algebraic definition. Thus we call an imaginary root α compact or nonsingular according as the three-dimensional simple group G_α over $\mathbb{R}$ determined by α is $\mathbb{R}$-isomorphic to $SU(2)$ or to $SL(2)$.

The imaginary roots of T in G are exactly the roots of T in the connected reductive subgroup $M_T = Cent(S_T, G)$ of G. The group M_T is defined over $\mathbb{R}$. We describe the Weyl group Ω_{M_T} of T in M_T, usually called the imaginary Weyl group of T, concretely as the group $Norm(T(\mathbb{C}), M_T(\mathbb{C}))/T(\mathbb{C})$. This Weyl group acts on the set of imaginary roots. We call an orbit for this action *totally compact* if each root in it is compact. This sets up our concrete algebraic description of adjacency.

16.18 General Picture

Suppose that the imaginary root α of T is not totally compact. Then we find an element s of $M_T(\mathbb{C})$ such that (i) $T' = sTs^{-1}$ is defined over $\mathbb{R}$ and (ii) $\sigma(s)s^{-1}$, which then normalizes T, acts on T as the Weyl reflection ω_α for α.

This ensures that T is adjacent to T'. Conversely, given adjacent pair T and T', we can find such an α.

If T does not contain a maximal $\mathbb{R}$-split torus of G then T has imaginary roots, and if at least one of these roots, say α, is not totally compact then there exists T' adjacent to T. Replacing α by a root in its imaginary Weyl group orbit does not change the stable conjugacy class of T'. Passing to a not totally compact imaginary root outside the Weyl orbit of α does change the stable conjugacy class of T'. Finally, suppose that T' is any given maximal torus over $\mathbb{R}$. Then, apart from the case T' is fundamental, there exists T adjacent to T'.

It is instructive to check how this view works in our examples above, but details are not included here.

16.19 Remark on Other Fields

How do things change when we replace $\mathbb{R}$ by other fields of interest to us here? Assume for the rest of this paragraph that F is nonarchimedean. Then a fundamental maximal torus over F in G is elliptic. On the other hand, stable conjugacy for maximal tori over F does not coincide with $G(F)$-conjugacy except in certain cases. We no longer have a unique fundamental (elliptic) stable conjugacy class. We do have a unique maximally F-split stable conjugacy class which is then a single conjugacy class.

16.20 $t_{st}(G)$ and Inner Forms

A lemma of Langlands [8, Lemma 3.2] shows that an inner twist $\psi : G \to G^*$, where the connected reductive group G^* is quasisplit over $\mathbb{R}$, determines a map $\psi^{(t)} : t_{st}(G) \to t_{st}(G^*)$. We see easily from our analysis above of familiar results on roots that $\psi^{(t)}$ maps $t_{st}(G)$ to an initial segment of $t_{st}(G^*)$. More precisely, $\psi^{(t)}$ is injective and maps the class of fundamental maximal tori over $\mathbb{R}$ in G to the corresponding class in G^*. Further, there is a unique maximal element in the image of $t_{st}(G)$, namely the image of the class of maximal tori containing a maximal $\mathbb{R}$-split torus in G. This image is the class of maximal tori containing a maximal $\mathbb{R}$-split torus in G^* only if G is quasisplit over $\mathbb{R}$ (and then ψ must be an isomorphism over $\mathbb{R}$). The notion of $t_{st}(G)$ as simply an initial segment of $t_{st}(G^*)$ is developed extensively as we go on.

16.21 Back to Endoscopic Transfer

How is $t_{st}(G)$ helpful in visualizing endoscopic transfer?

First, we recall the original goal in endoscopic transfer of orbital integrals. For some groups at least, the (finite) set of conjugacy classes in the stable conjugacy classes of a strongly regular element in $G(\mathbb{R})$ has the structure of a finite abelian group, a sum of Z/2s. An immediate difficulty for us is that this group structure is not uniquely determined by the stable conjugacy class. Nevertheless, first attempts at endoscopic transfer involved picking families of structures and showing that certain combinations of orbital integrals associated with these families could be identified with stable orbital integrals on a certain lower-dimensional group, an endoscopic group. This point of view prevailed a long time despite the fact, already emphasized, that it was clear that this does not lead to a well-defined notion of endoscopic transfer.

Another difficulty is that for other groups, only parts of the mentioned finite abelian groups appear in the considerations for a single group G. That is resolved by a variant of an already-discussed Vogan technique. It is not really significant for our present concerns, so we will just assume that we may work with a single group G.

While it is hardly surprising that a detailed analysis of the structure of $t_{st}(G)$ played an important role in the original point of view, can it really matter in the proof of existence of a well-defined geometric endoscopic transfer?

Our answer is that the only way we know to prove the existence is to explicitly construct it... in this one case $F = \mathbb{R}$. Moreover, our approach gives us much more: indeed, we see the form the transfer statement must take in the nonarchimedean case in order to satisfy local–global compatibility demands, although of course our methods do not offer a proof of the existence for the nonarchimedean case. We will see that these constructive methods are heavily influenced by the structure of $t_{st}(G)$.

16.22 Our Results for Endoscopic Transfer for $F = \mathbb{R}$

Our primary datum for endoscopic transfer is (on ignoring an easily handled technical modification) a pair $(\mathbf{s}, {}^L H \hookrightarrow {}^L G)$, where ${}^L G = G^\vee \rtimes W_\mathbb{R}$ is the L-group of a given extended group $\{(G_i, \psi_{ij})\}$, $\mathbf{s}$ is a semisimple element of $G^\vee$, and ${}^L H = Cent(\mathbf{s}, G^\vee)^0 \rtimes W_\mathbb{R}$.

Consider the product of the set of strongly regular stable conjugacy classes in $H(\mathbb{R})$ with the set of strongly regular conjugacy classes in $G(\mathbb{R})$. We identify a certain subset of this product as the set of *very regular pairs*. For each very regular pair $(\Gamma_H^{st}, \Gamma_G)$ we define a complex number $\Delta(\Gamma_H^{st}, \Gamma_G)$ such that for each nice measure m_G on $G(\mathbb{R})$ there exists a nice measure m_H on $H(\mathbb{R})$ satisfying

$$O^{st}(\Gamma_H^{st}, m_H) = \sum \Delta(\Gamma_H^{st}, \Gamma_G) O(\Gamma_G, m_G)$$

for all Γ_H^{st} contributing to very regular pairs (i.e. for all strongly G-regular stable classes in $H(\mathbb{R})$). The proof, while firmly based on only elementary consequences of the Harish-Chandra theory, is long and quite complicated.

The result is sufficient to establish our main goal, a well-defined geometric transfer in the endoscopic setting.

What about the attached dual transfer... does it behave as desired regarding representations? The answer is *yes* in the *HCS* case, also called the *tempered* case. Our constructive methods for the orbital integral matching greatly simplify the arguments on the spectral side; this will be explained in Part II of the present project. We will also describe some progress we have made for the C_c^∞-case, partially by recasting some results of others. For example, motivated by important work of Waldspurger, we see that the dual transfer builds in a natural way on the *elliptic* representations. We note in passing that Knapp–Zuckerman decomposition of unitary principal series builds from a wider, more complicated family of representations [7].

16.23 Algebraic Point of View Again

As we have indicated already, we have used a simple algebraic method for the normalization of Haar measures. Starting instead with Cartan involutions, usually called the geometric method, we get very different normalizations. This is proved by an elaborate calculation already available in the 1960s in the work of Harish-Chandra (see [6]) using different language. The algebraic approach clearly works better for our intended applications, including geometric ones. We will say no more about this in the present paper.

16.24 Getting Started on *Stable–Stable Transfer*

We come now to the main, and entirely different, type of transfer. Before we start, we ask again about what endoscopic transfer has achieved. It tells us that orbital integrals along conjugacy classes in $G(\mathbb{R})$ can be expressed in terms of orbital integrals along *stable* conjugacy classes from a certain related finite collection of groups $H(\mathbb{R})$. These groups include a quasisplit inner form of G, and all other groups in the collection are of lower dimension. For the attached dual transfer, if we consider the HCS-case then we know that all tempered characters on $G(\mathbb{R})$ are nicely expressed in terms of stable characters on the $H(\mathbb{R})$. Consider what we will call the *trivial case of endoscopic transfer*, that is, where the endoscopic group is a quasisplit inner form, say G^*, of G. It shows that stable orbital integrals on $G(\mathbb{R})$ may be viewed as stable orbital integrals on $G^*(\mathbb{R})$, and stable tempered characters on $G(\mathbb{R})$ as stable tempered characters on $G^*(\mathbb{R})$. This will allow us to reduce our new transfer involving only stable orbital integrals to the case where both groups, say G_1 and G_2, are quasisplit over $\mathbb{R}$. A more elaborate reduction will then bring us to the case that G_1 and G_2 have same split rank over $\mathbb{R}$. That is the only case we will investigate here.

Our primary datum is (up to a technicality we continue to ignore here) an L-homomorphism from ${}^L G_1$ to ${}^L G_2$. What if G_1 is endoscopic for G_2? Have we already solved the second transfer problem by doing endoscopic transfer? Emphatically, no... unless we are in the trivial case... where, because both groups are quasisplit over $\mathbb{R}$, the L-homomorphism determines an $\mathbb{R}$-isomorphism from G_1 to G_2.

16.25 Working Concretely

Consider the following example. We take G_1 to be a one-dimensional torus anisotropic over $\mathbb{R}$ and G_2 to be $SL(2)$. Before starting, we remark that we will apply to the general case a principle of Harish-Chandra that pervades his work on real groups. We have called it the *Semiregular is sufficient* Principle, and, very roughly, it tells us how this little example can be applied over and over, along with various elementary arguments, to generate the general case.

Now, for details in the example, we identify $G_1(\mathbb{R})$ as the group of rotation matrices $r(\theta) = \left[\begin{smallmatrix} \cos\theta & -\sin\theta \\ \sin\theta & \cos\theta \end{smallmatrix}\right], \theta \in \mathbb{R}$. We notice that $r(\theta)$ is strongly $SL(2)$-regular if and only if $\theta \neq 0 \bmod \pi$. The endoscopic transfer tells us that, as function of θ, the suitably normalized unstable combination of orbital integrals of nice measure $m_{SL(2)}$ along the stable conjugacy class of $r(\theta)$ extends smoothly across the points $\theta = 0 \bmod \pi$ on the real line. In the language of transfer, the stable orbital integrals, i.e. the point values at $r(\theta)$ for $\theta \neq 0 \bmod \pi$, of the function so defined match the unstable combination of orbital integrals of $m_{SL(2)}$. Here we have been careless in notation when switching back and forth between nice measures and nice functions.

This smoothness is a simpler preliminary version of the second formula in [5, page 40], well-known even longer than the formula itself. Notice that translation of our language to that of Harish-Chandra's F_f requires a change from our difference of two terms to the sum of his two terms.

Harish-Chandra's first formula in [5, page 40] points us towards the statement of stable–stable transfer. His formula tells us for $\theta \neq 0 \bmod \pi$ how to write a stable orbital integral on $SL(2, \mathbb{R})$ at $r(\theta)$ in terms of stable tempered characters on $SL(2, \mathbb{R})$. To review, we denote by $Ch(\Pi_n, *)$ the stable discrete series character attached to the positive integer n, and by $Ch(\Pi_{\lambda, +}, *), Ch(\Pi_{\lambda, -}, *)$ the two unitary prinicipal series characters attached to the positive real number λ. Here we are following, as closely as our different conventions allow, Harish-Chandra's notation in [5]. Then we write the first formula as

$$\widehat{O}(\Gamma_\theta, m) = \sum_{n>0} \Delta(\Gamma_\theta, \Pi_n)Ch(\Pi_n, \Gamma_\theta)$$

$$+ \int_0^\infty \Delta(\Gamma_\theta, \Pi_{\lambda,+})Ch(\Pi_{\lambda,+}, \Gamma_\theta)d\lambda$$

$$+ \int_0^\infty \Delta(\Gamma_\theta, \Pi_{\lambda,-})Ch(\Pi_{\lambda,-}, \Gamma_\theta)d\lambda,$$

where Γ_θ denotes the stable conjugacy class of $r(\theta)$ for $\theta \neq 0 \bmod \pi$. We may use Harish-Chandra's simple explicit formulas for the coefficients $\Delta(\Gamma_\theta, \Pi_n)$, $\Delta(\Gamma_\theta, \Pi_{\lambda,+})$ and $\Delta(\Gamma_\theta, \Pi_{\lambda,-})$, along with the temperedness of the representations $\Pi_n, \Pi_{\lambda,\mp}$ to see that convergence of the series is absolute, uniform on compact subsets of $\theta \neq 0 \bmod \pi$, and similarly for the integrals.

16.26 Some Heuristics

Now we explain some elementary and rather crude heuristics that do, however, lead us on from the general version of Harish-Chandra's first formula (namely, *Fourier inversion for stable orbital integrals*) to our final statement of stable–stable transfer. Without stating explicitly what we mean by the space Γ of stable orbital integrals nor describing the measure $d\Gamma$ on it, we write $O(\Gamma, m)$ for the (stable orbital) integral of a nice measure m along the stable conjugacy class Γ, and $\widehat{O}(\Gamma, m)$ for the normalized version via the usual discriminant function. On the spectral side we similarly use a space Π of tempered packets with measure $d\Pi$. Then $Tr(\Pi, *)$ denotes the stable trace for the packet Π, and $Ch(\Pi, *)$ is the real analytic function on the regular semisimple elements of $G(\mathbb{R})$ that represents the stable trace (via Harish-Chandra's Regularity Theorem). The normalized version is $\widehat{Ch}(\Pi, *)$.

Instead of G_1 and G_2, we label our two groups H and G, and attach subscript H or G to $\Pi, \boldsymbol{\Pi}, \Gamma$ and $\boldsymbol{\Gamma}$, as needed. We do not assume that H is endoscopic for G but do insist that H and G have same rank and that we have primary datum ξ embedding ^LH in ^LG. This determines a map $\boldsymbol{\Pi}_H \to \boldsymbol{\Pi}_G$, and we will write $\Pi_{H \to G}$ for the image of the packet Π_H.

What we seek are transfer identities of the following "shape":

for each nice measure m_G on $G(\mathbb{R})$ there exists a nice measure $m_H = (m_G)_H$ on $H(\mathbb{R})$ such that

$$\widehat{O}(\Gamma_H, (m_G)_H) = \int_{\boldsymbol{\Gamma}_G} \Theta(\Gamma_H, \Gamma_G)\widehat{O}(\Gamma_G, m_G)d\Gamma_G$$

and

$$\widehat{Ch}(\Pi_{H \to G}, \Gamma_G) = \int_{\boldsymbol{\Gamma}_H} \widehat{Ch}(\Pi_H, \Gamma_H)\Theta(\Gamma_H, \Gamma_G)d\Gamma_H$$

for all (strongly) regular semisimple Γ_H, Γ_G.

Assume this is true (in some sense!). We will also change order of integration freely. Using the Weyl integration formula on $G(\mathbb{R})$ we write $Tr(\Pi_{H\rightarrow G}, m_G)$ as

$$\int_{\Pi_G} \widehat{Ch}(\Pi_{H\rightarrow G}, \Gamma_G)\widehat{O}(\Gamma_G, m_G)d\Gamma_G.$$

Then $Tr(\Pi_{H\rightarrow G}, m_G)$ is given by

$$\int_{\Pi_G}\int_{\Pi_G} \widehat{Ch}(\Pi_H, \Gamma_H)\Theta(\Gamma_H, \Gamma_G)\widehat{O}(\Gamma_G, m_G)d\Gamma_H d\Gamma_G$$

$$= \int_{\Pi_H} \widehat{Ch}(\Pi_H, \Gamma_H)\int_{\Pi_G} \Theta(\Gamma_H, \Gamma_G)\widehat{O}(\Gamma_G, m_G)d\Gamma_G d\Gamma_H$$

$$= \int_{\Pi_H} \widehat{Ch}(\Pi_H, \Gamma_H)\widehat{O}(\Gamma_H, (m_G)_H)d\Gamma_H$$

$$= Tr(\Pi_H, (m_G)_H).$$

This is stable–stable transfer at the level of traces, which we do expect as our final, emphatically not our initial, transfer formula.

Continuing in the same spirit, we also see functoriality emerging.

Given a composition $^L J \rightarrow {}^L H \rightarrow {}^L G$ of L-homomorphisms, assume there are attached stable–stable transfer. Then m_G determines both $(m_G)_J$ and $((m_G)_H)_J$, and we see that these two measures are stably equivalent (same $\widehat{O}(\Gamma_J, *)$ for all Γ_J) provided

$$\Theta(\Gamma_J, \Gamma_G) = \int \Theta(\Gamma_J, \Gamma_H)\Theta(\Gamma_H, \Gamma_G)d\Gamma_H.$$

As a final comment, we write down our proposed "shape" for $\Theta(\Gamma_H, \Gamma_G)$, and concern ourselves just with our particular example, to make sense of this $\Theta(\Gamma_H, \Gamma_G)$ and verify the geometric stable–stable transfer.

Set

$$\Theta(\Gamma_H, \Gamma_G) = \int_{\Pi_H} \Delta(\Gamma_H, \Pi_H)\widehat{Ch}(\Pi_{H\rightarrow G}, \Gamma_G)d\Pi_H,$$

where $\Delta(\Gamma_H, \Pi_H)$ is the coefficient in Fourier inversion of the stable orbital integral $\widehat{O}(\Gamma_H, *)$ on $H(\mathbb{R})$:

$$\widehat{O}(\Gamma_H, m_H) = \int_{\Pi_H} \Delta(\Gamma_H, \Pi_H)Tr(\Pi_H, m_H)d\Pi_H$$

for all nice measures m_H on $H(\mathbb{R})$.

16.27 Back to the Example

Here we write the proposed $\Theta(r(\theta), \Gamma_G)$ as $\sum_{n \in \mathbb{Z}} e^{in\theta} \widehat{Ch}(\Pi_{n^*}, \Gamma_G)$, where $n^* > 0$ is attached to $n \in \mathbb{Z}$ using the Langlands' classification. We can make sense of $\Theta(r(\theta), \Gamma_G)$ as a distribution or as a generalized function. Calculation shows that we then get the desired transfer of orbital integrals, our stable–stable transfer. This example is included in the cases studied by Thomas [10].

References

[1] Adams, J. and Johnson, J. Endoscopic groups and packets of nontempered representations, *Compositio Math.* 64 (1987), 271–309.

[2] Adams, J., Barbasch, D., and Vogan, D. *The Langlands Classification and Irreducible Characters of Real Groups,* Birkhauser, 1992.

[3] Arthur, J. On the transfer of distributions: weighted orbital integrals, *Duke Math. J.*, 99 (1999), 209–283.

[4] Bouaziz, A. Sur les caractères des groupes de Lie réductifs non connexes, *J. Funct. Analysis*, 70 (1987), 1–79.

[5] Harish-Chandra Characters of semi-simple Lie groups, *Some Recent Advan. Basic Sciences*, Vol. 1, Academic Press Inc., New York 1966, 35–40.

[6]   Harmonic analysis on real reductive groups I, *J. Funct. Analysis*, 19 (1975), 104–204.

[7] Knapp, A. and Zuckerman, G. Classification of irreducible tempered representations of semisimple groups, *Annals of Math.*, 116 (1982), 389–455.

[8] Langlands, R. On the classification of irreducible representations of real algebraic groups, in *Representation Theory and Harmonic Analysis on Semisimple Lie Groups*, AMS Math Surveys and Monographs, 31, 1989, 101–170.

[9] Shelstad, D. Characters and inner forms of a quasisplit group over $\mathbb{R}$, *Compositio Math.* 39 (1979), 11–45.

[10] Thomas, J. *Towards Stable-stable Transfer Involving Symplectic Groups*, Ph.D. thesis, Rutgers-Newark, 2020.

[11] Waldspurger, J.-L. Stabilisation de la formule des traces tordue IV: transfert spectral archimédien, *Arxiv* 1403.1454.

PART V

Langlands' Contributions to Mathematical Physics

17

Robert Langlands' Work in
Mathematical Physics

Thomas Spencer

17.1 Introduction

Robert Langlands has had a profound influence on mathematics. His results and conjectures combine analysis, representation theory and number theory, in a surprising and elegant manner. The Langlands program is still one of the most active and exciting research areas today.

This article will describe some of Langlands' research in mathematical physics. Most of this work was done in collaboration with Yvan Saint-Aubin, whose chapter (Chapter 18) in this volume [YSA] will give a much more authoritative presentation of their joint work. My aim is to reinforce his description of the influence of this work and to give an overview of some related aspects of statistical mechanics from my perspective.

Langlands became interested in mathematical physics in the early 1980's. As Yvan suggests, Langlands is a mathematical adventurer striving to explore and understand new domains. This note will especially focus on his paper with Pouliot and Saint-Aubin [LPSA94] on critical percolation in two dimensions. The geometric perspective presented in this work had a major impact on our mathematical understanding of conformal field theory. It led the way to a rigorous mathematical analysis of both percolation and Ising models.

17.2 Some Background

Before explaining the ideas of [LPSA94], it will be helpful to discuss some background about statistical mechanics and scaling limits from the view point of both mathematics and theoretical physics.

17.2.1 Brownian Motion as the Scaling Limit of Random Walk

A well-understood example of a scaling limit is Brownian motion. Let $W(n) = (W_1(n), W_2(n))$, $n \in \mathbb{Z}$ be a random walk with values in $\mathbb{Z}^2$. For simplicity assume that W_1 and W_2 are composed of steps $s_1(n) = W_1(n+1) - W_1(n) \in \mathbb{Z}$ and $s_2(n) = W_2(n + 1) - W_2(n) \in \mathbb{Z}$, which are independent, identically distributed random variables with mean 0 and variance 1. Higher moments of s_1, s_2 are assumed to be finite. Brownian motion B(t) is the scaling limit of W given by

$$B(t) = (B_1(t), B_2(t)) = \lim_{r \to \infty} r^{-1/2} W([rt]) \in \mathbb{R}^2$$

where $[rt]$ is the integer closest to rt for real t. $B(t)$ is known to be a Gaussian process that is Hölder continuous in t and rotation invariant. It scales so that $\alpha^{-1/2} B(\alpha t)$ has the same distribution as $B(t)$. In addition, if F denotes a holomorphic function on $\mathbb{C} \simeq \mathbb{R}^2$ then $F(B(t))$ has the same law as $B(\tau(t))$, a time-rescaled Brownian motion. The limiting process is universal. This means that there is a wide class of random walks whose scaling limit is Brownian motion. The steps need not be independent and may even have short-range correlations. These are classical results of probability theory.

17.2.2 Two-Dimensional Ising Model

One of the most studied and basic models in statistical mechanics is the Ising model on $\mathbb{Z}^d, d \geq 2$. This is a model of interacting spins $s_j = \pm 1, j \in \Lambda \subset \mathbb{Z}^d$, where Λ is a large box of side L centered at the origin. At inverse temperature $\beta \geq 0$ the partition function is given by

$$Z_\Lambda(\beta) = \sum_{s_j = \pm 1, \, j \in \Lambda} e^{\beta \sum_{j,k \in \Lambda} J_{j,k} s_j s_k}, \quad j, k \in \mathbb{Z}^2, \tag{17.1}$$

where $J_{j,k} \geq 0$ is the short-range coupling depending only on $|j - k|$. In a box Λ, m-point correlations are given by

$$C(j_1, j_2, \ldots j_m)_\Lambda(\beta) = \langle s_{j_1} s_{j_2} \ldots s_{j_m} \rangle_\Lambda(\beta)$$
$$= Z_\Lambda(\beta)^{-1} \sum_{s_j = \pm 1, \, j \in \Lambda} s_{j_1} s_{j_2} \ldots s_{j_m} e^{\beta \sum_{j,k \in \Lambda} J_{j,k} s_j s_k}. \tag{17.2}$$

They vanish by symmetry if m is odd. By Griffiths inequalities, the correlations are monotone in L and thus the limit as $L \to \infty$ exists and is translation invariant. In the limit, the subscript Λ will be dropped.

When the temperature is high (β small), the spins are nearly independent. The spin–spin correlation $\langle s_0 s_j \rangle(\beta)$ decays exponentially fast for large $|j|$. On the other hand if β is large, the main contribution to the exponential (Gibbs Weight) in (17.2) occurs when the spins are aligned. It can then be shown that for $d \geq 2$, $\langle s_0 s_j \rangle(\beta) \approx M(\beta)^2 > 0$ for large $|j|$. The symbol $M(\beta)$ denotes the spontaneous magnetization. This is the ordered or magnetized phase of the Ising model. There is a unique intermediate value of $\beta = \beta_c = \beta_{critical}$ depending on the coupling J in (17.2) at which the pair correlation has a power law decay. In two dimensions, for the special *nearest-neighbor* case (J_{kj} vanishes for $|j - k| > 1$) it is known from work of L. Onsager [Ons44] and T. T. Wu [Wu66] that $\langle s_0 s_j \rangle (\beta_c) \approx C |j|^{-1/4}$ along lattice directions. The exponent 1/4 is called a critical exponent. It is conjectured to be universal. In particular, this means that it is expected to be independent of the short range coupling $J \geq 0$. Moreover, the scaling limit

$$\lim_{r \to \infty} r^{m/8} \langle s_{[rx_1]} s_{[rx_2]} \cdots s_{[rx_m]} \rangle (\beta_c)$$

$$= \langle \phi(x_1) \phi(x_2) \ldots \phi(x_m) \rangle_{EQFT} \quad x_1, x_2 \ldots x_m \in R^2, \quad (17.3)$$

is conjectured to exist and to be universal up to a scalar multiple. The expression $[rx_j]$ denotes lattice point closest to $rx_j \in R^2$. This limit defines the correlation functions for a Euclidean quantum field theory of the scalar fields ϕ which are invariant under rotations, translations. Furthermore they should satisfy natural relations under scaling and *conformal transformations*.

17.2.3 Theorems about the Ising Model

Although the high- and low-temperature behavior of the Ising has been quite well understood for some time, the proof of the above conjectures at β_c has only recently been established in the nearest-neighbor case by Smirnov [Smi10] and Chelkak, Hongler, and Izyurov [CHI15]. The renormalization group ideas of Kadanoff and Wilson [Wil75] suggest that universality holds for general short-range coupling and one may even add small generic local four-spin perturbations. But there is no proof of such a result. There are some rigorous renormalization group results for local energy correlations in [GGM12] based on earlier work with Pinson sketched in [S⁺00]. These results state that for small symmetric perturbations of the integrable nearest neighbor interaction, the local energy correlation decays like Clr^2. In three dimensions a similar scaling limit should be conformally invariant and universal. This conjecture seems to be beyond the reach of today's mathematical analysis. However, there are theorems that support such conjectures in four and five dimensions where the scaling limit is Gaussian. See [BBS19] for advances in rigorous perturbative renormalization group methods applied to many four-dimensional models. In very recent work [ADC19], Aizenman and Duminil-Copin established the Gaussian fixed point of the four-dimensional Ising model nonperturbatively. There is also very interesting work of [BDH98] establishing the existence of quantum field theories with a non-Gaussian fixed point.

17.2.4 Renormalization and Conformal Field Theory

In 1983, Langlands organized a mathematical physics seminar with A. Borel, A. Knapp, and L. Gross. The seminar focussed on understanding the renormalization group in hierarchical models and lattice Ising models. Daniel Amit, a physicist visiting the School of Natural Sciences, helped to explain some of the concepts. The renormalization group is a multiscale technique developed in theoretical physics for understanding universality at the critical temperature and estimating critical exponents. Langlands' later papers with Saint-Aubin often seem to have been motivated by renormalization group ideas.

In 1984, Belavin, Polyakov, and Zamolodchikov [BPZ84] published a foundational paper in which they described possible conformal field theories in two dimensions. Conjecturally, these included the scaling limit of the two-dimensional Ising model and the q state Potts model for $1 \leq q \leq 4$ at the critical temperature. The authors worked directly with the continuum limit and assumed conformal invariance of the correlations. From the abstract: "Their (correlations) basic property is their invariance under an infinite-dimensional group of conformal (analytic) transformations. It is shown that the local fields forming the operator algebra can be classified according to the irreducible representations of Virasoro algebra..." Shortly after this paper appeared, Friedan, Qiu, and Shenker [FQS84] strengthened this result by classifying certain unitary representations. These unitary representations should describe Ising models but not percolation or self-avoiding walk in two dimensions. In [Lan88], Langlands gave a very clear independent proof of this result. Details also appeared in [FQS86]. I remember attending his 1986 lecture series at IAS on this work.

Although [BPZ84] was a major conceptual advance in theoretical physics, mathematicians had serious problems relating this work to well-known lattice models such as the Ising model and percolation. The two-dimensional nearest-neighbor Ising model is in some sense soluble. Its partition function was exactly computed by Onsager in 1944. However, the scaling limit at β_c and conformal invariance were not rigorously established until recently. One of the first questions I recall discussing with Langlands was about proving that the spin–spin correlation of the Ising model at β_c was asymptotically rotation invariant at long distances. The proof of rotation invariance had to wait until the work of [Smi10, Pin12, CHI15], which were influenced by [LPSA94]. There is still no proof of rotational or conformal invariance for interactions J that are not nearest neighbor.

17.3 Two-Dimensional Percolation – Langlands, Pouliot, and Saint-Aubin

As Saint-Aubin has described in his chapter [YSA], he, Langlands, and Pouliot began to test the ideas of universality as predicted by the renormalization group for the case of two-dimensional percolation. I believe it was the apparent simplicity of the percolation model that drew their attention to it. However, unlike the Ising model, two-dimensional percolation is not a solvable model. It is in some ways more mysterious than the Ising model.

17.3.1 Definitions and Crossing Probabilities

Let's recall the basic definitions of *bond* percolation on the square lattice $S = \mathbb{Z}^2$. By a bond or edge we mean an unordered adjacent pair $\{j, j'\}$, with $|j - j'| = 1$, $j, j' \in S$. Given p, $0 < p \leq 1$ we define a random configuration of open (1) and closed (0) edges. With independent probabilities each edge of S is open with probability p and closed with probability $1 - p$. So we have a family of 0,1 independent random variables indexed by edges of the graph S. Two edges are said to be connected if they share a vertex j. We are interested in the geometry of the connected components formed by the open edges as we vary p.

For a large rectangular box B with horizontal side L and vertical side ρL parallel to the axes in $\mathbb{Z}^2$ we consider the probability that there is a connected set of open edges connecting the vertical sides of the box. Such an event is called a horizontal crossing. Equivalently, (as in [YSA]) we may consider a lattice $\delta\mathbb{Z}^2$ of small mesh size δ in a box of sides 1 and ρ in $\mathbb{R}^2$. With this notation $\delta \approx L^{-1}$. If p is small, the connected components of open edges will form small clusters and the probability that the sides of the box are connected is exponentially small for large L. When p is near 1 the crossing probability approaches 1 for large L. Figure 18.1 of [YSA] shows a path of open edges connecting two intervals of the boundary of a box. The two diagrams have different mesh sizes δ. By a classical theorem of Kesten [Kes82], when $p = p_c = 1/2$ the probability does not go to 0 or 1. This result uses duality in which one considers percolation of closed edges.

For *site* percolation, the vertices (sites) are (independently) open with probability p and closed with probability $1 - p$. Two open vertices are said to be connected if they are adjacent. As in bond percolation, we are interested in connected components of open vertices. On the square lattice, p_c for site percolation exists by Kesten's theorem but its exact value is not known.

17.3.2 Univerality Conjecture for Two-Dimensional Percolation

In [LPPSA92], the authors performed numerical tests for bond and site percolation on the square and triangular lattices. Although the value of p_c is not universal – the crossing probabilities at p_c which are a function of ρ were found to be universal, this means independent of the lattice as $\delta \downarrow 0$. As the authors were careful to explain, this was not a proof but is was clear evidence that the scaling limit exists ($\delta \downarrow 0$) and universality holds in this natural geometric framework. Note that these geometric crossing probabilities were not described in [BPZ84]. The conclusions are compellingly stated in [LPPSA92]: our result

> establishes conclusively that the crossing probabilities are universal, and therefore suitable coordinates for the fixed point, and that several basic models, to be described later, fall into the same universality class. The mathematical consequence is that attention is focussed not on the critical indices, which are from a mathematical viewpoint both literally and figuratively derived objects, since they are given hypothetically by eigenvalues of the jacobian matrix of the renormalization group at its fixed point, but on an object with a more direct mathematical significance, the fixed point itself.

This statement represents a fundamental change in perspective from quantum field correlations and critical exponents to probabilities of geometric crossings.

17.3.3 Geometric Conformal Invariance and Cardy's Formula

The Langlands, Pouliot, and Saint-Aubin paper in the *Bulletin of the AMS* entitled "Conformal invariance and two-dimensional percolation," [LPSA94], gave a broad review of percolation, Ising models, renormalization, and universality. The authors had numerous discussions with Michael Aizenman about their numerical work on crossing probabilities at p_c. Aizenman then formulated a key conjecture about the conformal invariance based on their simulations. To formulate this conjecture it is natural to phrase it on a fine grid $\delta \mathbb{Z}^2 \subset \mathbb{R}^2 = \mathbb{C}$ in the limit as $\delta \to 0$. Let $F(z)$ be a holomorphic function and suppose D and D' be two bounded domains of $\mathbb{C}$ defined by four disjoint boundary arcs A_1, A_2, A_3, A_4 and A'_1, A'_2, A'_3, A'_4 respectively. If D' is the image of D under the mapping F and the boundary arcs also images under F then the conjecture states the following.

In the limit $\delta \to 0$, the crossing probability from A_1 to A_3 exists and coincides with that from A'_1 to A'_3. We assume that A_1 and A_3 are not contiguous. See [YSA] for a discussion and a more precise formulation.

This conjecture was carefully checked in [LPSA94]. As this numerical work was being completed, Aizenman and I contacted John Cardy, who is an expert in conformal field theory and had studied boundary effects [Car86]. Motivated by the numerical work of Langlands and conjectures about geometric conformal invariance, Cardy quickly understood how to predict the crossing probabilities for an $L \times \rho L$ rectangle in terms of a hypergeometric function, [Car92]. This result, now known as Cardy's formula, matched numerical calculations of [LPSA94]. In [Car05] Cardy comments on the geometric view of Langlands et al. from a theoretical physics perspective:

Essentially, the RG (renormalisation group) programme of classifying all suitable renormalisable quantum field theories in two dimensions has been carried through to its conclusion in many cases, providing exact expressions for critical exponents, correlation functions, and other universal quantities. However, the geometrical, as opposed to the algebraic, aspects of conformal symmetry are not apparent in this approach.

There was also later work of Pinson [Pin94] motivated by [LPSA94] that formally calculated the probability that a percolation cluster belongs to a homology class on the torus. This work was based on the conjectured relation of critical percolation to the free field models and the Coulomb gas as explained in [DFSZ87].

17.3.4 Mathematical Results and Influence of Langlands

The elegant geometric perspective provided by Langlands et al. [LPSA94], together with Cardy's formula was very appealing to mathematicians and inspired intense research in percolation and Ising models. This included work by [Pin94], [Wat96], [BS+96] [AB+99], [ABNW99]. However, a proof of the conjectures of [LPSA94] remained open.

In 2000, Oded Schramm's breakthrough paper [Sch00] proposed an extraordinary stochastic geometric equation now called the Schramm–Loewner equation (SLE). Roughly speaking, this stochastic equation governs the random interface between open and closed bonds in critical percolation. The key idea of his paper was the assumption that these interfaces are conformally invariant. Schramm writes:

> It is a celebrated conjecture that critical Bernoulli percolation on lattices in R^2 exhibits conformal invariance in the scaling limit [LPSA94]. Assuming such a conjecture, we plan to prove in a subsequent work that a process similar to SLE describes the scaling limit of the outer boundary of the union of all critical percolation clusters in a domain D which intersect a fixed arc on the boundary of D.

To get a more concrete picture of the interface consider the upper half plane. On the boundary $\mathbb{R}$ suppose that the edges to the right of 0 are open bonds while those to the left are closed. There is a natural random lattice path starting from the origin and going into the upper half plane that separates the open and closed bonds. See [YSA], Figure 18.6. In the continuum limit this path can be described by the SLE stochastic differential equation. A large family of conformal models can conjecturally be described by the $SLE(\kappa)$, the Schramm–Loewner equation driven by $\sqrt{\kappa}B(t)$, where $B(t)$ is the standard Brownian motion. The value of κ depends on the model. These models include the q-state Potts, for $1 \leq q \leq 4$, Ising model, and self-avoiding walk. See [YSA] for further discussion of SLE.

The first proof of conformal invariance of percolation and Cardy's formula was given by S. Smirnov [SW01] for site percolation on the triangular lattice. He also proved that in the continuum limit, $\delta \to 0$, the random curve described above, is governed by $SLE(6)$. In this case $p_c = 1/2$ and Cardy's formula has a particularly simple expression due to a crucial observation of L. Carleson. Let T denote an equilateral triangle in $\mathbb{C}$ of side 1 with vertices $A_1 = 0, A_2 = 1, A_3 = 1/2 + i\sqrt{3/2}$. The probability that the interval $[0, x] \subset (A_1, A_2)$ is connected within T to the boundary (A_2, A_3) by adjacent open sites is exactly equal to x. Here is the key to proving this result: Let z be a point inside the triangle and let $h(z)$ denote the probability that as $\delta \to 0$ there is a cluster of open vertices which connects $[0, 1]$ to (A_2, A_3) and which separates z from (A_1, A_3). Smirnov proved that $h(z)$ is a harmonic function. He used the geometry of the triangular lattice in a crucial way to construct a related lattice holomorphic function. Intuitively we note that as $h(z)$ goes to zero as z approaches the boundary (A_1, A_3) and $h(z)$ goes 1 as z approaches the vertex A_2. Building on this work, [SW01] calculated various critical exponents exactly. In [CN06] additional details of the proof and the full scaling limit are given. The corresponding result for bond percolation on the square lattice is still open.

Smirnov's proof [Smi10] of conformal invariance of the two-dimensional Ising model with nearest-neighbor interaction was also partly motivated by the geometric perspective of Langlands et al. [LPSA94]. In this case Smirnov works on a square lattice (other planar lattices can also be analyzed). At β_c consider a large box B with $+$ spins on one half of the boundary of the box and $-$ spins on the other half. Let b_1 and b_2 be the vertices at the boundary that separate these spins. There is random path (interface) connecting b_1 to b_2 separating the $\pm$ spins inside B. In the continuum limit this path has the statistics of $SLE(3)$. See Section 18.4 of [YSA] and the expository review [DCS12].

17.4 Concluding Remarks

In summary, Langlands' work on percolation and Ising models played a very important early role in mathematical analysis of these models. His emphasis on universal geometric features contrasted with the conformal field theoretic perspective of Belavin–Polyakov–Zamolodchikov [BPZ84]. Today the relation between these perspectives is much better understood.

Two of Langlands' other ambitious contributions to mathematical physics include his papers with Lafortune [LL94] and later [Lan05] on real space renormalization and with Saint-Aubin [LSA95] on completeness of the Bethe-Ansatz. In [LL94] he presents a complex real space numerical renormalization approach to obtain the critical exponent v for two-dimensional percolation. His work was a serious effort to make the renormalization group mathematically controlled and thereby establish universality. Universality is still one of the major open questions in mathematical physics. For certain hierarchical models, the existence of a fixed point for the renormalization group had been controlled numerically [KW91] but here the hierarchical approximation is essential. His work with Saint-Aubin concerns the Bethe-Ansatz which is one of the basic nonperturbative tools of theoretical physics. Even in its simplest form, the ansatz for the eigenstates has a complicated expression, which involves solutions to certain algebraic equations. Completeness of the eigenstates, as well as their structure, are still difficult challenging mathematical problems. In [LSA95] geometric methods, such as the Lefschetz formula, were applied to give partial answers to these questions.

Finally, I would like to acknowledge the central role Langlands has played in the School of Mathematics. He always embraced a broad view of mathematics. In particular, he brought applied mathematics to IAS, especially in the field of partial differential equations. He was the driving force behind bringing a special program on fluids to IAS from 1991 to 1992. A. Chorin, B. Engquist, T. Hou, A. Majda, and V. Rokhlin were prominent members of the program, which combined theory with numerical analysis and computation. In addition, Weinan E was a long-term Member from 1991 to 1994, and George Papanicolaou from 1990 to 1992. Although Luis Caffarelli and I played a role in supporting applied mathematics at the IAS, Langlands was certainly the leader of this endeavor.

Acknowledgements

I thank John Cardy and Yvan Saint-Aubin for their help. Yvan kindly sent me a preview of his chapter so that I could refer to his detailed mathematical description and figures.

References

[AB^{+}99] Michael Aizenman, Almut Burchard, et al. Hölder regularity and dimension bounds for random curves. *Duke Mathematical Journal*, 99(3):419–453, 1999.

[ABNW99] Michael Aizenman, Almut Burchard, Charles M. Newman, and David B. Wilson. Scaling limits for minimal and random spanning trees in two dimensions. *Random Structures & Algorithms*, 15(3-4):319–367, 1999.

[ADC19] Michael Aizenman and Hugo Duminil-Copin. Marginal triviality of the scaling limits of critical 4d ising and ϕ_4^4 models. *arXiv preprint arXiv:1912.07973*, 2019.

[BBS19] Roland Bauerschmidt, David C. Brydges, and Gordon Slade. *Introduction to a Renormalisation Group Method*, volume 2242. Springer Nature, 2019.

[BDH98] D. Brydges, J. Dimock, and T. R. Hurd. A non-gaussian fixed point
         for φ 4 in 4- ε dimensions. *Communications in Mathematical Physics*,
         198(1):111–156, 1998.

[BPZ84] Alexander A. Belavin, Alexander M. Polyakov, and Alexander B.
         Zamolodchikov. Infinite conformal symmetry in two-dimensional quan-
         tum field theory. *Nuclear Physics B*, 241(2):333–380, 1984.

[BS$^+$96] Itai Benjamini, Oded Schramm, et al. Percolation beyond z^d many
         questions and a few answers. *Electronic Communications in Probability*,
         1:71–82, 1996.

[Car86] John L. Cardy. Effect of boundary conditions on the operator content
         of two-dimensional conformally invariant theories. *Nuclear Physics B*,
         275(2):200–218, 1986.

[Car92] John L. Cardy. Critical percolation in finite geometries. *J. Phys. A*,
         25(4):L201–L206, 1992.

[Car05] John Cardy. SLE for theoretical physicists. *Ann. Physics*, 318(1):81–118,
         2005.

[CHI15] Dmitry Chelkak, Clément Hongler, and Konstantin Izyurov. Conformal
         invariance of spin correlations in the planar ising model. *Annals of
         Mathematics*, pages 1087–1138, 2015.

[CN06]   Federico Camia and Charles M. Newman. Two-dimensional critical per-
         colation: the full scaling limit. *Communications in Mathematical Physics*,
         268(1):1–38, 2006.

[DCS12] Hugo Duminil-Copin and Stanislav Smirnov. Conformal invariance of
         lattice models. *Probability and Statistical Physics in Two and More
         Dimensions*, 15:213–276, 2012.

[DFSZ87] Philippe Di Francesco, Hubert Saleur, and Jean-Bernard Zuber. Rela-
         tions between the coulomb gas picture and conformal invariance of
         two-dimensional critical models. *Journal of Statistical Physics*, 49(1–2):
         57–79, 1987.

[FQS84] Daniel Friedan, Zongan Qiu, and Stephen Shenker. Conformal invariance,
         unitarity, and critical exponents in two dimensions. *Physical Review
         Letters*, 52(18):1575, 1984.

[FQS86] Daniel Friedan, Zongan Qiu, and Stephen Shenker. Details of the non-
         unitarity proof for highest weight representations of the virasoro algebra.
         Communications in Mathematical Physics, 107(4):535–542, 1986.

[GGM12] Alessandro Giuliani, Rafael L. Greenblatt, and Vieri Mastropietro. The
         scaling limit of the energy correlations in non-integrable ising models.
         Journal of Mathematical Physics, 53(9):095214, 2012.

[Kes82] Harry Kesten. *Percolation Theory for Mathematicians*, volume 423.
         Springer, 1982.

[KW91]   Hans Koch and Peter Wittwer. On the renormalization group transfor-
         mation for scalar hierarchical models. *Communications in Mathematical
         Physics*, 138(3):537–568, 1991.

[Lan88] Robert P. Langlands. On unitary representations of the virasoro algebra.
         Infinite-Dimensional Lie Algebras and Their Applications. World Scien-
         tific, Singapore, New Jersey, Hong Kong, pages 141–159, 1988.

[Lan05] Robert P. Langlands. The renormalization fixed point as a mathematical object. *Twenty Years of Bialowieza: a Mathematical Anthology*, 8:185–216, 2005.

[LL94] R. P. Langlands and M.-A. Lafortune. Finite models for percolation. In *Representation Theory and Analysis on Homogeneous Spaces (New Brunswick, NJ, 1993)*, volume 177 of *Contemp. Math.*, pages 227–246. Amer. Math. Soc., Providence, RI, 1994.

[LPPSA92] Robert P. Langlands, Claude Pichet, Ph. Pouliot, and Yvan Saint-Aubin. On the universality of crossing probabilities in two-dimensional percolation. *Journal of Statistical Physics*, 67(3–4):553–574, 1992.

[LPSA94] Robert Langlands, Philippe Pouliot, and Yvan Saint-Aubin. Conformal invariance in two-dimensional percolation. *Bull. Amer. Math. Soc. (N.S.)*, 30(1):1–61, 1994.

[LSA95] Robert P. Langlands and Yvan Saint-Aubin. Algebro-geometric aspects of the Bethe equations. In *Strings and Symmetries*, pages 40–53. Springer, 1995.

[Ons44] Lars Onsager. Crystal statistics. I. A two-dimensional model with an order–disorder transition. *Physical Review*, 65(3-4):117, 1944.

[Pin94] Haru T. Pinson. Critical percolation on the torus. *J. Statist. Phys.*, 75(5-6):1167–1177, 1994.

[Pin12] Haru Pinson. Rotational invariance of the 2d spin–spin correlation function. *Communications in Mathematical Physics*, 314(3):807–816, 2012.

[S+00] Thomas Spencer et al. A mathematical approach to universality in two dimensions. *Physica A: Statistical Mechanics and its Applications*, 279(1):250–259, 2000.

[Sch00] Oded Schramm. Scaling limits of loop-erased random walks and uniform spanning trees. *Israel J. Math.*, 118:221–288, 2000.

[Smi10] Stanislav Smirnov. Conformal invariance in random cluster models. I. Holmorphic fermions in the Ising model. *Annals of Mathematics*, pages 1435–1467, 2010.

[SW01] Stanislav Smirnov and Wendelin Werner. Critical exponents for two-dimensional percolation. *arXiv preprint math/0109120*, 2001.

[Wat96] G. M. T. Watts. A crossing probability for critical percolation in two dimensions. *Journal of Physics A: Mathematical and General*, 29(14):L363, 1996.

[Wil75] Kenneth G. Wilson. The renormalization group: Critical phenomena and the Kondo problem. *Reviews of Modern Physics*, 47(4):773, 1975.

[Wu66] Tai Tsun Wu. Theory of toeplitz determinants and the spin correlations of the two-dimensional ising model. I. *Physical Review*, 149(1):380, 1966.

[YSA] Yvan Saint-Aubin. L'invariance conforme et l'universalité au point critique des modèles bidimensionnels, this volume.

18

L'invariance conforme et l'universalité au point critique des modèles bidimensionnels

Yvan Saint-Aubin

Résumé Des quelques articles publiés par Robert P. Langlands en physique mathématique, c'est celui publié dans le *Bulletin of the American Mathematical Society* sous le titre *Conformal invariance in two-dimensional percolation* qui a eu, à ce jour, le plus d'impact : les idées d'Oded Schramm ayant mené à l'équation de Loewner stochastique et les preuves de l'invariance conforme de modèles de physique statistique par Stanislav Smirnov ont été suscitées, au moins en partie, par cet article. Ce chapitre rappelle sommairement quelques idées de l'article original ainsi que celles issues des travaux de Schramm et Smirnov. Il est aussi l'occasion pour moi de décrire la naissance de ma collaboration avec Robert Langlands et d'exprimer ma profonde gratitude pour cette fantastique expérience scientifique et humaine.

Extended Abstract Of all mathematical physics contributions by Robert P. Langlands, the paper *Conformal invariance in two-dimensional percolation* published in the *Bulletin of the American Mathematical Society* is the one that has had, up to now, the most significant impact: Oded Schramm's ideas leading to the stochastic Loewner equation and Stanislav Smirnov's proof of the conformal invariance of percolation and the Ising model in two dimensions were at least partially inspired by it. This chapter reviews briefly some ideas of the original paper and some of those by Schramm and Smirnov.

This chapter is also for me the occasion to reminisce about the extraordinary scientific and human experience that working with Robert Langlands was. It started in the late 1980s when Langlands would spend Summers at the *Centre de recherches mathématiques* in Montreal. The "Langlands program" was already launched and many colleagues were devoting their career to it. Beside his steady efforts in automorphic forms, Langlands was already exploring new fields, mathematical physics being one of them. He studied conformal field theory, just then introduced, and started thinking about the renormalization group. He presented some of these ideas in a study workshop in Montreal and this is when our collaboration took off. This collaboration concentrated on problems related to conjectures of universality and conformal invariance of two-dimensional discrete systems on compact domains, and on the Bethe Ansatz. Discussing, bouncing ideas, and simply collaborating with Langlands was a fantastic experience. I had a hard time understanding his more-formal presentations. But one-on-one discussions at the blackboard were always concrete, instructive, and fruitful. My barrage of questions never seemed to frazzle him. Whenever he understood where I was blocked, his answer would often be "Let me give you an example." I had imagined that he would prefer the loftier way of mathematical communication through abstraction. But it was a nice surprise to discover that he knew so many concrete examples that revealed the crux of difficult mathematical concepts. I am deeply indebted to him for this collaboration that lasted about 10 years and for his friendship that remains very much alive today.

18.1 Prologue: réminiscences et gratitude

Robert P. Langlands commença à visiter le Centre de recherches mathématiques à Montréal durant les étés des années quatre-vingt. Le programme de Langlands, à l'intersection des théories des formes automorphes, de la représentation et de l'analyse, était déjà enclenché et plusieurs mathématiciens y travaillaient. Le présent recueil y est principalement consacré. Sans l'intention de délaisser ses premiers amours, l'aventurier Langlands était curieux d'affronter de nouveaux défis. Il explora d'autres domaines, loin des formes automorphes, dont certains problèmes liés à la physique. Il consacra d'ailleurs à ses nouvelles explorations la Chaire Aisenstadt du Centre de recherches mathématiques qu'il détint en 1988-1989. Notre intérêt commun pour les théories des champs conformes qui venaient d'être proposées aurait pu être le point de départ d'une collaboration. Mais ce sont ses efforts pour exposer, lors d'ateliers d'étude estivaux, les idées de Kesten sur la percolation et les siennes sur le groupe de renormalisation qui lancèrent nos discussions. Pendant les dix années qui suivirent, nous travaillâmes sur des problèmes liés aux hypothèses d'universalité et d'invariance conforme, principalement sur des domaines compacts, ainsi que sur l'Ansatz de Bethe. Je ne suis pas le seul à avoir profité de cette collaboration : des étudiants de l'Université de Montréal, surtout parmi les miens, ont eu la chance de travailler avec lui.

Discuter, échanger des idées et collaborer avec Robert Langlands fut une expérience fantastique. Plusieurs collègues prirent le soin, au début, de m'informer de la prestance scientifique de mon nouveau collaborateur. Je devrais leur être reconnaissant, mais ma relative ignorance des formes automorphes n'obscurcissait pas complètement mon panorama mental des mathématiques de la seconde moitié du vingtième siècle. Les exposés de Robert ne font que faire miroiter ses buts et les chemins qu'il pressent pour les atteindre. Après plus d'un, j'ai été pris du doute de n'avoir contemplé qu'un mirage. Mais nos discussions, seul à seul, furent toujours concrètes, instructives et fertiles. Jamais le déluge de mes questions n'a été reçu par une quelconque impatience. Et, lorsque Robert comprenait l'endroit où je butais, sa réponse était souvent, à ma grande surprise : « Je te donne un exemple ! ». Chaque membre de notre communauté possède sa personnalité mathématique. J'avais pensé que celle de Robert se limitait à penser et à communiquer les mathématiques en les termes les plus abstraits. Ce fut donc une révélation que de reconnaître que ses outils ne se limitent pas à l'abstraction, mais qu'ils incluent moults exemples bien tangibles. Je lui suis donc profondément reconnaissant pour ces échanges, parfois abstraits parfois concrets, pour cette collaboration fructueuse de près de dix ans, et pour cette chaleureuse amitié qui, elle, demeure toujours vivante.

J'ai choisi de ne discuter qu'un seul des articles que Langlands a écrits en physique mathématique. En fait, cette étude n'occupera même qu'une partie des pages qui me sont allouées. La raison de ce choix est simple : cet article, *Conformal invariance in two-dimensional percolation* [1], est son article dans nos domaines d'intérêts communs qui a eu le plus grand impact en mathématiques et en physique. Modestement, je crois que ce texte a permis d'éclairer sous un angle nouveau des sujets déjà bien établis dans les deux communautés, la percolation et les transitions de phase, suffisamment pour que de jeunes mathématiciens audacieux y percent de nouvelles avenues. La première section sera donc consacrée à cet article que nous avons écrit avec Philippe Pouliot, les deux suivantes aux travaux pionniers d'Oded Schramm et de Stanislav Smirnov, respectivement.

18.2 Deux hypothèses sur les probabilités de traversée critiques

Soit $D \subset \mathbb{R}^2$ un domaine connexe de frontière ∂D le long de laquelle deux segments disjoints I et J sont choisis. Dans ce même plan $\mathbb{R}^2$, les points $\delta(m,n)$ où $m,n \in \mathbb{Z}$ et $\delta \in (0,\infty)$ sont identifiés au graphe $\mathbb{Z}^2$ dont les arêtes lient les voisins immédiats. La maille δ sera choisie suffisamment petite pour qu'aucune arête du plongement de $\mathbb{Z}^2$ dans $\mathbb{R}^2$ n'intersecte simultanément les deux segments I et J et l'intersection $D \cap \mathbb{Z}^2$ soit un graphe connexe. Seuls les points de $\mathbb{Z}^2$ à l'intérieur de D et ceux liés par une arête à un sommet à l'intérieur joueront un rôle. Une *configuration* de ces points *intérieurs* est obtenue en donnant à chacun de ces points un statut, soit *ouvert*, soit *fermé*. Si $p \in [0,1]$ est la probabilité qu'un de ces sommets intérieurs soit ouvert, alors la configuration se voit accorder la probabilité $p^{\#(\text{sommets ouverts})}$ $(1-p)^{\#(\text{sommets fermés})}$. Une *traversée entre I et J à l'intérieur de D* existe si les conditions suivantes sont remplies : d'abord il existe une arête chevauchant le segment I et une le segment J dont les extrémités soient des sommets ouverts, puis il est possible de joindre ces arêtes chevauchant I et J par des arêtes contiguës dont les extrémités sont ouvertes. L'existence d'une telle traversée dépend évidemment de la configuration. Il est donc naturel de définir la *probabilité* $\pi(D, I, J; \delta; p)$ *d'une traversée* allant de I à J à l'intérieur de D; c'est donc la somme des probabilités des configurations possédant une telle traversée. Cette définition rappellera le passage d'un liquide dans un milieu aléatoire qui donne le nom « percolation » à ce domaine des probabilités. Dans cette interprétation, le paramètre δ mesure les dimensions relatives du réseau $\mathbb{Z}^2$ (de la mouture du café) et du domaine D (le filtre où l'eau percole). Cette fonction $\pi(D, I, J; \delta; p)$ est un polynôme en p définissant une bijection de l'intervalle $[0, 1]$ sur lui-même. Le diagramme gauche de la figure 18.1 présente un domaine D carré avec les deux segments I et J de sa frontière identifiés par un trait gras. La configuration choisie possède une traversée de I à J, indiquée en pointillés. Cependant, cette même configuration n'a pas de traversée entre les segments disjoints I' et J' dont l'union est le complément dans la frontière de $I \cup J$.

L'étude de ces probabilités $\pi(D, I, J; \delta; p)$ est malaisée et d'intérêt limité. Notre objet d'étude sera plutôt la limite $\pi(D, I, J; p) = \lim_{\delta \to 0} \pi(D, I, J; \delta; p)$ quand la maille du réseau $\mathbb{Z}^2$ tend vers zéro, alors que le domaine D est tenu fixe. Le diagramme droit de la figure 18.1 décrit le même domaine, mais sur lequel est superposé un plongement de $\mathbb{Z}^2$ avec une maille plus fine que celle du diagramme de gauche. Il est clair que d'autres définitions, pour les conditions de départ et d'arrivée de la traversée, pour la position relative du domaine par rapport au réseau plongé, pour le processus limite lui-même, etc.,

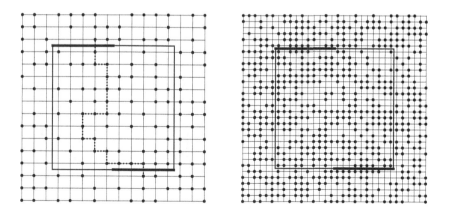

Figure 18.1 Deux configurations pour le même domaine D, mais deux mailles distinctes. Les deux configurations contiennent une traversée de I à J; celle du diagramme de gauche est indiquée en pointillés.

pourraient être choisies. Ces précisions s'avéreront avoir peu d'importance pour notre propos. Kesten [2] (voir également [3]) a montré qu'il existe une probabilité $p_c \in (0, 1)$, dite *critique*, telle que

$$\pi(D, I, J; p) = \begin{cases} 0, & p < p_c, \\ 1, & p > p_c, \end{cases} \tag{18.1}$$

et telle que

$$0 < \liminf_{\delta \to 0} \pi(D, I, J; \delta; p_c) \leq \limsup_{\delta \to 0} \pi(D, I, J; \delta; p_c) < 1. \tag{18.2}$$

Le résultat de Kesten laisse ouverte la détermination de la limite en p_c. En fait il n'est pas clair qu'elle existe ou que son existence soit indépendante des diverses variations des définitions ci-dessus. Mais, intuitivement, si cette limite existe, elle devrait dépendre du domaine D et des segments I et J choisis. Ainsi les probabilités de traversée horizontale entre les côtés opposés d'un rectangle devraient dépendre du rapport des longueurs de ses côtés. Par exemple, il semble raisonnable de supposer que, en p_c, la probabilité d'une traversée horizontale dans le rectangle à gauche de la figure 18.2 devrait être plus grande que celle dans le rectangle de droite.

Concentrons-nous pour l'instant sur les domaines rectangulaires, comme ceux de la figure 18.2 où les côtés gauche et droit sont les segments I et J. Notons par r le rapport de la longueur des côtés verticaux sur celle des horizontaux. Supposons enfin que les limites $\lim_{\delta \to 0} \pi(D, I, J; \delta; p_c)$ existent. Les probabilités de traversée ont été introduites ci-dessus sur le réseau carré $\mathbb{Z}^2$. Mais tout autre réseau périodique aurait pu être utilisé. En fait, notre formulation (18.1) et (18.2) du résultat de Kesten ne lui rend pas justice : il

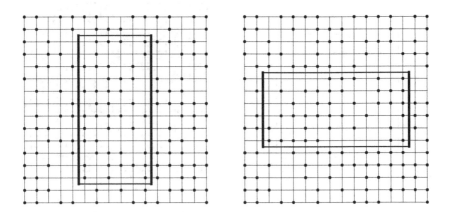

Figure 18.2 La probabilité de traversée horizontale dans le domaine rectangulaire de gauche devrait être supérieure à celle du domaine de droite.

a démontré ce résultat pour une large famille de graphes périodiques définis comme suit. Un graphe $\mathcal{G}$, plongé dans $\mathbb{R}^2$, est dit *périodique* si

(i) $\mathcal{G}$ ne contient pas de boucles (au sens des graphes);

(ii) $\mathcal{G}$ est périodique sous translation par les éléments d'un réseau $L \subset \mathbb{R}^2$ de rang deux;

(iii) le nombre d'arêtes attachées à un site est borné;

(iv) la longueur d'une arête est finie et tout ensemble compact de $\mathbb{R}^2$ intersecte un nombre fini d'arêtes et

(v) $\mathcal{G}$ est connexe.

Clairement les réseaux carré, triangulaire et hexagonal sont périodiques dans ce sens. Même si ses résultats valent hors p_c, ils soulèvent la question naturelle : les probabilités $\pi_\square(r)$, $\pi_\triangle(r)$ et $\pi_\bigcirc(r)$ de traversée horizontale en p_c, à l'intérieur d'un rectangle de rapport r, sur les réseaux carré, triangulaire et hexagonal, sont-elles reliées entre elles ? C'est d'un débat sur cette question qu'est née notre première collaboration : Langlands arguait que ces trois fonctions $\pi_\square(r), \pi_\triangle(r)$ et $\pi_\bigcirc(r)$ devaient coïncider, alors que je pensais que ces probabilités pouvaient appartenir à des classes d'universalité distinctes. (Ces positions reposaient sur notre compréhension des arguments physiques liés à la description des phénomènes critiques par la théorie des champs conformes. Dans cette théorie, les phénomènes critiques, telles les traversées horizontales en p_c, sont groupés en classes d'universalité étiquetées, entre autres, par la valeur de l'élément central c de l'algèbre de Lie qui gouverne ces théories, l'algèbre de Virasoro. Ces arguments physiques associaient à la percolation la valeur $c = 0$.) Notre premier article [4], écrit conjointement avec C. Pichet et P. Pouliot, présente des simulations sur ordinateur déterminant numériquement ces probabilités de traversée sur des rectangles pour plusieurs rapports r sur les trois réseaux ci-dessus ainsi que pour trois autres modèles de percolation où ce sont les arêtes qui sont ouvertes ou fermées, plutôt que les sommets. Ces simulations nous convainquirent que ces limites existent et que, tel que Langlands l'avait pressenti, ces probabilités vues comme fonction de r sont égales pour les six modèles considérés. Puisque toutes ces simulations étaient menées sur un réseau fini (et donc pour une maille plus grande que zéro), nous dûmes proposer une approximation des probabilités critiques p_c propices aux mesures à faire. (Même si les probabilités de traversée $\pi_\square, \pi_\triangle, \dots$, coïncident, les probabilités critiques p_c dépendent, elles, du réseau. La notice *percolation threshold* sur l'encyclopédie en ligne Wikipedia collige les meilleures valeurs connues de ces probabilités critiques pour une multitude de réseaux, périodiques ou non, en deux dimensions ou plus.) Plusieurs des techniques que nous utilisâmes étaient probablement connues, mais la coïncidence des fonctions probabilités de traversée était l'observation nouvelle. Même si ce premier travail observe numériquement l'universalité de ces probabilités de

traversée, la formulation générale d'une hypothèse d'universalité n'apparaît
dans toute sa généralité que dans notre second travail [1], écrit conjointement
avec P. Pouliot.

S'il est possible de changer le réseau sur lequel la percolation a lieu, il
est également possible de changer la probabilité que telle ou telle partie du
système soit ouverte. Par exemple, pour le réseau $\mathbb{Z}^2$, les sites dont la somme
des coordonnées est paire pourraient être ouverts avec une probabilité p_p
distincte de celle, disons p_i, des sites où cette somme est impaire. Soit donc
une fonction $p : S \to [0, 1]$ de l'ensemble des sites S du graphe périodique
$\mathcal{G}$ qui soit périodique sous les translations du réseau L de la définition (ii)
ci-dessus. Comme précédemment, les probabilités de traversée peuvent être
définies sur ce graphe où chaque site peut avoir une probabilité d'être ouvert
distincte de ces voisins. Cette paire $M = M(\mathcal{G}, p)$ définit donc un nouveau
modèle de percolation et nous noterons par π_M les probabilités de traversée
qui s'y rattachent. Puisque $\mathcal{G}$ est plongé dans $\mathbb{R}^2$, tout élément g de $GL(2, \mathbb{R})$
transforme ce graphe périodique en un graphe périodique $\mathcal{G}' = g\mathcal{G}$ par la
simple action $s \to gs$ sur les sites et similairement sur les arêtes. Soit donc gM
le modèle défini sur ce nouveau graphe. Si la même transformation linéaire est
appliquée au domaine D et aux segments I et J, les probabilités de traversée
vérifient trivialement la relation $\pi_M(D, I, J) = \pi_{gM}(gD, gI, gJ)$. Cependant,
les probabilités $\pi_M(D, I, J)$ et $\pi_{gM}(D, I, J)$ sont en général différentes.
L'hypothèse d'universalité relie les modèles M et M' associés à deux paires
$(\mathcal{G}, p)$ et $(\mathcal{G}', p')$ distinctes.

Hypothèse d'universalité — *Soient M et M' deux modèles de percolation
critiques. Il existe un élément $g \in GL(2, \mathbb{R})$ tel que*

$$\pi_{gM}(D, I, J) = \pi_{M'}(D, I, J)$$

pour tous les domaines D et segments I et J.

La vérification numérique offerte par notre premier article [4] est directe, mais
quelque peu limitée : les réseaux utilisés sont plongés de façon régulière,
c'est-à-dire de façon à préserver l'invariance sous rotation autour des sommets
par un angle de $\frac{\pi}{2}$, $\frac{\pi}{3}$ et $\frac{2\pi}{3}$ pour les réseaux carré, triangulaire et hexagonal
respectivement, et l'élément $g \in GL(2, \mathbb{R})$ faisant le lien entre les modèles
étudiés est toujours (un multiple de) l'identité. Le second article donne
cependant un exemple d'un modèle $M(\mathcal{G}, p)$ où le graphe est $\mathbb{Z}^2$, mais p n'est
pas la fonction constante. Pour ce modèle $M(\mathcal{G}, p)$, l'élément g le reliant au
modèle sur le réseau carré usuel n'est ni diagonal, ni une matrice orthogonale.

Il est probable que cet énoncé vaille pour des modèles qui ne sont pas définis
à l'aide d'un réseau périodique dans le sens de Kesten : viennent à l'esprit des
modèles définis à partir des pavages non périodiques de Penrose ou même ceux
dont le graphe est lui-même aléatoire. Quoiqu'il en soit, la conjecture demeure
à ma connaissance ouverte.

Deux collègues, M. Aizenman et J. Cardy, nous ont aidés à façonner la seconde hypothèse et à en faire un défi mathématique excitant. Après que nos premières données numériques sur les probabilités de traversée eurent été rendues publiques, Langlands eut la chance de discuter avec Aizenman qui lui proposa comment l'invariance conforme pouvait peut-être se manifester. À cette époque, la théorie des champs conformes avait déjà montré sa puissance pour prédire les exposants critiques qui caractérisent les transitions de phase en deux dimensions en leurs points critiques, et pour obtenir des formes explicites des fonctions de corrélation de diverses quantités physiques de ces modèles statistiques. Malgré que la percolation était parmi les modèles physiques étudiés par cette théorie des champs conformes, les probabilités de traversée étaient fort loin des quantités qu'elle avait permis de décrire. C'est pourquoi j'avais été étonné par l'audacieuse suggestion d'Aizenman.

Soit j une transformation linéaire du plan $\mathbb{R}^2$ dont le carré est moins l'identité. Elle définit donc une structure complexe sur le plan et la notion de fonctions j-holomorphes. Si le triplet (D, I, J) est la donnée d'un domaine et de deux segments disjoints le long de sa frontière, une fonction $\phi : \mathbb{R}^2 \to \mathbb{R}^2$ qui est j-holomorphe à l'intérieur de D, et continue et bijective jusqu'à sa frontière, définit un nouveau triplet $(\phi(D), \phi(I), \phi(J))$ d'un domaine avec deux segments disjoints $\phi(I)$ et $\phi(J)$ le long de sa frontière. Cette observation vaut également si ϕ est j-antiholomorphe à l'intérieur de D.

Hypothèse d'invariance conforme — *Pour tout modèle de percolation critique $M = M(\mathcal{G}, p)$, il existe une transformation linéaire j définissant une structure complexe telle que*

$$\pi_M(\phi(D), \phi(I), \phi(J)) = \pi_M(D, I, J)$$

pour tout triplet (D, I, J) et pour tout ϕ qui soit j-(anti)holomorphe à l'intérieur de D, continue et bijective jusqu'à sa frontière.

Cette hypothèse veut dire en particulier que la probabilité de traversée pour le triplet (D', I', J') présenté au diagramme droit de la figure 18.3 devrait être égale, dans la limite $\delta \to 0$, à celle du carré utilisé à la figure 18.1. Malgré son audace, l'hypothèse est possiblement vraie pour une famille plus grande de fonctions. Notre second article [1] présente une vérification numérique pour la fonction complexe $z \to z^2$ appliquée à un domaine D carré incluant l'origine. Sur un tel domaine, cette fonction n'est certainement pas bijective. Pour contrer cette difficulté, le réseau utilisé dans la mesure numérique est une paire de graphes $\mathbb{Z}^2$ joints le long d'une coupure longeant les points dont la première coordonnée est positive et la seconde nulle. On peut voir intuitivement cette paire de graphes $\mathbb{Z}^2$ comme couvrant le double recouvrement $\mathbb{X}$ du plan complexe qui rend bijective la fonction $\phi : \mathbb{C} \to \mathbb{X}$ donnée par $z \to z^2$. Sur ce « double recouvrement », l'hypothèse d'invariance conforme est remarquablement vérifiée.

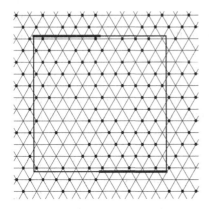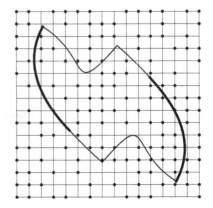

Figure 18.3 Deux probabilités de traversée qui devraient être égales, dans la limite $\delta \to 0$, à celle du triplet de la figure 18.1 : celle de gauche sur le réseau triangulaire (hypothèse d'universalité) et celle de droite pour un triplet (D', I', J') obtenu de celui utilisé à la figure 18.1 par une fonction holomorphe (hypothèse d'invariance conforme).

Fort de ces observations numériques, Langlands contacta Cardy, lui demandant s'il était possible d'extraire, de la théorie des champs conformes, une prédiction théorique pour ces probabilités de traversée. La réponse fut rapide et convaincante. Maintenant connue sous le nom de *formule de Cardy*, la probabilité de traversée prédite par cette théorie des champs conformes est

$$\pi_\square(r) = \frac{3\Gamma\left(\frac{2}{3}\right)}{\Gamma\left(\frac{1}{3}\right)^2} \sin^{\frac{2}{3}}(\theta(r)) \, _2F_1\left(\tfrac{1}{3}, \tfrac{2}{3}; \tfrac{4}{3}; \sin^2(\theta(r))\right) \qquad (18.3)$$

où Γ et $_2F_1$ sont les fonctions gamma et hypergéométrique usuelles, et $\theta(r)$ est déterminé comme suit. Par le théorème de Riemann, il existe une fonction holomorphe envoyant l'intérieur du rectangle de rapport de côtés r vers l'intérieur du disque de rayon 1. Cette fonction peut être choisie de façon à ce que les sommets du rectangle aient pour images les point $z = e^{i\theta}$, $\bar{z}$, $-z$ et $-\bar{z}$. Cet angle $\theta = \theta(r)$ est celui apparaissant dans la formule ci-dessus. La courbe de la figure 18.4 montre cette prédiction de Cardy à laquelle nous avons ajouté les mesures pour les probabilités de traversée sur le réseau carré obtenues pour des rectangles contenant approximativement un million de sites. L'échantillon choisi est tel que l'erreur sur ces mesures est plus petite que la grosseur des points sur la figure. On s'inquiètera peut-être du léger écart entre la prédiction et les mesures pour les valeurs de r extrêmes; cet écart est probablement dû au fait que, pour ces valeurs de r, le nombre de sites d'un des côtés du rectangle était trop petit pour atteindre la qualité de mesure des autres points. L'accord

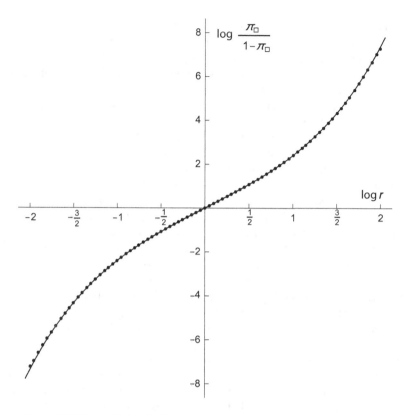

Figure 18.4 La prédiction de Cardy (trait continu) avec les mesures numériques pour le réseau carré de [1].

saisissant entre la prédiction et les mesures a été probablement un des éléments attirant de jeunes mathématiciens à ces conjectures.

Il existe une forme remarquablement simple de la formule de Cardy que Smirnov [5] attribue à L. Carleson. Elle est donnée pour le triplet (D, I, J) où D est un triangle équilatéral de côté unité, I un des côtés et J est le segment le long d'un des deux autres côtés commençant au sommet opposé à I et de longueur x. Alors $\pi(D, I, J) = x$.

Avant de clore cette section, voici quelques mots sur une autre avenue explorée par Langlands. Une des idées-clés de la physique statistique et de la théorie des champs quantiques est de remplacer la description ab initio d'un phénomène donné en « intégrant » les interactions à petite échelle, ne gardant ainsi que le comportement à plus grande échelle. Cette opération, nommée *renormalisation*, est souvent décrite par son action sur les divers paramètres décrivant la théorie. Par exemple, les atomes d'un cristal sont organisés sur un réseau de maille δ. Un ensemble de paramètres est alors nécessaire pour décrire l'interaction entre deux atomes à distance δ, entre deux atomes à distance 2δ, et ainsi de suite. Les points fixes sous la renormalisation sont liés aux points critiques du phénomène. L'étude rigoureuse de cette opération de renormalisation, si elle est possible, requiert usuellement l'étude d'un nombre infini de paramètres. Une des idées de Langlands était de mener (rigoureusement) cette étude de la renormalisation sur une famille de modèles, chacun possédant un nombre fini N d'états ou d' « événements ». Dans ses modèles l'opération de renormalisation agit sur le cube $[0, 1]^N$ dont les coordonnées sont les probabilités de chacun des événements.

Pour comprendre les modèles qu'il introduisit [6, 7], les triplets (D, I, J) sont remplacés par les événements $E = (D, (I_{i_1}, I_{j_1}), (I_{i_2}, I_{j_2}), \ldots, (I_{i_n}, I_{j_n}))$. Tous ces *événements* E auront lieu à l'intérieur d'un domaine D carré. Cependant, les côtés sont maintenant divisés en m segments, produisant ainsi $4m$ segments disjoints $I_1, I_2, \ldots, I_{4m}$. Parmi ces segments disjoints, certaines paires $(I_{i_1}, I_{j_1}), (I_{i_2}, I_{j_2}), \ldots, (I_{i_n}, I_{j_n})$ sont choisies et E est caractérisé par l'exigence, pour chacune de ces paires, qu'elle soit ou ne soit pas reliée. Par exemple, un événement E pourrait ne contraindre que deux paires : la paire (I_3, I_5) devrait être reliée par des sites ouverts mais la paire (I_3, I_8) devrait ne pas l'être; les autres paires sont libres d'être ou non reliées. L'opération de renormalisation introduite par Langlands regroupe alors k^2 de ces événements pour former un nouveau carré avec $4mk$ segments à sa frontière. Des règles sont alors données pour fusionner ces segments par ensembles de k segments contigus, redonnant ainsi un événement décrit par $4m$ segments. Les résultats numériques présentés dans [7] indiquent que ces modèles finis possèdent un point critique et qu'un des exposants critiques en ce point est proche de celui prédit pour la percolation. Clairement ces modèles finis devraient être explorés plus avant.

18.3 L'équation de Schramm–Loewner

Les hypothèses d'invariance conforme et d'universalité telles que formulées dans notre article dans le *Bulletin of the AMS* n'étaient que des conjectures dont les conséquences demeuraient à être explorées. Deux pionniers, Oded Schramm et Stanislav Smirnov, se mirent rapidement à pied d'œuvre pour explorer, formuler et prouver mathématiquement ces hypothèses. Les travaux de Schramm répondent à la question : existe-t-il des espaces de probabilité, permettant de mesurer les probabilités de traversée et respectant une condition d'invariance conforme similaire à celle que nous avions formulée ? Sa réflexion le mena à lier l'existence de tels espaces à celle du mouvement brownien. Le but de la présente section est modeste : il est d'expliquer, au moins intuitivement, ce lien. La présentation ci-dessous tient donc pour acquis qu'un espace de probabilité existe capturant les propriétés des plages de sites ouverts et de sites fermés et permettant de répondre à la question : quelle est la probabilité d'une traversée pour le triplet (D, I, J) en $p = p_c$? De plus, notre réflexion pourra utiliser le modèle de percolation $M = M(\mathcal{G}, p)$ de notre choix puisque l'hypothèse d'universalité suppose que l'espace de probabilité obtenu à la limite n'en dépend pas.

Les probabilités de traversée ont été introduites sur des domaines compacts, tels les rectangles, mais si l'hypothèse d'invariance conforme tient, ces probabilités de traversée devraient être reliées aux probabilités d'aller d'un segment à un autre de la frontière du demi-plan supérieur. Plutôt que de tracer une configuration de sites ouverts ou fermés, nous nous concentrerons maintenant sur une seule interface, c'est-à-dire la courbe séparant une plage de sites ouverts de la plage de sites fermés contiguë.

La figure 18.5 présente deux interfaces dans un domaine carré (deux diagrammes du haut) et leur image sous l'application conforme envoyant ce domaine sur le demi-plan supérieur (les deux du bas). Ces interfaces démarrent au sommet A du domaine carré. Puisque $p_c > 0$, dans la limite lorsque la maille tend vers zéro, une interface démarrant aussi proche de A que désiré existe presque sûrement. Les interfaces, parcourues en s'éloignant de A, ont la plage ouverte à leur droite et la fermée à leur gauche. Ces dessins ne tracent l'interface que jusqu'au moment où elle atteint soit le côté BC soit le côté CD. L'information dessinée est effectivement suffisante pour dire qu'il n'y aura pas de traversée horizontale dans le dessin de gauche, alors qu'il y en a une dans celui de droite. En effet, dans le dessin de gauche, les deux parties les plus à droite de l'interface (celle allant d'un point de AD à un de AB et celle de AB à BC) délimitent une plage de sites fermés qui bloque toute traversée. Au contraire, dans le dessin de droite, la dernière partie de l'interface, celle allant de AB à CD a, à sa droite, une plage de sites ouverts qui assure une traversée horizontale. Ainsi, suivre l'interface démarrant au coin A permet de décider si la configuration sous-jacente possède une traversée horizontale : elle en aura une si et seulement si elle atteint le côté CD avant le côté BC.

Soit ϕ l'inverse de l'application de Schwarz-Christoffel qui envoie l'intérieur d'un rectangle $ABCD$ sur le demi-plan supérieur. Les figures sous celles des domaines carrés (figure 18.5) représentent l'image des côtés AB, BC, CD et DA le long de la frontière horizontale du demi-plan, ainsi que les méandres de l'interface tracée dans le carré au-dessus. Notons par la lettre minuscule correspondante l'image par ϕ des sommets A, B, C et D. L'observation qu'une traversée dans le domaine carré existe si et seulement si l'interface atteint le côté CD avant le côté BC se traduit, pour les « traversées » dans le demi-plan supérieur comme suit : une « traversée » du segment ab au segment allant de $-\infty$ à d existera si et seulement si l'interface partant de a atteint la demi-droite de $-\infty$ à d avant qu'elle n'atteigne la demi-droite allant de b à $+\infty$. Le problème de déterminer la probabilité de traversée horizontale

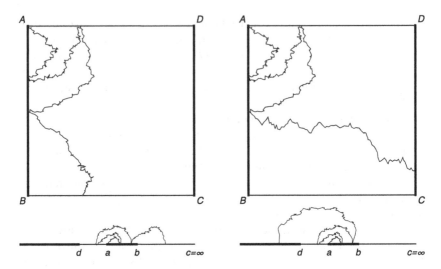

Figure 18.5 Deux interfaces à l'intérieur d'un domaine carré (en haut) et leur image dans le demi-plan supérieur (en bas). Les lettres a, b, c et d étiquettent les images par ϕ des sommets A, B, C et D.

(ou de tout événement décrit par le triplet (D, I, J)) est donc équivalent à déterminer la probabilité que l'interface partant de a atteigne la demi-droite $-\infty$ à d avant l'autre demi-droite.

L'argument précédent permet de réduire l'étude des probabilités de traversée à celle des probabilités des interfaces attachées à l'origine du demi-plan complexe $\mathbb{D}$. Quoique basé sur l'hypothèse d'invariance conforme, il ne révèle pas comment cette hypothèse relie les probabilités de diverses interfaces. À cette fin, il est utile de considérer l'interface, non pas tracée dans sa totalité, mais plutôt comme un processus dynamique où elle croît à partir de cette origine, c'est-à-dire du point a. Cette interface croît en séparant les sites ouverts immédiatement à sa droite, marqués d'un $+$, et les fermés à sa gauche, marqués d'un $-$. La figure 18.6 représente deux captures de ce processus sur un réseau triangulaire où chaque site est représenté par un hexagone. Dans la figure de gauche, l'interface a cru jusqu'au point e. Pour décider de la prochaine étape de cette croissance, l'état, ouvert ou fermé, de l'hexagone du site « devant » e doit être choisi. Si cet hexagone est ouvert $(+)$, l'interface ira vers la gauche et, s'il est fermé $(-)$, vers la droite. Ce processus de croissance est facile à comprendre pour la percolation puisque l'état des sites est décidé indépendamment des autres sites. Cette indépendance est perdue dans d'autres systèmes physiques tel le modèle d'Ising qui sera discuté à la prochaine section; heureusement cette indépendance probabiliste n'est pas nécessaire aux arguments qui suivent ci-dessous. Le diagramme droit de la figure capture

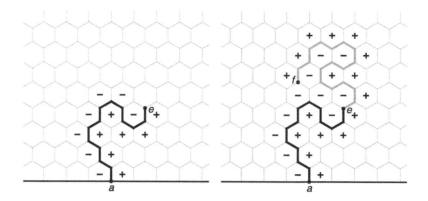

Figure 18.6 La croissance d'une interface de l'origine a à un point e (à gauche), puis jusqu'à f (à droite).

la croissance quelques instants plus tard. L'hexagone devant e est fermé et l'interface a bifurqué vers la droite, puis s'est rabattue vers la gauche. Du point f, il ne sera pas nécessaire de choisir l'état de l'hexagone devant f puisque ce site a déjà été fixé comme étant fermé à une étape précédente. Ainsi, même si le processus de croissance est simple, il dépend de toute l'histoire des choix précédents et est donc non-markovien. L'interface est appelée un *chemin auto-évitant*.

Cette propriété non-markovienne a une conséquence simple : l'interface croît à partir de f dans le demi-plan supérieur $\mathbb{D}$ duquel a été retirée l'interface de l'origine jusqu'à ce point f. La frontière de cette région va de $-\infty$ à l'origine a, suit alors le côté gauche de l'interface jusqu'à f, puis revient par son côté droit à l'axe horizontal qu'elle suit alors jusqu'à $+\infty$. Si l'interface est continue, cette région est un sous-ensemble ouvert simplement connexe du demi-plan et, par le théorème de Riemann, il existe donc une bijection holomorphe qui l'envoie sur ce même demi-plan. Si, de plus, un paramétrage $\gamma : [0, t_f] \to \mathbb{D}$ de l'interface est choisi, avec $\gamma(0) = a$ et $\gamma(t_f) = f$, alors il existe une famille de fonctions holomorphes $\psi_t, 0 \le t \le t_f$, chacune « retirant » la partie $\gamma([0,t])$ de l'interface de l'origine jusqu'à $\gamma(t)$. La figure 18.7 capture trois étapes de ce processus pour une interface (ou fente) très simple : la première montre l'interface complète et la dernière l'image de ψ_{t_f} où l'interface a été absorbée par l'axe réel. Des droites avec partie réelle ou imaginaire constante (et leur image) ont été tracées. Ces courbes montrent que les angles sont conservés par les fonctions ψ_t. L'extrémité f et l'origine de la fente (ici dédoublée puisqu'elle apparaît sur les deux côtés de cette interface) sont marquées par des points. Enfin l'interface elle-même, ou son image sous les ψ_t, est tracée en gris. Ces points et l'interface

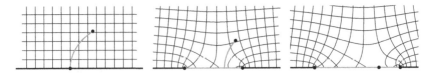

Figure 18.7 Le demi-plan supérieur duquel une interface a été retirée est envoyé holomorphiquement sur le demi-plan supérieur.

montrent la « fonte » progressive de l'interface dans l'axe réel. Cette famille ψ_t de fonctions holomorphes satisfait $\psi_{t=0}(z) = z$ pour tout $z \in \mathbb{D}$ et est uniquement déterminée si la condition suivante sur le comportement à l'infini est ajoutée : $\psi_t(z) = z + 2t/z + \mathcal{O}(z^{-2})$. Cette dernière condition fixe un paramétrage $\gamma(t)$ privilégié sur l'interface qui sera maintenant utilisé.

Cette bijection entre interfaces dans le demi-plan complexe et familles ψ_t de fonctions holomorphes peut être poussée un cran plus loin. L'*équation différentielle de Loewner*

$$\frac{\partial}{\partial t}\psi_t = \frac{2}{\psi_t - \xi(t)}$$

lie cette famille ψ_t à une fonction (continue) réelle $\xi : [0, t_f] \to \mathbb{R}$. Le point $\xi(t)$ est l'image du point $\gamma(t)$ de l'interface par ψ_t, c'est-à-dire $\psi_t(\gamma(t)) = \xi(t) \in \mathbb{R}$. C'est donc le point de rencontre avec l'axe réel de la partie restante de l'interface au temps t. En inversant le paramètre d'évolution $t \to -t$, la famille ψ_t peut être vue comme donnant naissance à une interface γ à partir du point ξ. Ainsi cette fonction est parfois nommée la *force motrice* (*driving force*) puisque ses mouvements le long de l'axe horizontal détermineront l'apparence de γ. Évidemment, si seule l'interface $\gamma([0, t_f])$ est donnée, ni la famille ψ_t, c'est-à-dire l'évolution de Loewner, ni la fonction $\xi(t)$ ne peuvent être aisément déterminées à partir de l'équation ci-dessus. L'importance de l'équation de Loewner réside plutôt dans le fait qu'elle établit une bijection entre les interfaces dans le demi-plan supérieur et les fonctions continues ξ décrivant le mouvement du point moteur sur l'axe réel auquel est attachée l'interface alors qu'elle est absorbée par cet axe (ou qu'elle en émane, si le paramètre t est inversé).

Cette façon d'appréhender les interfaces, par le processus dynamique de leur croissance ou par l'évolution de Loewner qui les fait disparaître, permet une formulation alternative de l'hypothèse d'invariance conforme. Supposons à nouveau qu'un espace de probabilité Ω existe dont les éléments sont les limites, lorsque la maille δ du réseau tend vers 0, des interfaces (infinies) construites comme à la figure 18.6. Notons par $\gamma : [0, +\infty) \to \mathbb{D}$ ces interfaces. Soit un ensemble mesurable ω de telles interfaces partageant leur partie initiale, disons jusqu'en un point e fixé atteint au temps t_e. Ainsi, si γ_1

et γ_2 appartiennent à ω, alors $\gamma_1(t) = \gamma_2(t)$ pour tout $t \in [0, t_e]$. Soit enfin ψ_{t_e} la fonction holomorphe retirant la partie commune $\gamma([0, t_e])$ de ces interfaces. Alors l'hypothèse d'invariance conforme requiert que la mesure de l'ensemble $\{\psi_{t_e}(\gamma) \mid \gamma \in \omega\}$ soit égale à celle de ω.

L'hypothèse d'invariance conforme contraint donc les espaces de probabilité Ω. En fait, la bijection obtenue par l'équation de Loewner entre les interfaces $\gamma : [0, +\infty) \to \mathbb{D}$ et les fonction $\xi : [0, +\infty) \to \mathbb{R}$ permet de les caractériser complètement. Si Ω est un tel espace de probabilité, alors il existe un nombre positif $\kappa > 0$ tel que la mesure sur ces interfaces dans $\mathbb{D}$ est induite par la bijection de Loewner si celle sur les fonctions réelles $\xi(t)$ est donnée par $\sqrt{\kappa}\, B(t)$ où B est le mouvement brownien.

C'est Schramm qui formula (rigoureusement) ces idées et démontra le théorème de classification de ces espaces de probabilité de courbes conformément invariants [8, 9]. Il nomma le processus de croissance des interfaces l'*équation de Loewner stochastique*, mais la communauté changea ce nom pour *équation de Loewner-Schramm* préservant ainsi l'acronyme anglais *SLE* qu'il avait choisi. En utilisant un résultat de Stanislav Smirnov [5] prouvant que les interfaces de la percolation sur un réseau triangulaire donnent lieu, dans la limite $\delta \to 0$, à l'espace Ω correspondant à $\kappa = 6$, Schramm put démontrer la formule de Cardy. L'équipe constituée de Greg Lawler, Oded Schramm et Wendelin Werner, à laquelle se joignirent leurs collègues et étudiants, développa ses idées en une théorie élégante, démontrant rigoureusement de nombreux résultats prédits par l'intuition des physiciens et, surtout, introduisant des outils mathématiques pour attaquer plusieurs autres questions physiques liées aux transitions de phase. Werner obtint la Médaille Fields en 2006 pour ses contributions à ce corpus.

18.4 L'invariance conforme du modèle d'Ising

Malgré la beauté des idées de Schramm et du grand nombre de résultats qui en découlèrent, une preuve de l'invariance conforme de la limite continue des modèles physiques existait que dans très peu de cas. À la publication du premier article de Schramm [8], il n'y avait à ma connaissance que la preuve de Smirnov pour la percolation sur un réseau triangulaire. Cette preuve, écrite dans un style compact, ne semblait pas pouvoir être étendue aisément à d'autres modèles. Il est donc naturel de se demander quels autres modèles sur réseau possèdent une limite décrite par l'espace de probabilité Ω, non seulement pour $\kappa = 6$, mais aussi pour toute autre valeur permise de ce paramètre. Les hypothèses d'invariance conforme et d'universalité peuvent être étendues à ces autres valeurs. Le modèle d'Ising fournit un exemple d'une telle limite qui, elle, va vers l'espace Ω en $\kappa = 3$. Le but de cette section, modeste comme celui de la précédente, est de présenter les outils intervenant dans la preuve de Smirnov [10] de l'invariance conforme de ce modèle statistique sur le réseau carré.

Il existe deux présentations du modèle d'Ising. La première que je présente d'abord est la définition originale et certainement la plus connue. J'introduirai après la seconde que Smirnov utilise dans sa preuve.

Soit R un rectangle contenant $M \times N$ points de $\mathbb{Z}^2$. À chacun de ces points est attachée une variable aléatoire σ_p, $p = (i, j)$, prenant les valeurs $+1$ ou -1. À un choix $\sigma = (\sigma_{p=(i,j)})_{1 \le i \le M, 1 \le j \le N}$ de ces MN variables, on associe une *énergie*

$$E(\sigma) = -J \sum_{\langle p, p' \rangle} \sigma_p \sigma_{p'} - H \sum_p \sigma_p$$

où la seconde somme porte sur tous les points p du rectangle R et la première sur les paires $\langle p, p' \rangle$ de voisins immédiats : si $p = (i, j)$, alors $p' \in \{(i + 1, j), (i - 1, j), (i, j + 1), (i, j - 1)\} \cap R$. Les constantes J et H ont une interprétation physique : J décrit le couplage ferromagnétique entre les sites (= spins) voisins et H est lié au champ magnétique extérieur dans lequel baignent les spins. L'*ensemble des configurations* $\{\sigma\} = \mathbb{Z}_2^{M \times N}$ est pourvu d'une mesure $P(\sigma) = e^{-E(\sigma)/kT}/Z$ où Z est un facteur de normalisation assurant que $\sum_\sigma P(\sigma) = 1$ et kT est le produit de la constante de Boltzmann et de la température T. On notera que seuls les rapports J/kT et H/kT interviennent; il est donc usuel de remplacer J/kT par β et H/kT par une autre constante η. Le modèle d'Ising est donc la famille d'espaces de probabilité $(\mathbb{Z}_2^{M \times N}, P_{\beta, \eta})$ étiquetée par l'inverse de la température β et la constante η. J'ai présenté le modèle sur un rectangle $\subset \mathbb{Z}^2$, mais la définition s'étend aisément à des (hyper-)rectangles dans $\mathbb{Z}^d$ pour $d \ge 1$. (La description du modèle en $d = 3$ demeure à ce jour un domaine de recherche actif.) Des conditions aux limites peuvent être ajoutées, par exemple en identifiant les deux côtés verticaux du rectangle R. Enfin, la limite de la maille du réseau est ici entendue comme une limite où le nombre de sites $M \times N$ tend vers l'infini.

Notons que, si la constante J est positive, alors une configuration σ est d'autant plus probable que le nombre de paires $\langle p, p' \rangle$ où les spins coïncident est grand. Cet effet sera d'autant plus marqué que la température $T > 0$ sera petite. Ainsi, lorsque J est positive, les spins tendent à s'orienter dans la même direction. Si H est positif, les configurations σ avec un grand nombre de sites $+1$ seront favorisées. Le champ extérieur H force donc les spins à s'aligner dans la direction (positive ou négative) où il pointe. Pour la suite, le champ extérieur H sera posé à zéro.

Il a été utile de concevoir les interfaces en percolation comme un processus de croissance dynamique où chaque site est décidé indépendamment des précédents. Cette façon d'appréhender les interfaces ne fonctionne pas dans le cas des interfaces entre les plages avec sites $+1$ et celles avec -1. En effet, ici, les variables aléatoires σ_p et $\sigma_{p'}$ ne sont pas indépendantes sous la mesure $P(\sigma)$. Les arguments de Schramm menant aux espaces Ω ne requièrent cependant pas cette indépendance.

La physique du modèle statistique est révélée par l'espérance de certaines variables, par exemple celle de l'énergie par site $E(\sigma)/MN$ ou par celle des fonctions de corrélations à n points, c'est-à-dire des produits $\sigma_{p_1}\sigma_{p_2}\ldots\sigma_{p_n}$ pour des points $p_i \in R$ distincts. Le signe révélateur d'une transition de phase pour un modèle physique comme le modèle d'Ising est la singularité d'une de ces valeurs moyennes ou d'une de leurs dérivées. Le calcul exact de l'énergie par site en $d = 2$ par Lars Onsager en 1944 démontra l'existence d'une telle singularité en un certain β_c, $0 < \beta_c < \infty$ en $H = 0$. (La valeur de l'inverse de la température critique est donnée implicitement par $\sinh 2\beta_c = 1$.) Ce calcul, considéré comme un tour de force, prouve donc l'existence d'une transition de phase pour le modèle d'Ising en deux dimensions.

Plutôt que d'utiliser un sous-ensemble rectangulaire de $\mathbb{Z}^2$ dont la base est horizontale, la seconde présentation opte pour un rectangle incliné à 45°. Le diagramme de gauche de la figure 18.8 montre une configuration σ où les sites p sont occupés par le signe σ_p et les paires de voisins immédiats sont liés par une diagonale en tirets. Une *configuration de boucles* sera maintenant construite. La première étape consiste à remplacer toutes les diagonales liant deux signes opposés par une tuile contenant deux quarts de cercle positionnés pour que ces arcs coupent la diagonale qu'ils remplacent. Le second diagramme de la figure 18.8 montre le résultat. La seconde étape remplace maintenant les diagonales liant des signes identiques. Les quarts de cercle pourront être positionnés pour couper la diagonale ou non. Ils la couperont avec probabilité $q = e^{-2\beta}$ et ne la couperont pas avec probabilité $p = 1 - q$. En ajoutant les demi-cercles à la frontière comme indiqué sur le dernier diagramme de la figure 18.8, la configuration ainsi obtenue ne contient que des boucles fermées.

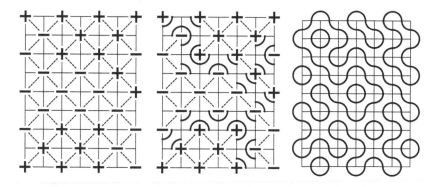

Figure 18.8 D'une configuration de spins à une configuration de boucles.

Fortuin et Kasteleyn [11] ont montré qu'une telle configuration de boucles γ apparaît alors avec la probabilité $q^{c(\gamma)}p^{d(\gamma)}2^{b(\gamma)}/Z$ où $c(\gamma)$ et $d(\gamma)$ sont respectivement les nombres de tuiles dont la diagonale est coupée ou préservée par les quarts de cercle, et où $b(\gamma)$ est le nombre de boucles fermées. (Pour la configuration de la figure 18.8, $b = 9$, $c = 26$ et $d = 22$, le nombre total de tuiles étant $c + d = 48$.) Le facteur de normalisation Z est alors identique à celui de la première présentation. Les valeurs critiques de p et q, c'est-à-dire celles en β_c, sont $2\sqrt{2}$ et $1 - \sqrt{2}$. En ces valeurs, une simplification remarquable a lieu pour les domaines tel celui considéré : la probabilité d'une configuration de boucles γ devient proportionnelle à $(\sqrt{2})^{b(\gamma)}$. C'est donc cet espace de probabilité qu'utilise Smirnov dans sa preuve : si R est un domaine connexe contenant N_R tuiles, l'espace des configurations est $\mathbb{Z}_2^{N_R}$ et la mesure d'une configuration, c'est-à-dire d'un choix parmi $\{\,\boxtimes,\ \boxtimes\,\}$ pour chaque tuile, est proportionnelle à $(\sqrt{2})^{b(\gamma)}$.

Il est ainsi possible de considérer un domaine quelconque du plan, comme celui représenté au diagramme gauche de la figure 18.9. Les tuiles de la région R sont maintenant celles qui intersectent l'intérieur de la courbe fermée. Des arcs frontières, tracés en gris, ont été ajoutés aux tuiles extérieures attenant au domaine R. Ces arcs ont été choisis de façon à ce que toute configuration γ sur R ne soit constituée que de boucles fermées (comme le diagramme droit de la figure l'indique) à l'exception d'une courbe entrant au point a et sortant au point b de R qui est mise en évidence sur le diagramme droit.

Le coeur de la preuve de Smirnov repose sur le choix d'une observable « physique » pour laquelle il est possible de montrer que la limite existe et qu'elle est conformément invariante. Soit C l'ensemble des côtés des tuiles qui sont à l'intérieur du domaine R. Voici un premier choix pour cette observable,

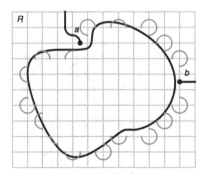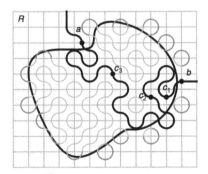

Figure 18.9 À gauche un domaine R et ses arcs frontières en gris; à droite une configuration de boucles, dans ce domaine et avec ces conditions aux limites, possédant une trajectoire allant des points frontières a et b avec trois points distingués sur cette trajectoire.

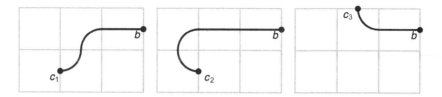

Figure 18.10 Les accroissement $w(b \to c_i)$ de l'angle du vecteur tangent de b aux points c_1, c_2 et c_3 sont $0, \pi$ et $-\pi/2$.

naturel quoique naïf. Soit $f : C \to [0, 1]$ la fonction définie par l'espérance $f(c)$ que la trajectoire $tr(\gamma)$ de a à b passe par le côté $c \in C$. Clairement f est une fonction réelle et, si elle possède une limite lorsque la maille du réseau tend vers zéro, cette limite sera également une fonction réelle. Si elle est holomorphe, elle sera une constante, ce qui n'est pas d'un grand intérêt physique. Ce premier choix doit être rejeté.

L'idée de Smirnov est de pondérer cette fonction $f(c)$ en fonction de l'accroissement $w(b \to c)$ de l'angle que fait le vecteur tangent à la trajectoire en allant du point b au point $c \in C$. Plus précisément soit $F : C \to \mathbb{C}$ la fonction donnée par $\mathbb{E}(\chi_{tr(\gamma)}(c) \cdot e^{-\frac{i\pi}{2} w(b \to c)})$ où $\chi_{tr(\gamma)}$ est la fonction caractéristique sur la trajectoire allant de a à b. Les accroissements $w(b \to c_i)$ de l'angle du vecteur tangent à la trajectoire jusqu'aux points c_i peuvent être lus aisément de la figure 18.10. Ce poids, ajouté au parcours de la trajectoire, n'était pas inconnu des physiciens. Il est la clé du calcul purement combinatoire de la fonction de partition du modèle d'Ising fait par Kac et Ward [12]. (Voir également le chapitre 5 du cours de mécanique statistique de Feynman [13].) Il n'en demeure pas moins inspiré. Cette fonction F de domaine C prend maintenant ses valeurs dans $\mathbb{C}$, et non pas seulement dans $\mathbb{R}$, et son caractère holomorphe (discret) peut être étudié. Les équations de Cauchy-Riemann discrète se lisent

$$F(c_{no}) - F(c_{se}) = i(F(c_{ne}) - F(c_{so}))$$

où les arêtes c_{no}, c_{se}, c_{ne} et c_{so} sont celles contiguës à c dans les directions nord-ouest, sud-est, et ainsi de suite. (Voir la figure 18.11.) Smirnov utilise une définition légèrement différente de cette discrétisation, mais il devra vérifier cette équation dans une preuve délicate et minutieuse.

La définition de F force également sa phase à la frontière du domaine R. En effet, si z est un point de la frontière de R ou très proche, l'espérance de l'accroissement $w(b \to z)$ sera, pour une maille du réseau très fine, mesurée avec un vecteur tangent à $tr(\gamma)$ arrivant en z perpendiculairement à la frontière ∂R. Cette condition aux limites, avec l'équation de Cauchy-Riemann discrète détermine donc un problème aux limites de Riemann. La solution du

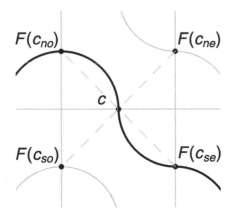

Figure 18.11 Les arêtes contiguës à c apparaissant dans l'équation de Cauchy-Riemann discrète.

problème continu correspondant est donnée par la fonction $(\Phi')^{\frac{1}{2}}$ où Φ est l'application conforme envoyant le domaine R sur le ruban horizontal infini de largeur unité, les points a et b étant envoyés aux extrémités.

Cette description ne fait que reformuler le problème de prouver l'invariance conforme du modèle d'Ising sur un réseau carré. De cette mise en place, la preuve de Smirnov procède alors en deux étapes : établir d'abord l'holomorphicité discrète de F, puis assurer que le passage à la limite peut être fait. Ces preuves établissent donc que, dans cette la limite, la fonction discrète F définie sur les arêtes du réseau tend vers la fonction analytique $(\Phi')^{\frac{1}{2}}$. Smirnov obtint la Médaille Fields en 2010 pour la preuve de l'invariance conforme pour les modèles de percolation et d'Ising en deux dimensions.

18.5 Conclusion

Les idées de Robert Langlands n'ont peut-être pas eu un impact aussi grand en physique mathématique qu'en théorie des formes automorphes. Il n'en demeure pas moins que, de par sa tribune privilégiée du *Bulletin of the American Mathematical Society*, son article *Conformal invariance in two-dimensional percolation* a permis de réorienter en partie les efforts de jeunes mathématiciens audacieux. Leurs travaux, et par conséquent les idées de Langlands, ont eu un impact important sur le domaine et de nombreux résultats ont suivi, réalisés par les équipes autour de Schramm et de Smirnov, mais aussi par la communauté mathématique.

La classification de Schramm des espaces de probabilité satisfaisant l'hypothèse d'invariance conforme donne une réponse claire aux questions soulevées par notre hypothèse d'invariance conforme. Mais l'hypothèse d'universalité demeure une conjecture, même si l'invariance conforme de la limite de plusieurs autres modèles physiques a maintenant été démontrée. Il est à espérer que les années prochaines sauront énoncer précisément les hypothèses sous lesquelles cette seconde conjecture est vraie.

Remerciements

L'auteur détient une subvention à la découverte du Conseil de recherches en sciences naturelles et en génie du Canada pour laquelle il est reconnaissant.

References

[1] R. Langlands, P. Pouliot, Y. Saint-Aubin, Conformal invariance in two-dimensional percolation. *Bull. of the Am. Math. Soc.*, **30**, 1–61, 1994.

[2] H. Kesten, *Percolation Theory for Mathematicians*. Birkhäuser, Boston, 1982.

[3] M. Aizenman, D. J. Barsky. Sharpness of the phase transition in percolation models. *Comm. Math. Phys.*, **108**, 489–526, 1987.

[4] R. Langlands, C. Pichet, P. Pouliot, Y. Saint-Aubin. On the universality of crossing probabilities in two-dimensional percolation. *Journ. Stat. Phys.*, **67**, 553–574, 1992.

[5] S. Smirnov. Critical percolation in the plane: conformal invariance, Cardy's formula, scaling limits. *C.R. Acad. Sci. Paris*, **333**, série I, 239–244, 2001.

[6] R. P. Langlands. Dualität bei endlichen Modellen der Perkolation. *Math. Nachr.*, **160**, 7–58, 1993.

[7] R. P. Langlands, M.-A. Lafortune. Finite models for percolation. *Contemp. Math.*, **177**, 227–246, 1994.

[8] O. Schramm. Scaling limits of loop-erased random walks and uniform spanning trees. *Israel Journal of Mathematics*, **118**, 221–288, 2000.

[9] O. Schramm. A percolation formula. *Electronic Communications in Probability*, **6**, 115–120, 2001.

[10] S. Smirnov. Conformal invariance in random cluster models. I. Holomorphic fermions in the Ising model. *Annals of Mathematics*, **172**, 1435–1467, 2010.

[11] C. M. Fortuin, P. W. Kasteleyn. On the random-cluster model – I. Introduction and relation to other models. *Physica*, **57**, 536–564, 1972.

[12] M. Kac, J. C. Ward. A combinatorial solution of the two-dimensional Ising model. *Physical Review*, **88**, 1332–1337, 1952.

[13] R. P. Feynman. *Statistical Mechanics, a set of lectures*. Addison-Wesley, Boston, 1972.

Printed in the United States
by Baker & Taylor Publisher Services